Decision Making Applications in Modern Power Systems

Decision Making Applications in Modern Power Systems

Edited by

Shady H.E. Abdel Aleem
Mathematical, Physical and Engineering Sciences Department,
15th of May Higher Institute of Engineering, Cairo, Egypt

Almoataz Youssef Abdelaziz
Ain Shams University, Faculty of Engineering, Cairo, Egypt

Ahmed F. Zobaa
College of Engineering, Design and Physical Sciences,
Brunel University London, Uxbridge, United Kingdom

Ramesh Bansal
Department of Electrical and Computer Engineering,
University of Sharjah, Sharjah, United Arab Emirates

ELSEVIER

ACADEMIC PRESS
An imprint of Elsevier

Academic Press is an imprint of Elsevier
125 London Wall, London EC2Y 5AS, United Kingdom
525 B Street, Suite 1650, San Diego, CA 92101, United States
50 Hampshire Street, 5th Floor, Cambridge, MA 02139, United States
The Boulevard, Langford Lane, Kidlington, Oxford OX5 1GB, United Kingdom

British Library Cataloguing-in-Publication Data
A catalogue record for this book is available from the British Library

Library of Congress Cataloging-in-Publication Data
A catalog record for this book is available from the Library of Congress

ISBN: 978-0-12-816445-7

For Information on all Academic Press publications
visit our website at https://www.elsevier.com/books-and-journals

Publisher: Joe Hayton
Acquisition Editor: Lisa Readings
Editorial Project Manager: Joanna Collett
Production Project Manager: R. Vijay Bharath
Cover Designer: Miles Hitchen

Typeset by MPS Limited, Chennai, India

Contents

List of contributors

Shady H.E. Abdel Aleem Mathematical, Physical and Engineering Sciences Department, 15th of May Higher Institute of Engineering, Cairo, Egypt

Almoataz Y. Abdelaziz Electrical Power and Machines Department, Faculty of Engineering, Ain Shams University, Cairo, Egypt

Hassan Haes Alhelou Department of Electrical and Computer Engineering, IUT, Isfahan, Iran; Department of Electrical Power Engineering, Tishreen University, Lattakia, Syria

C.C. Anthony Transmission, Chhattisgarh State Power Transmission Company Ltd., Raipur, India

Bhargav Appasani School of Electronics Engineering, KIIT Deemed University, Bhubaneswar, India

V. Ashok Department of Electrical Engineering, National Institute of Technology, Raipur, India

Murat E. Balci Department of Electrical and Electronics Engineering, Balikesir University, Balikesir, Turkey

Ramesh Bansal Department of Electrical and Computer Engineering, University of Sharjah, Sharjah, United Arab Emirates

Doug Black Lawrence Berkeley National Laboratory, Berkeley, CA, United States

Benedito Donizeti Bonatto Institute of Electrical Systems and Energy, Federal University of Itajuba, UNIFEI, Itajubá, Brazil

Ganesh Kumar Budumuru Department of Electrical Engineering, Veer Surendra Sai University of Technology, Burla, India

Tirso Lorenzo Reyes Carvajal Research Department, Institute of Technology and Education Galileo of Amazon - ITEGAM, Manaus, AM, Brazil

Cy Chan Lawrence Berkeley National Laboratory, Berkeley, CA, United States

Kalpana Chauhan Department of Electrical and Electronics Engineering, Galgotias College of Engineering and Technology, Greater Noida, India

Rajeev Kumar Chauhan Department of Electronic and Instrumentation Engineering, Galgotias College of Engineering and Technology, Greater Noida, India

David Barbosa de Alencar Research Department, Institute of Technology and Education Galileo of Amazon - ITEGAM, Manaus, AM, Brazil

Jorge de Almeida Brito Júnior Research Department, Institute of Technology and Education Galileo of Amazon - ITEGAM, Manaus, AM, Brazil

Carlos Alberto Oliveira de Freitas Research Department, Institute of Technology and Education Galileo of Amazon - ITEGAM, Manaus, AM, Brazil

Antonio Carlos Zambroni de Souza Institute of Electrical System and Energy, Federal University of Itajuba, UNIFEI, Itajubá, Brazil

Yan Deng College of Electrical Engineering, Zhejiang University, Hangzhou, P.R. China

Carlos Augusto Duque UFJF—Federal University of Juiz de Fora, Juiz de Fora, Brazil

Soham Dutta Department of Electrical Engineering, Indian Institute of Technology (Indian School of Mines), Dhanbad, Jharkhand, India

Milton Fonseca Júnior Mauá Generation Department, Generation Eletrobras Amazonas GT Manaus, Amazonas, Brazil

Saeid Ghassemzadeh Faculty of Electrical and Computer Engineering, University of Tabriz, Tabriz, Iran

M.E.H. Golshan Department of Electrical and Computer Engineering, IUT, Isfahan, Iran

Josep M Guerrero Department of Energy Technology, Aalborg University, Aalborg, Denmark

Xiangning He College of Electrical Engineering, Zhejiang University, Hangzhou, P.R. China

Seyed Hossein Hosseini Faculty of Electrical and Computer Engineering, University of Tabriz, Tabriz, Iran; Engineering Faculty, Near East University, North Cyprus, Mersin 10, Turkey

Mohammad Hemmati Faculty of Electrical and Computer Engineering, University of Tabriz, Tabriz, Iran

Sherif M. Ismael Electrical Engineering Division, Engineering for the Petroleum and Process Industries (ENPPI), Cairo, Egypt

Alp Karadeniz Department of Electrical and Electronics Engineering, Balikesir University, Balikesir, Turkey

Abhishek Kumar College of Electrical Engineering, Zhejiang University, Hangzhou, P.R. China

Praveen Kumar Department of Electronics and Electrical Engineering, Indian Institute of Technology (IIT) Guwahati, Guwahati, India

Jandecy Cabral Leite Research Department, Institute of Technology and Education Galileo of Amazon - ITEGAM, Manaus, AM, Brazil

Behnam Mohammadi-Ivatloo Faculty of Electrical and Computer Engineering, University of Tabriz, Tabriz, Iran

Dusmanta Kumar Mohanta Department of Electrical and Electronics Engineering, Birla Institute of Technology, Mesra, Ranchi, Jharkhand, India

Maíra R. Monteiro Institute of Electrical System and Energy, Federal University of Itajuba, UNIFEI, Itajubá, Brazil; School of Engineering, The University of British Columbia, Kelowna, BC, Canada

Nadime Mustafa Moraes Research Department, University of the State of Amazonas (UEA), Manaus, Amazonas, Brazil

Yahya Naderi Faculty of Electrical and Computer Engineering, University of Tabriz, Tabriz, Iran; Department of Energy Technology, Aalborg University, Aalborg, Denmark

Manoel Henrique Reis Nascimento Research Department, Institute of Technology and Education Galileo of Amazon - ITEGAM, Manaus, AM, Brazil

Mohammad Pazoki School of Engineering, Damghan University, Damghan, Iran

Lênio O. Prado, Jr IF—Federal Institute of Science and Technology Education, Minas Gerais, Brazil

Papia Ray Department of Electrical Engineering, Veer Surendra Sai University of Technology, Burla, India

Maddikara Jaya bharata Reddy Department of Electrical and Electronics Engineering, National Institute of Technology, Tiruchirapalli, Tamil Nadu, India

Luiz Carlos Ribeiro, Junior Institute of Electrical Systems and Energy, Federal University of Itajuba, UNIFEI, Itajubá, Brazil

Paulo Fernando Ribeiro Institute of Electrical System and Energy, Federal University of Itajuba, UNIFEI, Itajubá, Brazil

Yuri R. Rodrigues School of Engineering, The University of British Columbia, Kelowna, BC, Canada

Pradip Kumar Sadhu Department of Electrical Engineering, Indian Institute of Technology (Indian School of Mines), Dhanbad, Jharkhand, India

Bikash Sah Department of Electronics and Electrical Engineering, Indian Institute of Technology (IIT) Guwahati, Guwahati, India

Harish Kumar Sahoo Department of Electronics & Telecommunication Engineering, Veer Surendra Sai University of Technology, Burla, India

Mehdi Savaghebi SDU Electrical Engineering, Mads Clausen Institute, University of Southern Denmark (SDU), Odense, Denmark

Arvind R. Singh School of Electrical Engineering, Shandong University, Jinan, P.R. China

Alireza Soroudi University College Dublin, Dublin, Ireland

Juan Carlos Vasquez Department of Energy Technology, Aalborg University, Aalborg, Denmark

Francinei Lucas Vieira Institute of Electrical Systems and Energy, Federal University of Itajuba, UNIFEI, Itajubá, Brazil

Bin Wang Lawrence Berkeley National Laboratory, Berkeley, CA, United States

Anamika Yadav Department of Electrical Engineering, National Institute of Technology, Raipur, India

K.K. Yadav Meter, Relay and Testing Division, 400 kV Sub-Station, Bhilai-3, Chhattisgarh State Power Transmission Company Ltd., Bhilai, India

Umakant Yadav Extra High Tension: Construction and Maintenance Division, Gudhiyari, Chhattisgarh State Power Transmission Company Ltd., Raipur, India

Rongxin Yin Lawrence Berkeley National Laboratory, Berkeley, CA, United States

Ahmed F. Zobaa College of Engineering, Design and Physical Sciences, Brunel University London, Uxbridge, United Kingdom

Chapter 1

Multicriteria decision-making methodologies and their applications in sustainable energy system/microgrids

Abhishek Kumar[1], Bikash Sah[2], Arvind R. Singh[3], Yan Deng[1], Xiangning He[1], Praveen Kumar[2] and Ramesh Bansal[4]

[1]College of Electrical Engineering, Zhejiang University, Hangzhou, P.R. China, [2]Department of Electronics and Electrical Engineering, Indian Institute of Technology (IIT) Guwahati, Guwahati, India, [3]School of Electrical Engineering, Shandong University, Jinan, P.R. China, [4]Department of Electrical and Computer Engineering, University of Sharjah, Sharjah, United Arab Emirates

1.1 Introduction

As per International Energy Organization, currently, more than 38% of the global population suffers from energy poverty [1]. Around 2792 million people are deprived of access to clean cooking and electricity access. This condition is even worse in the case of India (834 million without clean cooking and 240 million without electricity) and sub-Saharan Africa (846 million without clean cooking and 588 million without electricity) where a combined population of the United States and Europe also do not have any access to energy for their general livelihood [1,2]. This situation is anticipated to become worse due to population growth, rapid industrialization, and economic development, especially in the case of India [3]. The overall development of a country is proportional to the increase in the requirements of energy; therefore, many policies are being framed by the governments to embrace on the suitable energy planning strategies leading to sustainable development [4,5]. Underscoring the importance of energy issues for overall development, United Nation (UN) general assembly has unanimously declared the decade 2014–24 as the "Decade of Sustainable Energy for All." Adding weight to the declaration, "2030 Agenda for Sustainable Development" was formulated and adopted in 2015 by the UN.

Decision Making Applications in Modern Power Systems. DOI: https://doi.org/10.1016/B978-0-12-816445-7.00001-3

Among the multiple goals in the agenda, three are focused on the energy. These are [6]:

1. to provide energy access for everyone by 2030;
2. prominent action plans in order to fight global climatic change; and
3. to reduce the emissions that cause air pollution.

Moreover, rising energy demand along with the depletion of fossil fuels and environmental issues associated with the use of conventional energy resources has motivated governments globally to use the renewable energy sources [7,8]. In the recent few decades, microgrids based on renewable energy technologies have become popular for electrifying the remote and isolated areas mostly in developing nations and can certainly provide a solution for providing energy access to such areas [9,10]. A number of methods and models exist in the literature for designing the microgrids based on the renewable energy technologies [11–14]. Sustainable energy design recently has become a tedious process due to the involvement of multiple performance indices having several targets and scenarios [15]. The participation of many actors having differing perceptions based on numerous aspects of sustainability has made the planning and analysis of systems more difficult [16].

The problem is no longer seen as a singular perspective; rather it is seen from multiple perspectives and needs to be evaluated based on several traits or key performance indicators to achieve the perspective of sustainability [17]. For a successful design of the energy systems with sustainability perspective, a wholesome cooperation is required between differing perspectives of stakeholders when various scenarios based on different criteria are pondered [18]. For such complex designs, multicriteria decision-making (MCDM) methods and tools can be effectual while solution accommodating multiple criteria, stakeholders and differing views in the same framework [19]. In this chapter a detailed illustration of MCDM methods, such as analytical hierarchical process (AHP), technique for order preference by similarity to ideal solutions (TOPSIS), elimination and choice translating reality (ELECTRE), fuzzy, and hybrid MCDM models such as AHP–TOPSIS and fuzzy–AHP is discussed, which can be efficiently utilized to renewable energy planning and design.

1.1.1 A general perspective

MCDM is a branch of operational research, which uses analytical methods to make proper decisions. It helps in addressing complex problems dealing with inconsistent objectives, heterogeneous data, interest, and uncertainty. A general classification of the various fields of the operational research is given in Fig. 1.1.

Multiattribute decision-making (MADM) aims to find distinct alternatives from a set of alternatives. Multiobjective decision-making (MODM), on the other hand, is inclined for decision problems that involve multiple objectives

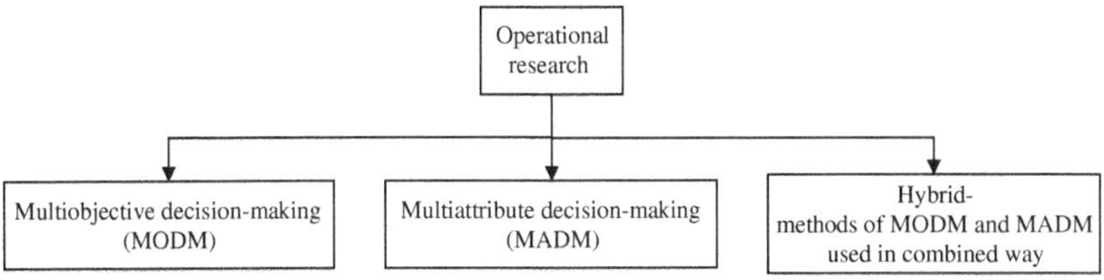

FIGURE 1.1 A general classification of operational research [20–23].

as well as alternatives. The hybrid extracts the best of the both methods to draw necessary conclusions for a decision maker. Energy management and policy making considering sustainability has become one of the topics of global benefits, and researchers and agencies (public and private) are inclining their interests to use MODM to get the best decision. Energy management and policy making involves multiple technical, economic, social, and environmental issues and objectives to attain. For example, in a process to select a site for a power plant, a decision maker needs to consider multiple objectives as well as alternatives. The objectives can be least capital investment, topographically suitable for generation, ease to access by all stakeholders, socially acceptable, environmentally safe for deploying the power plants and its supporting services, etc. Each of the objectives is arranged into various categories and indicators. The techniques in MCDM allow us to consider all the hues and cries to be taken care by the decision makers.

Apart from energy management and policy making, the MCDM is also applicable in formulating suitable strategies for management and control on the generation and distribution of electrical power as well. The decision science has grown deep roots in the power sector with microgrids and distributed energy resources utilization gaining popularity. Numerous studies are endeavored using MCDM techniques for energy and resource saving objectives. Apart from the objectives related to technical, social, environmental, and economic, sustainability analysis has also become a part of MCDM [20–23].

Renewable energy is playing a major role in the power and energy management. Its numerous benefits include local availability, environment friendliness, and it can be widely distributed depending on the topography and demography of the place. The projects based on renewable energy are running all round the globe. MCDM has been applied in various fields such as planning, integration, and management of renewable energy. MCDM is used to define the priority of projects to be funded by the agencies, deciding location, effectiveness, and feasibility analysis of the projects.

1.2 Multicriteria decision-making in energy planning

Two types of methods are discussed in this chapter: general and hybrid. The following general methods are used in the energy planning:

- weighted sum method (WSM);
- weighted product method (WPM);
- AHP;
- TOPSIS;
- ELECTRE; and
- preference ranking organization method for enrichment evaluation (PROMETHE).

The hybrid methods are based on fuzzy, and two types of hybrid methods are discussed in the chapter: fuzzy with AHP and fuzzy with TOPSIS. Each method is described with detailed description of steps and a flowchart to briefly describe the process to the readers. Each method and its utilization in energy planning are cited from the literature.

1.2.1 Weighted sum method

This method is most commonly used for single-dimension problem. This method is widely used in the structural optimization and energy planning. The ease in the computation to select alternatives brings in the significance of this method for utilization. The selection of alternatives is made by using the following mathematical equation:

$$W_s = \sum_{n}^{m} v_{nm} w_m \tag{1.1}$$

where W_s is the value of weighted sum, n is the number of alternative, and m is the number of criteria. v_{nm} is the function of the formulated vector and calculated as the normalized value for nth alternative and mth criteria. w_m is the weight of mth criteria. W_s is the score calculated for each alternative. The best alternative is determined as $\max(W_s)$. This value of W_s can also be used for ranking the alternatives based on the scores. This method provides a basic estimate of the function and fails to integrate multiple preferences [24].

1.2.2 Weighted product method

Alike WSM, this method performs multiplication. The comparison of each alternative is done by multiplying the ratios based on the number of criteria. While multiplying, the ratio is raised to the power of a value equivalent to the relative weight of the corresponding criteria. The comparison of the two alternatives can be defined by the following mathematical expression:

$$W\left(\frac{P_k}{P_L}\right) = \prod_{i=1}^{m} \left(\frac{p_{K_i}}{p_{L_i}}\right)^{w_i} \tag{1.2}$$

where the number of criteria is represented by m, and p_{K_i} and p_{L_i} are the actual values of the Kth and Lth alternative w.r.t. the ith criteria. w_i is the weight of the ith criteria. The value of $W(P_k/P_L)$ determines the desirability of an alternative [25].

If $W(P_k/P_L) \geq 0$ or 1, alternative P_k is considered better than P_L. The best alternative is the one whose $W(P_k/P_L)$ is found to be closer to "1." The value closer to "1" signifies that the selected alternative is closest to all other alternatives. An alternative approach to this is considered by avoiding the use of ratios. This is given by

$$W(P_k) = \prod_{i=1}^{m} \left(p_{K_i}\right)^{w_i} \tag{1.3}$$

The value of $W(P_k)$, which is the performance index of $(P_K)p_K$ considering all the criteria, determines the best alternatives. The ranking is made based on the value of $W(P_k)$.

It is to be noted that the first method of WPM is a dimensionless approach, whereas the later has dimensions. The first method that is based on the relative method is found to be beneficial than the latter [8].

1.2.3 Analytic hierarchy process

The AHP, proposed by R.W. Saaty in 1977 [26], is widely used by the energy planners to manage resources, making public policies, logistics, and transportation engineering worldwide. The method was later revised and updated in 1980. This method converts the operational problem to a hierarchical model. The detailed explanation of AHP model has been illustrated thoroughly with a real case study for the design of rural microgrids based on renewable energy sources in Ref. [27]. Table 1.1 provides a summary of few studies based on AHP in energy planning.

1.2.4 Technique for order preference by similarity to ideal solutions

TOPSIS, originally developed by Hwang and Yoon in 1981, is one of the popular MCDM models, which has the ability and applicability to provide solutions to real-world problems [40,41]. This method has a wide application area ranging from microgrid/energy planning, energy management, supply chain and logistics, water and waste resource management, manufacturing and design engineering, business and industrial management, etc. [41]. This MCDM method is based on the concepts of geometric distances where the best alternatives have a very short distance from the positive ideal solution (PIA^+) and a precisely longer distance from the negative ideal solution (NIA^-) [42].

Fig. 1.2 shows the detailed process for the implementation of TOPSIS method. It consists of major six steps that are as follows [27,43]:

TABLE 1.1 A summary of few studies based on AHP in energy planning.

Sl. no.	Year	Objective	Contribution	Reference
1	2018	To design a rural sustainable microgrid for developing nation	Development of a detailed three-step framework combining decision-making and multiobjective optimization. Detailed illustration of AHP methodology and implementation with a real case study	[27]
2	2017	Site selection for installation of solar power plants	Illustration for sustainable site selection using AHP method. Formulation of key performance factors based on social, technical, economic, environmental, and institutional aspects	[28]
3	2017	To provide a framework for evaluation and selection of energy projects	Application of AHP and other MCDM models for determination of suitable criteria as well as ranking of four major energy projects on social, environmental, and economic indicators	[29]
4	2016	To determine suitable power generation technologies with a case study of Lithuania	Deliberates an integrated approach of AHP and additive ratio assessment for the selection of six power generation technologies based on five dimensions of sustainability	[30]
5	2016	Design of electrification system for rural areas	A detailed methodology based on AHP method for rural electrification system based on locally available renewable sources, showing a clear stepwise approach to clearly proceed with real case study for system design	[16]
6	2015	To determine an optimal location to set up solar power plants	Developing a web-based spatial decision support system to determine the location	[31]

7	2015	To improve the use of combined cooling and heating plant in Japanese buildings and to measure the energy utilization	Environmental and economic criteria are studied to develop the model	[32]
8	2014	Controlling the rotation of crops that are involved in affecting the energy sources or being used as energy resource	A methodology is proposed for evaluating the economic and environmental criteria w.r.t. crop rotations	[33]
9	2014	To develop a model to analyze the performance of energy systems in buildings and to increase their efficiency	A building energy performance calculation method is developed	[34]
10	2014	To model a design support system for the energy supply management in any production plant	A case study is done by the developed model, and the results are found to be better than existing processes	[35]
11	2013	Ranking of power generation technologies	A framework is developed for decision makers to rank renewable and nonrenewable energy generation	[36]
12	2013	To determine best location for photovoltaic power plants in Spain	A location is determined considering the location, climate, terrain, and environmental variables to install photovoltaic plant	[37]
13	2012	Optimal distribution of energy among the projects and selection of renewable technologies	An analytical model for selecting the renewable energy technology in a research and development center of oil industry	[38]
14	2011	To design an evaluation method for the impact of excessive energy consumption in urban areas	Considering four indicators, a method is proposed to evaluate the impact	[39]

AHP, Analytical hierarchical process; *MCDM*, multicriteria decision-making.

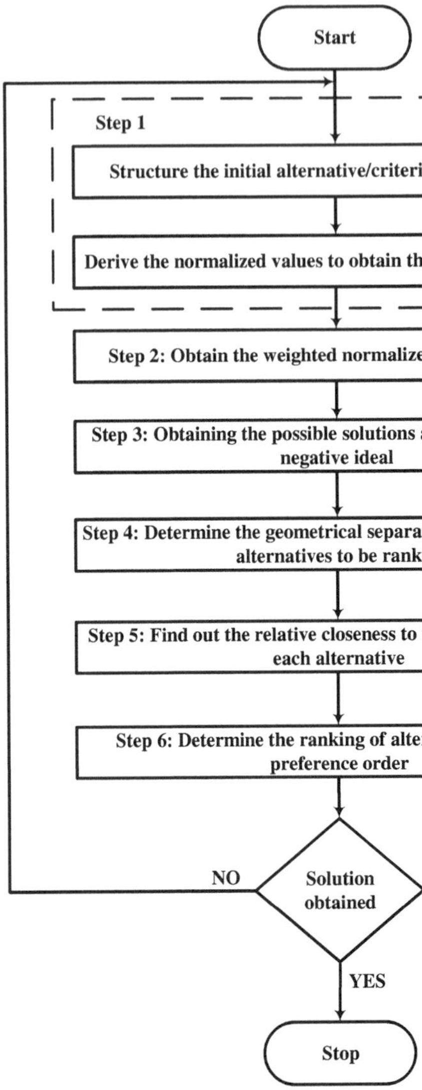

FIGURE 1.2 Flowchart to rank of alternatives using TOPSIS method [27,43]. *TOPSIS*, Technique for order preference by similarity to ideal solutions.

Step 1. Structure the initial alternatives/criteria decision matrix and derive the normalized matrix

Articulate an initial alternative/criteria decision matrix having a dimension of $a \times c$, where a is the number of alternatives and c is the number of performance indictors/criteria.

Calculate normalized values from the initial decision matrix to get normalized matrix. The normalized value (acv_{gh}) is given by

$$acv_{gh} = \frac{f_{gh}}{\sqrt{\sum_{h=1}^{a} f_{gh}^2}} \tag{1.4}$$

where f_{gh} is the value of the gth criterion function for the alternative Ah ($h = 1, \ldots, a$ and $g = 1, \ldots, c$).

Step 2. Obtain the weighted normalized decision matrix

Calculate the weighted normalized value v_{gh} as

$$v_{gh} = w_{c_g} acv_{gh} \tag{1.5}$$

where the weight of the criterion/performance indicator g is such that

$$w_{c_g} = \sum_{g=1}^{c} w_{c_g} = 1 \tag{1.6}$$

The weight of the criteria/performance indicators can be determined by using AHP method as described in the previous subsection or using specialized software packages such as Triptych or Expert Choice as given in [8].

Step 3. Obtain the possible solutions as positive ideal and negative ideal

First, the classification of the selected essential performance indicators (EPIs)/criteria into two groups has to be done, which is as follows:

Maximum value/benefit criterion: The EPI whose maximum value is sought among the alternatives is termed benefit criterion. The criteria, such as renewable fraction, total electrical energy production, storage autonomy, and grid sells whose maximum values are advantageous for the microgrid alternatives, fall under this category.

Minimum value/performance/economic criterion: Minimum value criterion is that whose low value is better for the microgrid alternatives. These consist of economic indicators such as capital cost, net present cost, and cost of electricity as well as performance criteria such as storage power losses, excess electricity, and grid energy purchase whose minimal values are sought.

After the classification of the EPIs/criteria, the PIA^+ and NIA^- can be obtained as

$$\mathrm{PIA}^+ = \{v_1^+, \ldots, v_n^+\} = \{(\max_h v_{gh}/g \in G'), (\min_h v_{gh}/g \in G'')\} \tag{1.7}$$

$$\mathrm{NIA}^- = \{v_1^-, \ldots, v_n^-\} = \{(\min_h v_{gh}/g \in G'), (\max_h v_{gh}/g \in G'')\} \tag{1.8}$$

where G' is the EPIs/criteria representing the benefit criterion whose maximum values are pursued. G'' is the EPIs associated with the performance/minimum value criterion.

Step 4. Determine the geometrical separation measures for alternatives

Compute the geometrical separation measures of the alternatives using the PIA^+ and NIA^- solutions obtained from Step 3.

The geometrical separation measure ($S^+_{h(PIA)}$) based on the PIA^+ of each alternative is given by the following equation:

$$S^+_{h(PIA)} = \sqrt{\sum_{g=1}^{c} \left(v_{gh} - v_g{}^*\right)^2} \tag{1.9}$$

Similarly, from the negative ideal, the separation measure ($S^-_{h(NIA^-)}$) is given as

$$S^-_{h(NIA^-)} = \sqrt{\sum_{g=1}^{c} \left(v_{gh} - v_g{}^-\right)^2} \tag{1.10}$$

Step 5. Find out the relative closeness to the ideal solution for alternatives

The relative closeness (CI^*_h) of the alternative from the ideal solution can be calculated using the following equation:

$$CI^*_h = \frac{S^-_{h(NIA^-)}}{\left(S^+_{h(PIA^+)} + S^-_{h(NIA^-)}\right)} \tag{1.11}$$

The numeric value of CI^*_h lies between 0 and 1, where 0 signifies the worst and 1, the best possible solution.

Step 6. Determine the ranking of alternatives based on preference order

Once the value of CI^*_h for each alternative has been determined, arrange them in the descending order and rank the alternatives. The alternative attaining the highest values of CI^*_h should be ranked first and proposed as the best possible solution. Table 1.2 illustrates a few studies based on TOPSIS with its applicability to design and evaluation of energy system.

1.2.5 Elimination and choice translating reality

ELECTRE also called elimination and choice translating reality was proposed by Benayoun et al. in 1966 [52]. The method finds its application in energy, financial and business management, information technology and communication, logistics and transportation engineering. This is an outranking method as it develops a binary association between alternatives w.r.t. to all criteria [53]. The literature gives several forms of ELECTRE. A brief description of this method is given in the following subsections [54].

TABLE 1.2 A few studies based on technique for order preference by similarity to ideal solutions applicable to energy system design.

Sl. no.	Year	Objective	Contribution	Reference
1	2013	To determine the best location for photovoltaic power plants in Spain	A location is determined considering the location, climate, terrain, and environmental variables to install photovoltaic plant	[37]
2	2013	To assess the small-scale combined heat and power plants–based technology in buildings	Determined the combined heat and power technologies that are more sustainable	[44]
3	2013	To analyze an electricity supply chain and propose a framework	The framework developed is used as a case study in Turkey	[45]
4	2013	To evaluate nine options of power generation w.r.t. seven criteria	The method is deployed to determine the best generation technology. Natural gas–fueled solid oxide fuel cells were found to be the best among all options	[46]
5	2012	To determine the location of a thermal power plant in India	Combined other methods of multicriteria decision-making and concluded that the cost and accessibility to resources are the most important factors to choose the location for a thermal power plant	[47]
6	2012	To give assistance to the policy makers to use linguistic variables in deciding priorities for sustainable energy technology	A two-stage methodological approach is developed for the decision makers. The sustainable energy technology priorities are also decided based on the country-specific requirements and available resources	[48]
7	2011	Selection of best energy generation technology considering technical, economic, environmental, and social aspects	The model is combined with fuzzy to determine the weights instead of using the conventional process of calculating distance. The results are later compared	[49]

(Continued)

TABLE 1.2 (Continued)

Sl. no.	Year	Objective	Contribution	Reference
			with other methods of multicriteria decision-making techniques as well	
8	2010	To illustrate that the linguistic variables can be used to study sustainable development and how energy policy objectives toward SD and RES options are related and assessed using linguistic variables	The method given in the article is country specific, although a similar framework can be used based on the criteria under study to energy policy and management	[50]
9	2007	To compare three techniques of multicriteria decision-making with specific focus on energy policy and management	The drawbacks and advantages of the method when applied in energy planning are presented	[51]

RES, Renewable energy sources; *SD*, Sustainable Development.

1.2.5.1 Elimination and choice translating reality I

This is based on the concordance and discordance indexes. The concept of concordance and discordance determines the outranking of alternatives. The concordance states and measures the intensity of favoring an argument placed for an alternative under analysis, whereas discordance gives the intensity of the opposition for the same. This method is designed for the selection of problems.

1.2.5.2 Elimination and choice translating reality II

This method is similar to the ELECTRE I. The only difference is the addition of a threshold value to the outranking matrices formed. This method is suitable for ranking problems.

1.2.5.3 Elimination and choice translating reality III

This method is considered an interaction method as it involves the direct participation of decision maker and the process. The quantitative and qualitative criterion can be analyzed. This method is used when there is a necessity to quantify a certain criterion.

1.2.5.4 Elimination and choice translating reality IV

The ELECTRE IV method allows the construction of several (nested) upgrade relationships when it is not possible to assign weights to each of the pseudocriteria. Instead, the decision maker must allow that none of the criteria is dominant or negligible (so able to deal with any grouping of one-half of the pseudocriteria).

Three basic steps are involved in the formulation:

1. finding the threshold function;
2. calculation of concordance and discordance indexes; and
3. determining the outranking degree.

A detailed flowchart illustrating the ELECTRE method is given in Fig. 1.3. A detailed explanation of the previous steps is as follows:

Step 1: Decision matrix

The decision matrix is formed similar to those mentioned in Section 1.2.4.

Step 2: Normalization

In this step the decision matrix formulated is normalized. This is done by making the entries of decision matrix as dimensionless by using the following equation:

$$x_{ij} = \frac{p_{ij}}{\sqrt{\sum_{k=1}^{m} p_{kj}^2}} \tag{1.12}$$

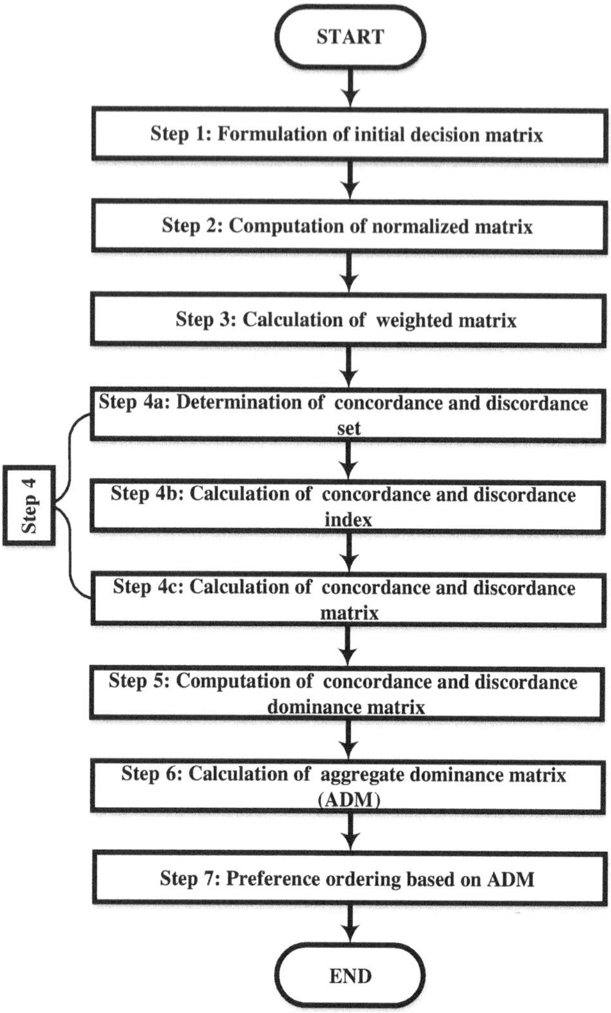

FIGURE 1.3 Flowchart for the implementation of ELECTRE method. *ELECTRE*, Elimination and choice translating reality.

where p_{kj} is the actual value of the kth alternative w.r.t. ith criteria, m is the number of alternatives, n is the number of criteria, x_{ij} is the normalized value of the ith alternative in terms of jth criteria. The normalized decision matrix is given by the following equation:

$$X = \begin{bmatrix} x_{11} & \cdots & x_{1n} \\ \vdots & \ddots & \vdots \\ x_{m1} & \cdots & x_{mn} \end{bmatrix} \tag{1.13}$$

Step 3: Calculation of weighted matrix

The decision matrix X mentioned in Eq. (1.13) is associated with the respective weights, which resembles the significance of corresponding criteria. Let the weight determined by the decision maker be denoted by $w_1, w_2, \ldots, w_n$, such that $\sum_{i=1}^{n} w_i = 1$, and W be the weighted matrix. W can be calculated as

$$W = XY \tag{1.14}$$

where Y is a diagonal matrix defined as

$$\begin{bmatrix} w_1 & \cdots & 0 \\ \vdots & \ddots & \vdots \\ 0 & \cdots & w_n \end{bmatrix} \tag{1.15}$$

Thus the weighted matrix W can be given by

$$\begin{bmatrix} y_{11} & \cdots & y_{1n} \\ \vdots & \ddots & \vdots \\ y_{m1} & \cdots & y_{mn} \end{bmatrix} = \begin{bmatrix} w_1 x_{11} & \cdots & w_n x_{1n} \\ \vdots & \ddots & \vdots \\ w_1 x_{m1} & \cdots & w_n x_{mn} \end{bmatrix} \tag{1.16}$$

Step 4: Concordance and discordance matrix formulation

This step involves three substeps.

Step 4a: Calculation of concordance and discordance set

Let P_k and P_l be two alternatives, $m \geq k$ and $l \geq 1$; the concordance set C_{kl} of the two alternatives is such that P_k desired over P_l is given by

$$C_{kl} = \{j, y_{kj} \geq y_{lj}\}, \quad \text{for } j = 1, 2, 3, \ldots, n \tag{1.17}$$

The discordance (D_{kl}) set is given by

$$D_{kl} = \{j, y_{kj} \geq y_{lj}\}, \quad \text{for } j = 1, 2, 3, \ldots, n \tag{1.18}$$

Step 4b: Calculation of concordance and discordance indexes

The concordance index presents the relative importance of one alternative w.r.t. other. It is calculated as a sum of weights associated with a criterion. Let c_{kl} be the concordance index.

$$c_{kl} = \sum_{j \in C_{kl}} w_j, \quad \text{for } j = 1,\ 2,\ 3,\ \ldots, n \tag{1.19}$$

Let d_{kl} be the discordance index. This measures the triviality of one alternative w.r.t. other. It is calculated as

$$d_{kl} = \frac{\max_{j \in D_{kl}} |y_{kj} - y_{lj}|}{\max_j |y_{kj} - y_{lj}|} \tag{1.20}$$

Step 4c: Calculation of concordance and discordance matrices

The concordance and discordance matrices are expressed in terms of the concordance and discordance indices. The concordance matrix C and discordance matrix D are given by

$$C = \begin{bmatrix} - & c_{12} & \cdots & c_{1m} \\ c_{21} & - & \cdots & c_{2m} \\ \vdots & \vdots & \vdots & \vdots \\ \vdots & \vdots & \vdots & \vdots \\ c_{m1} & c_{m2} & \cdots & - \end{bmatrix} \tag{1.21}$$

$$D = \begin{bmatrix} - & d_{12} & \cdots & d_{1m} \\ d_{21} & - & \cdots & d_{2m} \\ \vdots & \vdots & \vdots & \vdots \\ \vdots & \vdots & \vdots & \vdots \\ d_{m1} & d_{m2} & \cdots & - \end{bmatrix} \tag{1.22}$$

It is to be noted that the value elements of D are undefined for $k = l$, and it is nonsymmetric square matrix.

Step 5: Calculating the concordance and discordance dominance matrices

This step involves the use of a threshold value. The threshold value is a critical value that is involved in the formation of dominance matrices. Let $\hat{c}$ and $\hat{d}$ be the threshold values that are calculated as an average value of the concordance and discordance indexes.

$$\hat{c} = \frac{1}{m(m-1)} \sum_{\substack{k=1 \\ k \neq l}}^{m} \sum\nolimits_{\substack{l=1 \\ l \neq k}}^{m} c_{kl} \tag{1.23}$$

$$\hat{d} = \frac{1}{m(m-1)} \sum_{\substack{k=1 \\ k \neq l}}^{m} \sum\nolimits_{\substack{l=1 \\ l \neq k}}^{m} c_{kl} \tag{1.24}$$

The concordance and discordance dominance matrices are calculated based on the comparison of concordance and discordance indexes and threshold values ($\hat{c}$ and $\hat{d}$). Let cd_{kl} and dd_{kl} be the elements of the matrices. Mathematically, it is given by

$$cd_{kl} = \begin{cases} 1 \text{ if } c_{kl} \geq \hat{c} \\ 0 \text{ if } c_{kl} \leq \hat{c} \end{cases} \tag{1.25}$$

and

$$dd_{kl} = \begin{cases} 1 \text{ if } d_{kl} \geq \hat{\hat{d}} \\ 0 \text{ if } d_{kl} \leq \hat{d} \end{cases} \tag{1.26}$$

Step 6: Calculation of the aggregate dominance matrix (ADM)

The elements of the ADM are calculated as the product of concordance and discordance indexes and threshold values (cd_{kl} and dd_{kl}).

$$ADM_{kl} = cd_{kl} \times dd_{kl} \tag{1.27}$$

Step 7: Preference ordering based on the value of the ADM

The best alternatives will be the one that dominates all the other alternatives. The dominance is determined based on the presence of "1." If any of the value in a column is equal to 1, it can be stated that the column with "1" is preferred over the corresponding row.

The ELECTRE is preferred over the other method as it can deal with the heterogeneous scales. The drawbacks of this method are related to its versatility, and it needs a very good understanding of the objective a decision maker is working with, especially for quantitative features. Few applications of ELECTRE in energy planning are given in Table 1.3.

1.2.6 Preference ranking organization method for enrichment evaluation

PROMETHE was developed by Brans et al. [61]. This method is dominantly utilized in the field of risk and structural analysis, and mining engineering. This method encompasses a group level involvement in the decision-making. It also deals with both qualitative and quantitative information. In this method, there are various categories of analysis in this method, and each of them is named as versions—PROMETHE I, PROMETHE II, PROMETHE III, PROMETHE IV, PROMETHE V, and PROMETHE VI. The type of analysis the method focuses on is given in Table 1.4 [61–64].

This method involves the selection of transfer function and fixing the value of threshold. The transfer function, also called preference, is given in Table 1.5 [61,65,66].

PROMETHE II is the most commonly used method to calculate the rank of various alternatives. The flowchart of the process for PROMETHE is given in Fig. 1.4. The steps that are described later have been followed while calculating the rank of the alternatives.

Step 1: Calculation of preference degree

To start with, a decision matrix with rows filled with alternatives and columns with criterion is prepared. This is followed by giving weights to the entire criterion and establishing the threshold values. The transfer function is also selected for each criterion. Let $c_i(a)$ and $c_i(b)$ be the value of criterion j for decisions a and b. The difference $d_i(a, b)$ is calculated between the values given in the following equation:

TABLE 1.3 Selected applications of ELECTRE in energy planning.

Sl. no.	Year	Objective	Contribution	Reference
1	2014	Selection of a macrosite for developing generating station for renewable energy sources—wind, solar, and hybrid	Literature is rich in site selection for individual power-generating plants. This work is focused on the site selection for a hybrid power generation plant	[54]
2	2014	Assessment of best sites for photovoltaic farms to obtain strong conclusions	Combined the geographic information system and ELECTRE Tri for evaluating the sites for photovoltaic farm	[55]
3	2012	To select the best concept of material selection, design, and manufacturing process considering sustainability	The sustainable solution in industrial process involving material selection, design, and manufacturing process is determined using ELECTRE II. It is concluded that the manufacturing process improvement can be used to increase the sustainability	[56]
4	2011	To use multicriteria decision-making to manage multiple sources–based system involving renewable energy	A generalized procedure is developed to find that the optimal solution is presented using ELECTRE III. Economic and environmental criteria are studied	[57]
5	2008	To evaluate various energy efficiency initiatives undertaken by various stakeholders ranging from electric utility company to private or public-funded company	Used ELECTRE to determine the effectiveness of all initiatives that are undertaken for an efficient utilization of energy	[58]

(*Continued*)

TABLE 1.3 (Continued)

Sl. no.	Year	Objective	Contribution	Reference
6	2008	To determine the penetration of renewable energy sources in the power generation of a provincial region that is out of the mainstream	ELECTRE III is used to evaluate the alternatives to exploit renewable energy into islands (isolated)—Karpathos and Kassos of Greece	[59]
7	1998	To develop a framework to work out from a diverse set of criterion and alternatives, thereby helping decision makers to understand complex problems and make appropriate choice	Used ELECTRE to develop the framework. The framework focused on developing hybrid energy resource planning using renewables. Comparison is made with fuzzy logic for the developed framework	[60]

ELECTRE, Elimination and choice translating reality.

TABLE 1.4 Categorization of preference ranking organization method for enrichment evaluation (PROMETHE) [61–64].

Sl. no.	Name	Type
1	PROMETHE I	Calculation of partial ranking
2	PROMETHE II	Calculation of complete ranking
3	PROMETHE III	Calculation of ranking based on intervals (fuzzy logic)
4	PROMETHE IV	Calculation of ranking for a continuous case or many alternatives
5	PROMETHE V	Calculation of ranking based on PROMETHE II and the integer linear programming
6	PROMETHE VI	Calculation of ranking based on the representation of human brain

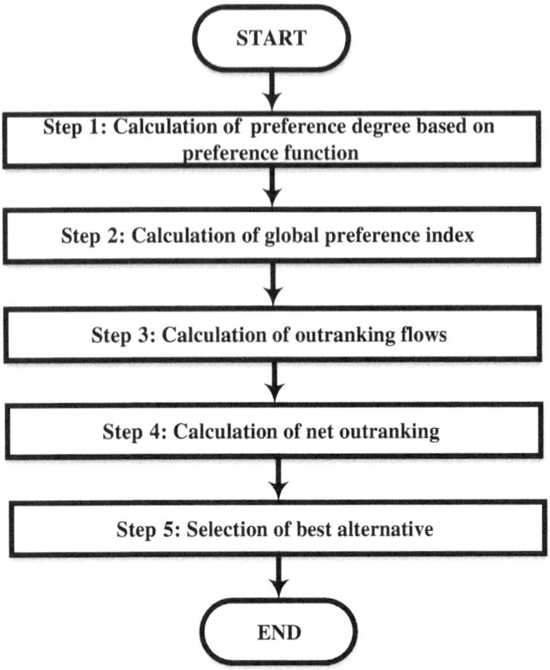

FIGURE 1.4 Steps for PROMETHE. *PROMETHE*, Preference ranking organization method for enrichment evaluation.

$$d_j(a,b) = c_j(a) - c_j(b) \tag{1.28}$$

The preference degree of a criterion j is calculated based on the preference function. Let $P_j(a,b)$ be the preference degree. It is defined as

$$P_j(a,b) = F(d_j(a,b)) \text{ such that } 0 < F(x) < 1 \tag{1.29}$$

Step 2: Calculation of global preference index

This is calculated by multiplying the weight associated with each criterion. Let w_j be the weight associated with the criterion j. The global preference index $gp(a,b)$ is computed as

$$gp(a,b) = \sum_{j \in C} w_j \times P_j(a,b) \tag{1.30}$$

Step 3: Calculation of outranking flows

Two outranking flows are calculated: positive and negative; let F^+ and F^- be the positive and negative outranking flows.

$$F^+(a) = \frac{1}{m-1} = \sum_{x \in A} gp(a,x) \tag{1.31}$$

TABLE 1.5 Transfer function for PROMETHE [61,65,66].

	Preference functions		
Sl. no.	**Name or type of creation**	**Graphical representation**	**Mathematical representation**
1	Usual	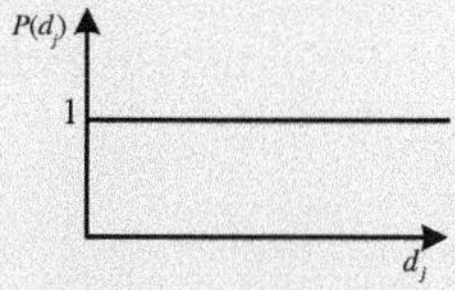	$P(d)_j = \begin{bmatrix} 0 \text{ if } d_j \le 0 \\ 1 \text{ if } d_j > 0 \end{bmatrix}$
2	Quasi	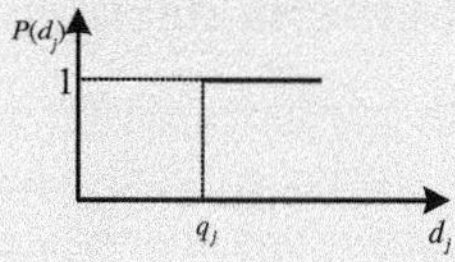	$P(d)_j = \begin{bmatrix} 0 \text{ if } d_j \le q_j \\ 1 \text{ if } d_j > q_j \end{bmatrix}$
3	Linear	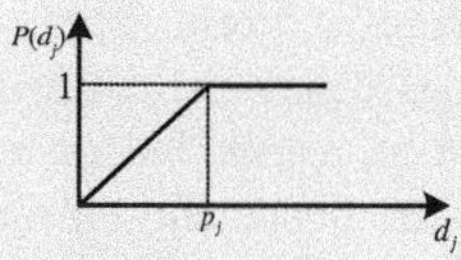	$P(d)_j = \begin{bmatrix} \frac{d_j}{p_j} \text{ if } d_j \le q_j \\ 0 \text{ if } d \le 0 \\ 1 \text{ if } d_j > q_j \end{bmatrix}$
4	Step or level	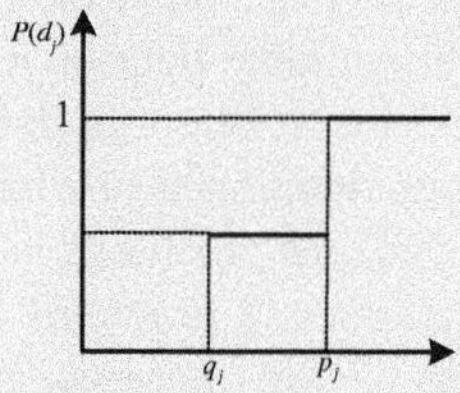	$P(d)_j = \begin{bmatrix} 0 \text{ if } d_j < q_j \\ 0.5 \text{ if } q_j < d_j \le p_j \\ 1 \text{ if } d_j > q_j \end{bmatrix}$
5	Indifference	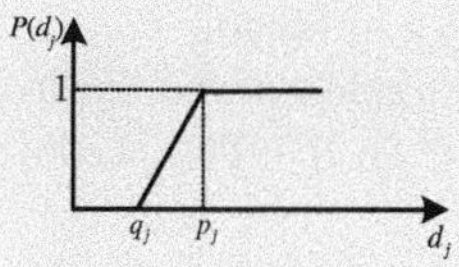	$P(d)_j = \begin{bmatrix} 0 \text{ if } d_j \le p_j \\ \frac{d_j - q_j}{p_j - q_j} \text{ if } q_j < d_j \le p_j \\ 1 \text{ if } d_j > q_j \end{bmatrix}$
6	Gaussian	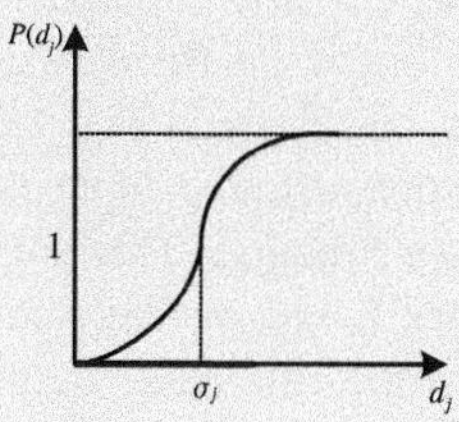	$P(d)_j = \begin{bmatrix} 1 - e^{\frac{-d_j^2}{2\sigma_j^2}} \\ 0 \text{ if } d \le 0 \end{bmatrix}$

and

$$F^-(a) = \frac{1}{m-1} = \sum_{x \in A} gp(x, a) \tag{1.32}$$

where m is the possible number of decisions.

Step 4: Calculation of net outranking

This is the final step of ranking all the alternatives or possible decisions. It is calculated as the difference of positive and negative outranking flows. $F(a)$ is the net outranking of the decision a.

$$F(a) = F^{+}(a) - F^{-}(a) \tag{1.33}$$

The selected applications of PROMETHE in the area of energy planning are given in Table 1.6.

1.3 Fuzzy logic in multicriteria decision-making

Conventional MCDM methods are based on the assigning values, which act as weights. These values are always a fixed number, generally called crisp values. The ranking and all other procedures are carried out based on the assigned crisp values. In a practical world application involving a real-world scenario, most of the quantities cannot be defined quantitatively in terms of numbers. They are expressed as a function or any linguistic variable. The real-world decision problems are dependent on multiple constraints of which the significance and consequences are not exactly defined and determined. Another situation is when the available data or information is not sufficient to judge or the crisp values are incompetent to determine the model of a real situation. Thus, in the situations mentioned previously, it becomes very difficult to use the classical MCDM methods. The MCDM models are suitable for dealing with situations and problems in which it is assumed that the performance or the outputs of any operation is known and can be further represented in the form of crisp numbers.

Fuzzy logic developed by Zadeh in the 1960s has established its application in various fields [76]. It is an IF–THEN rule–based controller. It is useful for solving complex systems whose behaviors cannot be understood. The solutions can be approximate, but a fast output solution is warranted [77]. The output of fuzzy logic is based on three basic processes—fuzzification, rule base, and defuzzification—as given in Fig. 1.5.

The fuzzification process converts the input variables in fuzzy variables, which is a form of membership function. This is given as an input to the rule base, which is the list of rules based on the number of inputs and outputs. Based on the logic defined in the rule base, output is given for defuzzification. The defuzzification process converts the fuzzy variables to crisp values [77,78].

The use of fuzzy in MCDM increases the complexity, but the use is justified for the cases when the goal and final output of the problem statement is not a crisp value. Bellman and Zadeh, and Zimmermann introduced the application of fuzzy in the field of MCDM in the years 1970 and 1978, respectively [79,80]. According to Bellman and Zadeh, the fuzzy goals and

TABLE 1.6 Selected applications of PROMETHE in energy planning.

Sl. no.	Year	Objective	Contribution	Reference
1	2003	To develop a decision-making support system for projects based on renewable energy	Used PROMETHE II to perform a case study in an island of Chios in Greece. The study included few new criteria such as resource exploitation	[67]
2	2015	Selection of renewable energy sources for the desalination plants	Five architectures for energy generation in the reverse osmosis desalination process are studied. The hybrid and direct connection architectures are found to be the best one	[68]
3	2014	To develop a multicriteria decision analysis system for the assessment of sustainability at national scale and rank 11 renewable energy technologies available in Scotland	Nine criteria are considered in the process, and a probabilistic ranking of the available renewable energy generation technologies is determined	[69]
4	2009	To analyze the planning for sustainable energy by considering the economic, technological, social, and environmental as indices for an island in Crete in Greece	Used PROMETHE I and PROMETHE II for the purpose	[70]
5	2009	To determine the best residential energy that is sustainable considering economic and environmental aspects	The work concluded that the competitiveness to use the renewable energy is not more if the environmental aspects are not compared	[71]
6	2009	An assessment of solar thermal energy technologies considering sustainable development	The work used PROMETHE I and PROMETHE II for the assessment. PROMETHE I is used to determine partial ranking, and PROMETHE II is used for the complete ranking of the technologies. The hybrid technologies are found to be better	[72]

(Continued)

TABLE 1.6 (Continued)

Sl. no.	Year	Objective	Contribution	Reference
7	2004	Evaluation and policy devising for replacing the cooking fuel with renewables	Analysis is done on nine energy alternatives for cooking available in India. The outranking nature of the method is exploited	[73]
8	2000	The multicriteria nature in utilizing the locally available resources is demonstrated to be studied using PROMETHE	The PROMETHE method is comprehensively utilized to contract with the fuzzy type data. The drawbacks and advantages are presented in the conclusion as a guideline to use fuzzy data type input to PROMETHE	[74]
9	1998	Design of a group decision support system to promote the utilization of renewable energy sources	PROMETHE II is used to evaluate the scenarios. The developed method is used in the planning of the renewable energy source utilization in Greece	[75]

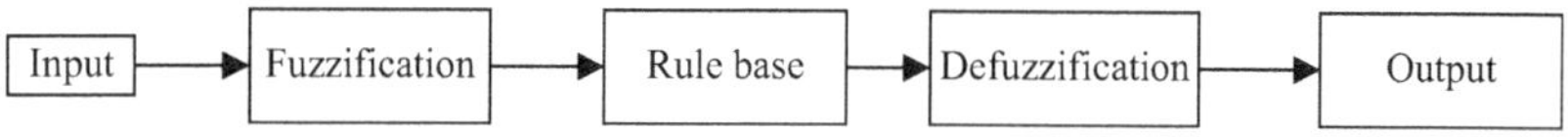

FIGURE 1.5 A fuzzy logic system.

constraints can be represented symmetrically as fuzzy sets. The decision is to be well defined such that the union between the constraints and the goals is satisfied. The maximum value of the membership function is to be defined at the point with maximized decision. Few other approaches are proposed by Bass and Kwakernaak (1977) and Yager (1978) [81,82]. The phase-wise approach was proposed by Dubois and Prade (1980), Zimmermann (1987), Chen and Hwang (1992), and Ribeiro (1996) [83–86]. The first phase is meant to determine the performance rating of the alternatives or calculating the degree of satisfaction w.r.t. all the attributes of respective alternatives. The second phase calculates the order of ranking of all alternatives based on the results of the first phase.

Although fuzzy-based MCDM solves a major issue of uncertainty or fuzziness in a decision-making problem, there are many drawbacks based on the literature. There is no standard solution technique to solve, mathematical model to represent a problem, increased complexity, and ambiguity. It is not possible to incorporate any quantitative factor, and the determined solution is very difficult to be analyzed.

1.3.1 Fuzzy–analytical hierarchical process

AHP is one of the most commonly used methods in the MADM procedure. It is considered to be one of the most systematic and logical approaches to determine the solution. It is noticed that decision maker finds it more relevant to give a conclusion that is in any range of values rather than exact numbers. There are various approaches proposed in the literature to jointly use fuzzy with AHP. Few of them are listed in Table 1.7.

The Buckley approach is described here in six steps.

Step 1: Fuzzification

Saaty proposed a scale to prioritize alternatives [26]. The scale is based on crisp values ranging between 1 and 8. Since fuzzy logic is based on linguistic variable, a range is defined for the same scale. The corresponding fuzzy triangular scale and the equivalent Saaty scale are given in Table 1.8.

The pairwise decision matrix is created using the fuzzy triangular scale. Let the elements of the matrix be f_{ij}^k that is a fuzzy number depicting the preference of the kth decision maker in selecting i over j. The complete matrix is given in the following equation:

TABLE 1.7 Fuzzy with AHP approaches with the respective years [87–92].

Sl. no.	Name	Year
1	Van Laarhoven and Pedrycz's approach	1983
2	Buckley's approach	1985
3	Boender's approach	1989
4	Extent analysis method by Chang	1992
5	Entropy-based fuzzy–AHP by Cheng	1996
6	Kahraman et al.	2004

AHP, Analytical hierarchical process.

$$F^k = \begin{matrix} f_{11}^k & f_{12}^k & \cdots & f_{1j}^k \\ f_{21}^k & \cdots & \cdots & f_{2j}^k \\ \cdots & \cdots & \cdots & \cdots \\ f_{j1}^k & f_{j2}^k & \cdots & f_{jj}^k \end{matrix} \tag{1.34}$$

Step 2: Normalization of the preferences of decision maker or averaged value calculation

Let N_{ij} be the averaged value.

$$N_{ij} = \frac{\sum_{k=1}^{m} f_{ij}^k}{m} \tag{1.35}$$

where m is the total number of decision makers.

Step 3: Update the pairwise decision matrix

Based on the average value calculated, the matrix is updated. It is given by

$$F = \begin{bmatrix} N_{11} & \cdots & N_{1j} \\ \vdots & \ddots & \vdots \\ N_{j1} & \cdots & N_{jj} \end{bmatrix} \tag{1.36}$$

Step 4: Calculation of geometric mean of the normalized values

Let G_i be the geometric mean.

$$G_i = \left[\prod_{j=1}^{n} N_{ij} \right]^{1/n} \tag{1.37}$$

TABLE 1.8 Fuzzy triangular scale for equivalent Saaty scale [87–92].

Definition	Saaty scale	Fuzzy triangular scale
Equally important	1	1,1,1
Weakly important	3	2,3,4
Fairly important	5	4,5,6
Strongly important	7	6,7,8
Absolutely important	9	9,9,9
The values between two adjacent scales	2	1,2,3
	4	3,4,5
	6	5,6,7
	8	7,8,9

Step 5: Calculation of fuzzy weight

The weight is calculated by the following formulae:

$$W_i = G_i \times (G_1 + G_2 + \ldots + G_n)^{-1} \tag{1.38}$$

$$W_i = x(G_i), y(G_i), z(G_i) \tag{1.39}$$

where x, y, and z are the three values of the triangular membership function.

Step 6: Defuzzification and normalization

The values determined in Step 5 are converted based on the center of area. Let the defuzzified value be D_i:

$$D_i = \frac{x(G_i) + y(G_i) + z(G_i)}{3} \tag{1.40}$$

The value is normalized by the following equation:

$$N_i = \frac{D_i}{\sum_{k=1}^{m} D_i} \tag{1.41}$$

The alternatives with the highest value of N_i are the most preferred ones and are suggested by the decision makers.

1.3.2 Fuzzy technique for order preference by similarity to ideal solutions

This method is based on the idea that the geometric distance between the best alternative should be the shortest from the PIA$^+$. Also, it should have the longest geometric distance from the NIA$^-$. A relative closeness or similarity index is determined considering the distance. This method was developed by Hwang and Yoon [93]. This method was later modified by Chen and Hwang [86]. Several versions of fuzzy–TOPSIS can be found in the literature [86,93]. In this chapter the method developed by Chen and Hwang [86] is described in detail considering the shape of the membership function trapezoidal. Most of the methods developed later have been possible due to this method with a minor modification in the ranking method.

Step 1: Formulation of decision matrix

Let D be the decision matrix formed by the decision makers. Let the elements of the matrix be represented by y_{ij} that is either a fuzzy or crisp value for a given value of i and j.

$$D = \begin{bmatrix} y_{11} & \cdots & y_{1n} \\ \vdots & y_{ij} & \vdots \\ y_{m1} & \cdots & y_{mn} \end{bmatrix} \tag{1.42}$$

Step 2: Normalization of the decision matrix

Each element of the decision matrix is normalized based on the following formula:

$$n_{ij} = \begin{bmatrix} \dfrac{y_{ij}}{y_j} \text{ for a ideal positive value of } j\text{th attribute} \\ \dfrac{y_j^-}{y_{jj}} \text{ for a ideal negative value of } j\text{th attribute} \end{bmatrix} \tag{1.43}$$

The normalized value calculated in the previous equation is also called benefit attribute for the positive value and cost attribute for the negative value.

Step 3: Calculation of weighted normalized decision matrix

For the case when the elements of decision matrix are a crisp value, it is easy to proceed, but if the values are fuzzy, each element is to be represented in terms of fuzzy values. Let $y_{ij} = (a_{ij}, b_{ij}, c_{ij}, d_{ij})$ and $y_j = (y_j, y_j, y_j, y_j)$. The weighted normalized decision matrix for the matrix is obtained as given by

$$WN_{ij} = n_{ij} w_j \tag{1.44}$$

where w_j is the weights of the fuzzy values, and n_{ij} is calculated as

$$n_{ij} = \begin{bmatrix} y_{ij}(+)y_j = \left(\dfrac{a_{ij}}{d_j}, \dfrac{b_{ij}}{c_j}, \dfrac{c_{ij}}{b_j}, \dfrac{d_{ij}}{a_j}\right) \\ y_j^-(+)y_{ij} = \left(\dfrac{a_j^-}{d_{jj}}, \dfrac{b_j^-}{c_{jj}}, \dfrac{c_j^-}{b_{jj}}, \dfrac{d_j^-}{a_{jj}}\right) \end{bmatrix} \tag{1.45}$$

The calculation of WN_{ij} becomes simple for crisp values, but when the values of n_{ij} are fuzzy, the formulae of WN_{ij} are updated as follows:

$$n_{ij} = \begin{bmatrix} y_{ij}(.)y_j = \left(\dfrac{a_{ij}}{d_j}\alpha_j, \dfrac{b_{ij}}{c_j}\beta_j, \dfrac{c_{ij}}{b_j}\gamma_j, \dfrac{d_{ij}}{a_j}\delta_j\right) & \text{for the benefit attribute} \\ y_j^-(.)y_{ij} = \left(\dfrac{a_j^-}{d_{jj}}\alpha_j, \dfrac{b_j^-}{c_{jj}}\beta_j, \dfrac{c_j^-}{b_{jj}}\gamma_j, \dfrac{d_j^-}{a_{jj}}\delta_j\right) & \text{for the cost attribute} \end{bmatrix} \tag{1.46}$$

The final weighted normalized decision matrix is given in the following form:

$$WN = \begin{bmatrix} WN_{11} & \cdots & WN_{1n} \\ \vdots & WN_{ij} & \vdots \\ WN_{m1} & \cdots & WN_{mn} \end{bmatrix} \tag{1.47}$$

Step 4: Calculation of positive and negative ideal solution

As calculated in the conventional TOPSIS, and hybrid technique, these values are calculated. The values are easily calculated for systems with the crisp values as given in the following equations:

$$P^* = [WN_1, \ldots, WN_n] \tag{1.48}$$

$$N^* = [WN_1^-, \ldots, WN_n^-] \tag{1.49}$$

where WN_n is equivalent to $\max(WN_{ij})$ and $WN_n^- = \min(WN_{ij})$. In case the values are fuzzy, the calculation becomes complex and is calculated as a generalized mean for the fuzzy number. This is calculated in numerous ways. The mathematical equation of the Lee and Li's ranking method is given as follows:

$$\text{Mean}\left(WN_{ij}\right) = \frac{-a_{ij}^2 - b_{ij}^2 + c_{ij}^2 + d_{ij}^2 - a_{ij}b_{ij} + c_{ij}d_{ij}}{[3(-a_{ij} - b_{ij} + c_{ij} + d_{ij})]} \tag{1.50}$$

The value of WN_n is equivalent to $\max(\text{Mean}(WN_{ij}))$ and $WN_n^- = \min(\text{Mean}\left(WN_{ij}\right))$.

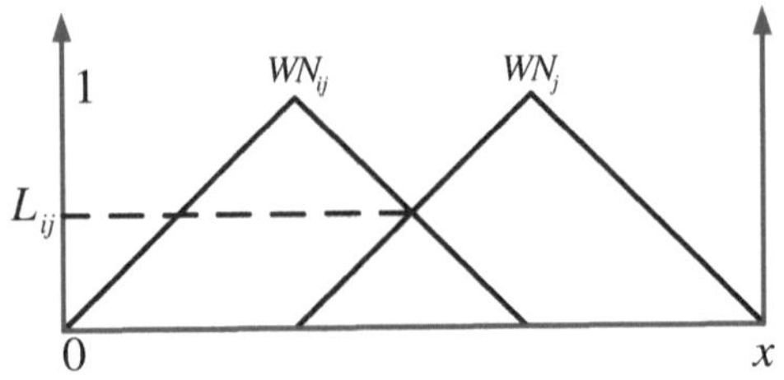

FIGURE 1.6 Pictorial representation of L_{ij}.

Step 5: Calculation of separation measures

Let S_p and S_n be the separation measures. Classically they are defined as

$$S_p = \sum_{j=1}^{n} d_{ij}, \quad i = 1, \ldots, m \text{ and} \tag{1.51}$$

$$S_i = \sum_{j=1}^{n} d_{ij}^{-}, \quad i = 1, \ldots, m \tag{1.52}$$

where d_{ij} and d_{ij}^{-} are the difference measures. For a crisp value, it is easily calculated as $d_{ij} = |WN_{ij} - WN_j|$ and $d_{ij}^{-} = |WN_{ij} - WN_j^{-}|$. With fuzzy numbers, the difference is made based on the fuzzy numbers. It is explained in the following equations:

$$d_{ij} = 1 - sup_x\left[\mu_{WN_{ij}}(x) \wedge \mu_{WN_j}(x)\right] = 1 - L_{ij} \tag{1.53}$$

$$d_{ij}^{-} = 1 - sup_x\left[\mu_{WN_{ij}}(x) \wedge \mu_{WN_{ij}^{-}}(x)\right] = 1 - L_{ij} \tag{1.54}$$

where L_{ij} is given in Fig. 1.6.

Step 6: Calculation of relative closeness of positive and negative ideals

Let C_i be the relative closeness. It is defined by the following equation:

$$C_i = \frac{S_n}{(S_p + S_n)} \tag{1.55}$$

The ranking is done based on the value of C_i. The higher the value, the higher the rank of the alternative. Few works in which hybrid techniques were utilized are given in Table 1.9.

A brief list of advantages, disadvantages, and applications of the methods mentioned in this chapter is given in Table 1.10.

TABLE 1.9 Hybrid techniques in energy planning.

Sl. no.	Year	Objective	Contribution	Reference
1	2014	To develop a model to analyze the performance of energy systems in buildings and to increase their efficiency	A building energy performance calculation method is developed. Fuzzy analytic network process with AHP and TOPSIS is used in this work	[34]
2	2001	To select suitable projects w.r.t. renewables—solar and wind energy systems	This work identifies the success factors for deploying solar and wind energy systems. Fuzzy and AHP are used in the work	[94]
3	2014	To propose a new preference scale in the intuitionistic fuzzy–AHP process and check the applicability in a practical case study related to solve the sustainable energy planning	The intuitionistic fuzzy is used to determine the weight to be used in AHP. A case study is done with nuclear, solar, wind, biomass, combined heat power, and conventional as alternatives	[95]
4	2011	To develop a model to evaluate the potential and present environmental performance of an urban region	AHP and fuzzy logic are used to formulate the model. The calculation of weight for AHP is claimed to be improved. In addition, an inclusive list of indicators is prepared for the evaluation of Beijing	[39]
5	2010	To analyze the list of factors/criteria that will affect the evaluation of "energy dissemination program" in the Republic of Korea	Five criteria (technical, economic, market associated, environmental, and policy making) are considered in the evaluation. The criteria "economic" is found to be the most significant factor affecting the dissemination program	[96]
6	2013	To develop a multicriteria decision support system to design and develop the hybrid energy systems	Fuzzy logic and TOPSIS are used to develop the system. 1D and 2D level figures are used with the fuzzy–TOPSIS to obtain the weight matrix. The	[97]

(*Continued*)

TABLE 1.9 (Continued)

Sl. no.	Year	Objective	Contribution	Reference
			technique can be utilized in solving other (not related to hybrid energy system) decision-making problems	
7	2012	To define the site for a thermal power plant in India	Fuzzy logic, AHP, and TOPSIS are used in this work. Fuzzy–AHP method is adopted to solve the issue of vagueness, and TOPSIS is used for ranking the locations	[47]
8	2012	An evaluation model is developed to check the benefits of combined cooling, heating, and power system plants	Fuzzy is combined with multicriteria decision-making process, weighing method, and gray system theory to evaluate the system	[98]
9	2012	Selection of prime mover for combined cooling, heating, and power system plants	Four criteria—technical, economic, social, and environmental—are used to check four options—conventional separate production, microgas turbine, internal combustion engine, and sterling engine. Fuzzy combined with gray incidence and AHP are used in this work for the analysis	[99]
10	2015	To develop a priority for hydrogen-producing technology to incline toward the growing hydrogen economy in China	Fuzzy–AHP and fuzzy–TOPSIS are used in the prioritization of technology. The developed model is capable of allowing many stakeholders to participate in the decision-making	[100]

AHP, Analytical hierarchical process; *TOPSIS*, technique for order preference by similarity to ideal solutions.

TABLE 1.10 Brief of various MCDM techniques.

Sl. no.	Name	Comparison and properties	Applications	Reference
1	Weighted sum method	*Advantage*: Weighted sum is the simplest method and involves fundamental calculations, whereas the weighted product method works on the basis of the relative quantities of alternatives	1. Energy planning 2. Work and organizational optimization 3. Bidding planning and policies	[8,24,25]
2	Weighted product method	*Disadvantage*: Both the methods are not suitable for multidimensional problem		
3	Analytical hierarchy process	*Advantage*: This method involves a hierarchical organization of criteria, which adds to the visibility of criteria *Disadvantage*: A major drawback of this method can be seen when a situation of interdependency arise between objective and alternatives	1. Policy making and energy planning 2. Transportation engineering and resource management 3. Logistics supply chain management 4. Water resource management 5. Communication engineering and information technology 6. Financial and business management 7. Civil and mining engineering	[27–31,39]
4	TOPSIS	*Advantage*: The problem of interdependency in AHP is not in this method. This method is suitable to work out the fundamental ranking of the alternatives *Disadvantage*: The use of the Euclidian distance in the calculation makes this method futile in some context		[37,44,51]
5	ELECTRE	*Advantage*: This method is valuable in dealing with heterogeneous measures. It can deal with both qualitative and quantitative attributes of the criteria, thus validating the results with proper reasons *Disadvantage*: Defining the quantitative scale or data is the only concern in this method		[54–60]
6	PROMETHE	*Advantage*: Unlike ELECTRE, this method also involves both qualitative and quantitative information in the decision-making. In addition, the involvement of group and fuzzy statistics makes this method preferred over other MCDM techniques *Disadvantage*: The assigning of the weight by the decision maker based on his wish makes the process unacceptable to many		[67–75]

(Continued)

TABLE 1.10 (Continued)

Sl. no.	Name	Comparison and properties	Applications	Reference
7	Fuzzy–AHP	*Advantage*: The drawback of AHP being subjective and based on a scale is overcome by using the fuzzy–AHP	1. Energy management 2. Civil engineering 3. Logistics 4. Risk assessment 5. Agriculture planning 6. Electric vehicles charging infrastructure	[94–100]
		Disadvantage: The uncertainty in the prioritization is concealed by this technique		
8	Fuzzy–TOPSIS	*Advantage*: The advantage of fuzzy–TOPSIS over TOPSIS is seen when the data is not precise or vague		
		Disadvantage: A considerable drawback of fuzzy–TOPSIS is the linguistic variable and their range, which is dependent on the decision maker		
		The advantage of fuzzy–TOPSIS will be visible in the cases when the importance of each of the criteria is also in the process of evaluation as fuzzy–AHP evaluates the alternatives in each of the criteria		

AHP, Analytical hierarchical process; *ELECTRE*, elimination and choice translating reality; *MCDM*, multicriteria decision-making; *PROMETHE*, preference ranking organization method for enrichment evaluation; *TOPSIS*, technique for order preference by similarity to ideal solutions.

1.4 Conclusion

MCDM is a revolutionary tool having a wide range of applications in energy planning covering the sustainable dimensions. For successful design and implementation of energy projects based on renewable energy sources, which involves a complex multiobjective traits, and differing views of stakeholders, MCDM can certainly play a crucial role. This chapter has iterated a few of the most popular techniques currently applicable not only in energy planning but also in other sectors, such as design and manufacturing, business models, and logistics management. A detailed process with the help of various flowcharts has been included for a better understanding of the readers.

References

[1] IEA, Energy Access Outlook 2017, 2017.

[2] IEA, IEA Statistics ©OECD/IEA, 2014.

[3] IEA, India Energy Outlook, 2015.

[4] S.R. Thangavelu, A.M. Khambadkone, I.A. Karimi, Long-term optimal energy mix planning towards high energy security and low GHG emission, Appl. Energy 154 (2015) 959–969.

[5] L. Proskuryakova, Updating energy security and environmental policy: energy security theories revisited, J. Environ. Manage. 223 (2018) 203–214.

[6] IEA, Introduction and Scope: Thinking About the Future of Energy, International Energy Outlook, 2017.

[7] A. Kumar, B. Sah, Y. Deng, X. He, P. Kumar, R.C. Bansal, Application of multi-criteria decision analysis tool for design of a sustainable micro-grid for a remote village in the Himalayas, J. Eng. 2017 (2017) 2108–2113.

[8] A. Kumar, B. Sah, A.R. Singh, Y. Deng, X. He, P. Kumar, et al., A review of multi criteria decision making (MCDM) towards sustainable renewable energy development, Renewable Sustainable Energy Rev. 69 (2017) 596–609.

[9] I. a. R. IRENA, Renewable energy policies in a Time of Transition, IRENA, OECD/IEA and REN21, 2018.

[10] F.S. Javadi, B. Rismanchi, M. Sarraf, O. Afshar, R. Saidur, H.W. Ping, et al., Global policy of rural electrification, Renewable Sustainable Energy Rev. 19 (2013) 402–416.

[11] S.S. Mohammad, S. Mishra, S.K. Sinha, V.K. Tayal, Integration of renewable energy resources for rural electrification, in: R. Singh, S. Choudhury (Eds.), Proceeding of International Conference on Intelligent Communication, Control and Devices: ICICCD 2016, Springer Singapore, Singapore, 2017, pp. 545–551.

[12] M.E. Khodayar, Rural electrification and expansion planning of off-grid microgrids, Electr. J. 30 (2017) 68–74.

[13] M.J. Herington, E. van de Fliert, S. Smart, C. Greig, P.A. Lant, Rural energy planning remains out-of-step with contemporary paradigms of energy access and development, Renewable Sustainable Energy Rev. 67 (2017) 1412–1419.

[14] J.F. Alfaro, S. Miller, J.X. Johnson, R.R. Riolo, Improving rural electricity system planning: an agent-based model for stakeholder engagement and decision making, Energy Policy 101 (2017) 317–331.

[15] T. Urmee, A. Md, Social, cultural and political dimensions of off-grid renewable energy programs in developing countries, Renewable Energy 93 (2016) 159–167.
[16] A. Kumar, Y. Deng, X. He, P. Kumar, A multi criteria decision based rural electrification system, in: Industrial Electronics Society, IECON 2016—42nd Annual Conference of the IEEE, 2016, pp. 4025–4030.
[17] J.C. Rojas-Zerpa, J.M. Yusta, Application of multicriteria decision methods for electric supply planning in rural and remote areas, Renewable Sustainable Energy Rev. 52 (2015) 557–571.
[18] A. Zomers, Remote access: context, challenges, and obstacles in rural electrification, IEEE Power Energ. Mag. 12 (2014) 26–34.
[19] B. Domenech, L. Ferrer-Martí, R. Pastor, Hierarchical methodology to optimize the design of stand-alone electrification systems for rural communities considering technical and social criteria, Renewable Sustainable Energy Rev. 51 (2015) 182–196.
[20] A.K. Basu, S.P. Chowdhury, S. Chowdhury, S. Paul, Microgrids: energy management by strategic deployment of DERs—a comprehensive survey, Renewable Sustainable Energy Rev. 15 (2011) 4348–4356.
[21] P. Rocha, A. Siddiqui, M. Stadler, Improving energy efficiency via smart building energy management systems: a comparison with policy measures, Energy Build. 88 (2015) 203–213.
[22] M. Rudberg, M. Waldemarsson, H. Lidestam, Strategic perspectives on energy management: a case study in the process industry, Appl. Energy 104 (2013) 487–496.
[23] J.C. Mourmouris, C. Potolias, A multi-criteria methodology for energy planning and developing renewable energy sources at a regional level: a case study Thassos, Greece, Energy Policy 52 (2013) 522–530.
[24] H.-C. Lee, C.-T. Chang, Comparative analysis of MCDM methods for ranking renewable energy sources in Taiwan, Renewable Sustainable Energy Rev. 92 (2018) 883–896.
[25] T. Evangelos, Multi-Criteria Decision Making Methods: A Comparative Study, Kluwer Academic Publication, The Netherlands, 2000.
[26] R.W. Saaty, The analytic hierarchy process—what it is and how it is used, Math. Modell. 9 (1987) 161–176.
[27] A. Kumar, A.R. Singh, Y. Deng, X. He, P. Kumar, R.C. Bansal, A novel methodological framework for the design of sustainable rural microgrid for developing nations, IEEE Access 6 (2018) 24925–24951.
[28] S. Sindhu, V. Nehra, S. Luthra, Investigation of feasibility study of solar farms deployment using hybrid AHP-TOPSIS analysis: case study of India, Renewable Sustainable Energy Rev. 73 (2017) 496–511.
[29] G. Büyüközkan, Y. Karabulut, Energy project performance evaluation with sustainability perspective, Energy 119 (2017) 549–560.
[30] D. Štreimikienė, J. Šliogerienė, Z. Turskis, Multi-criteria analysis of electricity generation technologies in Lithuania, Renewable Energy 85 (2016) 148–156.
[31] T. Wanderer, S. Herle, Creating a spatial multi-criteria decision support system for energy related integrated environmental impact assessment, Environ. Impact Assess. Rev. 52 (2015) 2–8.
[32] Q. Wu, H. Ren, W. Gao, J. Ren, Multi-criteria assessment of combined cooling, heating and power systems located in different regions in Japan, Appl. Therm. Eng. 73 (2014) 660–670.

[33] A. Werner, A. Werner, R. Wieland, K.-C. Kersebaum, W. Mirschel, H.-P. Ende, et al., Ex ante assessment of crop rotations focusing on energy crops using a multi-attribute decision-making method, Ecol. Indic. 45 (2014) 110–122.

[34] M. Kabak, E. Köse, O. Kırılmaz, S. Burmaoğlu, A fuzzy multi-criteria decision making approach to assess building energy performance, Energy Build. 72 (2014) 382–389.

[35] A. Mattiussi, M. Rosano, P. Simeoni, A decision support system for sustainable energy supply combining multi-objective and multi-attribute analysis: an Australian case study, Decis. Support Syst. 57 (2014) 150–159.

[36] E.W. Stein, A comprehensive multi-criteria model to rank electric energy production technologies, Renewable Sustainable Energy Rev. 22 (2013) 640–654.

[37] J.M. Sánchez-Lozano, J. Teruel-Solano, P.L. Soto-Elvira, M. Socorro García-Cascales, Geographical information systems (GIS) and multi-criteria decision making (MCDM) methods for the evaluation of solar farms locations: case study in south-eastern Spain, Renewable Sustainable Energy Rev. 24 (2013) 544–556.

[38] H. Davoudpour, S. Rezaee, M. Ashrafi, Developing a framework for renewable technology portfolio selection: a case study at a R&D center, Renewable Sustainable Energy Rev. 16 (2012) 4291–4297.

[39] L. Wang, L. Xu, H. Song, Environmental performance evaluation of Beijing's energy use planning, Energy Policy 39 (2011) 3483–3495.

[40] G.-H. Tzeng, J.-J. Huang, Multiple Attribute Decision Making: Methods and Applications, Chapman and Hall/CRC Press, 2011.

[41] M. Behzadian, S. Khanmohammadi Otaghsara, M. Yazdani, J. Ignatius, A state-of the-art survey of TOPSIS applications, Expert Syst. Appl. 39 (2012) 13051–13069.

[42] Y.-J. Lai, T.-Y. Liu, C.-L. Hwang, TOPSIS for MODM, Eur. J. Oper. Res. 76 (1994) 486–500.

[43] J.R.S.C. Mateo, TOPSIS, Multi Criteria Analysis in the Renewable Energy Industry, Springer, 2012, pp. 43–48.

[44] D. Streimikiene, T. Baležentis, Multi-criteria assessment of small scale CHP technologies in buildings, Renewable Sustainable Energy Rev. 26 (2013) 183–189.

[45] E. Bas, The integrated framework for analysis of electricity supply chain using an integrated SWOT-fuzzy TOPSIS methodology combined with AHP: the case of Turkey, Int. J. Electr. Power Energy Syst. 44 (2013) 897–907.

[46] A. Sarkar, A TOPSIS method to evaluate the technologies, Int. J. Qual. Reliab. Manage. 31 (2013) 2–13.

[47] D. Choudhary, R. Shankar, An STEEP-fuzzy AHP-TOPSIS framework for evaluation and selection of thermal power plant location: a case study from India, Energy 42 (2012) 510–521.

[48] R. Dapkus, D. Streimikiene, Multi-criteria assessment of electricity generation technologies seeking to implement EU energy policy targets, Proc. Econ. Dev. Res. 55 (2012) 50–56.

[49] T. Kaya, C. Kahraman, Multicriteria decision making in energy planning using a modified fuzzy TOPSIS methodology, Expert Syst. Appl. 38 (2011) 6577–6585.

[50] H. Doukas, C. Karakosta, J. Psarras, Computing with words to assess the sustainability of renewable energy options, Expert Syst. Appl. 37 (2010) 5491–5497.

[51] M.-T. Chu, J. Shyu, G.-H. Tzeng, R. Khosla, Comparison among three analytical methods for knowledge communities group-decision analysis, Expert Syst. Appl. 33 (2007) 1011–1024.

[52] R. Benayoun, B. Roy, B. Sussman, "ELECTRE: Une méthode pour guider le choix en présence de points de vue multiples", Note de travail, vol. 49 (1966).

[53] M. Mousavi, H. Gitinavard, S.M. Mousavi, A soft computing based-modified ELECTRE model for renewable energy policy selection with unknown information, Renewable Sustainable Energy Rev. 68 (2017) 774–787.

[54] D. Jun, F. Tian-tian, Y. Yi-sheng, M. Yu, Macro-site selection of wind/solar hybrid power station based on ELECTRE-II, Renewable Sustainable Energy Rev. 35 (2014) 194–204.

[55] J.M. Sánchez-Lozano, C. Henggeler Antunes, M.S. García-Cascales, L.C. Dias, GIS-based photovoltaic solar farms site selection using ELECTRE-TRI: evaluating the case for Torre Pacheco, Murcia, Southeast of Spain, Renewable Energy 66 (2014) 478–494.

[56] S. Vinodh, R.J. Girubha, Sustainable concept selection using ELECTRE, Clean Technol. Environ. Policy 14 (2012) 651–656.

[57] T. Catalina, J. Virgone, E. Blanco, Multi-source energy systems analysis using a multi-criteria decision aid methodology, Renewable Energy 36 (2011) 2245–2252.

[58] L.P. Neves, A.G. Martins, C.H. Antunes, L.C. Dias, A multi-criteria decision approach to sorting actions for promoting energy efficiency, Energy Policy 36 (2008) 2351–2363.

[59] A. Papadopoulos, A. Karagiannidis, Application of the multi-criteria analysis method ELECTRE III for the optimisation of decentralised energy systems, Omega 36 (2008) 766–776.

[60] M. Beccali, M. Cellura, D. Ardente, Decision making in energy planning: the ELECTRE multicriteria analysis approach compared to a FUZZY-SETS methodology, Energy Convers. Manage. 39 (1998) 1869–1881.

[61] J.P. Brans, Ph Vincke, A preference ranking organisation method: (the PROMETHE method for multiple criteria decision-making), Manage. Sci. 31 (1985) 647–656.

[62] A.T. de Almeida Filho, T.R.N. Clemente, D.C. Morais, A.T. de Almeida, Preference modeling experiments with surrogate weighting procedures for the PROMETHE method, Eur. J. Oper. Res. 264 (2018) 453–461.

[63] J.-P. Brans, B. Mareschal, PROMETHEE methods, Multiple Criteria Decision Analysis: State of the Art Surveys, Springer, 2005, pp. 163–186.

[64] N. Munier, A Strategy for Using Multicriteria Analysis in Decision-Making: A Guide for Simple and Complex Environmental Projects, Springer Science & Business Media, 2011.

[65] J.R.S.C. Mateo, Multi Criteria Analysis in the Renewable Energy Industry, Springer Science & Business Media, 2012.

[66] S. Vinodh, R. Jeya Girubha, PROMETHEE based sustainable concept selection, Appl. Math. Modell. 36 (2012) 5301–5308.

[67] D.A. Haralambopoulos, H. Polatidis, Renewable energy projects: structuring a multi-criteria group decision-making framework, Renewable Energy 28 (2003) 961–973.

[68] D. Georgiou, E.S. Mohammed, S. Rozakis, Multi-criteria decision making on the energy supply configuration of autonomous desalination units, Renewable Energy 75 (2015) 459–467.

[69] M. Troldborg, S. Heslop, R.L. Hough, Assessing the sustainability of renewable energy technologies using multi-criteria analysis: suitability of approach for national-scale assessments and associated uncertainties, Renewable Sustainable Energy Rev. 39 (2014) 1173–1184.

[70] T. Tsoutsos, M. Drandaki, N. Frantzeskaki, E. Iosifidis, I. Kiosses, Sustainable energy planning by using multi-criteria analysis application in the island of Crete, Energy Policy 37 (2009) 1587–1600.
[71] H. Ren, W. Gao, W. Zhou, K.I. Nakagami, Multi-criteria evaluation for the optimal adoption of distributed residential energy systems in Japan, Energy Policy 37 (2009) 5484–5493.
[72] F. Cavallaro, Multi-criteria decision aid to assess concentrated solar thermal technologies, Renewable Energy 34 (2009) 1678–1685.
[73] S.D. Pohekar, M. Ramachandran, Multi-criteria evaluation of cooking energy alternatives for promoting parabolic solar cooker in India, Renewable Energy 29 (2004) 1449–1460.
[74] M. Goumas, V. Lygerou, An extension of the PROMETHEE method for decision making in fuzzy environment: ranking of alternative energy exploitation projects, Eur. J. Oper. Res. 123 (2000) 606–613.
[75] E. Georgopoulou, Y. Sarafidis, D. Diakoulaki, Design and implementation of a group DSS for sustaining renewable energies exploitation, Eur. J. Oper. Res. 109 (1998) 483–500.
[76] L. Zadeh, Fuzzy sets, Inform. Contr. 8 (1965) 421–427.
[77] T.J. Ross, Fuzzy Logic With Engineering Applications, John Wiley & Sons, 2005.
[78] M. Singh, P. Kumar, I. Kar, Implementation of vehicle to grid infrastructure using fuzzy logic controller, IEEE Trans. Smart Grid 3 (2012) 565–577.
[79] R.E. Bellman, L.A. Zadeh, Decision-making in a fuzzy environment, Manage. Sci. 17 (1970) B-141–B-164.
[80] H.-J. Zimmermann, Fuzzy programming and linear programming with several objective functions, Fuzzy Sets Syst. 1 (1978) 45–55.
[81] R.R. Yager, Fuzzy decision making including unequal objectives, Fuzzy Sets Syst. 1 (1978) 87–95.
[82] S.M. Baas, H. Kwakernaak, Rating and ranking of multiple-aspect alternatives using fuzzy sets, IFAC Proc. Vol. 9 (1976) 585–600.
[83] R.A. Ribeiro, Fuzzy multiple attribute decision making: a review and new preference elicitation techniques, Fuzzy Sets Syst. 78 (1996) 155–182.
[84] D.J. Dubois, Fuzzy Sets and Systems: Theory and Applications, vol. 144, Academic Press, 1980.
[85] H.-J. Zimmermann, Fuzzy Sets, Decision Making, and Expert Systems, vol. 10, Springer Science & Business Media, 2012.
[86] S.-J. Chen, C.-L. Hwang, Fuzzy multiple attribute decision making methods, Fuzzy Multiple Attribute Decision Making, Springer, 1992, pp. 289–486.
[87] P. Van Laarhoven, W. Pedrycz, A fuzzy extension of Saaty's priority theory, Fuzzy Sets Syst. 11 (1983) 229–241.
[88] J.J. Buckley, Fuzzy hierarchical analysis, Fuzzy Sets Syst. 17 (1985) 233–247.
[89] C. Boender, J. De Graan, F. Lootsma, Multi-criteria decision analysis with fuzzy pairwise comparisons, Fuzzy Sets Syst. 29 (1989) 133–143.
[90] D. Chang, Extent Analysis and Synthetic Decision, Optimization Techniques and Applications, vol. 1, World Scientific, Singapore, 1992.
[91] D.-Y. Chang, Applications of the extent analysis method on fuzzy AHP, Eur. J. Oper. Res. 95 (1996) 649–655.
[92] C. Kahraman, U. Cebeci, D. Ruan, Multi-attribute comparison of catering service companies using fuzzy AHP: the case of Turkey, Int. J. Prod. Econ. 87 (2004) 171–184.

[93] C.-L. Hwang, K. Yoon, Methods for multiple attribute decision making, Multiple Attribute Decision Making, Springer, 1981, pp. 58–191.

[94] H.H. Chen, H.-Y. Kang, A.H.I. Lee, Strategic selection of suitable projects for hybrid solar-wind power generation systems, Renew. Sustain. Energy Rev. 14 (2010) 413–421.

[95] L. Abdullah, L. Najib, Sustainable energy planning decision using the intuitionistic fuzzy analytic hierarchy process: choosing energy technology in Malaysia, Int. J. Sustainable Energy (2014) 1–18.

[96] E. Heo, J. Kim, K.-J. Boo, Analysis of the assessment factors for renewable energy dissemination program evaluation using fuzzy AHP, Renewable Sustainable Energy Rev. 14 (2010) 2214–2220.

[97] A.T.D. Perera, R.A. Attalage, K.K.C.K. Perera, V.P.C. Dassanayake, A hybrid tool to combine multi-objective optimization and multi-criterion decision making in designing standalone hybrid energy systems, Appl. Energy 107 (2013) 412–425.

[98] Y.-Y. Jing, H. Bai, J.-J. Wang, A fuzzy multi-criteria decision-making model for CCHP systems driven by different energy sources, Energy Policy 42 (2012) 286–296.

[99] M. Ebrahimi, A. Keshavarz, Prime mover selection for a residential micro-CCHP by using two multi-criteria decision-making methods, Energy Build. 55 (2012) 322–331.

[100] J. Ren, S. Gao, S. Tan, L. Dong, A. Scipioni, A. Mazzi, Role prioritization of hydrogen production technologies for promoting hydrogen economy in the current state of China, Renewable Sustainable Energy Rev. 41 (2015) 1217–1229.

Chapter 2

Uncertainty management in decision-making in power system operation

Mohammad Hemmati[1], Behnam Mohammadi-Ivatloo[1] and Alireza Soroudi[2]
[1]*Faculty of Electrical and Computer Engineering, University of Tabriz, Tabriz, Iran,* [2]*University College Dublin, Dublin, Ireland*

2.1 Introduction

The development of power system and the emergence of new energy concepts such as microgrids and smart grids have caused various challenges in the scheduling and operation of these networks. Modern power system scheduling such as conventional networks can be performed for short-, medium-, and long-term periods [1,2]. However, the need for accurate decision-making for these periods is essential. Managing the challenges of the power system faces a set of decision problems affiliated to different parts (e.g., scheduling, investment, and operation) where decision makers must distinguish all alternatives from cost, revenue, or risk point of views.

Apart from all of the modern power system challenges, the growth of total installed renewable energy resources with probabilistic nature causes the complex planning of power system. It is forecasted that the renewable energy share can reach 36% by 2030 [3]. This massive power generated by renewable energy is associated with uncertainties. Renewable energy resource power depends on initial sources such as wind and solar. Dependence of these resources on climate condition causes an increase of uncertainty in generated power [4]. However, as renewable energy penetration increases, there will be an increase in the uncertainty associated with power system. Hence, uncertainty modeling and suitable addressing in planning and operation of a power system are essential [5,6].

The uncertainty handling is one of the main issues of decision makers in power system [7]. All of the uncertain parameters faced by power system can be classified into economical and technical parameters. According to

Decision Making Applications in Modern Power Systems. DOI: https://doi.org/10.1016/B978-0-12-816445-7.00002-5

TABLE 2.1 Classified uncertain parameters in power system [7].

Technical parameters		Economical parameters
Operational parameters	**Topological parameters**	
Load demand	Line outage	Economic growth
Generation output	Generator outage	Price levels
		Governmental regulation
		Unemployment rate
		Fuel price

Ref. [7], technical uncertain parameters are related to a topological network such as failure rate of transmission lines or generators. The other technical uncertain parameters that affect the operating decisions are generation and demand values in system. The economical uncertain parameters contain energy price, economic growth, environmental policies, etc., which face the decision-making process with multiple challenges [8]. Table 2.1 categorizes all the possible uncertain parameters in the power system that should be handled for suitable operation of the system.

However, in scheduling and operation of power system, the main objective is cost minimization or profit maximization [9]; therefore the main uncertain parameters faced by us include generated power of renewable energies, load demand of consumption, and energy price. So, we focus on the modeling of these parameters. In the following section a comprehensive review on uncertainty handling in power system will be provided. Then, we will choose one of them and provide the example to describe the uncertainty modeling in power system.

2.2 Uncertainty management in power system: a review

There are multiple methods for handling uncertainty in power systems. The main feature that causes the discrepancy between multiple methods is in line with the different techniques used to describe the uncertain input data and parameters [10] and the degree of uncertainty. In the following section, some of the main uncertainty handling approaches will be described.

2.2.1 Probabilistic method

One of the simple and earliest methods, which assumes all the input uncertain parameters as random variables, is a probabilistic approach [11]. In this method, each variable has a known probability density function (PDF), and the uncertain parameters are modeled according to the corresponding PDF. The probabilistic programming was first introduced by Dantzig [12].

There are three different probabilistic approach techniques that are used for uncertainty management: Monte Carlo simulation (MCS), point estimation method (PEM), and scenario-based optimization method [13].

In MCS, n_s samples for each uncertain parameter are generated according to the corresponding PDF of each one. Assuming $n_s = \{n_1, n_2, \ldots, n_s\}$, $y^s = f(n_s)$ is calculated. The process is repeated for a lot of iterations, until the average value of each uncertain parameter is obtained. Some of the valuable works that use MCS in the probabilistic programming can be found in Refs. [13–17].

The PEM is based on the concept of statistical moments of input uncertain parameters. Unlike the MCS, the PEM only generates two samples for each parameter. Some of the valuable works that use MCS in the probabilistic programming can be found in Ref. [18].

The decision-making on the basis of scenarios is another technique that is based on probability theory. A list of scenarios is generated using the corresponding PDF of each uncertain parameter. In Section 2.3 an example of modern power system and scenario-based optimization approach is implemented to uncertainty modeling. The details of this technique are presented [19]. Some of the valuable works that use the scenario-based decision-making method can be found in Ref. [20].

2.2.2 Information gap decision theory

In some cases the uncertain parameters do not follow a PDF or the PDF is not known by the decision maker. In such cases the information gap decision theory (IGDT) is used to model the uncertainty [7,21]. In IGDT the robustness is defined as the inviolability of euphoria of a predefined constraint [21]. Suppose X is the set of uncertain parameters and all the components of this set are equal to its predicated value $\left(X = \overline{X}\right)$, so the predicated value for objective function $(\overline{y})$ is obtained. When the value of the uncertain parameter is not equal to the predicated value and is unknown, IGDT is implemented to find a good solution that gives the robustness feature to the value of the objective function against the predicting error. Some of the valuable works that use IGDT can be found in Refs. [22–25].

2.2.3 Robust optimization

The robust optimization technique is another important uncertainty modeling tool in the power system studies. In this method, it is assumed that the uncertain parameter belongs to an uncertainty set, and it is tried to make the optimal decision considering this fact. In other words the decision variables are found in a way that the objective function remains optimal even if the uncertain parameter takes its worst case value [26,27]. For example, consider a power system that includes wind and solar sources and load demand, which are all associated with uncertainty. The robust optimization first finds the worst case for the realization of uncertain parameters and then finds the optimal scheduling of system flexibilities. The use of this method makes the system robust against the uncertain parameters, and the value of objective function remains reasonable even if the uncertain parameters do not take their predicted values. In this way the risk of decision in scheduling and operation of system can be controlled [28]. Some of the valuable works that use the robust optimization method can be found in Refs. [29,30].

2.3 Problem formulation

In this section an example of modern power system scheduling problem is presented, which involves the uncertainty. The microgrid concept as a futuristic distribution network is one of the modern power systems that has attracted much interest in recent years. Each microgrid includes multiple components: distributed generation (generation units and energy storage system), different loads (controllable and noncontrollable load), switches, etc. The penetration of renewable energy, such as photovoltaic (PV) and wind in the microgrid with probabilistic nature, may cause the complexity for the scheduling and operation of these networks. Fig. 2.1 shows that the 33-bus microgrid consists of the wind and PV units as renewable energy resources,

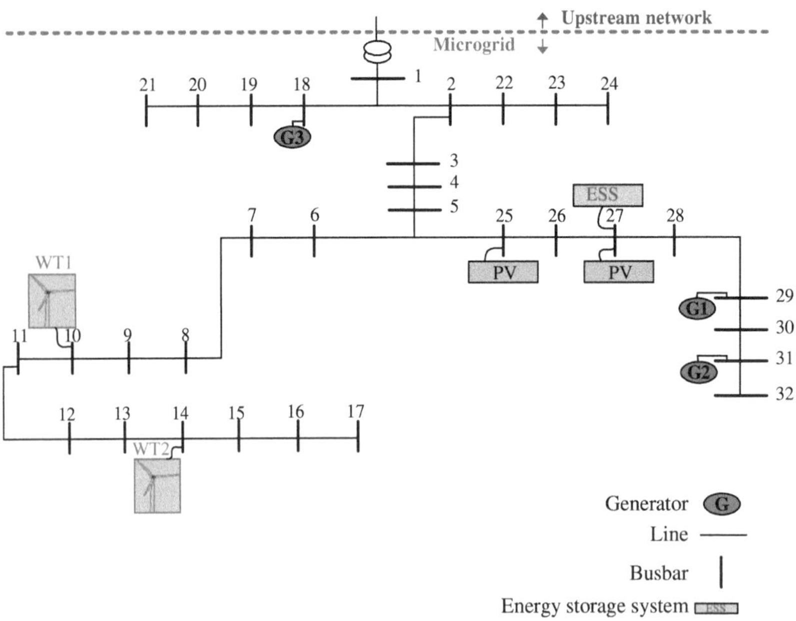

FIGURE 2.1 Schematic of microgrid with multiple units [31].

along with the dispatchable units (diesel generators and energy storage system), which can exchange the power with the upstream network.

In this problem the main objective is profit maximization, which is obtained by the revenues minus the system costs. The objective function can be formulated as follows:

$$\text{Maximize} \quad \text{Profit} = \text{RV} - \text{TC} \tag{2.1}$$

where RV is the microgrids (MGs) revenue ($), TC is the total cost ($) of MG. The revenue for MGs is calculated according to

$$\text{RV} = \sum_{\omega=1}^{N_\omega} \pi_\omega \sum_{t=1}^{T} \sum_{l=1}^{L} \lambda_t^L (P_{l,t,\omega}) + r \sum_{t=1}^{T} \lambda_t^N P_{t,\omega}^N, \quad \omega = 1, 2, \ldots, N_\omega \tag{2.2}$$

where λ_t^L is the power market price that the consumers pay at t^{th} time, $P_{l,t,\omega}$ is the active power demand of lth load at t^{th} time in ω^{th} scenario, $P_{t,\omega}^N$ is the power sold to the upstream network at tth time, λ_t^N is the price of selling/purchasing power with the upstream network at t^{th} time in ω^{th} scenario, r is the binary variable that separates selling or purchasing state; if r is 1 MG sells power to the upstream network, t is the index of time, i is the index of dispatchable unit, l is the index of load, ω is the index of the scenario, and N_ω is the number of scenarios.

The total operation cost of MG includes fuel cost, startup and shutdown cost, emission cost of the dispatchable units, and the cost of purchasing power from the upstream network that is calculated as follows:

$$\begin{aligned} \text{TC} = \sum_{\omega=1}^{N_\Omega} \pi_\omega \sum_{i=1}^{N} \sum_{t=1}^{T} &(F(P_{i,t,\omega}) X_{i,t,\omega} + \text{SU}_{i,t,\omega} + \text{SD}_{i,t,\omega} \\ &+ (C^{\text{emi}} \cdot \tau_i \cdot P_{i,t,\omega})) + (1 - r) \sum_{t=1}^{T} \lambda_t^N P_{t,\omega}^b \end{aligned} \tag{2.3}$$

where π_ω is the probability of the ωth scenario, and N_Ω is the number of total scenarios. $P_{i,t,\omega}$ is the power output of ith generator at t^{th} time and ω^{th} scenario. $X_{i,t,\omega}$ is the commitment state of ith generator at t^{th} time and ω^{th} scenario. $F(P_{i,t})$ is the fuel cost consumption function of the ith generator at t^{th} time that is calculated as follows:

$$F(P_{i,t}) = a_i + b_i P_{i,t} + c_i (P_{i,t})^2, \quad i \in N, t \in T \tag{2.4}$$

where a_i, b_i, and c_i are the cost coefficients of ith generator.

The second and the third terms of (2.2) show the technical cost of generator named startup and shutdown cost, respectively. The fourth term represents the emission cost of ith generator, τ_i is the emission factor (kg/kWh) of ith generator, and C^{emi} is the emission cost of ith generator ($/kg) [32]. The fifth term represents the cost of purchased power, and $P_{t,\omega}^b$ is the purchased power from the upstream network at t^{th} time and ω^{th} scenario.

2.3.1 Constraints

2.3.1.1 Power balance constraint

The sum of generated power by dispatchable (generators and energy storage system (ESS)) and nondispatchable (PV and wind turbine (WT)) units and exchanged power with the upstream network must be greater or equal to the sum of forecasted power demand and power loss as follows:

$$\sum_{i=1}^{N} P_{i,t,\omega} \cdot X_{i,t,\omega} \pm \sum_{e=1}^{E} P_{\mathrm{ESS}_{e,t,\omega}} \cdot X_{e,t,\omega} + \sum_{k=1}^{K} P_{\mathrm{PV}_{k,t,\omega}} + \sum_{w=1}^{W} P_{\mathrm{WT}_{w,t,\omega}} \pm P_{\mathrm{grid}_{t,\omega}} \geq P_{\mathrm{loss}_{t,\omega}} + \sum_{l=1}^{L} P_{l,t,\omega}, \quad t \in T, \ \omega \in N_{\omega} \tag{2.5}$$

where $P_{\mathrm{ESS}_{e,t,\omega}}$ is the charged/discharged power by the eth ESS at t^{th} time in ω^{th} scenario, $X_{e,t,\omega}$ is the commitment state of the eth ESS at t^{th} time in ω^{th} scenario. $X^{c}_{e,t,\omega}$ and $X^{d}_{e,t,\omega}$ are the binary variables denoting the eth ESS charging and discharging modes at t^{th} time in ω^{th} scenario, respectively. $P_{\mathrm{PV}_{k,t,\omega}}$ is the power output of kth PV at t^{th} time in ω^{th} scenario, $P_{\mathrm{WT}_{w,t,\omega}}$ is the power output of wth WT at t^{th} time in ω^{th} scenario, and $P_{\mathrm{grid}_{t,\omega}}$ is the exchanged power with the upstream network at t^{th} time and ω^{th} scenario, if r is 1, $P_{\mathrm{grid}_{t,\omega}} = P^{N}_{t,\omega}$ and power sold to the upstream network, otherwise, $P_{\mathrm{grid}_{t,\omega}} = P^{b}_{t,\omega}$ and power purchased from the upstream network. $P_{\mathrm{loss}_{t,\omega}}$ is the total power loss at t^{th} time in ω^{th} scenario.

2.3.1.2 Inequality constraints

The generated power by generators and exchanged power are bounded by upper and lower limits as follows:

$$P_i^{\min}(t) \leq P_i(t) \leq P_i^{\max}(t), \quad t \in T \tag{2.6}$$

$$P_{\mathrm{grid}}^{\min}(t) \leq P_{\mathrm{grid}}(t) \leq P_{\mathrm{grid}}^{\max}(t), \quad t \in T \tag{2.7}$$

where $P_i^{\min}$ and $P_i^{\max}$ are the minimum and maximum power outputs of ith generator, respectively, $P_{\mathrm{grid}}^{\min}$ and $P_{\mathrm{grid}}^{\max}$ are minimum and maximum exchanged powers with the upstream network, respectively.

2.3.1.3 Energy storage constraints

The power that is charged or discharged by ESS is bounded by the upper and lower limits:

$$P_{\mathrm{dis}_e}^{\min}(t) X^{d}_{e,t} \leq P_{\mathrm{ESS}_e}(t) \leq P_{\mathrm{dis}_e}^{\max}(t) X^{d}_{e,t}, \quad \forall t \in T, \ \forall e \in E \tag{2.8}$$

$$P_{\text{char}_e}^{\min}(t)X_{e,t}^{c} \le P_{\text{ESS}_e}(t) \le P_{\text{char}_e}^{\max}(t)X_{e,t}^{c}, \quad \forall t \in T, \ \forall e \in E \tag{2.9}$$

where $P_{\text{dis}}^{\min}$ and $P_{\text{char}}^{\min}$ are the minimum power discharged and charged by the eth ESS, respectively. $P_{\text{dis}}^{\max}$ and $P_{\text{char}}^{\max}$ are the maximum power discharged and charged by the eth ESS, respectively. $X_{e,t}^{d}$ and $X_{e,t}^{c}$ are binary variables for discharging and charging modes of the eth ESS, when charging, the charging state X_e^c is 1 and the discharging state X_e^d is 0, so the minimum and maximum limits of charging mode are imposed (2.9). When discharging, the discharging state X_e^d is 1 and the charging state X_e^c is 0, so the minimum and maximum limits of discharging are imposed (2.8). To separate the ESS operation modes (ESS cannot operate in both charging and discharging modes, simultaneously), another constraint is considered as follows:

$$X_e^c + X_e^d \le 1, \quad \forall e \in E \tag{2.10}$$

Energy storage system state-of-charge (SOC) constraint is calculated as

$$0 \le \text{SOC}_{e,t} \le \text{SOC}_e^{\max}, \quad \forall e \in E \tag{2.11}$$

where $\text{SOC}_e^{\max}$ is the maximum SOC of the eth ESS.

The energy storage system is subject to minimum charging and discharging time limits, respectively [33]:

$$\begin{aligned} &T_e^{\text{char}} \ge \text{MC}_e\left(X_{e,t}^{c} - X_{e,t-1}^{c}\right) \\ &T_e^{\text{dis}} \ge \text{MD}_e\left(X_{e,t}^{d} - X_{e,t-1}^{d}\right) \end{aligned} \quad \forall e \in E, \ \forall t \in T \tag{2.12}$$

where MC_e and MD_e are the minimum charging and discharging time of the eth ESS, respectively. T_e^{char} and T_e^{dis} are the number of continuous charging and discharging time (per hour) of the eth ESS, respectively.

2.3.1.4 Minimum up/down time constraint

As we know, a dispatchable unit is limited by minimum up and down times. The time during which a unit must be on/off after being startup/shutdown is called minimum up/down time constraints, respectively, which are as follows [34]:

$$\begin{aligned} &\text{MUT}_i\left(X_{i,t} - X_{i,t-1}\right) \le T_i^{\text{on}} \\ &\text{MDT}_i\left(X_{i,t-1} - X_{i,t}\right) \le T_i^{\text{off}} \end{aligned} \tag{2.13}$$

where MUT_i and MDT_i are the minimum up and down time of ith generator, respectively, $X_{i,t}$ and $X_{i,t-1}$ are the commitment status of ith generator at $t-1^{\text{th}}$ and t^{th} time, respectively. T_i^{on} and T_i^{off} are the numbers of on and off hours of ith generator, respectively.

2.3.1.5 Ramp up/down constraint

Any increase or decrease in power output of ith generator for two consecutive periods of time must be limited by ramp up and ramp down, respectively, which are as follows:

$$P_{i,t} - P_{i,t-1} \leq \mathrm{UR}_i \tag{2.14}$$

$$P_{i,t-1} - P_{i,t} \leq \mathrm{UD}_i \tag{2.15}$$

where $P_{i,t}$ and $P_{i,t-1}$ are the power outputs of ith generator at t^{th} and $t-1^{th}$ time, respectively. UR_i and UD_i are the up/down ramp rates of ith generator, respectively.

2.3.2 Uncertainty modeling

As mentioned earlier, the penetration of renewable energy sources (RESs) into MG can influence the scheduling and operation of MG. PVs and WTs are one of the common types of RESs in active distribution networks and MGs. The generated powers by PVs and WTs are coming from the solar irradiation and wind speed as a prime energy source, respectively [35]. Due to the probabilistic nature of wind speed and sun irradiance, the generated power of those resources causes a significant amount of uncertainties. Furthermore, daily load behavior is considered as an uncertain parameter. Therefore the proposed optimal profit maximization of MG scheduling consists of a large number of uncertain parameters. The probabilistic analysis in the presence of multiple uncertain parameters is a mighty tool for scheduling and operation of power network. To address the uncertain parameters a probabilistic scenario-based framework is presented in this section.

2.3.2.1 Scenario generation

2.3.2.1.1 WT power output

The power output of WT depends on the speed of wind. To model the uncertainty of wind speed, it is assumed that the wind speed follows the Weibull distribution [7]. If V_{mean} and σ are the mean and standard deviations of forecasted wind speed, respectively, the parameters of the Weibull distribution are calculated as [36].

$$r = \left(\frac{\sigma}{V_{\text{mean}}}\right)^{-1.086}, \quad c = \frac{V_{\text{mean}}}{\mathrm{Gamma}\left(1 + (1/r)\right)} \tag{2.16}$$

According to the Weibull parameters (r, c), the Weibull probability distribution function (PDF) is calculated as

$$f(V) = \frac{r}{c}\left(\frac{V}{c}\right)^{r-1} \exp\left[-\left(\frac{V}{c}\right)^{r}\right] \tag{2.17}$$

The MCS generates a high number of scenarios subject to the Weibull distribution, each of which is assigned a probability value that is equal to 1 divided by the number of generated scenarios [37]. In each scenario a random value for the wind speed is considered for the current hour. According to the assigned PDF, in each scenario, the hourly random wind speed is generated; therefore, according to the random wind speed, WT power generation is calculated as

$$P_W(V) = \begin{cases} 0 & 0 \le V \le V_{\text{cut-in}} \\ (k_1 + k_2 V + k_3 V^2) P_{W_{\text{rated}}} & V_{\text{cut-in}} \le V \le V_{\text{rated}} \\ P_{W_{\text{rated}}} & V_{\text{rated}} \le V \le V_{\text{cut-out}} \\ 0 & V_{\text{cut-out}} \le V \end{cases} \tag{2.18}$$

where k_1, k_2, and k_3 are the coefficients of wind turbine, $P_{W_{\text{rated}}}$ is the rated power output of WT, $V_{\text{cut-in}}$ and $V_{\text{cut-out}}$ are the minimum and maximum allowable wind speeds, and V_{rated} is the rated wind speed.

2.3.2.1.2 PV power output

The generated power of PV depends on air temperature and solar radiation. To model the uncertainty of irradiation and air temperature, it is assumed that those parameters are following the normal distribution. If μ and σ are the mean and standard deviation of forecasted irradiation (air temperature), respectively, then the normal distribution PDF for irradiation (G_{ING})and air temperature (T_r)is calculated as

$$f(G_{\text{ING}}, T_r) = \frac{1}{\sqrt{2\pi} \times \sigma} \exp\left(-\frac{((G_{\text{ING}}, T_r) - \mu)^2}{2 \times \sigma^2} \right) \tag{2.19}$$

The same as the wind speed simulation, a large number of scenarios are generated by MCS to model the PV unit outputs.

According to the assigned PDF, in each scenario, hourly random air temperature and irradiation are generated. Therefore PV power generation output is calculated as

$$P_{\text{PV}} = P_{\text{STC}} \times \frac{G_{\text{ING}}}{G_{\text{STC}}} \times (1 + K(T_c - T_r)) \tag{2.20}$$

where G_{ING} is the hourly irradiation, G_{STC} is the standard irradiation (1000 W/m^2), T_c and T_r are cell and air temperatures, respectively, P_{STC} is the rated power of PV, and K is the maximum power temperature coefficient [38].

2.3.2.1.3 Demand variation modeling

Due to the load variation during the day the probabilistic behavior of load is considered as the uncertain parameter. The uncertainty of load demand is subject to normal distribution [4] like (2.19). Therefore MCS generates a large number of scenarios. Similar to the previous uncertain parameters, in each scenario, hourly random load demand is generated according to the assigned PDF.

2.3.2.2 Scenario reduction

Initially, a large number of scenarios are generated by MCS. To simplify the computation requirements the generated scenarios should be reduced. Some of the different scenario reduction techniques are presented in Refs. [19,39]. In this chapter the fast forward selection algorithm is used. The base of this method is to calculate the distance between the scenarios; therefore the most possible scenarios with more probability are selected. The fast forward selection algorithm works as per the following steps [19]:

Step 1: Consider Ω as the initial set of the scenarios: $\Omega = \{1, \ldots .\omega_1, \omega_2, \ldots .\omega', \ldots .N_\Omega\}$. Compute the cost function $\upsilon(\omega, \omega')$ for each pair of scenarios ω and ω' in Ω. For example, two simulated wind speeds corresponding to ω^{th} and ω'^{th} scenarios are 15 and 10 m/s, respectively; therefore the cost function (υ) for these two scenarios is $\upsilon(\omega, \omega') = 15 - 10 = 5$.

Step 2: Compute the distance between each pair of the scenarios as follows:

$$d_\omega = \sum_{\substack{\omega'=1 \\ \omega' \neq \omega}}^{N_\Omega} \pi_{\omega'} \upsilon(\omega, \omega'), \quad \forall \omega \in \Omega \tag{2.21}$$

where $\pi_{\omega'}$ is the probability of ω'^{th} scenario. The scenario with minimum d_ω is selected (e.g., ω_1) and $\Omega_s^{[1]} = \{\omega_1\}$. $\Omega_s^{[1]}$ demonstrates the new set of the most probable scenarios in the first iteration. When ω_1 is selected, $\Omega_j^{[1]}$ is defined. $\Omega_j^{[1]}$ is equal to the initial set of the scenarios (Ω) except ω_1; therefore $\Omega_j^{[1]} = \{1, 2, \ldots, N_\omega\}/\omega_1$.

Step 3: Compute $\upsilon(\omega, \omega')$ for the new set of scenarios as

$$\upsilon^{[2]}(\omega, \omega') = \min\left\{\upsilon^{[i-1]}(\omega, \omega'), \upsilon^{[i-1]}(\omega, \omega_{i-1})\right\}, \quad \forall \omega, \omega' \in \Omega_j^{[1]} \tag{2.22}$$

According to $\upsilon^{[2]}$, the distance between each pair of scenarios is computed as (2.21). Like Step 2, the scenario with minimum d_ω is selected (e.g., ω_2); therefore Ω_s and Ω_j are updated as

$$\begin{gathered} \Omega_s^{[2]} = \{\omega_1, \omega_2\} \\ \Omega_s^{[2]} = \frac{\{1, 2, \ldots, N_\omega\}}{\omega_1, \omega_2} \end{gathered} \tag{2.23}$$

Step 4: Repeat Steps 2 and 3 until N_{Ω_s} (the number of scenarios in Ω_s^{th} set) is equal to the desired number of scenarios.

Step 5: If Ω_s^* and Ω_j^* are the final sets of selected and deleted scenarios, respectively, and $\omega \in \Omega_s^*$, calculate the probabilities of selected scenarios as

$$\pi_\omega^* = \pi_\omega + \sum_{\omega' \in J(\omega)} \pi_{\omega'} \tag{2.24}$$

where $J(\omega)$ is defined as the set of scenarios $\omega' \in \Omega_j^*$ such that $\omega \in \arg\min_{\omega'' \in \Omega_s^*} v(\omega'', \omega')$.

2.4 Case study

The proposed microgrid problem that is presented in the previous section is implemented on the 33-bus microgrid (Fig. 2.1). The characteristic of three generators is given in Table 2.2. Tables 2.3–2.5 show the characteristic of PV, wind, and energy storage system, respectively.

Based on the uncertainty modeling for different parameters (PV and wind power generation and load demand), which is described in Section 2.3.2.1, the forecasted values of these parameters are depicted in Figs. 2.2–2.4.

The stochastic framework models the power output of WT and PV and the load consumption using the corresponding PDF that is described in Section 2.2. To model the uncertainties, 1000 scenarios are generated for each variable, which is reduced to 10 scenarios using the scenario generation algorithm (Section 2.3.2.2).

TABLE 2.2 The characteristic of generators of system.

Generator	G1	G2	G3
P_{min}(kW)	25	75	25
P_{max}(kW)	300	150	300
a ($)	25	10	20
b ($/kW)	0.15	0.85	0.25
c ($/kW)^2	0.0023	0.012	0.003
Startup/shutdown cost ($)	0.96	1.9	0.96

TABLE 2.3 PV characteristics.

Technology	T_c (°C)	P_{STC} (kW)	G_{STC} (W/m^2)	K
PV	25	250	1000	0.001

TABLE 2.4 Wind turbine characteristic.

	$V_{\text{cut-in}}(m/s)$	$V_{\text{cut-out}}(m/s)$	$V_{\text{rated}}(m/s)$	$P_{\text{wind}}^{\text{min}}(\text{kW})$	$P_{\text{wind}}^{\text{max}}(\text{kW})$	k_1	k_2	k_3
WT	3	25	12	0	100	0.123	−0.096	0.0184

TABLE 2.5 Energy storage system characteristic.

ESS	Minimum/maximum charge/ discharge power (kW)	Minimum charge/ discharge time (h)	Capacity (kWh)
	−50/+50	2	100

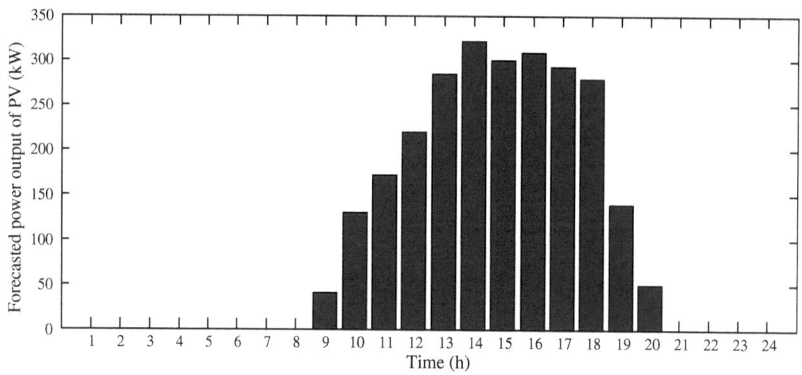

FIGURE 2.2 The forecasted value of PV power generation.

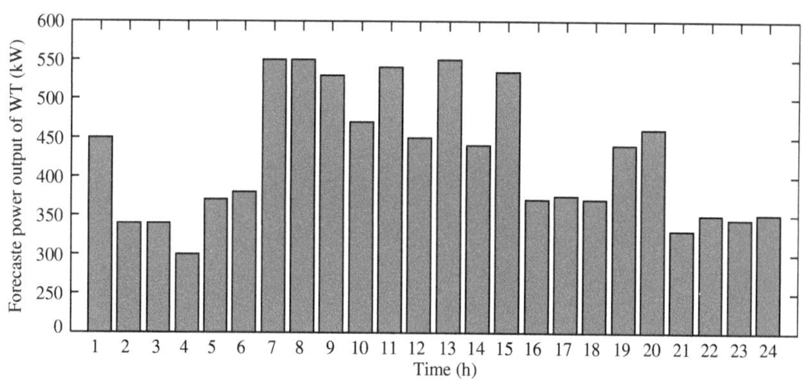

FIGURE 2.3 The forecasted value of WT power generation.

Table 2.6 shows the power market and power exchanged prices for each hour. The values of the emission factor and the emission cost are fixed at 0.003 kg/kWh and 0.02$/kg, respectively [32].

The optimal scheduling of the microgrid problem under high level of uncertainties for profit maximization will be solved by the time-varying acceleration coefficient particle swarm optimization (TVAC-PSO) algorithm. It has been discovered that parameter adapting is a key factor in PSO to find an accurate solution [41]. In the TVAC-PSO algorithm, unlike the

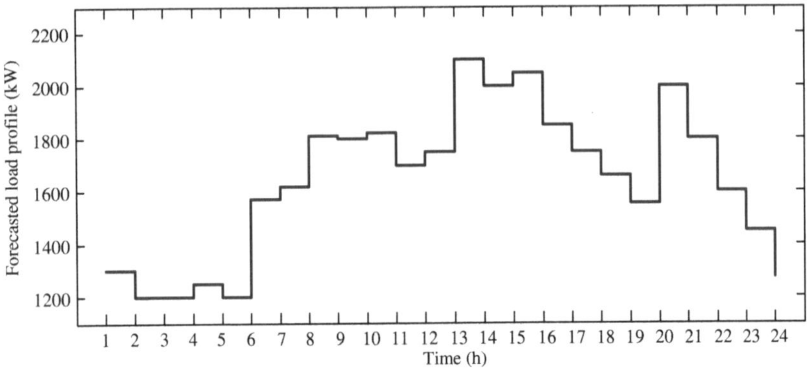

FIGURE 2.4 The forecasted value of load demand.

TABLE 2.6 Hourly energy price [40].

Hour	Power market price ($/kWh)	Power exchanged price ($/kWh)
1	0.6	1.1
2	0.6	1.1
3	0.6	1.1
4	0.6	1.1
5	0.6	1.1
6	0.6	1.1
7	0.6	1.1
8	0.6	1.1
9	0.9	1.3
10	0.9	1.3
11	0.9	1.3
12	1.2	1.4
13	1.45	1.7
14	1.6	1.7
15	1.7	1.95
16	1.75	1.8
17	1.7	1.8

(Continued)

TABLE 2.6 (Continued)

Hour	Power market price ($/kWh)	Power exchanged price ($/kWh)
18	1.4	1.6
19	1	1.3
20	0.8	1.3
21	0.8	1.25
22	0.8	1.3
23	0.7	1.2
24	0.6	1.1

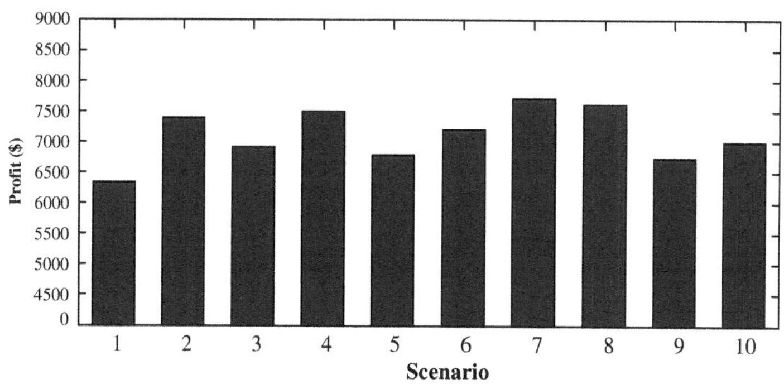

FIGURE 2.5 MG profit for different scenarios.

conventional PSO that considered the fixed coefficient, the acceleration coefficients are changed and updated in the search proceeds [42]. So, unlike conventional PSO, the acceleration coefficients are updated. More details can be found in Refs. [31,42]. All computer simulations and required coding are carried out in MATLAB software and using a CPLEX 11.2 solver.

2.4.1 Simulation and results

In this section the optimal scheduling of MG for profit maximization is analyzed. Based on daily power market price and exchanged power price that are shown, the proposed algorithm finds the decision variables $(X_{i,t}, X_{e,t})$. To observe the impact of the proposed scheduling, we execute microgrid optimal scheduling for the reduced scenarios (10 scenarios) and describe one of them with details (power dispatch and hourly cost). Fig. 2.5 shows the MG

TABLE 2.7 Results of the three generators and ESS states based on decision variable for scenario number 7.

H	1	2	3	4	5	6	7	8	9	10	11	12	13	14	15	16	17	18	19	20	21	22	23	24
G1	1	1	1	1	1	1	1	1	1	1	1	1	1	1	1	1	1	1	1	1	1	1	1	1
G2	0	0	0	0	0	0	0	0	0	0	0	0	1	1	1	1	1	0	0	0	0	0	0	0
G3	1	1	1	1	1	1	1	1	1	1	1	1	1	1	1	1	1	1	1	1	1	1	1	1
ESS	−1	−1	−1	−1	−1	−1	0	0	0	0	1	1	1	1	0	0	0	0	1	1	1	0	0	0

TABLE 2.8 Dispatch schedule and monetary value for scenario number 7.

Hour	Power dispatch (kW)			Monetary value				
	G1	G2	G3	Gen cost ($)	E mission cost ($)	Energy import cost ($)	Total revenue ($)	Profit
1	300	0	250	547	79.2	246	930	57.8
2	300	0	250	547	79.2	252	1005	126.8
3	300	0	250	547	79.2	252	1005	126.8
4	300	0	250	547	79.2	298.8	942	17
5	300	0	250	547	79.2	241.8	1320	452
6	300	0	250	547	79.2	465	1038	–
7	300	0	250	547	79.2	357	1082	98.8
8	300	0	250	547	100.8	540	1359.5	171.7
9	300	0	300	642	115.2	415.8	1340	167
10	300	0	300	642	115.2	234	1379	387.8
11	300	0	300	642	115.2	0	1560	802.8
12	300	0	300	642	115.2	0	1304.15	546.95
13	300	100	300	857	194.4	0	1970	918.6
14	300	150	300	1049.5	219.6	0	1600.4	331.3
15	300	150	300	1049.5	219.6	0	1433.75	164.65
16	300	75	300	783.25	190.8	0	1797.5	823.45

(Continued)

TABLE 2.8 (Continued)

Hour	Power dispatch (kW)			Monetary value				
	G1	G2	G3	Gen cost ($)	E mission cost ($)	Energy import cost ($)	Total revenue ($)	Profit
17	300	0	300	642	115.2	0	1442	684.8
18	300	0	300	642	115.2	83	1145	304.8
19	300	0	300	642	115.2	66.4	1015	191.4
20	300	0	300	642	115.2	238.4	1500	504.4
21	300	0	300	642	115.2	352	1340	230.8
22	300	0	300	642	115.2	294	1220	168.8
23	300	0	250	547	79.2	322.5	1095	146.3
24	300	0	250	547	79.2	210	1186	349.8
Total				Total cost ($)			31,009.3	7721.35
				23,287.95				

profit for 10 scenarios. The scenario number 7 with the maximum profit is selected and described in the following.

The optimal generators and ESSs state for 24 hours are given in Table 2.7. ESS charging, discharging, and idle states are represented by -1, 1, and 0. As can be seen in Table 2.2, generator number 3 is a high-cost generator, so it is committed in peak-load hours (13–16 hours).

The proposed algorithm detects that the price of exchanged power is low at some hours (1–8 hours), so purchasing power from the upstream network is rational for MG's owner ($r=0$). In this interval, most of the ESS is charged. At the higher exchanged power price (10–16 hours), it is beneficial for MG to sell surplus power to the upstream network to achieve significant benefits. In this interval, ESS is discharged and because the exchanged power price exceeds the cost coefficients of all generators, generator number 2 (high-cost generator) is committed at 13–16 hours. While MG supplies their loads, the surplus power is sold to the upstream network ($r=1$).

The monetary result of the problem consists of MG's costs, revenues, and profits, which are given in Table 2.8.

It is shown in Table 2.8, at the peak-load hours (11–16), in which the value of the power price exceeds generators cost coefficients, MG sells the surplus power (that generated by the high-cost generator) to the upstream network and receives significant revenues; at this interval, r that shows the exchanged power direction with the upstream network is equal to 1. Based on the proposed optimal scheduling, the profit of MG is determined as \$7721.35 for scenario number 7.

Finally, the expected profit of 10 reduced scenarios based on the probability of each scenario can be calculated as follows:

$$Expected\ profit = \sum_{\omega=1}^{10} \pi_\omega f_\omega \quad \text{where} \quad \sum_{\omega=1}^{10} \pi_\omega = 1 \tag{2.25}$$

According to (2.25), the value of expected profit is \$7528.9.

2.5 Conclusion

In this chapter a brief overview of uncertainty management in the modern power system is studied. The penetration of renewable energy resources, as well as load demand deviation, causes different challenges in the operation and scheduling of modern power systems. After classifying a variety of uncertain parameters in the power system, some useful methods for uncertainty management are discussed. To demonstrate the uncertainty modeling in the power system decision-making process, we considered that a microgrid consists of different generation units (dispatchable units and renewable units such as PV and wind). The MG scheduling for profit maximization in the presence of multiple uncertain parameters is examined. After the formulation of the problem, generated power by renewable energy and load demand are considered as the uncertain parameters, and the stochastic scenario-based approach is used for solving the problem. The proposed method was examined on the 33-bus microgrid test system. The result of decision variables of the problem consists of the status of the storage and dispatchable units and power output of dispatchable units, exchanged power with the upstream network, and the monetary value that was studied for the best scenario. Finally, the expected profit of reduced scenarios was calculated. The results show that by modeling the uncertainty in a lot of scenarios, the operator of the system can decide with a better view, about the conditions of the network. However, the high volume of computations in this method requires an examination of the other methods that will be studied in future works.

Acknowledgments

The work done by Alireza Soroudi is supported by a research grant from Science Foundation Ireland (SFI) under the SFI Strategic Partnership Programme Grant No. SFI/15/SPP/E3125. The opinions, findings, and conclusions or recommendations expressed in this material are those of the author(s) and do not necessarily reflect the views of the Science Foundation Ireland.

References

[1] H.M. Merrill, A.J. Wood, Risk and uncertainty in power system planning, Int. J. Electr. Power Energy Syst. 13 (1991) 81–90.

[2] M. Hemmati, S. Ghassemzadeh, B. Mohammadi-Ivatloo, A new framework for optimal scheduling of smart reconfigurable neighbour micro-grids, IET Gen. Trans. Distrib. 13 (2018).

[3] http://www.ieso.ca/en/Power-Data/Supply-Overview/Transmission-Connected-Generation.

[4] N. Nikmehr, S. Najafi-Ravadanegh, A. Khodaei, Probabilistic optimal scheduling of networked microgrids considering time-based demand response programs under uncertainty, Appl. Energy 198 (2017) 267–279.

[5] P.H. Henault, R.B. Eastvedt, J. Peschon, L.P. Hajdu, Power system long-term planning in the presence of uncertainty, IEEE Trans. Power Appar. Syst. PAS-89 (1970) 156–164.

[6] S.A. Malcolm, S.A. Zenios, Robust optimization for power systems capacity expansion under uncertainty, J. Oper. Res. Soc. 45 (1994) 1040–1049.

[7] A. Soroudi, T. Amraee, Decision making under uncertainty in energy systems: state of the art, Renew. Sustain. Energy Rev. 28 (2013) 376–384.

[8] L. Shi, Y. Luo, G. Tu, Bidding strategy of microgrid with consideration of uncertainty for participating in power market, Int. J. Electr. Power Energy Syst. 59 (2014) 1–13.

[9] A. Soroudi, Power System Optimization Modeling in GAMS, Springer, 2017.

[10] S.M. Mohseni-Bonab, A. Rabiee, B. Mohammadi-Ivatloo, S. Jalilzadeh, S. Nojavan, A two-point estimate method for uncertainty modeling in multi-objective optimal reactive power dispatch problem, Int. J. Electr. Power Energy Syst. 75 (2016) 194–204.

[11] A. Narayan, K. Ponnambalam, Risk-averse stochastic programming approach for microgrid planning under uncertainty, Renew. Energy 101 (2017) 399–408.

[12] G.B. Dantzig, Linear programming under uncertainty, Stochastic Programming, Springer, 2010, pp. 1–11.

[13] G. Carpinelli, P. Caramia, P. Varilone, Multi-linear Monte Carlo simulation method for probabilistic load flow of distribution systems with wind and photovoltaic generation systems, Renew. Energy 76 (2015) 283–295.

[14] R. Dufo-López, E. Pérez-Cebollada, J.L. Bernal-Agustín, I. Martínez-Ruiz, Optimisation of energy supply at off-grid healthcare facilities using Monte Carlo simulation, Energy Convers. Manage. 113 (2016) 321–330.

[15] E.J. da Silva Pereira, J.T. Pinho, M.A.B. Galhardo, W.N. Macêdo, Methodology of risk analysis by Monte Carlo method applied to power generation with renewable energy, Renew. Energy 69 (2014) 347–355.

[16] U. Arnold, Ö. Yildiz, Economic risk analysis of decentralized renewable energy infrastructures—a Monte Carlo simulation approach, Renew. Energy 77 (2015) 227–239.

[17] L. Bin, M. Shahzad, Q. Bing, M.R. Zafar, R. Islam, M.U. Shoukat, Probabilistic power flow model to study uncertainty in power system network based upon point estimation method, Am. J. Electr. Power Energy Syst. 6 (2017) 64–71.

[18] A. Soroudi, R. Caire, N. Hadjsaid, M. Ehsan, Probabilistic dynamic multi-objective model for renewable and non-renewable distributed generation planning, IET Gener. Transm. Distrib. 5 (2011) 1173–1182.

[19] A.J. Conejo, M. Carrión, J.M. Morales, Decision Making Under Uncertainty in Electricity Markets, vol. 1, Springer, 2010.

[20] A. Soroudi, Possibilistic-scenario model for DG impact assessment on distribution networks in an uncertain environment, IEEE Trans. Power Syst. 27 (2012) 1283–1293.

[21] A. Soroudi, M. Ehsan, IGDT based robust decision making tool for DNOs in load procurement under severe uncertainty, IEEE Trans. Smart Grid 4 (2013) 886–895.

[22] A. Soroudi, A. Rabiee, A. Keane, Information gap decision theory approach to deal with wind power uncertainty in unit commitment, Electr. Power Syst. Res. 145 (2017) 137–148.

[23] S. Nojavan, M. Majidi, K. Zare, Risk-based optimal performance of a PV/fuel cell/battery/grid hybrid energy system using information gap decision theory in the presence of demand response program, Int. J. Hydrogen Energy 42 (2017) 11857–11867.

[24] A. Rabiee, S. Nikkhah, A. Soroudi, E. Hooshmand, Information gap decision theory for voltage stability constrained OPF considering the uncertainty of multiple wind farms, IET Renew. Power Gener. 11 (2016) 585–592.

[25] A. Soroudi, P. Maghouli, A. Keane, Resiliency oriented integration of DSRs in transmission networks, IET Gener. Transm. Distrib. 11 (2017) 2013–2022.

[26] D. Bertsimas, E. Litvinov, X.A. Sun, J. Zhao, T. Zheng, Adaptive robust optimization for the security constrained unit commitment problem, IEEE Trans. Power Syst. 28 (2013) 52–63.

[27] S. Nojavan, B. Mohammadi-Ivatloo, K. Zare, Robust optimization based price-taker retailer bidding strategy under pool market price uncertainty, Int. J. Electr. Power Energy Syst. 73 (2015) 955–963.

[28] R. Wang, P. Wang, G. Xiao, A robust optimization approach for energy generation scheduling in microgrids, Energy Convers. Manage. 106 (2015) 597–607.

[29] A. Soroudi, P. Siano, A. Keane, Optimal DR and ESS scheduling for distribution losses payments minimization under electricity price uncertainty, IEEE Trans. Smart Grid 7 (2016) 261–272.

[30] A.J. Conejo, J.M. Morales, L. Baringo, Real-time demand response model, IEEE Trans. Smart Grid 1 (2010) 236–242.
[31] M. Hemmati, B. Mohammadi-Ivatloo, S. Ghasemzadeh, E. Reihani, Risk-based optimal scheduling of reconfigurable smart renewable energy based microgrids, Int. J. Electr. Power Energy Syst. 101 (2018) 415–428.
[32] Z. Wang, J. Wang, Self-healing resilient distribution systems based on sectionalization into microgrids, IEEE Trans. Power Syst. 30 (2015) 3139–3149.
[33] A. Khodaei, Microgrid optimal scheduling with multi-period islanding constraints, IEEE Trans. Power Syst. 29 (2014) 1383–1392.
[34] H. Sadeghian, M. Ardehali, A novel approach for optimal economic dispatch scheduling of integrated combined heat and power systems for maximum economic profit and minimum environmental emissions based on Benders decomposition, Energy 102 (2016) 10–23.
[35] N. Nikmehr, S.N. Ravadanegh, Optimal power dispatch of multi-microgrids at future smart distribution grids, IEEE Trans. Smart Grid 6 (2015) 1648–1657.
[36] R. Jabbari-Sabet, S.-M. Moghaddas-Tafreshi, S.-S. Mirhoseini, Microgrid operation and management using probabilistic reconfiguration and unit commitment, Int. J. Electr. Power Energy Syst. 75 (2016) 328–336.
[37] J. Wang, M. Shahidehpour, Z. Li, Security-constrained unit commitment with volatile wind power generation, IEEE Trans. Power Syst. 23 (2008) 1319–1327.
[38] N. Nikmehr, S.N. Ravadanegh, Reliability evaluation of multi-microgrids considering optimal operation of small scale energy zones under load-generation uncertainties, Int. J. Electr. Power Energy Syst. 78 (2016) 80–87.
[39] L. Wu, M. Shahidehpour, T. Li, Stochastic security-constrained unit commitment, IEEE Trans. Power Syst. 22 (2007) 800–811.
[40] M. Honarmand, A. Zakariazadeh, S. Jadid, Integrated scheduling of renewable generation and electric vehicles parking lot in a smart microgrid, Energy Convers. Manage. 86 (2014) 745–755.
[41] A. Salhi, D. Naimi, T. Bouktir, TVAC based PSO for solving economic and environmental dispatch considering security constraint, in: Renewable and Sustainable Energy Conference (IRSEC), 2013 International, 2013, pp. 396–401.
[42] B. Mohammadi-Ivatloo, A. Rabiee, A. Soroudi, M. Ehsan, Iteration PSO with time varying acceleration coefficients for solving non-convex economic dispatch problems, Int. J. Electr. Power Energy Syst. 42 (2012) 508–516.

Chapter 3

Uncertainty analysis and risk assessment for effective decision-making using wide-area synchrophasor measurement system

Bhargav Appasani[1] and Dusmanta Kumar Mohanta[2]
[1]*School of Electronics Engineering, KIIT Deemed University, Bhubaneswar, India,*
[2]*Department of Electrical and Electronics Engineering, Birla Institute of Technology, Mesra, Ranchi, Jharkhand, India*

Abbreviations

CPU	central processing unit
CT	current transformer
GPS	global positioning system
ILP	integer linear programming
MCS	Monte Carlo simulation
MTTF	mean time to failure
MTTR	mean time to repair
OPP	optimal placement of PMUs
PDC	phasor data concentrator
PMU	phasor measurement unit
PT	potential transformer
SPCS	synchrophasor communication system
TTF	time to failure
TTR	time to repair
WASMS	wide area synchrophasor measurement system

3.1 Introduction

The current power system is a complex network that caters to the demands of several applications with diverse energy requirements. Such a complex network is susceptible to faults caused due to several reasons such as failure

Decision Making Applications in Modern Power Systems. DOI: https://doi.org/10.1016/B978-0-12-816445-7.00003-7

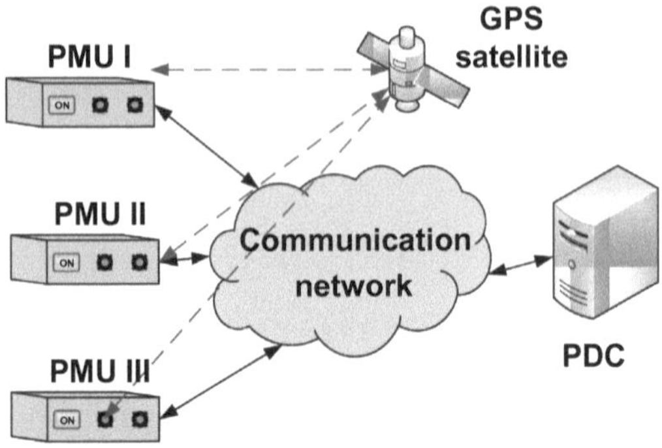

FIGURE 3.1 A typical WASMS. *WASMS*, Wide area synchrophasor measurement system.

of the equipment and hostile weather conditions. These faults, if not detected in the real time, may lead to cascading failures resulting in a blackout [1]. These blackouts have catastrophic consequences, which may result a huge loss of resources. For example, a blackout in 2003 caused an economic loss of 10 billion US dollars as per the report of the Electricity Consumers Resource Council [2]. Subsequent investigation of the blackout revealed that the catastrophe could have been prevented if there was an early warning system [3]. Similar other blackouts across the globe forced the power system engineers to devise an effective solution for detection and control of the power system abnormalities. The consequence of these efforts is the wide area synchrophasor measurement system (WASMS) [4]. The WASMS consists of several measurement devices that are termed phasor measurement units (PMUs) that collect the real-time information concerning the health of the power grid in the form of time-stamped voltage and current phasors. These measurements are known as synchrophasors and are communicated to the central control center or the phasor data concentrator (PDC) for the detection of power system anomalies.

In the past few decades the PMUs have carved their niche as important sensors for estimating the state of the power system [5] and are being widely installed at the electrical nodes, separated by large geographical distances. The communication of the synchrophasor data from these geographically separated PMUs to the PDC is the responsibility of the synchrophasor communication system (SPCS) [6]. Thus the PMUs, the PDCs, and the SPCS together constitute the WASMS and are illustrated in Fig. 3.1.

3.1.1 Phasor measurement unit

The PMU is defined by the Institute of Electrical and Electronic Engineers as "a device that produces synchronized phasor, frequency, and rate of change of frequency (ROCOF) estimates from voltage and/or current signals

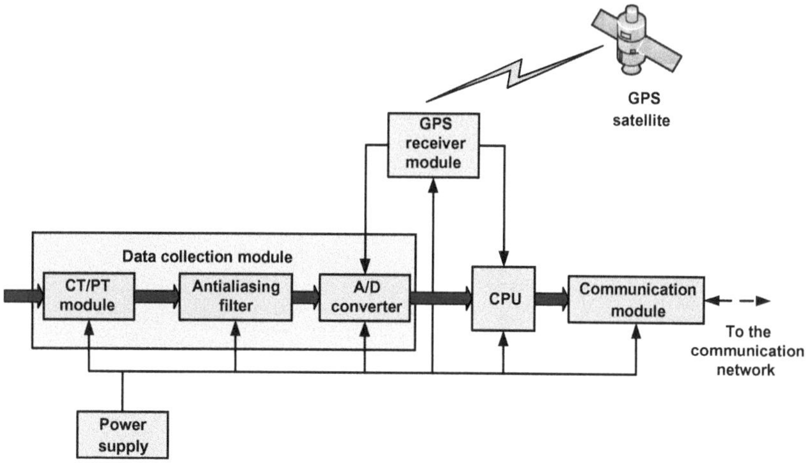

FIGURE 3.2 Various functional blocks in the PMU. *PMU*, Phasor measurement unit.

and a time synchronizing signal." In a nutshell the PMU is a high-speed signal processor capable of estimating current and voltage phasors. These measurements are time marked using the signals acquired from the global positioning system (GPS). The PMU is a combination of seven subsystems or modules, namely, the central processing module (M_1), the current transformer (CT) and potential transformer (PT) module or the transformer board (M_2), the preprocessing filter (M_3), the converter module (M_4), communication module (M_5), the GPS module (M_6), and the supply module (M_7) [7]. These modules and their functionalities are described later and are depicted in Fig. 3.2.

1. The transformer board is responsible for stepping down the magnitudes of the transmission line voltages and currents.
2. To prevent aliasing, the bandwidth of the signal is limited by using an antialiasing filter.
3. The aliased signal is then fed to an A/D converter for sampling the analog signal and to convert it into a digital signal.
4. The central processing unit (CPU) uses the discrete Fourier transform for computing the magnitude and phase of the voltage and current phasors. It also time-stamps the data using the one-pulse-per-second provided by the GPS receiver module.
5. The communication module communicates the synchrophasor data to the PDC, and the power supply module provides power to all the modules of the PMU.

3.1.2 Synchrophasor communication system

A single PMU can only observe a small number of electrical buses in the entire network. To observe the entire power system, phasor measurements

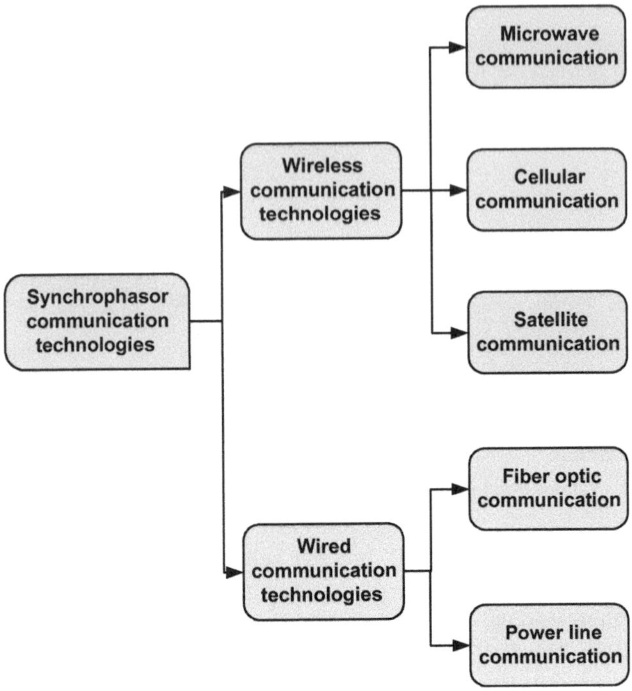

FIGURE 3.3 Various synchrophasor communication technologies.

from several PMUs are required. In a practical power system network the electrical buses are scattered over a span of several hundred kilometers and to communicate their phasor measurements to the PDC in real time is the responsibility of the SPCS. The SPCS should be reliable, have low latency, and have sufficient bandwidth to support the PMU data rates. The key synchrophasor communication technologies are categorized into two groups: wired technologies and wireless technologies [8]. The wired communication technology offers high reliability, huge bandwidth, and protection against interference. On the other hand, the wireless technology enjoys superiority in the matter of rapid deployment, low installation and maintenance costs, access to remote geographic locations, etc. The various synchrophasor communication technologies are depicted in Fig. 3.3.

The power line communication uses the existing power transmission lines to transfer the synchrophasor data [9]. This technology is cost-effective as no additional infrastructure is needed to set up the communication between the devices. However, the noisy environment of the transmission cables increases the bit error rate, thereby limiting its application for intersubstation communication. While the wireless technologies, such as microwave communication and cellular communication, are less reliable due to the signal

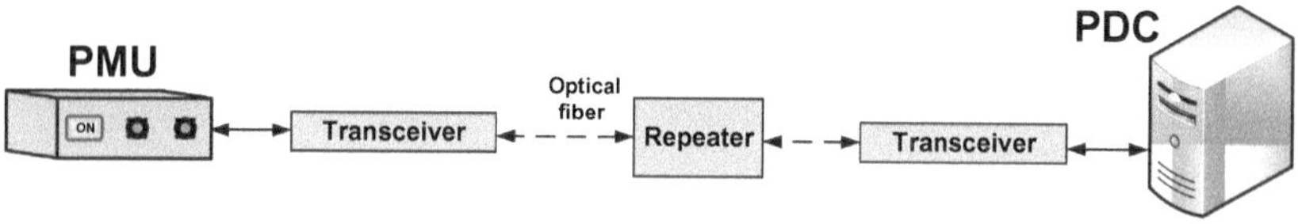

FIGURE 3.4 Optical fiber synchrophasor communication system.

propagation in open environment and multipath fading, the satellite communication has a large propagation delay [10]. Thus the wireless technologies are not reliable for real-time applications.

The fiber optic technology has low signal attenuation and is suitable for long-distance communication [11,12]. It offers high bandwidth and is a widely adopted synchrophasor communication technology. The communication channel can be shared by other data applications or it may be dedicated solely for the purpose of synchrophasor data transfer. Sharing the communication resources increases the chances of packet losses and decreases the reliability of the system. In this work the communication channels between the PMUs and the PDC are considered to be dedicated only for synchrophasor applications. Such an assumption simplifies the analysis as the loss of synchrophasor data is due to the failure of the communication infrastructure.

A dedicated optical fiber synchrophasor communication network between a PMU and a PDC consists of several components that are shown in Fig. 3.4.

The data from the PMU is sent to an optical transceiver (C_{PMU}) to transform the data into optical signals, which are then sent to the PDC through the fiber optic cables (C_2). As the PMU and the PDC are separated by several hundred kilometers, optical repeaters (C_3) are placed at intermediate distances to maintain the quality of the signal. The optical transceiver (C_{PDC}) at the PDC converts these optical signals back into electrical signals and sends the synchrophasor data to the PDC. Thus the optical transceivers, the optical repeaters, and the fiber optic cables are the main constituent components of the synchrophasor communication networks.

3.1.3 Phasor data concentrator

The PDC creates a time-aligned output data stream from the phasor measurements received from several PMUs, which can be used by other applications. A PDC can also exchange this time-aligned phasor data with other PDCs. Use of multiple PDCs results in hierarchical communication architecture, as shown in Fig. 3.1. To record the phasor measurements of all the PMUs, a PDC has a maximum wait time of 1–4 seconds. If the phasor data is received within this maximum wait time, the output data is created immediately; else, the PDC waits until the time limit expires to create the output data. A quality check is then performed on the PMU data, and suitable flags are inserted in the output data stream. The output from the PDC can be used to provide visualizations, which enhance the situational awareness of power system operator.

3.2 Risk assessment and uncertainty analysis of wide area synchrophasor measurement system

The failure of a PMU or its communication network results in the loss of observability of a certain portion of the electrical network. Thus it is important to assess the impact that the failure of the PMU or its communication network has on the monitoring capability of the WASMS. This impact is measured in terms of risk, which is defined as the loss incurred due to the probable failure of a system [13,14].

Consider an individual event k, whose risk r_k is measured as the product of the probability of failure (p_k) with the severity of failure (s_k) and is given by the following equation:

$$r_k = p_k \times s_k \tag{3.1}$$

The composite risk O when a system involves N distinct events is the sum of the risk pertaining to the individual events and is given by the following equation:

$$O = \sum_{k=1}^{N} r_k \tag{3.2}$$

In the framework of risk assessment for WASMS, the severity of loss is taken as the number of buses that become unobservable due to the failure of a PMU or its communication network, while the probability of failure depends on the failure rates and repair rates of the constituent components of a system. The uncertainties involved with these parameters pose a formidable challenge for the accurate determination of the risk. Hence, uncertainty analysis must be incorporated into the risk assessment of the WASMS.

3.2.1 Basics of estimating the probability of failure

Reliability analysis is an important aspect of system design. It provides the probabilistic estimate of the system's capability to function for the designated time period under certain operating conditions [15]. The mathematical representation of reliability (R) is given by the following equation:

$$R(t) = e^{-\int_0^t \lambda(t)dt} \tag{3.3}$$

where $\lambda(t)$ is the system's failure rate and t represents the time.

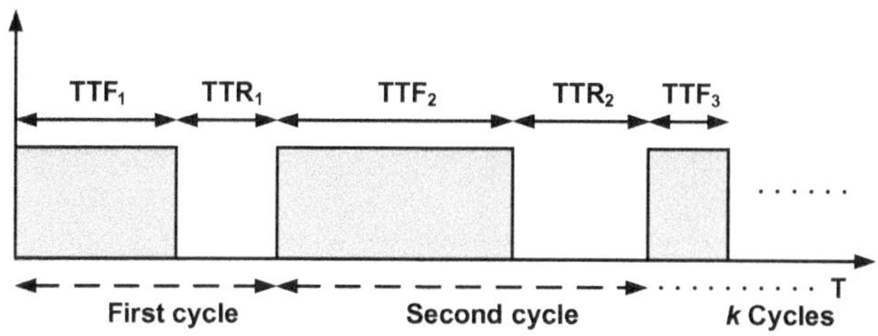

FIGURE 3.5 Operation cycle of a repairable system.

For a system having a constant failure rate, the reliability is given by the following equation:

$$R(t) = e^{-\lambda t} \tag{3.4}$$

Unavailability or the probability of failure is another parameter that is often used as a measure of system's reliability. Unavailability is the steady-state probability that the given system does not deliver the desired functionality. For a repairable system whose operation cycle is shown in Fig. 3.5, the mean time to failure (TTF) (MTTF), mean time to repair (TTR) (MTTR), the availability (A), and the unavailability or the probability of failure (U) are given by the following equations:

$$\text{MTTF} = \frac{1}{\lambda} = \frac{\sum_{i=1}^{k} \text{TTF}_i}{k} \tag{3.5}$$

$$\text{MTTR} = \frac{1}{\mu} = \frac{\sum_{i=1}^{k} \text{TTR}_i}{k} \tag{3.6}$$

$$A = \frac{\mu}{\mu + \lambda} \tag{3.7}$$

$$U = \frac{\lambda}{\mu + \lambda} \tag{3.8}$$

where μ is the repair rate of the component, and λ is the failure rate of the component.

An engineering system is a combination of several components that are interconnected with one another. These components may be connected in series, in parallel, or in standby. The components in a group are said to be in series if the system fails due to the failure of any one of the constituent components. The components are said to be connected in parallel if the group fails only when all the components fail. Two components are connected in standby if upon the failure of the first component, the second component comes into operation. These three connections are shown in Fig. 3.6.

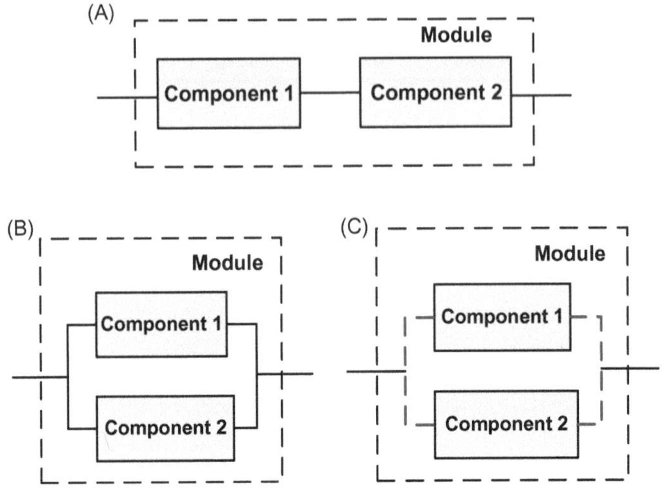

FIGURE 3.6 Module with (A) serially connected components, (B) parallel connected components, and (C) components in a standby mode.

The PMU can also be modeled in terms of these interconnections and is shown in Fig. 3.7.

The PMU consists of seven modules connected in series as any one of their failures leads to the failure of the PMU. Some of these modules have a standby component that is brought into operation when the principal component fails. The transformer board module (M_2), the filter module (M_3), the A/D module (M_4), communication module (M_5), and the power module (M_7) consist of main components M_{2a}, M_{3a}, M_{4a}, M_{5a}, and M_{7a} along with their corresponding standby components M_{2b}, M_{3b}, M_{4b}, M_{5b}, and M_{7b}, respectively, while the GPS receiver module (M_6) consists of a main GPS receiver component (M_{6a}) along with a crystal oscillator (M_{CO}) that acts as a standby component. The CPU module (M_1) requires both hardware and software to deliver the desired functionality. Thus it is modeled as a series combination of hardware (M_{hw}) and software (M_{sw}) components [16]. The switching operating between the original component and their standby is assumed to be instantaneous. The failure data and the repair data for the modules are given in Table 3.1 [16].

The communication network between the PMU and the PDC can also act as a series combination of the optical transceiver modules (C_{PMU} and C_{PDC}), the repeater modules (C_3), and the optical fiber (C_2). Standby components (C_{PMU2}, C_{PDC2}, and C_{3b}) that are similar to the original components (C_{PMU1}, C_{PDC1}, and C_{3a}) are provided for the transceiver modules and the repeater modules. The model for a single-repeater optical fiber synchrophasor communication network is shown in Fig. 3.8, and the failure rates for these components are given in Table 3.2 [12,17,18].

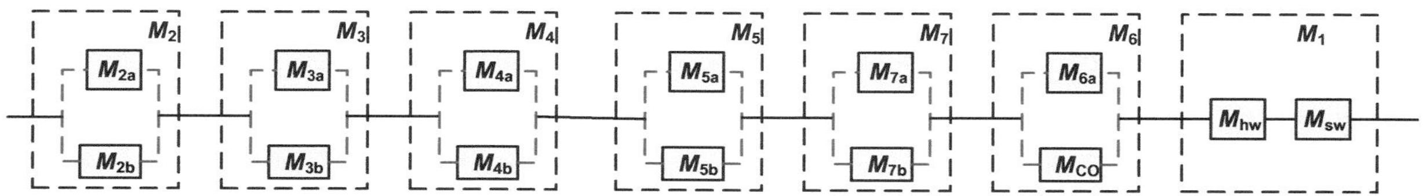

FIGURE 3.7 Modeling of the PMU. *PMU*, Phasor measurement unit.

TABLE 3.1 Failure and repair rates of various constituent components of the phasor measurement unit.

Component	λ (failures/year)	μ (repairs/year)
M_{hw}	$\lambda_{hw}=0.2368$	$\mu_{hw}=365$
M_{sw}	$\lambda_{sw}=0.0657$	$\mu_{sw}=1460$
M_{2a} and M_{2b}	$\lambda_{2a}=\lambda_{2b}=0.4155$	$\mu_{2a}=\mu_{2b}=673.85$
M_{3a} and M_{3b}	$\lambda_{3a}=\lambda_{3b}=0.1923$	$\mu_{3a}=\mu_{3b}=547.5$
M_{4a} and M_{4b}	$\lambda_{4a}=\lambda_{4b}=0.4155$	$\mu_{4a}=\mu_{4b}=673.85$
M_{5a} and M_{5b}	$\lambda_{5a}=\lambda_{5b}=0.1383$	$\mu_{5a}=\mu_{5b}=438$
M_{7a} and M_{7b}	$\lambda_{7a}=\lambda_{7b}=0.2751$	$\mu_{7a}=\mu_{7b}=365$
M_{6a}	$\lambda_{6a}=0.7727$	$\mu_{6a}=365$
M_{CO}	$\lambda_{CO}=0.0188$	$\mu_{CO}=312.88$

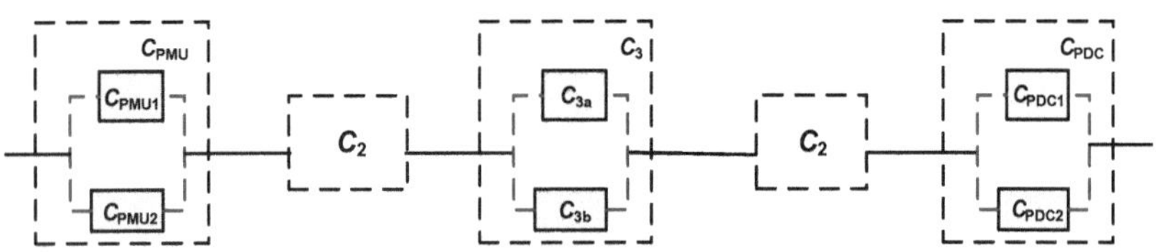

FIGURE 3.8 Modeling of a single-repeater optical fiber communication network for PMUs. *PMU*, Phasor measurement unit.

TABLE 3.2 Failure and repair rates of various constituent components of the phasor measurement unit communication network.

Component	λ (failures/year)	μ (repairs/year)
C_{PMU1} and C_{PMU2}	$\lambda_{PMU1}=\lambda_{PMU2}=0.01752$	$\mu_{PMU1}=\mu_{PMU2}=4380$
C_{PDC1} and C_{PDC2}	$\lambda_{PDC1}=\lambda_{PDC2}=0.01752$	$\mu_{PDC1}=\mu_{PDC2}=4380$
C_{2a} and C_{2b} (per 100 km)	$\lambda_{2c}=0.00438$	$\mu_{2c}=104.244$
C_{3a} and C_{3b}	$\lambda_{3ca}=\lambda_{3cb}=0.03504$	$\mu_{3ca}=\mu_{3cb}=4380$

The intermediate optical repeaters required for reliable communication among the end devices depend on the geographical distance between them. Usually, repeaters are placed after every 100–150 km so as to maintain the quality of the signal [12].

3.2.2 Monte Carlo simulation models for phasor measurement unit and their communication networks

Reliability evaluation methods, such as Markov models [19], fault trees [20], reliability block diagrams [21], and Petri nets [22], assume constant failure rates and constant repair rates for obtaining the probability of failure. Different versions of the same components may have different failure and repair rates, and thus Monte Carlo simulation (MCS) is often used to obtain a better estimate of the failure probability [23]. It is a probabilistic method that uses random values of TTR and TTF for the various components of the system. These random values are generated, and the entire cycle of real-time events is simulated in real time. Several such cycles are iterated to obtain the failure distribution for the various components of the system. Using this approach, one can effortlessly model different component interconnections and can simulate simultaneous component failures. The random values for TTF and TTR used by the MCS are given by Eqs. (3.9) and (3.10). These values are based on the nominal values of the component's failure rate and repair rate [16].

$$TTF = -\frac{1}{\lambda}\ln(\theta_1) \tag{3.9}$$

$$TTR = -\frac{1}{\mu}\ln(\theta_2) \tag{3.10}$$

where θ_1 and θ_2 are random numbers between 0 and 1.

The modules M_2, M_3, M_4, M_5, M_6, and M_7 of the PMU and the modules C_{PMU}, C_{PDC}, and C_3 of the optical fiber SPCS are two component modules, with the components connected in standby mode. The MCS model for these modules is shown in Fig. 3.9.

The modules fail when the standby component fails before the main component is repaired and is indicated by the checkered line (system downtime is the duration for which the system is unavailable) in Fig. 3.9. The variables MT and CT stand for mission time and current time, respectively. In order to generate sufficient failure data the MT taken should be sufficiently large. All the simulations performed in this work consider a mission time of 200 million years. Each simulation is run for 1000 iterations (indicated by the variable N) in order to obtain consistent failure data. In every iteration, initially the values of TTF, TTR, and CT are taken to be zero. Random values of failure times (TTF_1 and TTF_2) and repair times (TTR_1 and TTR_2) are generated

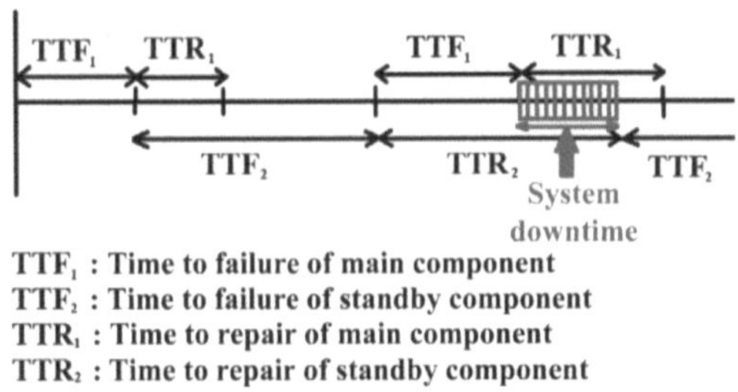

FIGURE 3.9 Monte Carlo model for the various modules of the PMU and its communication network. *PMU*, Phasor measurement unit.

for both the components using Eqs. (3.9) and (3.10), respectively. The whole system fails only when the second component fails prior to the first component becoming operative (i.e., $TTF_2 < TTR_1$) or the first component fails prior to the second component becoming operative (i.e., $TTF_1 < TTR_2$). The simulation time (CT) is updated from time to time until the termination criterion (i.e., $CT > MT$) is met. The number of failures, the frequency of failures, the MTTF of the composite system, and the MTTR of the composite system are thus obtained. The flowchart for calculating the failure rates and the repair rates of these modules using the MCS is shown in Fig. 3.10.

The module M_1 of the PMU is a two-component module with the components in series. The Monte Carlo model for this module is shown in Fig. 3.11, and the flowchart for calculating the failure rate and the repair rate of this module is shown in Fig. 3.12.

The failure rates and the repair rates are obtained for the various modules shown in Table 3.3. The failure distribution histograms for these modules are shown in Fig. 3.13. As these modules are in series, the failure probability of the PMU and its communication network combined as a single system is the sum of the individual failure probabilities, which is calculated for a varying number of repeaters and is shown in Table 3.4. The failure probabilities were obtained using the MCS with uncertainty included in the analysis.

3.2.3 Risk assessment of a sample wide area synchrophasor measurement system

Risk assessment of WASMS can be explained with the help of a sample network shown in Fig. 3.14. The diagram also depicts the position of the PMUs, the PDC, and the repeaters.

The network consists of six buses with the PMUs placed at buses 3 and 4. The locations for the PMU placement have been selected to achieve complete system observability. The PDC is placed on bus 6. The optical fiber SPCS between the buses 3 and 6 requires three repeaters for communication feasibility. Similarly, the optical fiber SPCS between buses 4 and 6 requires

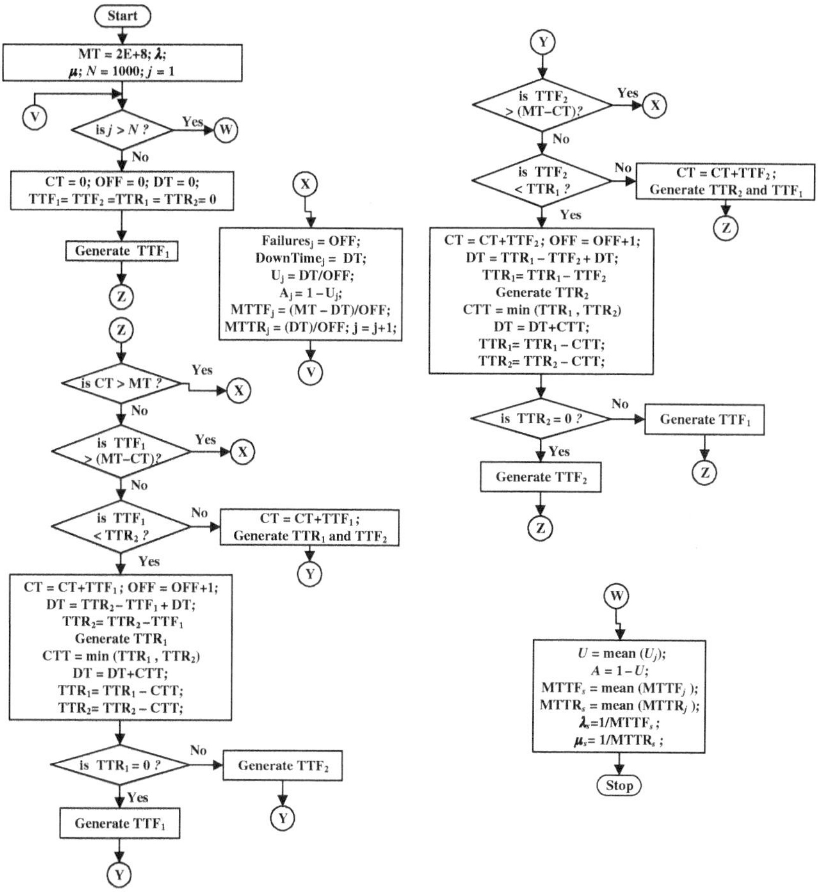

FIGURE 3.10 Flowchart for generating the failure rates and repair rates of the various modules using MCS. *MCS*, Monte Carlo simulation.

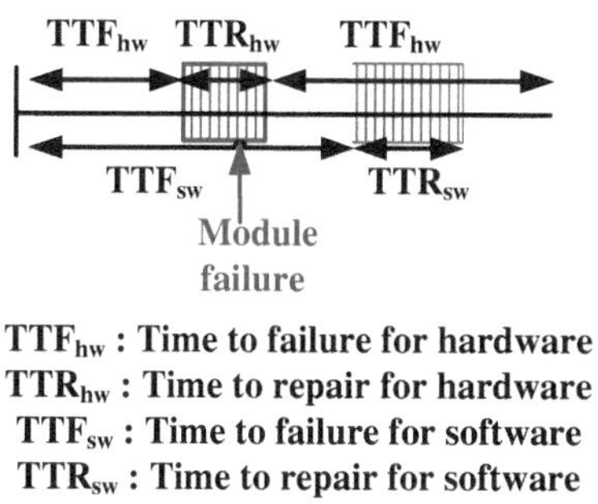

FIGURE 3.11 Monte Carlo model for the CPU module of the PMU. *CPU*, Central processing unit; *PMU*, phasor measurement unit.

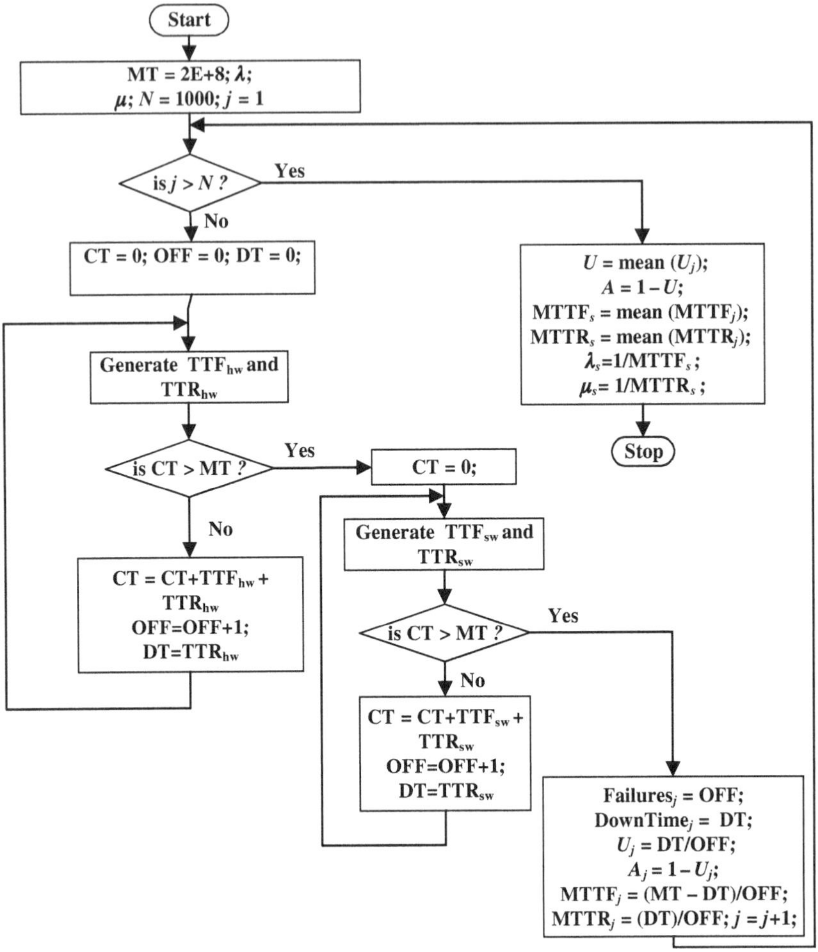

FIGURE 3.12 Flowchart for generating the failure rates and repair rates of the CPU module using MCS. *CPU*, Central processing unit; *MCS*, Monte Carlo simulation.

two repeaters for communication feasibility. When the PMU at bus 3 or its communication network fails, buses 1 and 2 cannot be observed. On the other hand, when the PMU at bus 4 or its communication network fails, buses 5 and 6 cannot be observed. Thus the severity of failures for both the communication networks is 2. The probabilities of failure, the severity, and the risk for each of these communication networks are shown in Table 3.5.

The overall system risk is the sum of their individual risks and is equal to 0.0034042. When the PDC is placed on bus 1, it happens that the communication network of both the PMUs requires only one repeater for communication feasibility. By placing the PDC at bus 1, the risk is reduced to

TABLE 3.3 Failure and repair rates of the various modules of the phasor measurement unit (PMU) and its communication system.

Module	λ (failures/year)	μ (repairs/year)	Failure probability
M_1	$\lambda_1 = 5.00347E-07$	$\mu_1 = 0.000721$	0.0006934
M_2	$\lambda_2 = 5E-06$	$\mu_2 = 8.838459224$	5.657E − 07
M_3	$\lambda_3 = 5E-06$	$\mu_3 = 26.8536$	1.862E − 07
M_4	$\lambda_4 = 5E-06$	$\mu_4 = 8.838459224$	5.657E − 07
M_5	$\lambda_5 = 5.00001E-06$	$\mu_5 = 3.335687$	1.499E − 06
M_6	$\lambda_6 = 5E-09$	$\mu_6 = 0.023224$	2.153E − 07
M_7	$\lambda_7 = 5E-06$	$\mu_7 = 5.847309$	8.551E − 07
C_{PMU}	$\lambda_{PMU} = 0.000000005$	$\mu_{PMU} = 210.4021$	2.376E − 11
C_{PDC}	$\lambda_{PDC} = 0.000000005$	$\mu_{PDC} = 210.4021$	2.376E − 11
C_2 (per 100 km)	$\lambda_{2c} = 0.00438$	$\mu_{2c} = 104.244$	4.202E − 05
C_3	$\lambda_{3c} = 0.000000005$	$\mu_{3c} = 51.72361$	9.667E − 11

0.0032914. Thus, by optimally choosing the location for placing the PDC, the overall risk can be minimized.

3.3 Optimal placement of phasor measurement unit for power system observability

PMUs have gained immense popularity due to their profound application in the real-time monitoring of the power system. By properly placing the PMUs and through the use of network equations, all the other buses in the network can be observed and the problem is known as optimal placement of PMUs (OPP) [24–26].

When a PMU is placed on a bus, it measures the voltage phasor of the bus and the current phasors of all the lines incident on the bus. Consider a simple 4-bus system depicted in Fig. 3.15. Placing the PMU on bus 1 provides voltage phasor at bus 1 and the current phasors of branch 1, 2, and 3. Now, by Ohm's law, one can easily evaluate the voltage phasors of the remaining buses as indicated by Eqs. (3.11)–(3.13), without the need for placing a PMU on these buses. Thus, when the PMU is placed on a bus, all the connected buses become observable [27].

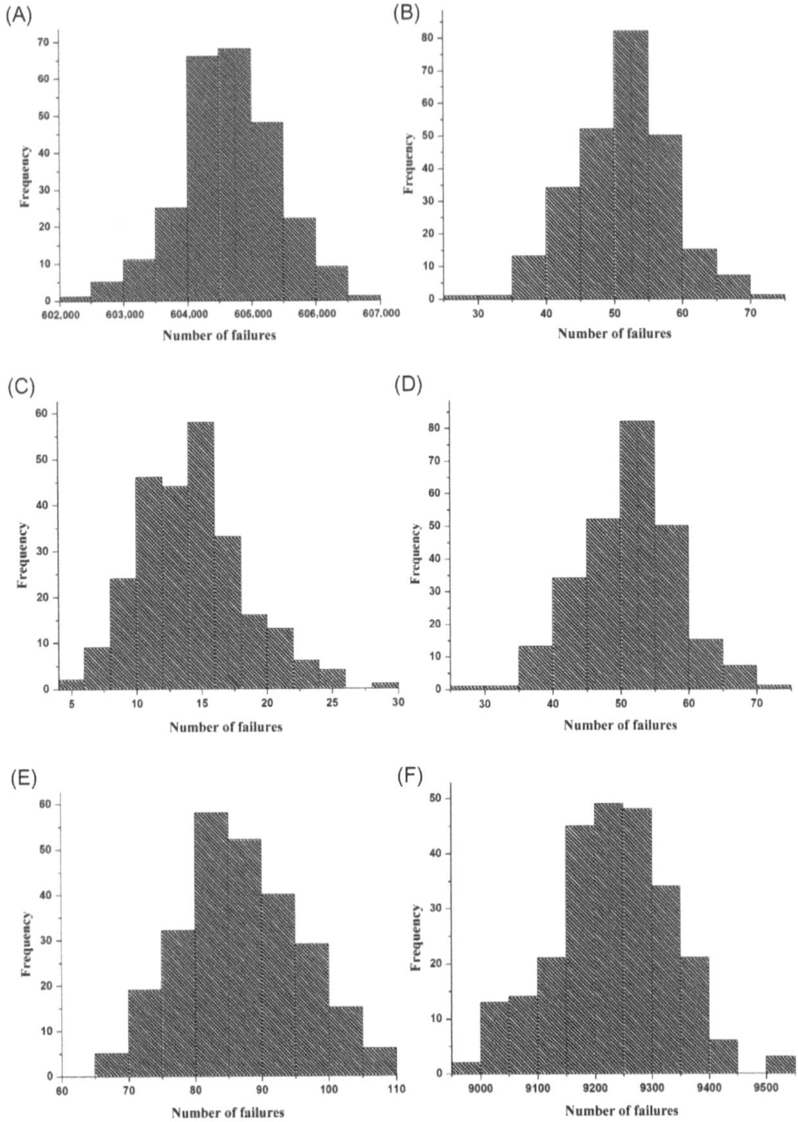

FIGURE 3.13 Failure distribution with uncertainty incorporated for (A) CPU module (M_1), (B) transformer module (M_2), (C) antialiasing filter module (M_3), (D) A/D converter module (M_4), (E) communication module (M_5), (F) GPS receiver module (M_6), (G) power supply module (M_7), (H) PMU optical transceiver module (C_{PMU}), (I) PDC optical transceiver module (C_{PDC}), (J) optical repeater module (C_3), (K) optical fiber (C_2). *CPU*, Central processing unit; *GPS*, global positioning system; *PMU*, phasor measurement unit.

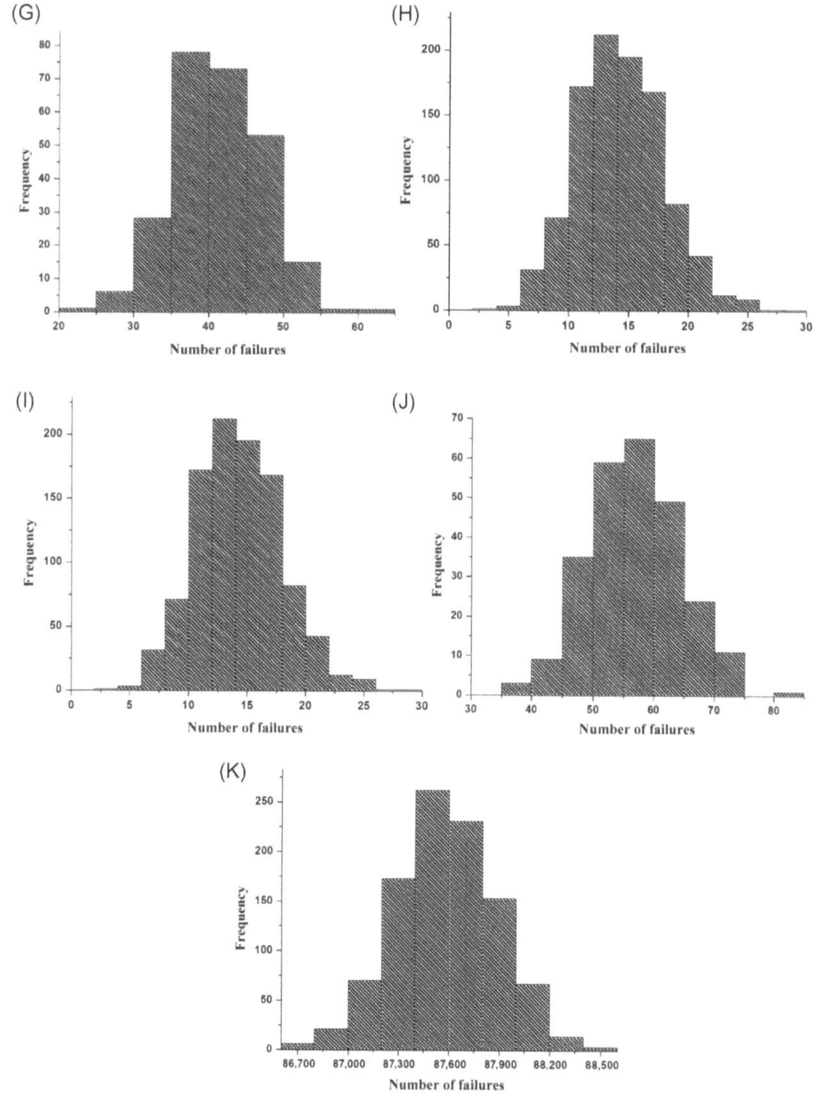

FIGURE 3.13 (Continued).

$$V_2 = V_1 - I_{12}Z_{12} \tag{3.11}$$

$$V_3 = V_1 - I_{13}Z_{13} \tag{3.12}$$

$$V_4 = V_1 - I_{14}Z_{14} \tag{3.13}$$

In this chapter, integer linear programming (ILP) is used for determining the optimal locations for the placement of PMUs [28]. Furthermore, the

TABLE 3.4 Failure probability of the phasor measurement unit and its communication network combined as a single system.

Number of optical repeaters (*l*)	λ_{sys} (failures/year)	μ_{sys} (repairs/year)	Failure probability *U* (*l*)
0	0.0044047	5.538669	7.9466E − 04
1	0.0087873	10.66897	8.2285E − 04
2	0.0131652	15.87640	8.2887E − 04
3	0.0175549	19.94771	8.7926E − 04
4	0.0219404	23.94447	9.1550E − 04

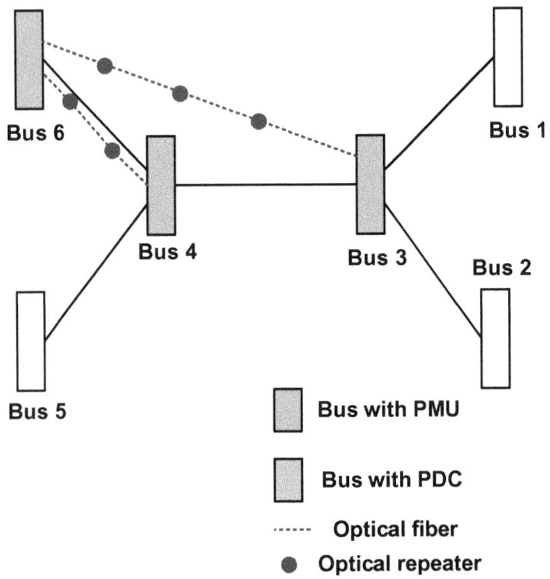

FIGURE 3.14 Sample network for risk assessment.

TABLE 3.5 Risk assessment for the sample network.

PMU	Severity (*s*)	Failure probability $p = U(R)$	Risk (*r*) $r = s \times p$
PMU at bus 3	2	*U* (3) = 8.7926E − 04	0.0017585
PMU at bus 4	2	*U* (2) = 8.2285E − 04	0.0016457

PMU, Phasor measurement unit.

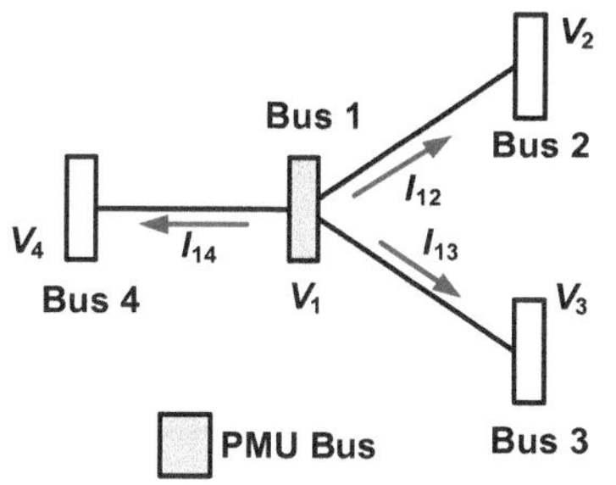

FIGURE 3.15 Placement of PMUs for observability. *PMU*, Phasor measurement unit.

location of the PDC is selected such that the risk involved is minimized. The OPP using the ILP is described in this section.

An electrical network consisting of p buses is considered to elucidate the approach. The connectivity amongst the buses can be represented with the aid of a $(p \times p)$ matrix (B). The elements of this matrix are binary values, which indicate the connectivity between the various buses and are given by Eq. (3.14). Another matrix Q of the order $(p \times 1)$ is used in order to determine the location of the PMUs in the network as indicated using Eq. (3.15).

$$B_{ij} = \left\{ \begin{array}{l} 1 \ \textit{if bus i is associated with bus j or if } i = j \\ 0 \ \textit{elsewhere} \end{array} \right\} \tag{3.14}$$

$$Q_i = \left\{ \begin{array}{l} 1 \ \textit{if the PMU is located on the } i^{th} \textit{ bus} \\ 0 \ \textit{if the PMU is not located on the } i^{th} \textit{ bus} \end{array} \right\} \tag{3.15}$$

The optimal locations for achieving complete observability can be determined using the ILP given by the following equation:

$$N_{\min} = \min\left(\sum_{u=1}^{p} q_u \right) \tag{3.16}$$

$$\text{Constraints:} \quad BQ \geq [1 \ 1 \ \ldots \ 1]^T_{(p \times 1)}$$

where at least $N_{\min}$ number of PMUs are needed for observing the complete system.

Depending on the location of the PDC, the number of optical repeaters and the number of optical links required by the PMU communication network may vary. Thus the overall system risk can be minimized by optimally placing the PDC.

3.4 Simulation results

The uncertainty analysis and the risk assessment are performed for the North-Eastern Power Grid of India consisting of 14 buses as shown in

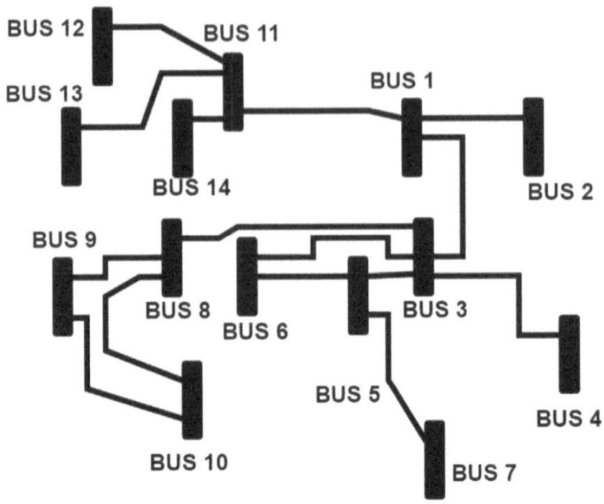

FIGURE 3.16 The electrical network of the North-Eastern Power Grid of India.

Fig. 3.16 [29]. The geographical information pertaining to these buses is given in Table 3.6.

The connectivity matrix B is constructed, and the optimal locations are determined for achieving complete observability of the power system. A minimum of five PMUs, placed at bus 1 (*Bongaigaon*), bus 3 (*Balipara*), bus 5 (*Biswanath Chariali*), bus 8 (*Kathalguri*), and bus 11 (*Silchar*), are required. For determining the optimum PDC location, the risk with uncertainty incorporated is calculated for all possible PDC locations.

The number of intermediate repeaters depends on the topological distance between the PMU and the PDC. If the latitude and longitude of the PMU and the PDC are (φ_1, ψ_1) and (φ_2, ψ_2), the geographical distance d between them can be calculated using the Haversine formula as given in Eqs. (3.17)–(3.19) [30]. The geographical distances between the PMUs and the PDC are shown in Table 3.7 along with the number of optical repeaters.

$$d = Rad \times Z \tag{3.17}$$

$$Z = 2 \times a\tan2\left(\sqrt{a}, \sqrt{(1-a)}\right) \tag{3.18}$$

$$a = \sin^2\left(\frac{\varphi_1 - \varphi_2}{2}\right) + \cos(\varphi_1)\cos(\varphi_2)\sin^2\left(\frac{\psi_1 - \psi_2}{2}\right) \tag{3.19}$$

where Rad is the radius of the Earth = 6371 km and $a$tan2 (x, y) is the principal value of arc tangent of y/x expressed in radians.

When the PMU at bus 1 or its communication network fails, bus 2 becomes unobservable. When the PMU at bus 3 or its communication

TABLE 3.6 Geographical data pertaining to the North-Eastern Power Grid of India.

Bus	Location	Latitude (°N)	Longitude (°E)
1	Bongaigaon	26.2626	90.2143
2	Bongaigaon Thermal Power Station (BTPS)	26.4440	90.3642
3	Balipara	26.8257	92.7755
4	Khupi	24.5800	89.2560
5	Biswanath Chariali	26.7267	93.1479
6	Ranganadi	27.3426	93.8168
7	Subansiri	27.5526	94.2588
8	Misa	26.4830	92.9330
9	Kathalguri	26.4932	92.6773
10	Mariani	26.6516	94.3283
11	Silchar	24.8333	92.7789
12	Imphal	24.8170	93.9368
13	Melriat	23.6455	92.7267
14	Pallatana	23.5097	91.4362

network fails, bus 4 becomes unobservable. When the PMU at bus 5 or its communication network fails, bus 7 becomes unobservable. When the PMU at bus 8 or its communication network fails, buses 9 and 10 become unobservable. Finally, when the PMU at bus 11 or its communication network fails, buses 12, 13, and 14 become unobservable. Thus the severity indices for the WASMS are 1, 1, 1, 2, and 3, respectively. The uncertainty analysis is carried out, and the risk is calculated for different locations of the PDC. These results are shown in Table 3.8.

It can be observed from Table 3.8 that the WASMS with the PMUs at buses 1, 3, 5, 8, 11, and with the PDC at bus 8 has a minimal risk of 0.00647606. The risk is also the minimum if the PDC is located on bus 9. Thus the PDC can either be placed at bus 8 or on bus 9 to minimize the system risk. The failure frequencies for these five systems with the PDC located at bus 9 obtained using the MCS are shown in Fig. 3.17. Thus, for practical power systems, the risk can be estimated and can be minimized through the OPP and the PDC.

TABLE 3.7 Geographical distance and the number of repeaters between phasor measurement unit (PMU) and the phasor data concentrator (PDC).

	Distance in km, number of repeaters				
	PMU at bus 1	PMU at bus 3	PMU at bus 5	PMU at bus 8	PMU at bus 11
PDC at bus 1	0, 0	262.35, 2	296.46, 2	271.94, 2	302.40, 3
PDC at bus 2	25.1, 0	243.4 2	278.6, 2	255.74, 2	301.11, 3
PDC at bus 3	262.35, 2	0, 0	38.57, 0	41.2, 0	221.55, 2
PDC at bus 4	210.4, 2	432.05, 4	457.29, 4	425.278, 4	356.97, 3
PDC at bus 5	296.46, 2	38.57, 0	0, 0	34.51, 0	213.75, 2
PDC at bus 6	377.15, 3	118.03, 1	95.29, 0	129.67, 1	297.65, 2
PDC at bus 7	425.89, 4	167.50, 1	143.24, 1	177.18, 1	336.49, 3
PDC at bus 8	271.94, 2	41.2, 0	34.51, 0	0, 0	184.08, 1
PDC at bus 9	246.69, 2	38.24, 0	53.51, 0	25.47, 0	184.85, 1
PDC at bus 10	411.80, 4	155.41, 1	117.57, 1	140.03, 1	254.87, 2
PDC at bus 11	302.4, 3	221.55, 2	213.75, 2	184.09, 1	0, 0
PDC at bus 12	406.57, 4	251.79, 2	226.56, 2	210.81, 2	116.87, 1
PDC at bus 13	385.77, 3	353.66, 3	345.22, 3	316.20, 3	132.18, 1
PDC at bus 14	329.98, 3	392.57, 3	397.05, 3	363.39, 3	200.5, 2

The failure distribution curves shown in Fig. 3.17 can be approximated by the normal distribution, with the peak of the curve corresponding to the average number of failures (i.e., the mean of the normal distribution). In Fig. 3.17 for the WASMS between the PDC and PMU at bus 1, this value is approximately equal to 263,000. For the WASMSs between the PDC and the PMUs at buses 3, 5, and 8, this value is approximately 90, and for the last WASMS between the PDC and PMU at bus 11, it is 17,600. In the first case due to the large separation between the PMU at bus 1 and the PDC (see Table 3.7), the WASMS requires a more number of components (i.e., repeaters and optical cable), thereby resulting in more failures. In the last case as only one repeater is needed, the number of failures is lower than that of the first case but higher than that of the remaining cases. Thus, by proper placement of the PMUs, the probability of failure can be minimized.

TABLE 3.8 Risk estimated for different locations of phasor data concentrator (PDC).

	Risk					Overall risk O
	PMU at bus 1	PMU at bus 3	PMU at bus 5	PMU at bus 8	PMU at bus 11	
PDC at bus 1	0.00079466	0.00082887	0.00082887	0.00165774	0.00263778	0.00674792
PDC at bus 2	0.00079466	0.00082887	0.00082887	0.00165774	0.00263778	0.00674792
PDC at bus 3	0.00082887	0.00079466	0.00079466	0.00158932	0.00248661	0.00649412
PDC at bus 4	0.00082887	0.0009155	0.0009155	0.001831	0.00263778	0.00712865
PDC at bus 5	0.00082887	0.00079466	0.00079466	0.00158932	0.00248661	0.00649412
PDC at bus 6	0.00087926	0.00082285	0.00079466	0.0016457	0.00248661	0.00662908
PDC at bus 7	0.0009155	0.00082285	0.00082285	0.0016457	0.00263778	0.00684468
PDC at bus 8	**0.00082887**	**0.00079466**	**0.00079466**	**0.00158932**	**0.00246855**	**0.00647606**
PDC at bus 9	**0.00082887**	**0.00079466**	**0.00079466**	**0.00158932**	**0.00246855**	**0.00647606**
PDC at bus 10	0.0009155	0.00082285	0.00082285	0.0016457	0.00248661	0.00669351
PDC at bus 11	0.00087926	0.00082887	0.00082887	0.0016457	0.00238398	0.00656668
PDC at bus 12	0.0009155	0.00082887	0.00082887	0.00165774	0.00246855	0.00669953
PDC at bus 13	0.00087926	0.00087926	0.00087926	0.00175852	0.00246855	0.00686485
PDC at bus 14	0.00087926	0.00087926	0.00087926	0.00175852	0.00248661	0.00688291

PMU, Phasor measurement unit. The bold values indicate the location of the PDC for which the risk is minimum.

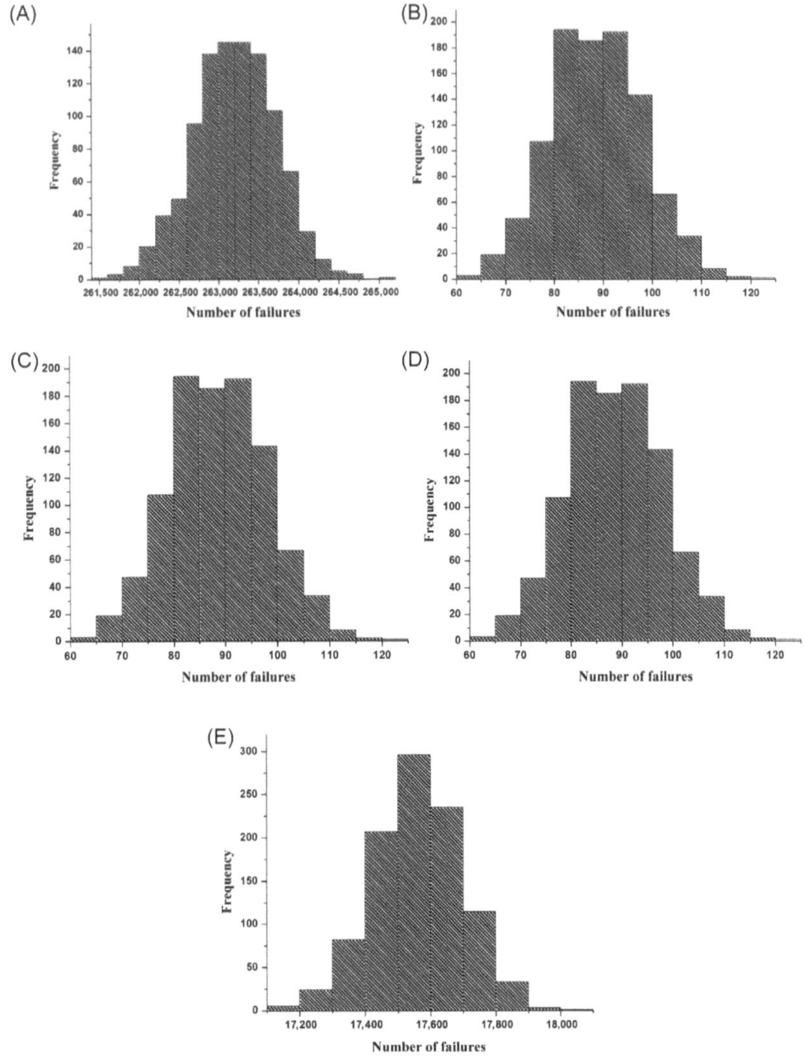

FIGURE 3.17 Failure distribution for WASMS of the North-Eastern Power Grid of India between the PDC at bus 9 and PMU (A) at bus 1, (B) at bus 3, (C) at bus 5, (D) at bus 8, and (E) at bus 11. *PDC*, Phasor data concentrator; *PMU*, phasor measurement unit; *WASMS*, wide area synchrophasor measurement system.

3.5 Conclusion

The failure of system components also affects the performance of the system that can be estimated as a risk. The effect of the failure of the PMU and its communication network on WASMS has been estimated using the risk parameter. MCS has been used to account for the uncertainty in the failure rates and the repair rates of the system components. Simulation models were constructed for the various modules of the PMU and for its optical fiber communication network, and the risk has been estimated. The proposed methodology was used for obtaining the risk metrics for a practical power grid of India, and it has been observed that by optimally placing the equipment, the overall risk can be minimized.

References

[1] P. Pourbeik, P.S. Kundur, C.W. Taylor, The anatomy of a power grid blackout – root causes and dynamics of recent major blackouts, IEEE Power Energy Mag. 4 (5) (2006) 22–29.

[2] The Economic Impacts of the August 2003 Blackout, Electricity Consumers Resource Council, 2004.

[3] A.G. Phadke, The wide world of wide-area measurement, IEEE Power Energy Mag. 6 (5) (2008) 52–65.

[4] A.G. Phadke, Synchronized phasor measurements in power systems, IEEE Comput. Appl. Power 6 (2) (1993) 10–15.

[5] A.G. Phadke, J.S. Thorp, Synchronized Phasor Measurements and Their Applications, Springer, New York, 2008.

[6] A.G. Phadke, J.S. Thorp, Communication needs for wide area measurement applications, in: Fifth International Conference on Critical Infrastructure, Beijing, China, 2010, pp. 1–7.

[7] D.K. Mohanta, M. Cherukuri, D.S. Roy, A brief review of phasor measurement units as sensors for smart grid, Electric Power Compon. Syst. 44 (4) (2016) 411–425.

[8] R. Zurawski, From wireline to wireless networks and technologies, IEEE Trans. Ind. Inform. 3 (2) (2007) 93–94.

[9] V.C. Gungor, D. Sahin, T. Kocak, S. Ergut, C. Buccella, C. Cecati, et al., Smart grid technologies: communication technologies and standards, IEEE Trans. Ind. Inform. 7 (4) (2011) 529–539.

[10] L. Berger, A. Schwager, J. Escudero-Garzás, Power line communications for smart grid applications, J. Electr. Comput. Eng. 2013 (2013) 1–16.

[11] B. Naduvathuparambil, M.C. Valenti, A. Feliachi, Communication delays in wide area measurement systems, in: Thirty-Fourth Southeastern Symposium on System Theory, Huntsville, Alabama, 2002, pp. 118–122.

[12] B. Appasani, D.K. Mohanta, Co-optimal placement of PMUs and their communication infrastructure for minimization of propagation delay in the WAMS, IEEE Trans. Ind. Inform. 14 (5) (2018) 2120–2132.

[13] W. Li, Risk Assessment of Power Systems: Models, Methods, and Applications, John Wiley & Sons, NJ, 2014, pp. 313–350.

[14] V.K. Verma, B. Appasani, D.K. Mohanta, Risk assessment of wireless communication network for PMUs, in: 2017 IEEE International Conference on Power, Control, Signals and Instrumentation Engineering, Chennai, India, 2017, pp. 2711–2714.

[15] R. Billinton, R.N. Allan, *Reliability Evaluation of Engineering Systems – Concepts and Techniques.*, Springer, New York, 1992.

[16] M. Cherukuri, D.S. Roy, D.K. Mohanta, Reliability evaluation of phasor measurement unit: a system of systems approach, Electric Power Compon. Syst. 43 (4) (2015) 437–448.

[17] I. Jurdana, R. Ivče, Availability model of communication network in connecting ship systems using optical fibre technology, Shipbuilding: Theor. Pract. Naval Archit. Naval Tech. 65 (3) (2014) 17–30.
[18] W.J. Bock, I. Gannot, S. Tanev, Optical Waveguide Sensing and Imaging., Springer, Netherlands, 2007.
[19] B. Appasani, D.K. Mohanta, A two-stage Markov model aided frequency duration approach for reliability analysis of PMU microwave communication networks, Proc. Inst. Mech. Eng., O: J. Risk Reliab. (2018). Available from: https://doi.org/10.1177/1748006X18785685.
[20] W.S. Lee, D.L. Grosh, F.A. Tillman, C.H. Lie, Fault tree analysis, methods and applications: a review, IEEE Trans. Reliab. 34 (3) (1985) 194–203.
[21] B. Appasani, D.K. Mohanta, Optimal placement of synchrophasor sensors for risk hedging in a smart grid, IEEE Sensors J. 17 (23) (2017) 7857–7865.
[22] S. Ghosh, D. Ghosh, D.K. Mohanta, Impact assessment of reliability of phasor measurement unit on situational awareness using generalized stochastic Petri nets, Int. J. Electr. Power Energy Syst. 93 (2017) 75–83.
[23] P. Zhang, K.W. Chan, Reliability evaluation of phasor measurement unit using Monte Carlo dynamic fault tree method, IEEE Trans. Smart Grid 3 (3) (2012) 1235–1243.
[24] N.M. Manousakis, G.N. Korres, P.S. Georgilakis, Taxonomy of PMU placement methodologies, IEEE Trans. Power Syst. 27 (2) (2012) 1070–1077.
[25] F. Aminifar, C. Lucas, A. Khodaei, M. Fotuhi-Firuzabad, Optimal placement of phasor measurement units using immunity genetic algorithm, IEEE Trans. Power Delivery 24 (3) (2009) 1014–1020.
[26] K.K. More, H.T. Jadhav, A literature review on optimal placement of phasor measurement units, in: IEEE International Conference on Power, Energy and Control, Tamil Nadu, India, 2013, pp. 220–224.
[27] B. Gou, Optimal placement of PMUs by integer linear programming, IEEE Trans. on Power Syst. 283 (3) (2008) 1525–1526.
[28] B. Gou, Generalized integer linear programming formulation for optimal PMU placement, IEEE Trans. Power Syst. 23 (3) (2008) 1099–1104.
[29] P. Gopakumar, G.S. Chandra, M.J.B. Reddy, D.K. Mohanta, Optimal redundant placement of PMUs in Indian power grid – northern, eastern and north-eastern regions, Front. Energy 7 (4) (2013) 413.
[30] [Online]. Available: <http://www.longitudestore.com/haversine-formula.html>.

Chapter 4

Power quality issues of smart microgrids: applied techniques and decision making analysis

Yahya Naderi[1,2], Seyed Hossein Hosseini[1,4], Saeid Ghassemzadeh[1], Behnam Mohammadi-Ivatloo[1], Mehdi Savaghebi[3], Juan Carlos Vasquez[2] and Josep M Guerrero[2]

[1]*Faculty of Electrical and Computer Engineering, University of Tabriz, Tabriz, Iran,* [2]*Department of Energy Technology, Aalborg University, Aalborg, Denmark,* [3]*SDU Electrical Engineering, Mads Clausen Institute, University of Southern Denmark (SDU), Odense, Denmark,* [4]*Engineering Faculty, Near East University, North Cyprus, Mersin 10, Turkey*

4.1 Introduction

It is Friday and you have been working all day on a project report that is due on Monday to be submitted; you are happy that you will be able to finish the task on time, and that your boss will be proud of you; however, you have not saved the report for several hours, and all of a sudden everything changes, a voltage disturbance in the electricity grid reboots your computer, what a disaster! You are upset, you have lost several hours of work, and you have to work on the weekend to meet the deadline. The problem is not just for you, it could be a costly problem for many electricity consumers all over the world. It was just a very small disturbance of electricity grid, a possible phenomenon that could happen once in a while if the quality of the power is low. Considering that, an outage with the duration of less than 100 ms will have the same effect that a general outage with the duration of minutes on some industrial process. Based on a research in 2006, it was estimated that power interruptions cost the United States an amount of $79 billion annually, which was updated in 2014 to be $110 billion annually [1–3]. The increasing cost of power quality disturbances is an obvious reason for electrical engineers to pay extra attention to the power quality issue. Other reasons for the increasing importance of power quality are inclusive application of more sensitive equipment to voltage disturbances, increasing the number of nonlinear loads, and increasing the awareness of the customers.

Decision Making Applications in Modern Power Systems. DOI: https://doi.org/10.1016/B978-0-12-816445-7.00004-9

The electrical engineers are aware of the importance of the power quality. They care about it, and they research how to improve it. They see everything that affects voltage, current, and frequency of the power that would be supplied to consumers. They have made standards on power quality at all levels of the power system, such as large generation units, distributed generation (DG) units, transmission lines, and ultimate consumers. Nowadays, not only the electricity generation units deal with power quality but also the consumers, small generation unit owners, utility owners, industrial units, and home appliance producers care about the same.

4.1.1 Power quality definition and standards

Power quality has different meanings from different points of view; it could be a problem to be solved, or it is a part of the product. If it is seen from the point of view of an electrical engineer or a power quality expert, it is a problem that should be solved; on the other hand, if it is seen by a power marketer, a producer of electricity, or a consumer, it is a part of the product. On both ends of the spectrum, power quality is an important part of the product that should best fit the needs.

From the technical point of view, reliability, availability, and power quality are the most important aspects of electrical power, which are somehow interconnected. The power quality concept could be studied posterior to having a reliable electricity source, which is available most of the time. As these qualitative definitions are not any proper evaluation criteria for power quality, there should be some quantitative standards to measure the quality of the power. During the last 30 years, several standards for power quality have been published and updated. Though the definitions are almost the same, some details are being added to each update, or the limits are becoming more rigorous. IEC, IEEE, ANSI, and NEMA have adapted several standards on power quality for different aspects of the issue [4]. IEC has developed a category of standards called EMC (electromagnetic compatibility) to deal with the power quality issues. Although EMC has been adopted in the European Union as a requirement of the equipment sold in Europe, their application in other countries varies, and few of them are used in the United States. In the United States, there are a number of standards developed by IEEE, ANSI, and some manufacturer companies such as NEMA on power quality, and most of these standards are application based. Among those the most important power quality standards are IEEE p1159 and IEEE 519 that have been revised several times and define limits of distortions for different levels of power system. An example of these definitions is shown in Table 4.1 [5–7]. For detailed definitions on voltage quality, current quality availability, etc., the provided references could be useful [8,9].

TABLE 4.1 Voltage distortion limits standard.

Bus voltage V_{PCC}	Individual harmonics (%)	THD (%)
$V \leq 1$ kV	5.0	8.0
1 kV $\leq V \leq$ 69 kV	3.0	5.0
69 kV $\leq V \leq$ 161 kV	1.5	2.5
161 kV $\leq V$	1.0	1.5

THD, Total harmonic distortion.

4.2 Smart microgrids

The ambiguous concept of a smart microgrid has been defined during the years that it was developed; although there are similarities between the definitions, some differences lie between them based on the institute providing the definitions. Here are four sample definitions of the smart grid:

> *A smart grid is an electricity grid that uses information and communications technology to gather and act on information, such as information about the behaviors of supplier and consumers, in an automated fashion to improve the efficiency, reliability, economics, and sustainability of the production and distribution of electricity [10].*

(US Department of Energy, 2012)

> *Smart Grids [concern] an electricity network that can intelligently integrate the actions of all users connected to it, generators, consumers and those that do both to efficiently deliver sustainable, economic and secure electricity supplies [11].*

> *A Smart Grid is an electricity network that can cost-efficiently integrate the behavior and actions of all users connected to it generators, consumers and those that do both in order to ensure economically efficient, sustainable power system with low losses and high levels of quality and security of supply and safety [12].*

> *Smart grids are networks that monitor and manage the transport of electricity from all generation sources to meet the varying electricity demands of end users. The widespread deployment of smart grids is crucial to achieving a more secure and sustainable energy future [13].*

Based on these definitions and the authors' opinion, smart microgrid could be described as a microgrid that has some special characteristics that would improve the overall efficiency of system to make it environment friendly, gain more functionality by increasing energy intensity, increasing the overall use and values of existing productions and transmission capacity, integrate greater levels of renewable energy sources (RESs), improve power quality to correspond to new digital demands, become more reliable, resilient, flexible, and sustainable. The key characteristics needed for these changes are listed as follows [13]:

- Intelligence (learning ability)
- Two-way communication
- Self-healing
- Advanced metering infrastructure (AMI)

4.2.1 Challenges in smart grid power quality

Similar to every technology, smart grids bring some challenges to traditional grids. Meanwhile, these also bring some new tools to improve the functionalities. The key challenges and tools that smart grids will bring are categorized in the following subsections.

4.2.1.1 Power electronic devices

The recent progress in the field of power electronics has led to increasing penetration of power electronic devices to the modern electrical grids; these devices are like a double-edged sword, improving the electrical grid performance in one side and bringing some new challenges such as injecting harmonics to the electricity grid on the other. Nowadays, most of the RESs need power electronic interfaces to connect to the main electricity grid, most of the home appliances use power electronic converters, also power electronic converters are used in industrial loads and many other applications. On the other hand, most of the power quality improvement (PQI) devices that are developed to mitigate the power quality problems are power electronic based. Power electronic–based PQI devices such as active power filter (APF), dynamic voltage restorer (DVR), static synchronous compensator (STATCOM), uninterruptible power supply (UPS) systems, smart impedances, electrical springs (ESs), and multifunctional DGs (MFDGs) are of this category [14].

4.2.1.2 Plug-in hybrid electrical vehicles integration

As is known, smart grid is green, meaning that it has the biggest potential to deliver carbon saving. By growing tendency to the use of environment-friendly vehicles, the future of the electricity grid will face a power quality challenge. Integration of a huge amount of storage units that use rectifiers to charge the batteries with different charge rates will greatly affect the power quality of the electricity grid. Also, the peak demand will increase significantly while injecting different orders of harmonic to the electrical grid. On the other hand, this challenge could become an opportunity to improve the reliability of the electrical system, if these storage units could be used as an active demand-side management (DSM) tool. This needs the enactment of ownership and utilization legislations for electrical vehicles storage units. Also, distributed DSM and smart charging methods could be used to improve the overall quality of the system.

4.2.1.3 Renewable energy sources integration

RES have changed the nature of the electricity generation from bulk generation units to DG units. This has helped one to improve the reliability of the system, voltage profile and decreased the transmission line costs, losses, and dependency on the main grid. These are the benefits of using RESs, while on the other hand, these energy sources are not fully reliable because of the probabilistic nature of the energy sources such as solar or wind power. Another drawback of RES integration is that most of these sources have power electronic–based interfaces to convert the power; as mentioned before, overusing of power electronic converters in the electricity grid will cause lots of harmonic pollutions. Recently, researchers are working on methods to make these RES multifunctional so that the integrated power electronic converters could improve the power quality of the grid [7,14].

4.2.2 New tools of smart grids

As mentioned earlier, smart grids technology will bring new tools as well as new challenges that are inevitable. These tools could be divided into several categories such as technologies, concepts, and novel control methods. A smarter grid requires the participation of the tools, which can deliver technology solutions to assist utilities and engage consumers. In this section a brief explanation of these tools and how they will affect the power quality of the grid will be introduced.

4.2.2.1 Advanced metering infrastructure

AMI enables the application of technologies, such as smart meters and other advanced metering devices, to enable a two-way flow of information between customers and utility and to provide customers and utility with data on consumption including time and amount of consumed energy and electricity price. This will give the smart grids a wide range of functionalities, such as remote consumption control, time-based pricing, consumption forecast, fault and outage detection, remote connection and disconnection of users, theft detection and loss measurements, and effective cash collection and debt management. Meeting these goals means the progress to a smarter grid that will have better control over power quality from different aspects. Logging and reporting of any kind of disturbance and outage in a very fast way will improve the power quality index in AMI-equipped grids.

4.2.2.2 *Modern monitoring devices*

Real-time monitoring and display of the devices and performance in a wide area is a key parameter to understand and optimize the system operation. Advanced system awareness avoids blackouts and reports the system logs to predict and prevent probable faults, generate data for future decision-makings, prevent wide-area disturbances, and improve the transmission capability and reliability of the grid. This feature is the beginning of the path that leads to PQIs. Without monitoring devices, smart grid is just a grid, but enabling this feature will grant more efficiency, quickness, and precision to PQI in smart microgrids.

4.2.2.3 *Information and communication technology*

Communication technology will play an important role in improving the power quality issues of smart microgrids. Previously, most of these devices were trying to become dependent on communication that will have some drawbacks such as uncertainty of data and latency. Other researchers suggested using low-bandwidth communication system that was not so efficient. Some research have been done to make the communication technology more reliable by predicting the lost bits in the communication links. However, after all, the lack of proper, fast, and high-bandwidth communication system is a drawback for the PQI devices. With a reliable, fast communication technology the power quality of smart microgrid will improve a lot.

4.2.2.4 *Smart appliances*

Nowadays, most of the home appliance manufacturers have started to include smart chips inside the home appliances to make it possible for each device to have a two-way communication with the grid, other than the ability to be controlled from almost everywhere while taking part in demand–response program. A category of smart appliances could be smart loads, which could directly influence the smart grid power quality; more details about the operation of these devices could be found in Ref. [15].

4.2.2.5 *Storage devices*

With the increasing penetration of probabilistic RESs, using storage devices is an inevitable part of the smart microgrids. Appearance of advanced electricity storage technologies has greatly influenced the vision for the future of this technology. Deployment and integration of advanced storage device technologies, such as the flywheel, super capacitors, advanced sodium–sulfur battery technologies, flow batteries along with compressed air, and pumped hydro and thermal storage technologies, have created a wide vision for the future of smart grids [16,17]. Developments in storage technologies will affect the plug-in hybrid vehicles (PHEV) technology and its influences on DSM and peak management scenarios.

4.2.2.6 Computational intelligence

An important component of smart grids is the computational intelligence that has progressed enormously during the last decade, making it possible to perform advanced control methods in real-time and forecasting applications. Deployment of this technology will greatly affect the power quality of smart microgrids since most of the real-time PQI methods need high computational capabilities; giving this ability to digital controllers will greatly affect the PQI process.

4.2.2.7 Advanced control methods

Advanced control methods can monitor and control the power system components and can make it possible for power electronics to give timely and rapid dynamic response to any event. These methods also involve in decision-making procedures of market pricing, enhancing asset management and a wide area of computer-based algorithms, such as data collecting, monitoring, and analyzing to provide innovative solutions from deterministic and predictive perspectives. As is mentioned in Section 4.2.2.6, improvements in computational capabilities have opened a path to power electronic experts to merge power-converting task with PQI capabilities for power electronic–based converters [14].

4.2.2.8 Active demand-side management and demand response

The DSM is a set of activities, which finally will lead to enhanced reliability, expense management, peak shaving, peak shifting, transmission and generation of cost reduction, and improved voltage quality. Different technologies are involved in DSM, including monitoring system, RESs, battery storage, smart appliances, computational intelligence, and almost all the smart grid technologies. Having a more reliable microgrid that has a high-quality voltage profile is a consequence of the implementation of DSM.

4.2.2.9 Multiagent technology

Multiagent technology is somewhat an umbrella term that encompasses several technologies for a common goal; this goal could be any achievement, such as PQI in smart microgrids. Multiagent system makes it possible for different sections to work in parallel to achieve the defined goal; it is a kind of hierarchical system that makes the use of different agents to perform a task, for example, to perform a demand response scenario; different agents should be employed, such as monitoring and metering agent, computational agent, decision-making agent, and, finally, the agent that is responsible to perform the actions regarding generation and consumption units.

4.2.2.10 Internet of things

The concept of Internet of Things is a very vast area of technology that could include almost everything, ranging from a patient who has implanted a heart monitor to a home appliance with an integrated chip that could connect to the internet. Nowadays, some home appliance companies have started to integrate the essential chipsets inside appliances to give them access to the internet; this will lead to an active participation of devices in DSM, better system planning, and cost-effective design of transmission systems. This will also improve the functionality of a two-way communication between grid and costumers.

4.3 Power quality improvement devices

It has been several decades since PQI devices have been introduced and installed all over the electricity grid. To introduce the PQI devices used in smart grids, a review of previously used PQI devices should be introduced. PQI devices could be categorized into three main generations based on their developing time plus a transition condition; a brief explanation about these categories is provided in the following sections. As the progress through conventional electrical systems to smart electrical systems is not sudden, a transitional condition should play the role of a bridge to pass through it to reach the smart electrical systems' standards. As a summary, before introducing different generations of PQI devices, a block diagram of classification of PQI devices is shown in Fig. 4.1 [14].

4.3.1 First generation of power quality improvement devices

The first generation of PQI devices is simple and reliable in structure, and they do not cost so much. This category includes passive, active, and hybrid

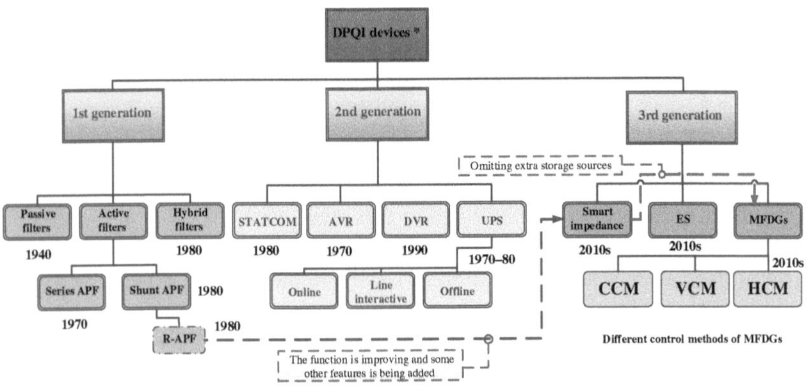

FIGURE 4.1 A classification of distributed power quality devices based on application and features.

filters [14]. Since the early 1970s, passive filters, a combination of capacitors and inductances, have been used in parallel or series to filter-defined harmonics. When it is installed in parallel with loads, it will make a detour for harmonic currents, by setting the inductance and capacitor values in a way that it has high impedance in fundamental frequencies and very low impedance facing desired harmonic frequencies to absorb the harmonic current. These devices are installed in series with a load to prevent harmonic currents from entering the load. Although passive filters are cheap and simple in structure, they have the drawback of need for redesign for each new case. The filters need to be tuned for a specific harmonic to act correctly and may lead to overvoltage at the end of line during low power demand. Passive filters are used in some specific applications nowadays despite their drawbacks for their simplicity and cost-effectiveness; it is worth noting that most of these applications are some hybrid applications of passive filters to reduce the costs and increase the total reliability of the system. To overcome the drawbacks of the passive filters, APFs were developed to compensate and improve power factor, to compensate current harmonics, unbalance and flicker, and to regulate voltage. APFs have been used in several applications with several topologies and control methods, and detailed comparison between different APFs and their application has been done in Ref. [14,18,19]. APFs could be divided into two main groups, shunt APFs and series APFs. Shunt APFs are used in parallel to compensate current harmonics by adding the harmonic current with the same magnitude and with 180 degrees phase difference to have nearly sinusoidal grid current. It could also be used to compensate reactive power if a proper control method is applied [20]. It is worth mentioning that improving the operation of control methods is an interesting ongoing topic about these devices [21]. Series APF is used in series with harmonic loads to compensate harmonic voltages by adding harmonic voltages with the same magnitude but with the opposite phase. The main drawback of these devices is that it could be in the same power rating with the load, so for high-power applications, it will be an expensive and unaffordable solution. To reduce the cost of using high-power APFs, hybrid power filters are an appropriate alternative that has the advantage of both active and passive power filters at the same time. Defining new applications for hybrid power filters has been popular for the last decade; another research field in the area of hybrid power filters is to find improved control strategies to enhance the performance of these hybrid power filters [14,22–24]. Fig. 4.2 shows a group of first- and second-generation PQI devices.

4.3.2 Second generation of power quality improvement devices

The second generation of PQI devices includes the most popular group of PQI devices used in electricity grid up to now. These devices are not as cost

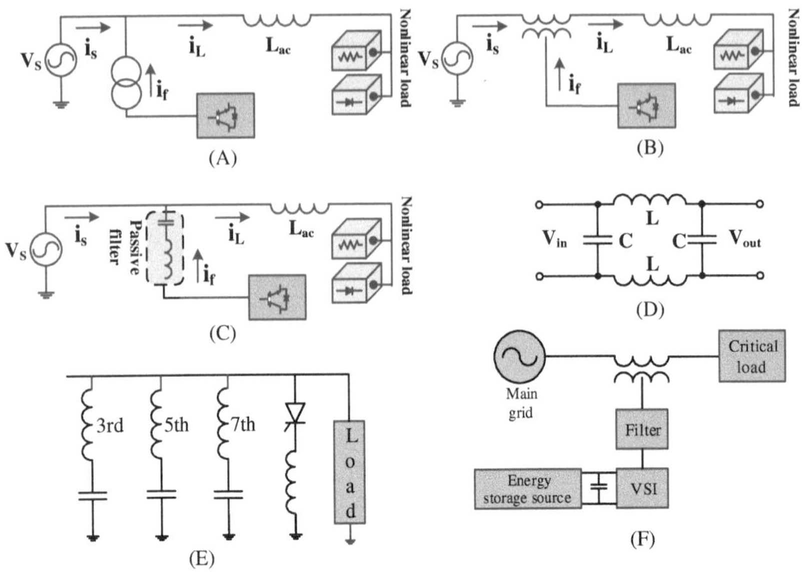

FIGURE 4.2 First and second generation of PQI devices: (A) shunt APF, (B) series APF, (C) hybrid APF, (D) passive filter, (E) DVR, and (F) SVC. *APF*, Active power filter; *DVR*, dynamic voltage restorer; *PQI*, power quality improvement; *SVC*, static VAR compensator.

effective as the first generation of devices, but they have more functionalities in comparison with the first generation. PQI devices, such as DVR, static volt amperes reactive (VAR) compensator (SVC), STATCOM, the automatic voltage regulator (AVR), and UPS, are categorized in the second generation of PQI devices, since they have more complex controllers. Since bringing a full definition and application case of these devices is out of the purpose of this chapter, it seems enough to give a brief explanation and refer the researchers to more detailed papers to read about these devices [14].

DVR is a power electronic–based device that includes an energy source, an isolated transformer, and a power electronic converter. It is connected in series with sensitive loads to protect them against voltage unbalances. For compensating voltage harmonics, it also needs an energy storage source such as a capacitor, an ultracapacitor, superconductor energy storage, or flywheel. Researching on improving the performance and emerging this technology with novel storage device technologies is popular among researchers [25,26].

AVR is a device that changes the output voltage that would be applied to critical loads to keep it at a sufficient voltage level. It could change the voltage using transformer taps continuously using servo motors or in a discrete way using power electronic relays. Although power electronic device–based AVRs are faster in comparison with rotary AVRs (servo motor based), they have less precision due to the discrete output voltage levels.

STATCOM and SVC are kind of power electronic–based devices that are used widely in industry to regulate the output voltage continuously and discretely by absorbing or giving reactive power to it [27].

UPS is a well-known power electronic–based device that can sense frequency and voltage unbalances and supply the sensitive load with a pure sinusoidal waveform with a fixed frequency that is generated by the included converters. From the physical point of view, UPSs could be categorized into two main groups, static UPSs that are power electronic based and dynamic UPSs that are based on some rotary elements, such as flywheels. The main drawback of the static UPSs is the requirement of large energy storage though it has to feed the whole load in the case of unbalance or blackout. Based on the application point of view, UPSs could be classified into three main groups: offline, line-interactive, and online. Researchers are trying to develop a method to use UPSs as APFs to actively compensate the unbalances instead of feeding the whole load; this will overcome the storage device problem of these kinds of UPSs [28,29]. For a more detailed comparison between different types of UPSs and their application, it is worth referring to Ref. [14].

4.3.3 Transition condition, a bridge between conventional and smart electrical systems

The operation and control of conventional electric power systems due to the increasing use of energy storage systems and deployment of variable generation technologies mostly as DG will become more complicated. On the other hand, power quality has always been a challenge in conventional electric power systems, and it is expected to become a vigorous task due to the strict standards and the ever increasing of variety of loads in upcoming modernized systems. Smart electric power systems discussed so far are the promising solution to overcome these challenges and difficulties arising from aging infrastructure, demand growth, and of course power quality issues.

But the problem with smart grids is that their development is faced with numerous economic, technical, and legal barriers, and it is not a sudden one-day event. In other words the time needed to equip conventional power systems with smart devices is much longer than the changes brought by the new conditions of power systems.

These conditions, which could be called "transition conditions," lead to more challenges in the operation and control of power grids. This "transition conditions" also affect the quality of power delivered to the consumer. Conventional power quality enhancement methods are no longer efficient, and the network has not reached the degree of intelligence that is needed for solving power quality problems.

This problem and the variety of power quality requirement by the consumer will further increase the challenges that utilities will face. Modern devices for improving power quality can play an effective role under these conditions. By using distributed configuration, these devices have required flexibility to work with various loads and generation technologies. Advanced control methods are also implemented in order to integrate these devices into conventional networks and provide the required degree of power quality. In this regard a detailed discussion about the third generation of PQI devices is presented in the following section to help the researchers build the missing bridge between the power quality issues of conventional electrical systems and the power quality issues of smart electricity systems.

4.3.4 Third generation of power quality improvement devices

The third generation of PQI devices is the pioneer in emerging smart grid technologies into PQI devices. These devices are mostly multifunctional, capable of performing several tasks at a time without adding any hardware to the structure of the device, which leads to increased cost-effectiveness besides being effective and reliable. Smart impedance, ES, and MFDGs are the main devices included in this category [30,31].

4.3.4.1 Smart impedance

A PQI device, which has the characteristics of all abovementioned second- and first-generation PQI devices, is named "smart impedance." From the physical point of view, smart impedance is a combination of an APF, a coupling transformer, a capacitor bank, and an appropriated control strategy. Smart impedance can solve the tuning process of passive filters, while compensating harmonic currents, harmonic unbalances, improving quality factor, tuned and displacement power factor (DPF). Smart impedance can improve voltage regulation and stability in weak systems such as small microgrids (smart grids), in which the source impedance is not negligible. Smart impedance, which is controlled by a proportional resonant (PR) control method, could perform as a series APF, shunt active filter, a tuned passive filter, a capacitor bank, and a combination of an active and passive filter to reduce the capacity of the filters. Smart impedance is able to mitigate the selected harmonics of interest, it can act as a short circuit (zero impedance) for load current harmonics, while acting as infinite impedance against undesired harmonics. Fig. 4.3 shows a simple view of smart impedance topology [32]. The power circuit of mart impedance includes a capacitor bank, which is connected to a power converter via a transformer. Three phase control blocks are on the basis of the proportional resonant (PR) converter to mitigate system current harmonics without need for phase-locked loop (PLL). The harmonic control block is used to eliminate harmonics using a PR controller. DPF block is controlling the injected reactive power

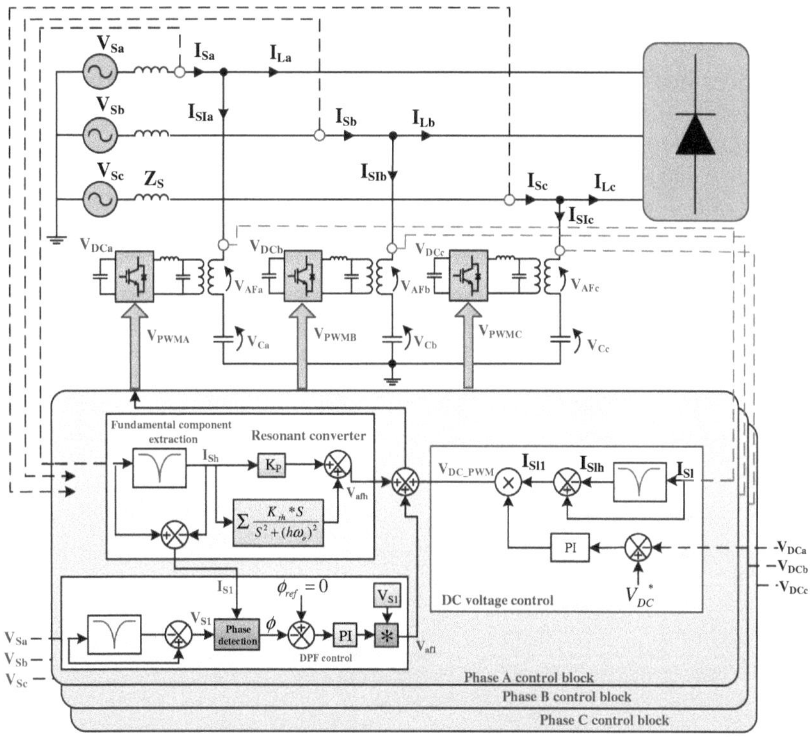

FIGURE 4.3 A scheme of smart impedance.

by capacitor bank, and DC voltage of each converter is controlled by DC voltage control block regulates.

4.3.4.2 Electrical spring

The concept of ES was developed on the basis of the mechanical spring principles to regulate the voltage in a distributed way; it could play the role of a smart load, which is able to follow the power generation profile in the case of integration into the noncritical electrical appliances. The application of distributed smart loads over the electricity grid will lead to increase the stability of the system independent of the communication system. As a mechanical spring prevents the subsidence of a mattress, in an identical way, the ES will prevent voltage drops over the electricity grid and will improve the voltage stability of the grid, the details could be better understood by referring to Ref. [14]. Since ESs have an integrated energy storage unit, it can save energy and export it back to the grid in the case of need; the most important advantage of this device is the distributed nature of ES, which means that even in the case of the failure of several devices, stability of the system remains untouched. A simplified connection diagram of an ES is shown in

Fig. 4.4. Like mechanical springs, ES can do three tasks in a power system, to store energy, support the grid with voltage regulation, to damp electrical oscillations and act as a capacitance and inductance. Improved types of ESs have also been introduced in some research works with some new capabilities such as P and Q compensation and harmonic reduction [15]. As is shown in Fig. 4.4, the energy storing capability of the ES is aided by the integrated storage unit. To damp the electrical oscillations a noncritical load is connected in series to an ES to form a smart load; this way the smart load could follow the generation profile of the power system, and this feature will make smart load an important device for the demand–response process of the smart grids. It will also improve the voltage stability of the power system. As was mentioned before, the ES plays a similar role to the mechanical springs used inside the mattresses to prevent mattress subsidence; this simulated mechanical feature of ES to lift the voltage dips is shown in Fig. 4.5 [14]. It is worth mentioning that the working principle of ES is similar to

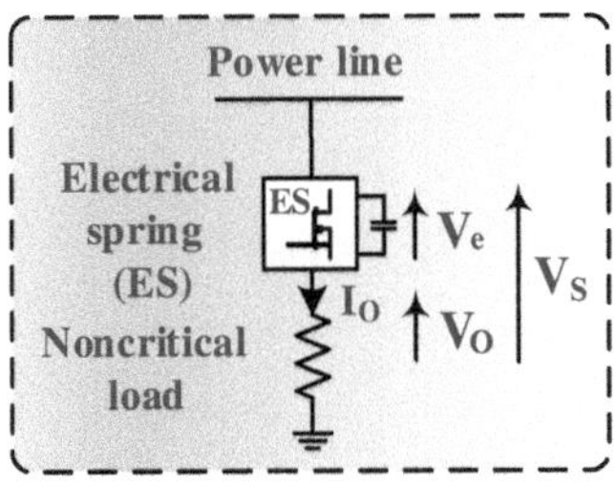

FIGURE 4.4 A simplified connection diagram of an ES. *ES*, Electrical spring.

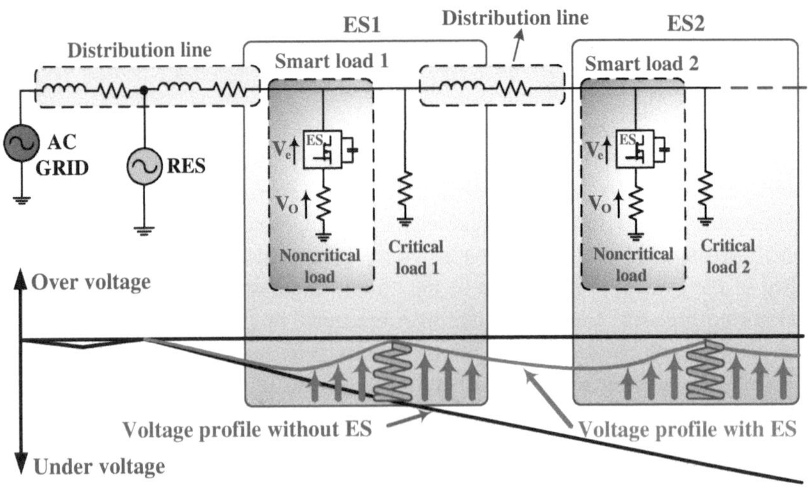

FIGURE 4.5 A simplified connection diagram of electric springs, and their simulated mechanical behavior in a microgrid.

reactive power controllers that control the input voltage instead of output voltage, but in contrary to other reactive power controllers, such as SVC and flexible alternating current transmission system (FACTS), which only participate in reactive power compensating, ES is capable of compensating active and reactive power. Recently, to take part in demand-side response of DC microgrids, the concept of DC-ES has been proposed; it has the storage units integrated and a bidirectional DC–DC converter in the structure that makes the DC-ES to perform the mentioned tasks in a DC microgrid [33].

4.3.4.3 Multifunctional distributed generations

The pioneer technology to improve the power quality in smart grids is the use of MFDGs to enhance the power quality locally and globally. These days the progress to a CO_2-free world, cheap and clean energy sources would accelerate the attention drawing to RESs. However, most of these energy sources use power electronic–based converters to output the desired AC voltage to the main electricity grid, which makes these energy sources a costly electricity generation. To make the technology more cost effective, other functionalities could be added to the power electronic–based interfacing converters, such as power quality enhancement capabilities. These converters could be used in harmonic compensation, voltage regulation, and as an energy storage source for different smart grid applications. These features could be empowered by means of several smart grid–enabled tools such as smart metering infrastructures and computational intelligence, and they will have a big portion in smart grid power quality enhancement. These devices could be categorized based on the controlled objects and the applied control methods. Since a critical part of MFDGs is the applied advanced control methods, it is worth dedicating a section to study the control methods and their secondary applications in smart grids.

4.3.4.4 Applied control methods to multifunctional distributed generations to enhance power quality

There are several control methods applied to MFDGs interfacing power electronic–based converters in the case of harmonic compensation in the literature; the most relevant control methods are PR controller and model-based predictive controller (MPC). Each of the mentioned methods has some advantages to previously proposed methods; in the case of a PR controller, the simplicity and the ability to control the voltage and current simultaneously are the advantages; and in the case of MPC, the advantages are the flexibility of the control method, fast dynamic response, and acceptable reference tracking operation in lower switching frequencies. The other merit of the MPC method when applied to MFDG interfacing inverters is the capability of multiobjective operation, which means the ability to control several objectives simultaneously regarding the priority of each objective; this would have a variety of applications in decision-making processes in smart grids that would be discussed in detail in Section 4.3.4.4.3. Another classification of control methods divides the methods into three groups of voltage-controlled method (VCM), current-controlled method (CCM), and hybrid control method (HCM), for which a detailed discussion will be done defining control objective. Based on the influence area of MFDGs, different control objectives should be pursued, that is, for local systems the harmonic-free output current of an MFDG is satisfactory, and CCM could be applied, while for a regional control of MFDGs, other objectives such as point of common coupling (PCC) voltage THD and harmonic-free PCC current should be considered as objectives; therefore different control methods could be used for each case [7,34–38].

4.3.4.4.1 The Proportional + Resonant control method

To overcome the harmonic reference tracking problems of a PI controller, the PR controller was developed; the main difference between PI and PR controllers is the parallel resonant loops that have the duty of tracking the harmonic references. It has been widely used to control MFDGs and hierarchical microgrid control [7,39,40]. In some references the PR control method has been used as a hierarchical control method; to have a higher supervision over the system, the included control levels could be energy management, voltage and current control at the first control level, power quality and power flow issues at the secondary control level, and economic dispatch, DSM, and microgrid supervision are done at the tertiary control level [36,41,42]. Since the third level is not the focus area, in this chapter, first and second levels of control methods will be discussed; in this regard, to have a better understanding of PR control method, it would be discussed in three parts: CCM, VCM, and HCM. In each case a solution for each control objective would be proposed; to make the following operation of these control methods easier, an overall view of harmonic compensation in microgrids is shown in Fig. 4.6. In this configuration, MFDGs are considered to play a role in PQI-related tasks, such as harmonic compensation, in addition to participating in load sharing.

Current-controlled method The CCM is the most used control method in grid-connected DGs; the main idea in CCM is to create an appropriate current reference due to the control objective and then track the current reference. Fig. 4.7 explains an overall view of the CCM applied to interfacing converters; as it could be seen in (4.1), the reference current is made up of two elements: the fundamental current and harmonic current. The fundamental reference current calculation is done based on P and Q exchange between microgrid and the main grid. The harmonic current calculation is as shown in Fig. 4.7; for three different control objectives, there are three harmonic

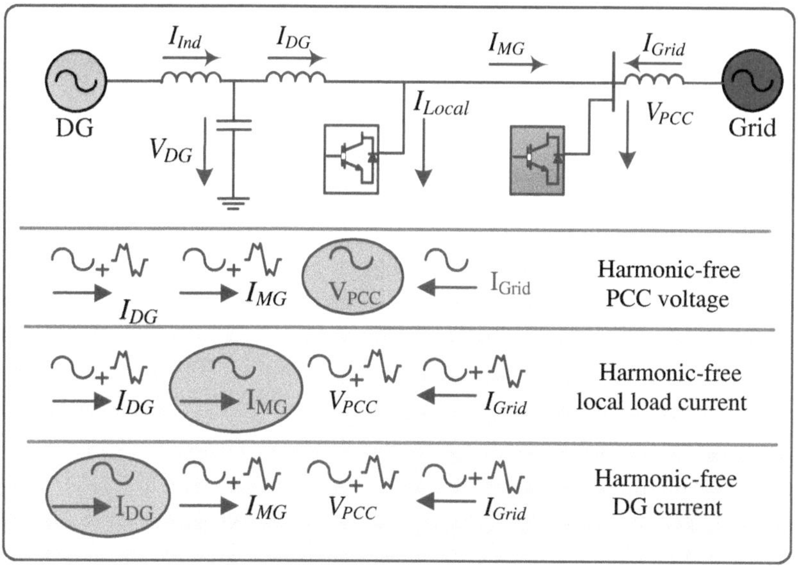

FIGURE 4.6 Overall view of harmonic compensation in microgrids.

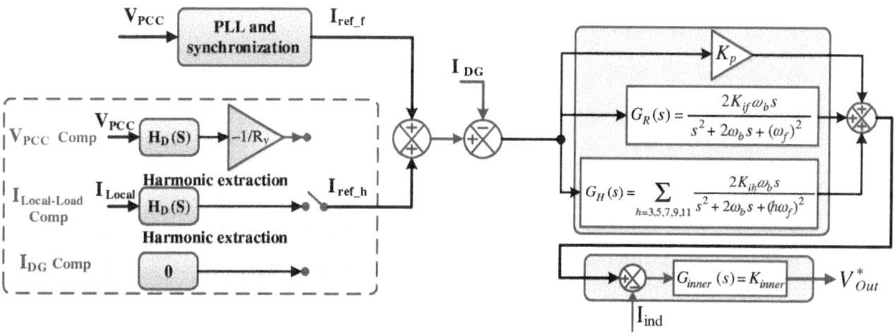

FIGURE 4.7 Overall view of CCM with harmonic current reference calculation. *CCM*, Current-controlled method.

reference calculation methods. To have a harmonic-free output current of MFDG, it is the default compensating objective of CCM, that is, it is simply done by setting the harmonic current reference to zero as in (4.2).

$$I_{ref} = I_{ref\_f} + I_{ref\_h} \tag{4.1}$$

where I_{ref} is the main current reference for CCM, I_{ref_f} is the fundamental current reference, and I_{ref_h} is the harmonic reference current for CCM.

$$I_{ref} = I_{ref\_f} \tag{4.2}$$

To compensate the local load harmonic current, as it could be seen in (4.3), the harmonic reference would be the harmonic elements of the local loads, and to calculate the harmonic elements, a harmonic extractor block is needed in this structure. To compensate the voltage harmonics of the PCC, since the PCC load current is not accessible for MFDG, PCC voltage compensation would be indirectly done by measuring the PCC voltage. The basic idea is to provide all the load currents inside the microgrid by MFDG so that the grid current would be almost harmonic free (4.4); as a result the PCC voltage will be harmonic free also. It is worth mentioning that, when compensating the PCC voltage, some other disturbances may appear inside the microgrid. Since the main objective is to compensate the PCC voltage harmonics, these disturbances could be neglected. This effect is called the whack-a-mole effect; it means that in some cases, compensating harmonics in a point will result in the appearance of harmonics in other points of microgrid; then when a compensating effort is being applied to a microgrid, all aspects of power quality issues should be considered and a trade-off between different objectives should be done.

$$\begin{aligned} I_{ref} &= I_{ref\_f} + I_{ref\_h} \\ &= I_{ref\_f} + H_D(s) \cdot I_{local} \end{aligned} \tag{4.3}$$

where $H_D(s)$ is the harmonic detector to extract the harmonic elements of the local load current.

$$\begin{aligned} I_{ref} &= I_{ref\_f} + I_{ref\_h} \\ &= I_{ref\_f} - H_D(s) \cdot \frac{V_{PCC}}{R_V} \end{aligned} \tag{4.4}$$

where R_V is the equivalent MFDG resistor at harmonic frequencies. It is worth mentioning that the only difference between PCC voltage harmonic and local load harmonic current compensation lies in calculating the harmonic current reference in Eqs. (4.3) and (4.4). As is shown in Fig. 4.7, the presented CCM is based on a stationary reference frame with PR (proportional and parallel resonant) controllers at harmonic and fundamental frequencies.

Voltage-controlled method Although CCM is used for most of the grid-connected MFDGs, VCM is increasingly being used in stand-alone applications of MFDGs to simulate the behavior of a synchronous generator; on the other hand for autonomous control of microgrids, for voltage and frequency control, VCM should be applied to interfacing converters. Another advantage of using VCM is the control of several MFDG units to share the power in a decentralized way by means of a droop controller without any need to communicate between MFDGs. Since there are no closed-loop line current regulators in the controller, it could be hardly used to regulate the MFDG output current; as a result, VCM is rarely used to address any harmonic compensating issues. However, PQI by means of VCM could be possible with virtual

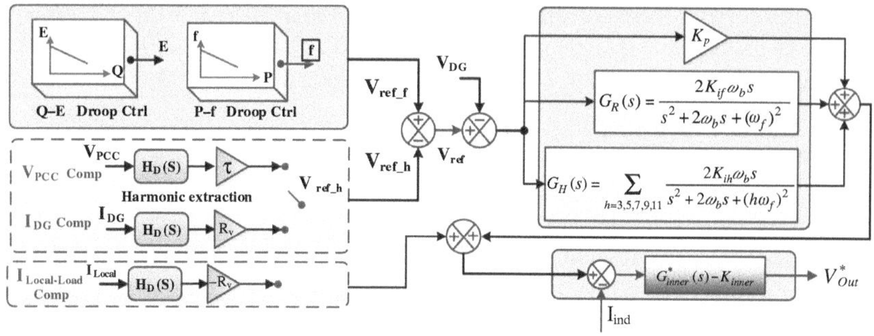

FIGURE 4.8 Overall view of VCM with harmonic current reference calculation. *VCM*, Voltage-controlled method.

harmonic impedance control. An overall view of the VCM applied to microgrids to fulfill the three objectives is shown in Fig. 4.8. It is worth mentioning that to control the reactive power in zero steady state, an integral control term should also be added to voltage magnitude reference. As is shown in Fig. 4.8, similar to CCM, a double-loop PR controller is used to control the voltage.

To compensate the PCC voltage harmonics, for voltage reference a feed-forward term could be used as

$$\begin{aligned} V_{ref} &= V_{ref\_f} + V_{ref\_h} \\ &= V_{ref\_f} - H_D(s) \cdot V_{PCC} \end{aligned} \tag{4.5}$$

where $H_D(s)$ is the harmonic detector block and τ is the feed-forward gain. This way, MFDG will play the role of small impedance in selected harmonic frequencies with equivalent harmonic impedance of that harmonic so that it could absorb selected harmonic currents.

$$Z_{DG,eq} = \frac{Z_{DG}}{1 + \tau} \tag{4.6}$$

To compensate the local load harmonic using VCM is not easy since, by default, it is not capable of compensating local load harmonic. It could be possible by adding a feed-forward high-bandwidth controller such as hysteresis band controller, model predictive controller, or multiple harmonic resonant controllers to the inner control loop that should replace G_{Inner} in Fig. 4.8. Although it would increase the complexity of the controller, it would make local load harmonic compensation by means of HCM. Using the same strategy in CCM, the harmonic elements of the local load current should be added to the current reference.

To have a harmonic-free output current of MFDG, it could not be controlled directly, instead current regulation could be done considering $\tau = -1$, so that virtually it would create a large closed-loop harmonic impedance at the terminal of MFDG, which would reject any harmonic current forcing them to the grid current. It could also be controlled by means of virtual series impedance control that could be implemented as shown in Fig. 4.8.

Hybrid control method In contrast to CCM and VCM, which control current or voltage output of the interfacing inverter at a time, the HCM is able to simultaneously control the voltage and current through the output Inductance + Capacitance + Inductance Filter (LCL) filter fundamental capacitor voltage and line harmonic current. The main difference between CCM, VCM, and HCM is that, unlike the two methods, it could control the fundamental and harmonic elements in a decoupled way, so that it will introduce some new characteristics to MFDG interfacing converter controllers. As a default, similar to VCM in HCM, output fundamental power is being controlled by output filter capacitor voltage; meanwhile, a closed-loop harmonic current controller regulates the line current harmonics. Instead of having a cascaded control loop structure as in Fig. 4.8, a more effective parallel structure of parallel converters is used with multiple control branches. As shown in (4.7), this parallel controller consists of three control branches: fundamental voltage, harmonic current, and an active damping term, which will provide damping to both fundamental voltage and harmonic current path.

$$\begin{aligned} V_{out}^{*} &= G_{power}(s)\cdot(V_{ref\_f} - V_C) \\ &\quad + G_{harmonic}(s)\cdot(I_{ref\_h} - I_{DG}) \\ &\quad + G_{damping}(s)\, I_{Ind} \end{aligned} \tag{4.7}$$

As is shown in Fig. 4.9, the parallel controller has three control branches, the first branch is a resonant controller in fundamental frequency to regulate the output filter capacitor voltage and to control the power flow. To minimize the interface between capacitor voltage and line harmonic current, the proportional gain of the voltage controller in VCM is removed without much affecting the HCM operation. The second term of (4.7) is a closed-loop harmonic current controller, which operates in harmonic frequencies, so multiple resonant controllers are adopted for different frequencies. To track the noncharacteristic harmonic currents a small proportional gain of K_p has been

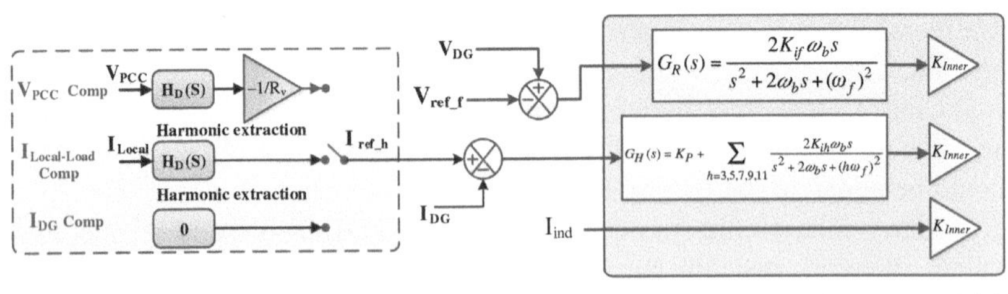

FIGURE 4.9 Overall view of HCM with harmonic current reference calculation. *HCM*, Hybrid control method.

added to the equation, since there is no fundamental frequency resonant controller involved, and the gain is small, tracking of line current harmonic would not affect fundamental voltage control of HCM. Finally, a proportional term is added to actively damp the fundamental voltage and harmonic current control oscillations. The previous PQI objectives could be satisfied easily by applying HCM to MFDGs interfacing converters. To have a harmonic-free line current, similar to CCM, I_{ref_h} should be set to zero; in this case, fundamental power flow is controlled through the first term and harmonic current tracking is done by the setting the second term of (4.7) to zero. It is worth mentioning that HCM could be used instead of CCM; in the case of fundamental current control, the first term of (4.7) should be replaced by a fundamental current control loop, and the harmonic current is being tracked using the second term of (4.7). In this case, fundamental and harmonic current elements are controlled in a decoupled way without affecting each other. If the mentioned method is used instead of traditional CCM, it has the advantages of HCM, such as no need for harmonic extraction block for local load harmonic compensation, since the small gain of the harmonic control loop would not affect the fundamental frequency current, local load current could be used as harmonic current reference without any harmonic extraction block.

To compensate local load harmonic current, in a similar way to CCM, but with a slight difference of not using the harmonic extractor, the local load harmonic current could be fed into the closed-loop harmonic current controller, without the need for a harmonic extractor block since it does not have any fundamental frequency controller elements to affect the fundamental frequency current, so that MFDG interfacing converter would provide most of the harmonic components produced by nonlinear load, leading to an improved PCC current, this advantage of HCM to previously introduced control methods would make it an appropriate, cost-effective option for low-power MFDG units with lower computing capabilities. To compensate the PCC voltage harmonics, similar to what has been done in CCM, the harmonic reference should be calculated based on PCC voltage harmonics as (4.4), which would be fed to the harmonic current loop controller; as a result, MFDG at selected harmonic frequencies will operate as a small virtual resistive impedance to improve the PCC voltage similar to what was done in CCM [14,43].

4.3.4.4.2 Model-based predictive control (MPC)

The other popular control method in PQI area is model predictive control or direct control that could be utilized in harmonic compensation, PQI, and APF applications because of its fast dynamic response and simplicity of the controller. The model-based predictive control (MPC) is a direct control method that uses the discrete model of a system to forecast behavior of system and has been popular with power converters' control recently for its prominent characteristics, such as robustness, fast and precise dynamic response, multiobjective control ability, and capability of application to nonlinear systems. MPC is classified into two main groups: continuous control set MPC and finite control set MPC (FCS−MPC). The first classification of MPC uses a modulator to generate switching signals based on a continuous output of the predictive control. The main advantage of this controller is the fixed switching frequency besides the common advantages of MPC controller. The first one has the advantage of dealing with a finite number of states in the optimization problem, which will lead to a lower amount of computational burden and a good solution for the control systems with limited computational abilities. Another advantage of FCS−MPC is the direct control of the converter without need for a modulation step that decreases the computational burden but has the drawback of variable switching frequency; if the control case is not sensitive to variable switching frequency, then FCS−MPC is a good solution. To apply the multiobjective MPC control, it is much convenient to use FCS−MPC because of the lower computational burden requirement; since it is the case in this study to have a multiobjective control over smart grids, the focused MPC method will be FCS−MPC [44,45]. The main idea behind FCS−MPC is to generate a discrete model of system to forecast the behavior of the system, then a cost function is formed, which best fits the control objectives. After forecasting the system behavior, it would be applied to the defined cost function; the control actions that minimize the cost function will be the selected control actions in each cycle. To ensure the optimum function of the system, this cycle would be done repeatedly; the flowchart view of this control method is shown in Fig. 4.10. In the next section, MPC would be applied to a prototype microgrid including an MFDG to validate the PQI characteristics as well as multiobjective operation capability of the control method. The simplest cost function for an MPC controller could be a current reference tracking such as

$$g[K+1] = \left| i_L[k+1] - i_L^{*}[k+1] \right| \tag{4.8}$$

which could be easily applied to an MFDG interfacing converter to track the current reference, which could represent a CCM as discussed in Section 4.3.4.4.1. Similar to PR-CCM, the current reference would be fed into the controller, so that the controller could track it with the minimum error. The advantage of simple MPC to PR-CCM is that, in MPC, there is no need for harmonic extractor block and the local load current could be directly used as a current reference when compensating local load harmonics.

4.3.4.4.3 Multiobjective model-based predictive control

The main difference between multiobjective model predictive control (MOMPC) and single-objective model predictive control lies in defining the cost function

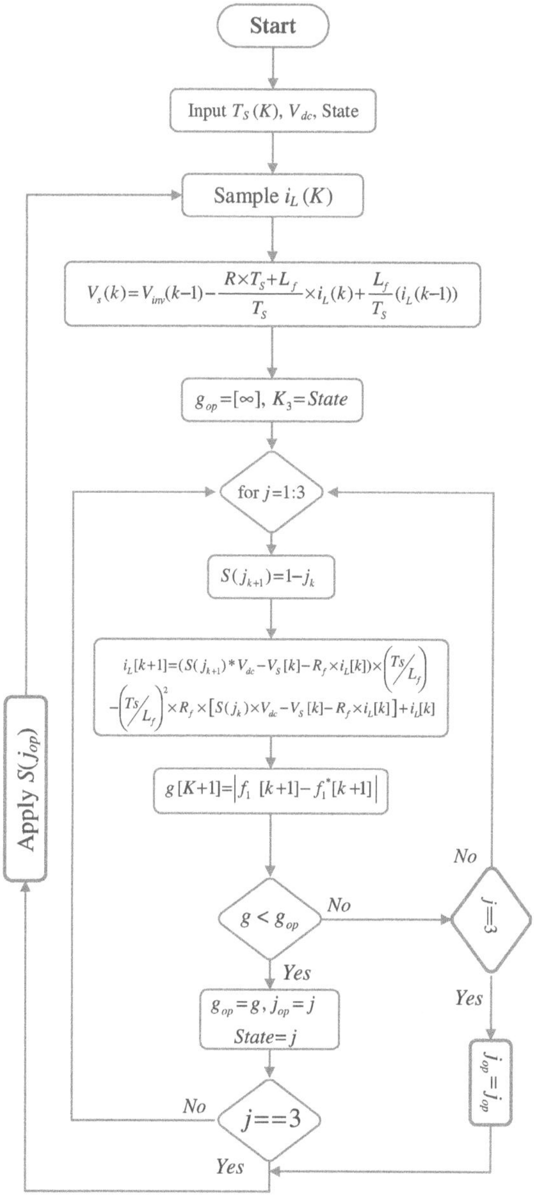

FIGURE 4.10 Flowchart view of the simple MPC.

and the weighting factors so that instead of using a simple cost function such as (4.8), a more complicated cost function will be used, which is as follows:

$$g[K+1]=\lambda_1\times\left|f_1[k+1]-f_1^*[k+1]\right|+\cdots+\lambda_n\times\left|f_n[k+1]-f_n^*[k+1]\right| \tag{4.9}$$

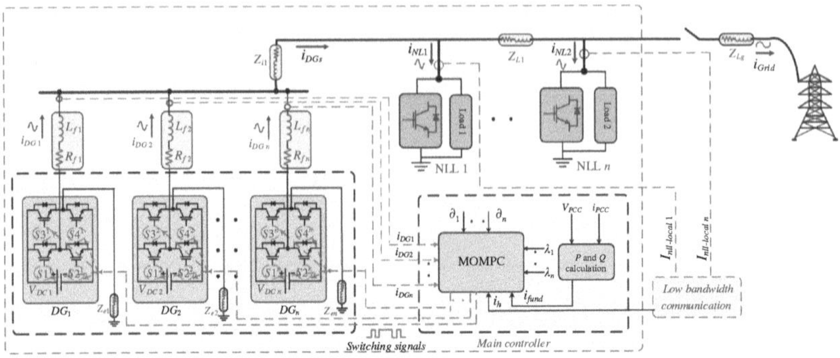

FIGURE 4.11 A prototype smart grid with computational intelligence and communication links.

where $f_1[k+1]$, $f_2[k+1]$, ... and $f_n[k+1]$ are related to different control objectives that should be minimized in the cost function in accordance with their weighting values (λ_1, $\lambda_2, \ldots,$ and λ_n) in the cost function. It should be mentioned that the application of mentioned weighting factors will define the priority of each control objective. MOMPC could be applied to several MFDGs that operate in parallel; this is the main advantage of the MOMPC to handle several objectives at a time for several converters. These objectives could be reference tracking, fundamental and harmonic power sharing, power management, output current THD minimization, switching frequency control, etc. Since the idea behind MOMPC is to fulfill multiple objectives in an acceptable way and not having the best performance over an objective and affecting the other objectives in an inverse way, to fulfill the control objectives in an acceptable way, a decision-making should be done over defining weighting factors. This decision-making could be based on heuristic methods or learning, which is dependent on system smartness. An example of MOMPC could be applied to the microgrid shown in Fig. 4.11; the control objectives would be tracking the defined current reference to compensate the nonlinear load harmonics, fundamental and harmonic power sharing, and switching frequency control.

The multiobjective cost function for this purpose will include three control terms: the first term for reference current tracking, the second one for fundamental and harmonic power sharing, and the third term will control the switching frequency. The cost function will be as follows:

$$g[K+1] = \lambda_1 \times \left|I_{DG1} + I_{DG2} - I_{ref}\right| + \lambda_2 \times \left|\partial_1 I_{DG1} - I_{ref}\right| + \lambda_3 \times f_{sw}(S_{sw}(k), S_{sw}(k+1)) \tag{4.10}$$

where λ_1, λ_2, and λ_3 are the weighting factors defining the priority of each control objective. And ∂_1 is the power-sharing factor that in this case defines the power-sharing ratio between first and second MFDG and is defined as

$$\partial_1 = \frac{\alpha_1 + \alpha_2}{\alpha_1} \tag{4.11}$$

where α_1 and α_2 are the coefficients defining the rated capacity of the DGs, so application of ∂_1 sharing factor will ensure the proportional sharing of power (current) between two MFDGs.

In (4.10), $S_{sw}(k)$ is the switching state of instant t_k, and it would be calculated in a different way for each converter, but the main idea is to sum the switching state of all the power electronic switches of the converter topology, that is, for a parallel operation of two, three-level H-bridge converter, $S_{sw}(k) = 2$ means that both converters are at their positive output state at instant t_k and $S_{sw}(k) = 0$, means that both converters are operating in zero output state, or one is operating in positive and the other one is operating in negative output state. Therefore $f_{sw}(S_{sw}(k), S_{sw}(k+1))$ is the mathematical function, relating the number of changes in switching states from instant t_k to t_{k+1}.

It is obvious that, in multiobjective operation of a controller, different priorities will result in different switching outputs; so to have an acceptable rate of satisfaction for all of the objectives, there should be an effort to set the best weighting factors for each application. This kind of controller is a very good solution for the cases with multiple objectives, which need a fast dynamic response, such as parallel MFDGs and modular APFs.

A comparison between different control methods applied to MFDGs is provided in Table 4.2. As can be observed, each method has its advantages and drawbacks, compared all together will make the selection of control method proper for the control cases.

A comprehensive comparison is done to clarify the advantages, disadvantages, and other PQI indexes between all generations of PQI devices in Table 4.3.

In Table 4.3 the devices based on cost-effectiveness factor are not only ranked based on the price but also on a price-to-capability ratio, so the passive filter is less cost effective in comparison to MFDGs, although passive filter will cost much lower.

4.4 Conclusion

To study the concept of power quality in smart grids, first, a definition of power quality and smart grid was proposed, new challenges and tools that smart grids will bring to traditional grids have been also discussed. There has been much research on power quality of smart grids, but less has been focused on smart grid concept; in this chapter, it has been tried to include the smart grid technology while studying power quality of smart grids. In this regard, almost all the PQI devices were discussed. The advantages, disadvantages, and applications in the case of each device were studied, and a

TABLE 4.2 Comparison of different control methods applied to multifunctional distributed generations (MFDGs).

Control method	Advantages	Disadvantages	Application case
PR-CCM	Control simplicity	Needs H_D block	Grid-connected MFDGs
PR-VCM	Decentralized power sharing without communication systems	Problem in I_{DG} comp	Stand-alone MFDGs
		Needs grid side info	
		Problem in I_{Local} comp	
		Needs H_D block	
PR-HCM	No need for H_D block	Control complexity	Grid-connected and stand-alone MFDGs
	Independent Ctrl of I_{DG} and V_{DG}	Slow dynamic response	
	Could replace CCM without H_D		
MPC	Fast dynamic response	Operation in grid-connected mode only	Active power filters
	Robustness		Grid-connected MFDGs
MOMPC	Fast dynamic response	Operation in grid-connected mode only	Modular active power filters
	Robustness		Parallel grid-connected MFDGs
	Multiobjective operation		
			Modular UPSs

CCM, Current-controlled method; *HCM*, hybrid control method; *MOMPC*, multiobjective model predictive control; *PR*, proportional resonant; *UPS*, uninterruptible power supply; *VCM*, voltage-controlled method.

comprehensive comparison has been provided between all PQI devices to clarify the level of effectiveness each device has when integrated to smart grids. Much focus has been dedicated to where the third generation of PQI devices is introduced, because the third generation of PQI devices has integrated some levels of smart grid technologies, such as computational intelligence, AMIs, and communications. In authors' opinion, "power quality issues of smart grids" include the integration of new technologies that smart

TABLE 4.3 A full comparison of all generations of power quality improvement devices.

PQI device	Year	Capabilities	Advantages	Disadvantages	Distributed PQI nature	Cost-effectiveness
Passive filters	The mid 1940s	Current harmonics and Q compensate	Simple, cheap, and highly reliable	Parameter design for each application	Yes	4
Shunt active filters	The mid 1970s	Improve PF, unbalances, flicker, VR and harmonics	Operating in various harmonic frequencies	Rather expensive	Yes	6
Series active filters	The mid 1980s	Compensate voltage harmonics, unbalances	Reduced cost in comparison to APF	High capacity and expensive	No	7
Hybrid filters	The mid 1980s	Advantages of active and passive filters together	Lower cost, reliable, efficient	Rather expensive	No	5
D-STATCOM	The mid 1980s	Voltage regulation harmonic compensation	Easy implementation	Control complexity, needs transformers	No	8
DVR	The mid 1990s	Voltage unbalances and regulation, harmonic compensation	Simple controller system	Control complexity, needs transformers	No	9
Smart impedance	The mid 2010s	All advantages of passive and active filters, improve stability, VR[a], selective harmonic elimination	One device instead of passive, active, and hybrid power filters	Increasing costs and control system complexity	No	3

(*Continued*)

TABLE 4.3 (Continued)

PQI device	Year	Capabilities	Advantages	Disadvantages	Distributed PQI nature	Cost-effectiveness
Electrical springs	The mid 2010s	Store energy, support voltage regulation, damp oscillations, and manage *P* and *Q*	Improve reliability stability and demand response, do not rely on communication	Expensive and needs storage	Yes	2
MFDGs	The mid 2010s	Eliminate harmonics, regulate voltage, lower power losses	Lower costs, power losses, harmonics	Control complexity	Yes	1

APF, Active power filter; *DVR*, dynamic voltage restorer; *MFDGs*, multifunctional distributed generations; *PQI*, Power quality improvement; *STATCOM*, static synchronous compensator.
[a] *Voltage regulation.*

grids bring to traditional grids, besides developing new devices that make use of these novel technologies to enhance power quality indexes in smart grids. The authors believe that some other characteristics of smart grids, such as learning and self-healing, would be possible discussion topics about power quality issues of smart grids for the next couple of years.

References

[1] K.H. LaCommare, J.H. Eto, Cost of power interruptions to electricity consumers in the United States (US), Energy 31 (2006) 1845–1855.

[2] B.W. Kennedy, Power Quality Primer, McGraw Hill Professional, 2000.

[3] K.H. LaCommare, J.H. Eto, L.N. Dunn, M.D. Sohn, Improving the estimated cost of sustained power interruptions to electricity customers, Energy 153 (2018) 1038–1047.

[4] International Standard IEC 61000-4-15, Flickermeter—functional and design specifications (IEC, Geneva, Switzerland, Edition 2.0, 2010-07), 2003.

[5] IEEE, IEEE recommended practice and requirements for harmonic control in electric power systems – redline, in: IEEE Std. 519-2014 (Revision of IEEE Std 519-1992) – Redline, 2014, pp. 1–213.

[6] J.C. Smith, G. Hensley, L. Ray, IEEE recommended practice for monitoring electric power quality, in: IEEE Std. 1995, 1159-1995.

[7] Y.N.S.H. Hosseini, S.G. Zadeh, B. Mohammadi-Ivatlo, J.C. Vasquez, J.M. Guerrero, Distributed power quality improvement in residential microgrids, in: 2017 10th International Conference on Electrical and Electronics Engineering (ELECO), 2017, pp. 90–94.

[8] M.H. Bollen, What is power quality? Electr. Power Syst. Res. 66 (2003) 5–14.

[9] R.C. Dugan, M.F. McGranaghan, H.W. Beaty, I c1996 Electrical Power Systems Quality, vol. 1, McGraw-Hill, New York, 1996.

[10] F. Hvelplund, P. Østergaard, B. Möller, B.V. Mathiesen, D. Connolly, A.N. Andersen, Analysis: smart energy systems and infrastructuresed Renewable Energy Systems, Elsevier Science, 2014, pp. 131–184.

[11] S. Grids, European Technology Platform for the Electricity Networks of the Future, European Commission, 2005.

[12] Employment, E. C. D.-G. f. and I. D. A., Employment and Social Developments in Europe, Publications Office of the European Union, 2011.

[13] A. B. IAB, International Energy Agency (IEA), 2006.

[14] Y. Naderi, S.H. Hosseini, S. Ghassem Zadeh, B. Mohammadi-Ivatloo, J.C. Vasquez, J.M. Guerrero, An overview of power quality enhancement techniques applied to distributed generation in electrical distribution networks, Renewable Sustainable Energy Rev. 93 (2018) 201–214.

[15] S. Yan, S.C. Tan, C.K. Lee, B. Chaudhuri, S.Y.R. Hui, Use of smart loads for power quality improvement, IEEE J. Emerg. Sel. Top. Power Electron. 5 (2017) 504–512.

[16] H. Abdi, B. Mohammadi-ivatloo, S. Javadi, A.R. Khodaei, E. Dehnavi, Energy storage systems, Distributed Generation Systems: Design, Operation and Grid Integration, Elsevier, 2017, p. 333.

[17] A.A. Gandomi, S. Saeidabadi, M. Sabahi, M. Babazadeh, Y.A. Gandomi, Design, engineering and optimization of a grid-tie multicell inverter for energy storage applications, arXiv preprint arXiv:1708.08008, 2017.

[18] H. Akagi, New trends in active filters for improving power quality, in: Proceedings of the 1996 International Conference on Power Electronics, Drives and Energy Systems for Industrial Growth, 1996, pp. 417–425.
[19] W. Yang-Wen, W. Man-Chung, L. Chi-Seng, Historical review of parallel hybrid active power filter for power quality improvement, in: TENCON 2015—2015 IEEE Region 10 Conference, 2015, pp. 1–6.
[20] L.B.G. Campanhol, S. A. O. d Silva, A. Goedtel, Application of shunt active power filter for harmonic reduction and reactive power compensation in three-phase four-wire systems, IET Power Electron. 7 (2014) 2825–2836.
[21] Y. Terriche, J.M. Guerrero, J.C. Vasquez, Performance improvement of shunt active power filter based on non-linear least-square approach, Electr. Power Syst. Res. 160 (2018) 44–55.
[22] E.L.L. Fabricio, S.C.S. Júnior, C.B. Jacobina, M. B.D.R. Corrêa, Analysis of main topologies of shunt active power filters applied to four-wire systems, IEEE Trans. Power Electron. 33 (2018) 2100–2112.
[23] R. Panigrahi, B. Subudhi, Performance enhancement of shunt active power filter using a Kalman filter-based H_∞ control strategy, IEEE Trans. Power Electron. 32 (2017) 2622–2630.
[24] J. Lu, P. Fu, J. Li, H. Mao, X. Shen, L. Xu, et al., A new hybrid filter based on differential current control method for low-order harmonic suppression in Tokamak power system, Int. J. Energy Res. 42 (2018) 82–90.
[25] G.A.D.A. Carlos, E.C.D. Santos, C.B. Jacobina, J.P.R.A. Mello, Dynamic Voltage restorer based on three-phase inverters cascaded through an open-end winding transformer, IEEE Trans. Power Electron. 31 (2016) 188–199.
[26] A.M. Gee, F. Robinson, W. Yuan, A superconducting magnetic energy storage-emulator/battery supported dynamic voltage restorer, IEEE Trans. Energy Convers. 32 (2017) 55–64.
[27] S.H. Hosseini, A. Ajami, Transient stability enhancement of AC transmission system using STATCOM, in: TENCON '02. Proceedings. 2002 IEEE Region 10 Conference on Computers, Communications, Control and Power Engineering, vol.3, 2002, pp. 1809–1812.
[28] J.M. Guerrero, L.G.D. Vicuna, J. Uceda, Uninterruptible power supply systems provide protection, IEEE Ind. Electron. Mag. 1 (2007) 28–38.
[29] M. Aamir, S. Mekhilef, An online transformerless uninterruptible power supply (UPS) system with a smaller battery bank for low-power applications, IEEE Trans. Power Electron. 32 (2017) 233–247.
[30] R.B. Gonzatti, S.C. Ferreira, C.H. da Silva, R.R. Pereira, L.E.B. da Silva, G. Lambert-Torres, Smart impedance: a new way to look at hybrid filters, IEEE Trans. Smart Grid 7 (2016) 837–846.
[31] C.H. Da Silva, R. Pereira, L. Silva, G. Lambert-Torres, R. Gonzatti, S. Ferreira, et al., Smart impedance: expanding the hybrid active series power filter concept, in: IECON2012—38th Annual Conference on IEEE Industrial Electronics Society, 2012, pp. 1416–1421.
[32] A. Baloi, A. Pana, F. Molnar-Matei, Contributions on harmonic impedance monitoring in smart grids using virtual instruments, in: 2011 Second IEEE PES International Conference and Exhibition on Innovative Smart Grid Technologies (ISGT Europe), 2011, pp. 1–5.
[33] K.T. Mok, M. Wang, S.C. Tan, S.Y. Hui, DC electric springs – a new technology for stabilizing DC power distribution systems, IEEE Trans. Power Electron. 32 (2016) 1088–1105.

[34] J. He, Y.W. Li, F. Blaabjerg, X. Wang, Active harmonic filtering using current-controlled, grid-connected DG units with closed-loop power control, IEEE Trans. Power Electron. 29 (2014) 642–653.

[35] J.C. Vasquez, J.M. Guerrero, A. Luna, P. Rodríguez, R. Teodorescu, Adaptive droop control applied to voltage-source inverters operating in grid-connected and islanded modes, IEEE Trans. Ind. Electron. 56 (2009) 4088–4096.

[36] J.M. Guerrero, M. Chandorkar, T.-L. Lee, P.C. Loh, Advanced control architectures for intelligent microgrids, Part I: Decentralized and hierarchical control, IEEE Trans. Ind. Electron. 60 (2013) 1254–1262.

[37] L. Meng, F. Tang, M. Savaghebi, J.C. Vasquez, J.M. Guerrero, Tertiary control of voltage unbalance compensation for optimal power quality in islanded microgrids, IEEE Trans. Energy Convers. 29 (2014) 802–815.

[38] J.M. Guerrero, P.C. Loh, T.-L. Lee, M. Chandorkar, Advanced control architectures for intelligent microgrids—Part II: Power quality, energy storage, and AC/DC microgrids, IEEE Trans. Ind. Electron. 60 (2013) 1263–1270.

[39] S.Y.M. Mousavi, A. Jalilian, M. Savaghebi, J.M.G. Ge, Autonomous control of current and voltage controlled DG interface inverters for reactive power sharing and harmonics compensation in islanded microgrids, IEEE Trans. Power Electron. 33 (2018) 9375–9386.

[40] M. Savaghebi, A. Jalilian, J.C. Vasquez, J.M. Guerrero, Secondary control for voltage quality enhancement in microgrids, IEEE Trans. Smart Grid 3 (2012) 1893–1902.

[41] J.M. Guerrero, M. Chandorkar, T.-L. Lee, P.C. Loh, Advanced control architectures for intelligent microgrids—Part I: Decentralized and hierarchical control, IEEE Trans. Ind. Electron. 60 (2013) 1254–1262.

[42] J.M. Guerrero, J.C. Vasquez, J. Matas, D. Vicuna, L. García, M. Castilla, Hierarchical control of droop-controlled AC and DC microgrids—a general approach toward standardization, IEEE Trans. Ind. Electron. 58 (2011) 158–172.

[43] Y.W. Li, J. He, Distribution system harmonic compensation methods: An overview of DG-interfacing inverters, IEEE Ind. Electron. Mag. 8 (2014) 18–31.

[44] J. Rodriguez, M.P. Kazmierkowski, J.R. Espinoza, P. Zanchetta, H. Abu-Rub, H.A. Young, et al., State of the art of finite control set model predictive control in power electronics, IEEE Trans. Ind. Inf. 9 (2013) 1003–1016.

[45] Y. Naderi, S.H. Hosseini, S.G. Zadeh, B. Mohammadi-Ivatloo, M. Savaghebi, J.M. Guerrero, An optimized direct control method applied to multilevel inverter for microgrid power quality enhancement, Int. J. Electr. Power Energy Syst. 107 (2019) 496–506.

Chapter 5

Modeling and simulation of active electrical distribution systems using the OpenDSS

Luiz Carlos Ribeiro, Junior, Francinei Lucas Vieira, Benedito Donizeti Bonatto, Antonio Carlos Zambroni de Souza and Paulo Fernando Ribeiro
Institute of Electrical System and Energy, Federal University of Itajuba, UNIFEI, Itajubá, Brazil

5.1 Introduction

The passive nature of the distribution grid is changing; since the number of distributed generation (DG) units in the world has been increasing exponentially, it is expected that this growth will continue for the next few years. New possibilities arise with the structural changes of the electric system, such as the development of smart grids with electric vehicles (EVs), storage systems, and a high interaction between utilities and customers, which is already something real.

The main sources of DG in the distribution grid, such as photovoltaic (PV) and wind power, have peculiar characteristics that bring to the planning of the distribution system some technical challenges. Several problems are reported in the literature, for example, overvoltage problems caused by reverse power flow, possible failures in the protection system, and an increase of harmonic distortion levels.

Some aspects not widely approached are the social issues caused by new technologies, because they may impact, directly or indirectly, the people's lives. Thus the social impact should always be a topic to be considered in the design of engineering projects.

How the modeling of the new distribution grids is treated has also been changed in recent times. Powerful distribution system analysis tools with distributed source models have been developed. In order to analyze the behaviors of the new distribution grid system, a powerful distribution analysis tool along with the distributed source models is essential for better accuracy and reliability of power system planning and operation. The software used in

Decision Making Applications in Modern Power Systems. DOI: https://doi.org/10.1016/B978-0-12-816445-7.00005-0

this chapter, the OpenDSS, has been used by researchers worldwide and will be presented with more details in the next topics.

5.2 Active electrical distribution systems

In recent years the characteristics of the current distribution networks are in transformation. A passive network is becoming active with the increasing penetration level of DG units, mainly wind and PV. The insertion of these new types of generation, near the consumer centers, brings significant benefits to the energy matrix of a country, to its economic, and its social and technological development. DG can also bring economic benefits for both customers and companies in the electricity sector if proper regulation is applied. On the customer side, the decrease in dependence on centrally generated energy, minimizing the generation from thermal power plants which, besides environmental impacts, usually has high electricity costs, for example, and brings advantages to economy [1].

On the side of companies in the electricity sector, the increase in DG reduces the demand of the traditional system, thus postponing investments in the construction of new lines and large power plants, provided that there is quality control in the energy injected by the DGs. The peak periods of power demand may also decrease, reducing costs of maintenance and replacement of equipment, besides the decongestion of transmission and distribution lines [1]. The losses in the system may also be reduced because the DG is preferably installed near the consumption center, which minimizes the power flow in the transmission lines and consequently the power losses. Although these changes in the distribution network bring benefits, there are also many challenges linked to the insertion of DG in the network.

The direction of power flow in the traditional distribution network is from a substation transformer to loads (residences, commercial centers, and industries). However, with an increasing number of DG systems, it might have moments on a day when the power flow reverses—that is, it goes from the low voltage to medium and high voltage—which increases complexity and uncertainty in the distribution system [2]. The reverse flow can lead the system to an overvoltage condition, may cause protection devices to malfunction, and can decrease system safety.

The following sections will present some problems that may arise with the high penetration of DGs in the distribution systems.

5.2.1 Impact of high penetration of distributed generation on power distribution systems

5.2.1.1 Voltage issues

Many texts in the literature report the problem of overvoltage due to the high penetration of distributed PV systems. The usual voltage profile

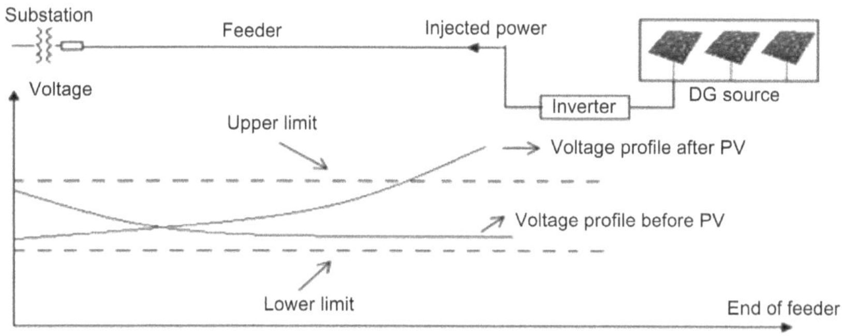

FIGURE 5.1 Voltage profile with PV-distributed generation at the end of the feeder [3]. *PV*, Photovoltaic.

characteristic of a feeder—that only contains passive loads and unidirectional power flow from the transformer to the load—is decreasing, and this means there is a typical voltage drop along the feeder. As the number of PV DG is significantly increasing, and the energy generated is often higher than that consumed, the power flow is inverted and flows from the low-voltage network to the medium and high voltage, causing the voltage variation to become positive as shown in Fig. 5.1.

5.2.1.2 The influence of protection

The distribution network protection usually consists of a simple overcurrent protection scheme, because of the radial scheme and the unidirectional power flow. The connection of the DG to the distribution network leads to multiple sources of fault current that can affect the detection of disturbances [4]. Thus several problems are reported in Refs. [2,5–8], such as fuse coordination, impact of DG on interrupting ratings of devices, fault detection–relay desensitizing, islanding problems, and false tripping.

5.2.1.3 Issues on the electric performance metrics (power quality)

The reason that the flicker effect occurs in DG distribution networks is the rapid generation change, for example, by a quick change of irradiation in a PV generation. Also, the interaction between the DG unit and the voltage feedback control device in the system may cause the flicker effect [2]. DG systems connected to the grid can cause voltage distortion, mostly related to the increasing insertion of electronic converters, such as power inverters and rectifiers. This distortion is generated by the electronic switching of semiconductor devices—for example, insulated gate bipolar transistor (IGBT) and gate turn-off thyristor (GTO) devices in inverters, which modulate a DC current produced by the generator to an AC current compatible to the grid frequency.

Small PV systems in homes are usually single-phased. Unless PVs are uniformly distributed among the phases, the imbalance may increase, resulting, for example, in an increase of the neutral current in the cables and a voltage increase at the point of common coupling (PCC) [9]. System imbalance can drastically alter and affect the operation of induction motors, electronic devices, and equipment for voltage regulation.

Finally, the clear majority of PV systems are connected to the system through inverters with unitary power factor (PF). The main reason is economical—the PV system should inject as much active power as possible for maximum revenue to the PV owner, and the grid has to supply the reactive power required by the load. The PF at the customer's PCC may be reduced, which is measured as a surplus of reactive power, subjecting the consumer to tariff charges for low PF.

5.2.1.4 Operation of the power grid

In order to adapt to the needs of the new electrical network with high DG penetration, additional investments are required, for example, the update of protection schemes and the necessity to make enough operational reserve capacity. Such investments should ensure the user's electricity reliability and suitability of lines and substations. Also, policies should promote a fair evaluation of the risks over all stakeholders and guarantee an efficient use of the power system for DG [10].

Procedures related to the connection control of DGs to the grid and the business models involving DG (e.g., demand response, feed-in, and time-of-use tariffs) are still subjects of ongoing discussions and research.

5.2.1.5 Socioeconomic impact problems

Not only technical problems arise with the change in the distribution system and dissemination of DG, socioeconomic impacts must also be analyzed in the context of changes. The consequences and costs of problems related to the increased level of penetration of distributed systems may affect a portion of the population that does not have financial access to such resources, so there is a concern that the new distribution network does not become an instrument to promote social balance.

On the side of utilities, there are also concerns about revenue losses and increased costs with network depreciation and maintenance [1]. The philosophy of technology discussed in Ref. [11] helps one to provoke a critical reflection about the socioeconomic impacts of engineering. Several issues and challenges are described in the literature and should be considered in the analysis of distribution networks.

5.2.2 Smart functions on power inverters

The growth and spread of distributed systems, especially PVs, suggest a future where utilities will need to accommodate high levels of DG penetration in their distribution and transmission systems. New forms of communication and interaction of distributed systems based on inverters have been developed to maximize the benefits of DG [1].

In 2009, EPRI—Electric Power Research Institute—along with the US Department of Energy, Sandia National Laboratories, and the Solar Electric Power Association launched a research project to begin a process of identifying and standardizing the set of capabilities that the inverter can perform to improve the performance of the network and thus increase the level of DG penetration without any damage to the system [12].

Based on studies carried out, EPRI published a document called *Common Functions for Smart Inverters* [8] with the description of the main functions for smart inverters, focused mainly on solar inverters and energy storage.

Smart inverter functions [8] are divided according to the control drives and their purpose. The division according to the control drives is described as follows:

- Functions driven by an operator require direct interaction with the operator and can be divided into the following:
  - *Basic functions*: basic operations of inverters, which include remote connection and disconnection from the grid, system monitoring through a supervisory and displaying parameters that allow the operator to understand the behavior of the inverter.
  - *Direct control*: functions that allow the operator to limit the output power and change the PF of the inverter also allow charging and discharging of an energy storage system, if available.
- *Autonomous functions*: functions that allow the inverter to decide on its own since it has been supplied with logic and previous parameters. The inverter collects measurement data at the control point for processing. Some examples of this type of function are based on curves, such as the Volt/Watt function, for example, which requires the inverter to decrease the injected power as the system voltage increases.
- *Functions driven by independent variables*: such as the autonomous functions, differentiated by the data collection, which is sent here to the inverter of some remote data stream, considering that the data can be electric or not.
  - *Electrical data*: electrical data is sent to the inverter from a load or generator that the inverter has been requested to follow, data from the PCC or another point in the network can also be sent.
  - *Indirect control*: functions performed by the inverter are performed through nonelectrical data, such as temperature, prices, and time-based data.

Regarding purpose or functionalities, smart inverters can be divided into five categories as follows:

- *Monitoring and scheduling*: functions that allow the operator to adjust and collect information from the inverter, including the connect and disconnect function.
- *Frequency support*: functions that provide frequency support for the grid, such as the Frequency/Watt function.
- *Real-power support*: functions that provide support for active power to the grid, such as dynamic real-power support.
- *Power factor support*: functions that act by adjusting the reactive power, such as the fixed PF function.
- *Voltage support*: functions that support the voltage to the grid, such as the dynamic Volt/Watt function.

5.2.3 Final considerations

Changes in the distribution system require more detailed engineering about the impacts that may arise; insertion of generation into distribution changes the concept of the unidirectional power flow and exposes customers and distributors to new challenges. Understanding and dealing with new problems will be critical to the future of distribution systems.

5.3 Modeling and simulation using OpenDSS

The changes in the distribution grid with the insertion of DG systems bring several technical challenges as presented in the previous topic. Also, the modeling of these new systems has become more complex, requiring sophisticated analyses and reliable algorithms for a true description of grid behavior.

According to [13], the distribution system analysis programs have evolved from simple voltage drop calculators of balanced loads to sophisticated systems with graphical interactions that allow one to know and quantify grid parameters. The best known methods, such as *Gauss–Seidel* and *Newton–Raphson*, may not show convergence in many distribution system analyses due to high R/X ratio and the radial structure of distribution grids. In addition, as the distribution grid has a strong tendency to be unbalanced among phases, such methods are not advised to work with positive sequence models only.

Thus this topic is intended to analyze the three-phase current injection method, based on the decomposition of the nodal admittance matrix that composes the OpenDSS software. A brief historical presentation of the software, PV systems, storage, and load models of an active distribution grid is presented.

5.3.1 The OpenDSS

The software came in 1997 still under the name of DSS—Distribution System Simulator. In 2004 the DSS was acquired by EPRI solutions, and in 2008, it is released with open-source license, named OpenDSS [14].

The OpenDSS software is a power flow simulation algorithm that performs the most varied analyses related to the planning of the electric distribution system and the quality of the energy. Also, the software performs analyses to meet the demands of future electric grids with DG. *Quasi-static* solution modes allow the execution of sequential simulations over time, and thus system analysis can be performed at any time of the day [15]. The main modes of software simulation are

- instantaneous power flow (*snapshot*),
- daily power flow (*daily mode*),
- annual power flow (*yearly mode*),
- harmonic analysis,
- dynamic analysis,
- fault study, and
- Monte Carlo fault study.

After being released in 2008 as open-source software, OpenDSS has become widely used around the world. One of the features that make it popular is the offered package of interfaces for simulation. The program was launched with a Component Object Model (COM) interface, and recently, the "Direct DLL" interface was released so that users could access program features on platforms incompatible with the COM interface [16].

The COM interface can be controlled by software such as Python, MATLAB, and MS Office tools, with an emphasis on Visual Basic for Applications. OpenDSS has recently come up with a version to run parallel computing on modern multicore computers. This version is called OpenDSS-PM (parallel machine) and is also freely available [16,17].

5.3.2 Power flow in OpenDSS: the current injection method

OpenDSS has been using two power flow algorithms. The standard method, described as *Normal method* or current injection method, and the *Newton method*, which should not be confused with the *Newton–Raphson method* [14]. The current injection method is an iterative method based on the theorems of *Thévenin* and *Norton* and the concept of the nodal admittance matrix.

Admittance matrix (Y) is assembled through the primitive nodal admittance matrix method. Such matrix represents the admittance matrix of a single element, and the various primitive matrices are specifically allocated in

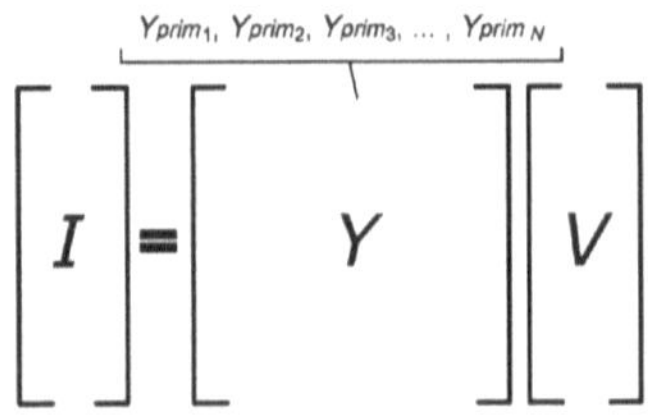

FIGURE 5.2 Construction of the Y matrix OpenDSS.

the complete nodal admittance matrix of the system as schematized in Fig. 5.2, adapted from Ref. [14].

The elements for circuit building on OpenDSS are divided into two categories. First, power delivery elements, where the main delivery elements are the lines and transformers. They are characterized by their primitive nodal admittance matrix. The second category is the power conversion (PC) elements, such as loads and generators. The characterization of nonlinear PC elements is performed in the OpenDSS through a compensation current, illustrated by the example shown in Fig. 5.3, adapted from Ref. [18], which presents a single-phase nonlinear load connected to a generic grid by the bus 1.

The absorbed current $\dot{I}_{\text{term}}$ is calculated as a function of the voltage applied to the load terminals. OpenDSS converts the model from Fig. 5.3 to a Norton equivalent, where linear and constant admittance Y_{linear} is calculated under nominal load voltage condition. The equivalent Norton model is shown in Fig. 5.4. The admittance of the Norton model is added to the complete grid matrix as a passive component of the system. Note that the linear part of the PC elements is also modeled as a nodal primitive matrix.

The compensation current shown in Fig. 5.4 includes the nonlinearity of the current absorbed by the load $\dot{I}_{\text{term}}$ and mathematically can be written as in (5.1).

$$\dot{I}_{\text{comp}} = \dot{V}_1 \times Y_{\text{linear}} - \dot{I}_{\text{term}} \tag{5.1}$$

The power flow algorithm is performed in four steps described as follows:

Step 1: Initial guess

As in most fixed-point iterations, the initial guess should be close to the final result, which is relatively simple to achieve by executing a direct solution of the complete nodal admittance matrix of the system, considering as zero the compensation current of the conversion elements.

Step 2: Calculation of injected currents and compensation currents

For each conversion element the compensation currents are calculated considering the various types of load models available. The calculated currents are organized into a specific vector of currents $\dot{I}_{\text{inj}}$.

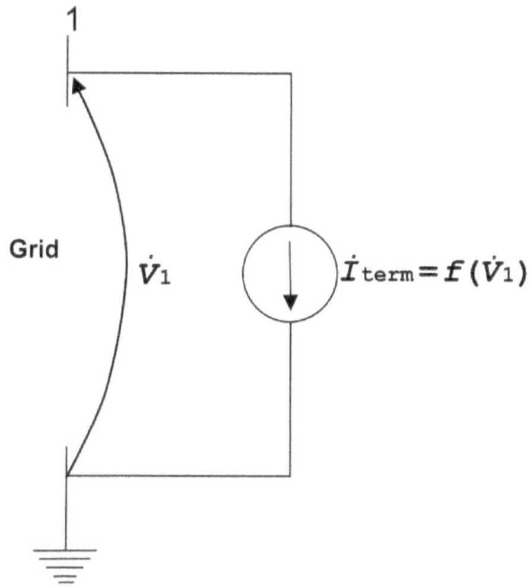

FIGURE 5.3 Single-phase nonlinear load model [18].

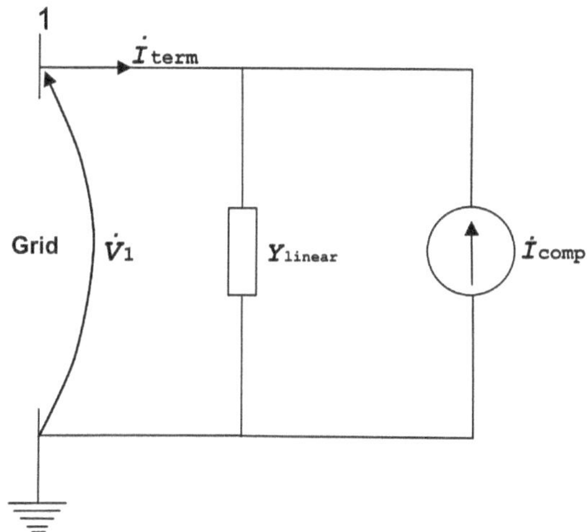

FIGURE 5.4 Norton equivalent model for single-phase nonlinear load [18].

Step 3: Calculation of nodal voltages for the first iteration

With the complete nodal admittance matrix of the previously calculated system and the injected currents, the nodal voltages of the system are calculated through the matrix (5.2).

$$\dot{V}_{nodal} = [Y_{Sist}]^{-1} \times \dot{I}_{inj} \tag{5.2}$$

Step 4: Convergence test

In this step the convergence of the method (5.3) is tested for any bus i, with values in p.u. or real values.

$$\begin{aligned} \text{error}_i^{(k)} &= \frac{||\dot{V}_i^{(k)}| - |\dot{V}_i^{(k-1)}||}{V_{\text{Base}_i}}, \text{in p.u.values} \\ \text{error}_i^{(k)} &= \frac{||\dot{V}_i^{(k)}| - |\dot{V}_i^{(k-1)}||}{|\dot{V}_i^{(k)}|}, \text{in real values} \end{aligned} \tag{5.3}$$

For either case (real or p.u. values) the power flow algorithm converges when $\text{error}_i = 0.0001 \forall i$.

5.3.3 The photovoltaic system model

The PV system model is modeled as a PC element, represented in Fig. 5.5. It combines the characteristics of a PV system and an inverter in its structure. The model assumes that the inverter can quickly find the maximum power point of the panel, a simplifying premise that considers an optimized computational time with suitable results for most impact studies [19].

In Fig. 5.5 the parameters necessary for system characterization in daily, annual, and cyclic modes are represented. Thus three basic parameters (power at the point of maximum power, temperature, and irradiation) are provided to the software. Also, some critical curves are necessary for the correct operation of the model.

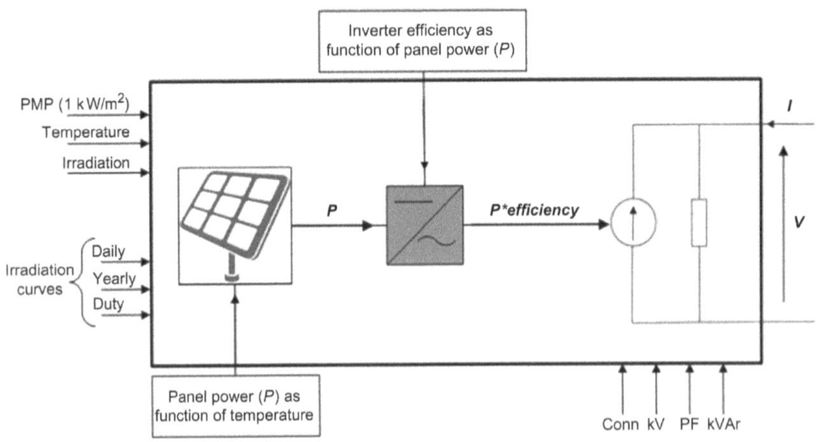

FIGURE 5.5 OpenDSS PV object diagram [19]. *PV*, Photovoltaic.

- *Irradiation curve*: The irradiation curve is in per unit values, based on a specified value.
- *Daily panel temperature*: The temperature curve is measured on the panel in Celsius over the simulated period.
- *Variation of panel power with temperature*: Temperature influences the performance of PV systems; thus a panel power reduction curve is provided as a function of its temperature.
- *Inverter efficiency*: The output power of the model considers the efficiency of the inverter. This gives an efficiency curve of the inverter as a function of the input power in per unit values.

Some parameters such as the connection topology of the PV system to the grid, the PV PF, and the output voltage should be provided. Then, the power supplied by the PV to the system is given by (5.4), where $Eff\ (P_{(t)})$ corresponds to the efficiency of the inverter for a given output power of the panel, and $P_{(t)}$ is the power of the panel provided by (5.5).

$$P(t)_{\text{out}} = P_{(t)} \times \text{Eff}(P_{(t)}) \tag{5.4}$$

$$P_{(t)} = P_{\text{mpp}} \times \text{irrad}_{(t)} \times \text{irrad}_{(\text{base})} \times P_{\text{mpp}}(T(t)) \tag{5.5}$$

where P_{mpp} corresponds to the power at the point of maximum power, $\text{irrad}_{(t)}$ is the irradiance in p.u. at time t, $\text{irrad}_{(\text{base})}$ is the base irradiance in kW, and $P_{\text{mpp}}(T(t))$ is the power correction factor with temperature.

The reactive power is defined separately and can be defined with a fixed kVAr or a fixed PF.

5.3.4 The OpenDSS storage

OpenDSS storage element is modeled as a generator that produces energy (discharge) or consumes it (charge) within its capacity as shown in Fig. 5.6.

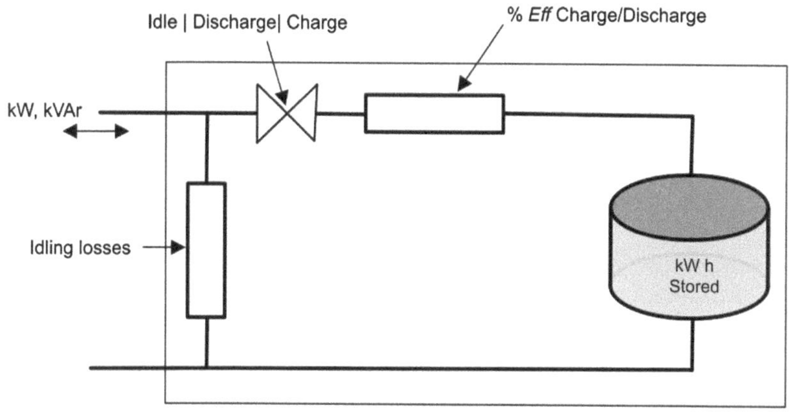

FIGURE 5.6 OpenDSS storage object diagram [20].

Generally, the studies are done in the time-varying simulation modes, controlled by control elements, the *StorageController*.

The control element performs tasks of energy charging dispatches in a controlled manner, and some loading and unloading modes can be found as follows and in the program [20]:

- *PeakShave*: Storage starts unloading when the monitored load exceeds an established value of active power. Thus it has attempted to maintain demand within a range.
- *Time*: In this mode, charging and discharging start from a preset time.
- *Follow*: This mode also tries to keep the load within a traditional value, but, unlike the *PeakShave* mode, the charging and discharging start from a time to be set.
- *Load shape*: Both charge and discharge follow an established curve, so when the values are negative, the storage unit is charging, and when the values are positive, the storage is in the discharge mode.

5.3.5 The load model

The load model in OpenDSS is a PC element that can be defined by its nominal power (kW, kVAr, kVA) and PF. It can be modified by load multipliers, daily or yearly load shapes, for example. By default, the load behaves as a current source, and its primitive Y matrix contains the impedance between the neutral of the load to the ground if existent [14]. However, for fault studies, it would be convenient to change the default behavior of the load to admittance, thus including it in the system Y matrix.

The load models can be characterized in different ways. There are eight options to define how the loads will vary with the voltage:

- Constant real and reactive powers ($P + jQ$). This is the default option of the software, and it is widely used for power flow studies.
- Constant impedance (Z) load.
- Constant real power (P) and quadradic reactive power (Q). Similar behavior of a motor load.
- Exponential. By default, it is a real linear power (P) and a quadratic reactive power (Q). Used for feeder mix studies or voltage optimization measures.
- Constant current magnitude (I). It may be used to represent rectified loads.
- Constant real power (P) and fixed reactive power (Q).
- Constant real power (P) and fixed reactive impedance.
- ZIP model. Array of eight coefficients for weighting factors and cutoff voltage.

In any case, new considerations should be done for load modeling on harmonic analysis. The harmonic load model in OpenDSS is a Norton

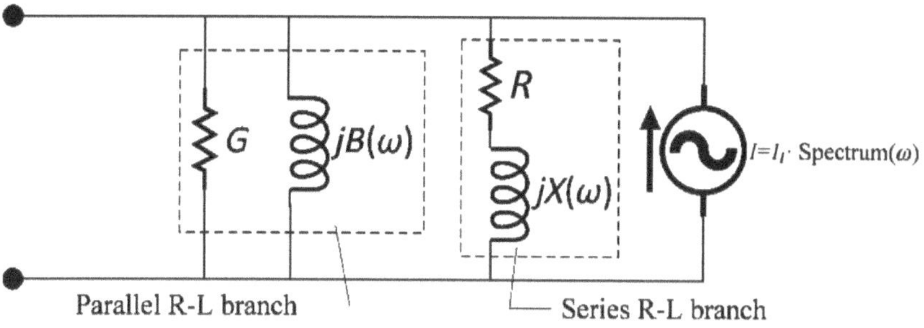

FIGURE 5.7 Harmonic load model for OpenDSS.

equivalent with a shunt admittance, as shown in Fig. 5.7. By default, the admittance is composed by a parallel R-L branch and a series R-L branch with half of the load specified to each one. Some options are available in the program to modify the load model in the harmonics mode, and the user can even neglect the load admittance branch to get nearly ideal current injections [21].

A harmonic case study is presented in the next section, focusing on the impacts of different load model representations and harmonic issues. Those issues derive from the changes in the power system impedance caused by switching of capacitor banks and PC inverters from distributed energy resources. The solution analysis begins with a power flow solution in OpenDSS (*solve snapshot*) following the harmonic solution (*solve mode = harmonics*) for all frequencies of interest.

Frequency scan is a well-known technique in harmonic studies that can reveal the existence of resonant points [22], used in equipment tests or specific buses in electrical power systems. A spectrum object with all the frequencies of interest in the scan should be defined and associated to a unitary current source connected to a node of the grid. A voltage monitor on the current source gets the frequency response (the impedance image as a function of frequency) of the system on that specific node.

5.4 Application in case studies

5.4.1 Case 1: Voltage control in distribution systems with high penetration of photovoltaics through smart functions

This first application is based on ref. [23] and presents the problem of overvoltage caused by high insertion of PV systems in distribution grids, one of the classic problems mentioned in Section 5.3. The solution proposed in this work is the use of some smart functions implemented in the inverters of PV systems for overvoltage mitigation.

The actual inverters that have the main function of DC-to-AC conversion are being adapted for an active operation in the new systems in such a way

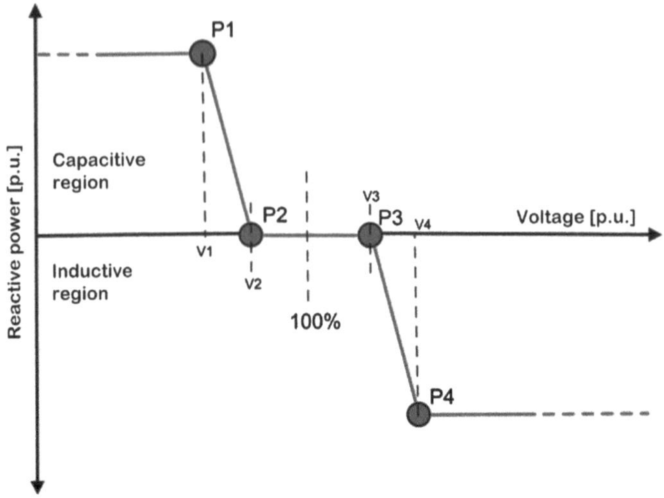

FIGURE 5.8 Volt/Var control.

that they control the PF, have support for voltage and frequency sags, interact with the operator, and possess control of the voltage via Volt/Var and Volt/Watt functions.

Volt/Var control for smart inverters operates using a custom curve that controls the voltage regulation at the PCC. If the voltage is below the established limit, the inverter acts on the capacitive region, injecting reactive power, and if the voltage is above the set point, the inverter actuation takes place in the inductive region, absorbing reactive power. The rated power of the inverter is a limiting boundary, and Volt/Var function will not work properly if it reaches this limit [24]. When the voltage is within the preestablished operational limits, the region of the curve described with *Dead Band*, no control action is taken by the inverter; a characteristic curve of the function is presented in Fig. 5.8.

In the characteristic of Fig. 5.9 (Volt/Watt control), when the voltage in the system is higher than *V*2, the output power decreases linearly to the point *P*3, where the reactive power reduces to zero. An application for the use of this function is the high penetration of PV systems in moments of light load in the system that is causing overvoltage [23].

Volt/Var and Volt/Watt features may be identical for all devices in a feeder, or they can be configured uniquely for each device. OpenDSS software has a command that performs the characteristics of Figs. 5.8 and 5.9 through the iterative power flow solution algorithm described earlier.

The system chosen for the work simulations [23] was based on the IEEE 13-bus test feeder developed by IEEE (The Institute of Electrical and Electronics Engineers) [24]. The original system was modified by connecting

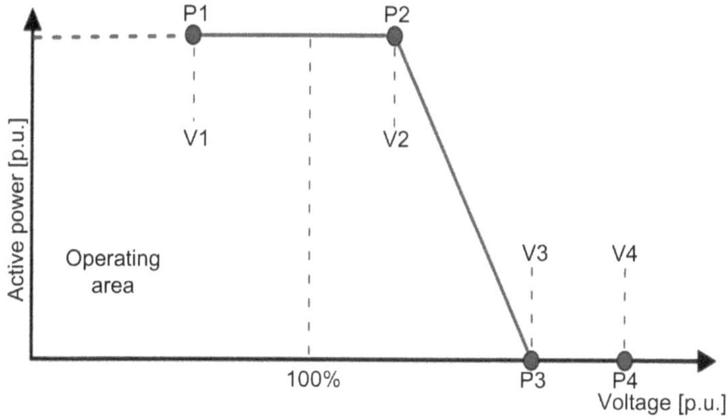

FIGURE 5.9 Volt/Watt control.

step-down transformers in bars where there are loads connected. Also, the values of active and reactive power of some loads have changed, as well as the length of the feeder lines.

Details of the system loads are shown in Table 5.1. The characteristics of the capacitor banks presented in buses 611 and 675, the switch between the sections 692 and 675, and the line voltage–reducing transformers have not been modified according to Ref. [23] and can be accessed at Ref. [25].

The loads of Table 5.1 follow typical residential and commercial demand curves as discussed in Ref. [23] and are discretized every 15 minutes in the software through the *loadshape* command, and such step is also used in the simulation of the system. The PV systems of this work are connected in the bars where there are loads (Table 5.1). Buses with three-phase loads have three-phase PVs, and the ones with single-phase loads have single-phase PVs.

The level of penetration of PV systems adopted in this work is a percentage ratio between the DG capacity and the nominal generation capacity of the system. Three penetration levels (50%, 75%, and 100%) are analyzed and simulated, and the PV generation is dimensioned in proportion to each local load in the respective buses.

Initially, the system condition is presented before the insertion of the PV generations for comparison purposes, and some solutions to the system overvoltage problem through the smart functions will be analyzed. The maximum deviation of the proposed base voltage at work is in line with the IEEE 1547 standard [26].

5.4.1.1 Simulation with and without distributed generation photovoltaic insertion

To analyze the overall active energy consumption required by the system, an energy meter is inserted into terminal 1 of the line connecting the buses 650 and 632.

TABLE 5.1 System loads of the 13-bus test feeder.

		Phases					
Bus	Model	1 kW	1 kVAr	2 kW	2 kVAr	3 kW	3 kVAr
634-2	PQ	160	110	120	90	120	90
645-2	PQ	0	0	170	125	0	0
646-2	PQ	0	0	0	0	230	112
652-2	Z	128	86	0	0	0	0
671-2	PQ	333	153	333	153	333	153
675-2	PQ	265	100	268	150	260	112
692-2	I	170	110	0	0	0	0
611-2	I	0	0	0	0	170	80
680-2	PQ	66	39	66	39	66	39
	Sum	1122	598	957	557	1179	586

Figs. 5.10 and 5.11 show the voltage in the phases along the feeder without the insertion of PV's and with 100% penetration at 12 a.m.

The full, dotted, star represent phases one, two, and three, respectively.

As can be seen in Fig. 5.11, voltage on phase 3 was impaired with the insertion of the PVs due to the reverse power flow generated in the system. The limit proposed by IEEE 1547 standard is exceeded at certain times of day discussed.

Fig. 5.12 represents the single-phase voltage of phase 3 of the single-phase bus 611 for the cases without PV penetration and with penetration of PVs at the different levels as previously described. Note that conventional tap control at the substation is unable to correct the voltage problem for this case.

5.4.1.2 *Simulation of smart controls*

To mitigate the overvoltage problem caused by the high penetration of PVs in the grid, the Volt/Var control is applied to the single-phase inverter of 611-bus. The voltage versus Vars curve created in the OpenDSS through the *VVcontrol* function and applied to the inverter is configured so that between 0.98 and 1.02 p.u. of voltage, the curve presents a dead band, not generating or absorbing reactive power. The control tested in this section is for the condition of maximum penetration of PVs. In the first case the inverter power is set at 432 kVA, 120% of the expected value for the maximum power generated (VV), and in the second case the inverter power is

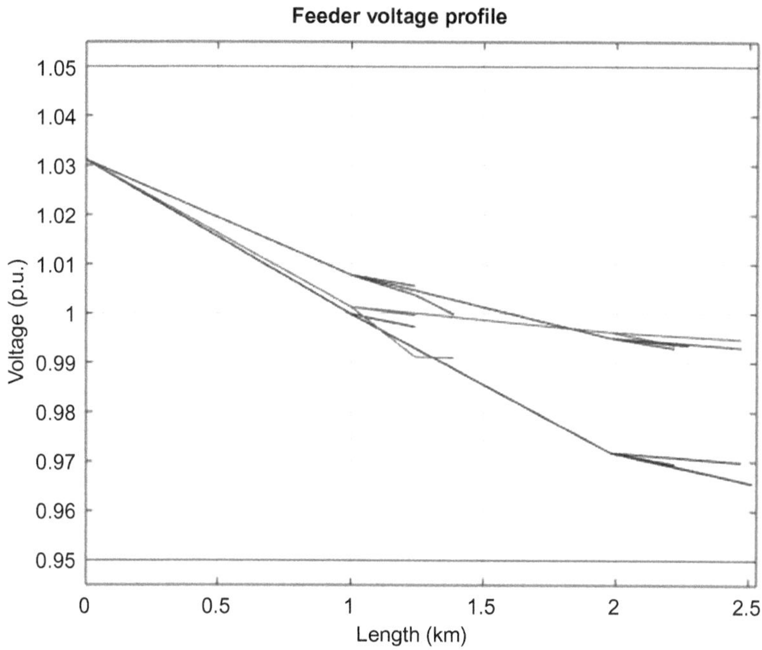

FIGURE 5.10 Voltage curve for the feeder at 12 h without PVs. *PV*, Photovoltaic.

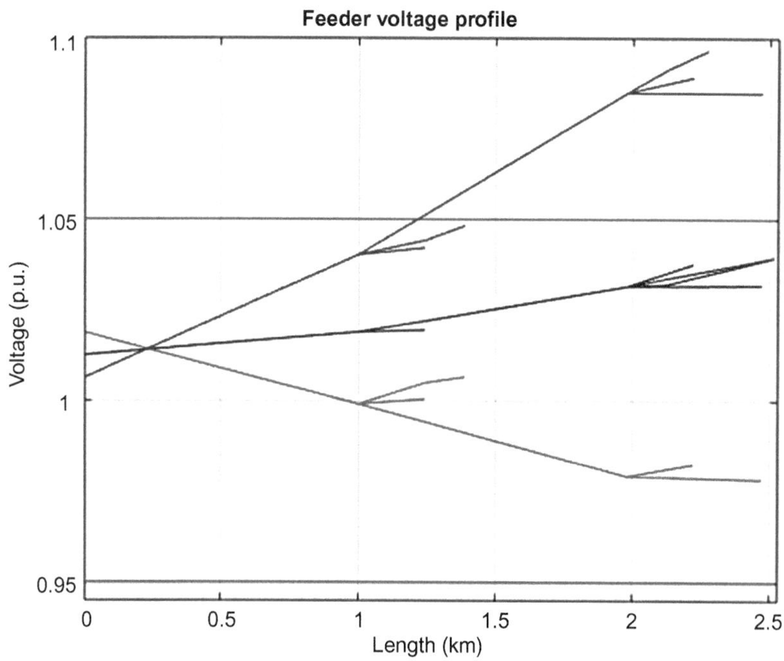

FIGURE 5.11 Voltage curve for the feeder at 12 h with PV. *PV*, Photovoltaic.

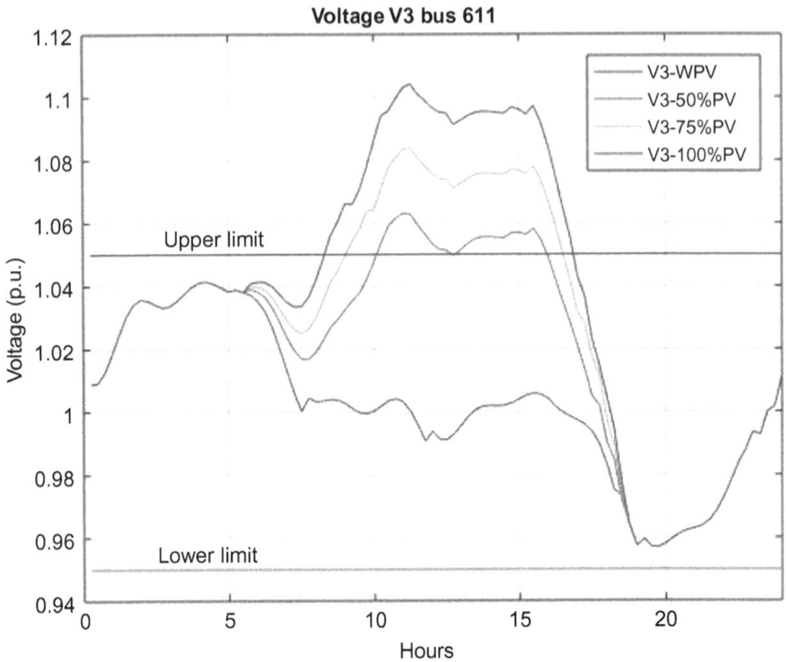

FIGURE 5.12 611-Bus voltages.

reduced to 360 kVA, the expected value for the maximum active power generated by the PV with unitary PF (*VVred*). Results can be seen in Fig. 5.13.

There are some possibilities to aid in voltage control when the Volt/Var control is not able to keep the voltage at the proper level.

The first alternative tested is the joint performance of the existing Volt/Watt and Volt/Var function, considering the inverter with a power of 360 kVA. For the study in question, when the nominal voltage measured at the PCC is greater than 1 p.u., the power output is reduced based on the straight slope, up to the limit of 1.07 p.u., when the power output becomes zero. Fig. 5.14 shows the voltage profile after application of the control (V3-VW) in comparison with the reduced Volt/Var case and 100% insertion of PVs. Power reduction for this case is shown in Fig. 5.15.

A second proposal for the voltage control aid is the Volt/Var control applied in more points of the grid. Thus the Volt/Var control is applied to other system buses where there is PV generation, and the result of the maximum voltage obtained for several combinations of application of the control throughout the day is presented in Table 5.2.

The last possibility presented in the work to aid the voltage control is the insertion of energy storage (batteries) to the PV system on bus 611, working

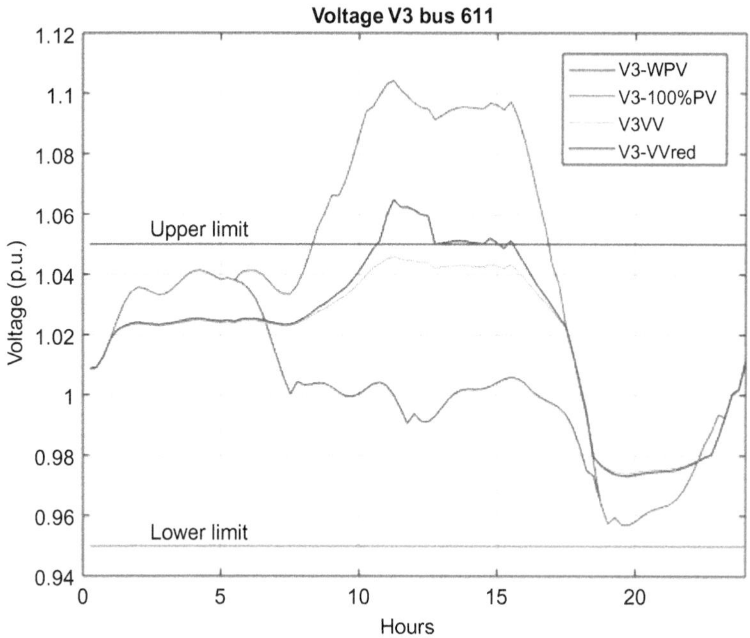

FIGURE 5.13 611-Bus voltages with VV control.

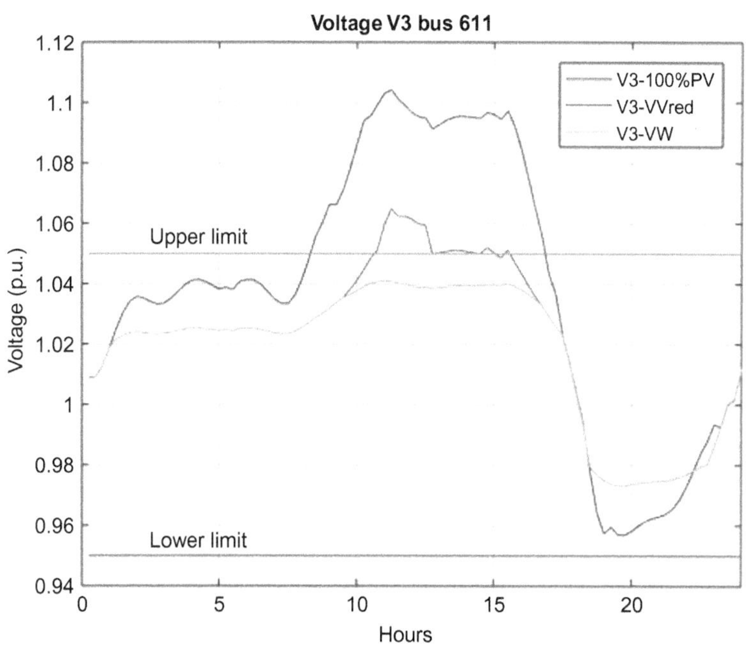

FIGURE 5.14 611-Bus voltage with Volt/Var and Volt/Watt control.

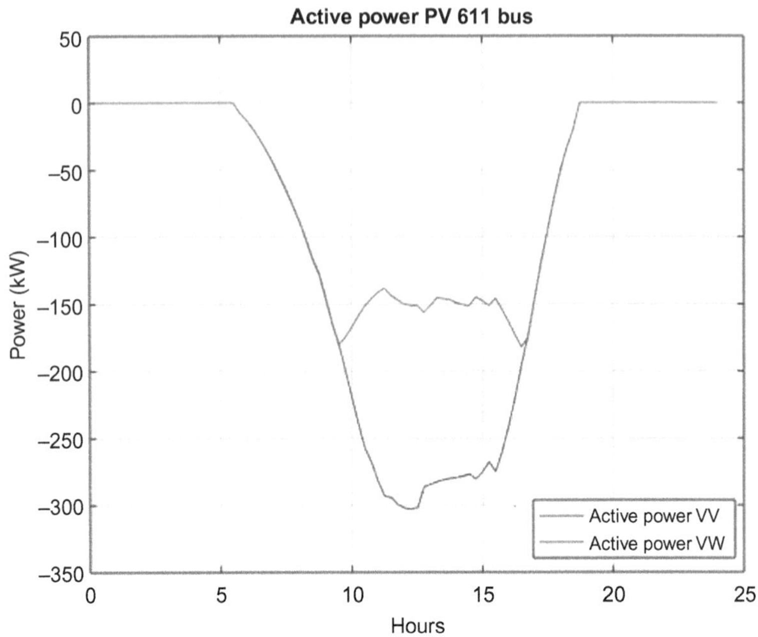

FIGURE 5.15 Active power injected by the PV of the 611-bus with and without Volt/Watt control. *PV*, Photovoltaic.

in conjunction with the Volt/Var control. The total power of the storage units is 150 kW. Charging occurs in the period of greatest power generation by the PV, and the stored energy consumption occurs in the night period, where the demand of the load connected to the bus is higher. The result of the voltage of bus 611 for the case in question is shown in Fig. 5.16.

5.4.2 Case 2: Harmonic studies in OpenDSS considering renewable distributed generation and aggregate linear load models

An increasing adoption of power electronic technologies—such as inverters for PV systems, variable-frequency drives, battery energy storage systems, and EVs—is expected to grow during the coming years, not only in quantities but also in size of installations as well [27]. However, high penetrations of these technologies may decrease the power performance metrics and amplify technical problems in distribution networks [28]. For example, overvoltage and harmonic distortion are operational circumstances of power distribution systems that are becoming more relevant in a progressive smarter grid scenario [29].

Many utility systems are not prepared for the increasing penetration of DG, taking the distribution grids to a new level of complexity. Thus the use

TABLE 5.2 Max 611-bus voltage with various Volt/Var controlling the system.

Bus with Volt/Var control	Maximum voltage 611-bus (p.u.)
611	1.0648
611–680	1.0618
611–675	1.0549
611–671	1.0562
611–680–675	1.0544
611–680–671	1.0554
611–671–675	1.0529
611–680–675–680	1.0528
611–646	1.0546
611–634	1.0641
611–646–634	1.0544
611–646–675	**1.0489**
611–646–671	**1.0495**
611–646–680	1.0528
All buses	1.0526

Note: Bold represents the identify the lowest voltage levels.

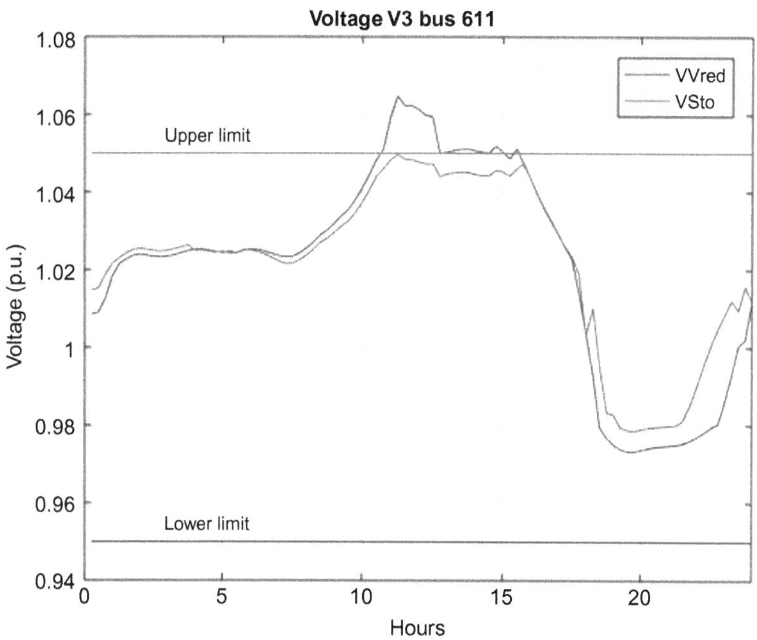

FIGURE 5.16 611-Bus voltage with storage insertion.

of system analysis tools for modeling this new scenario is required to achieve a better comprehension of grid behaviors under different levels of DG [30].

5.4.3 Harmonic studies

Loads play an important role in harmonic studies. Induction motors, synchronous motors, and most commercial and residential loads present a linear behavior, while electronic-based PC technologies are usually nonlinear loads since they generate harmonics [31]. The following case is an example of how aggregate linear load models impact harmonic studies, since their type, magnitude, and composition may change the resonance conditions and the voltage distortion in a power distribution system. Simulations were performed in a COM environment involving OpenDSS and MATLAB.

Three load models proposed in the literature [31,32] are analyzed—a series model, a parallel model, and a model dedicated for a more specific characterization of induction motors on an aggregated load. Their respective parameters are indicated in Table 5.3.

The first test feeder is shown in Fig. 5.17, and an extension of this system, with additional lines and loads connected, is displayed in Fig. 5.18. Their parameters can be seen in Table 5.4 [33]. The source equivalent is the same for both systems. Symmetrical parameters of power lines describe a typical 185 mm^2 primary network conductor. Power factor correction (PFC) is a three-phase capacitor bank without losses. Current source represents harmonic injection of electronic-based generation (e.g., inverters of PV generators). The dashed box represents where the load models of Table 5.3 are placed and analyzed. Skin effect has not been considered.

TABLE 5.3 Aggregated linear load models.

Model type	Series	Parallel	Induction motors
	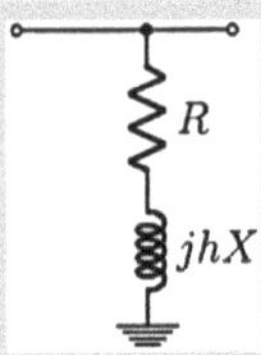	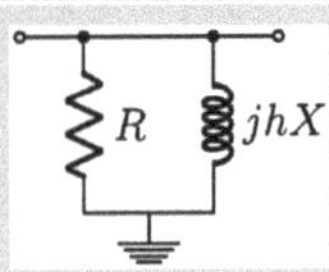	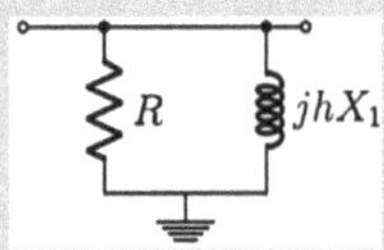
Parameters	$R = P \cdot \frac{V^2}{P^2+Q^2}$	$R = \frac{V^2}{P}$	$R = \frac{V^2}{(1-K) \cdot P}$
	$X = Q \cdot \frac{V^2}{P^2+Q^2}$	$X = \frac{V^2}{Q}$	$X_1 = X_M \cdot \frac{V^2}{K_m \cdot K \cdot P}$

P and Q are the active and reactive power, respectively; V is the rated line-to-line voltage, for a three-phase system; K_m is the install factor ($\approx$1.2 p.u.); X_M is the motor-locked rotor reactance ($\approx$0.15 − 0.25 p.u.); and K is the fraction of motor load into the total load demand.

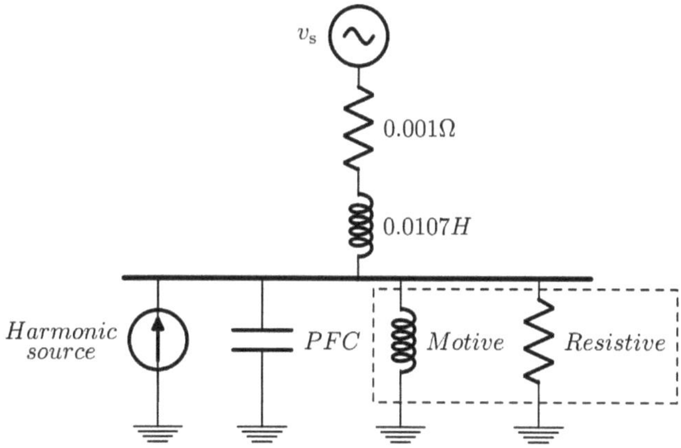

FIGURE 5.17 Test system 1 for harmonic studies.

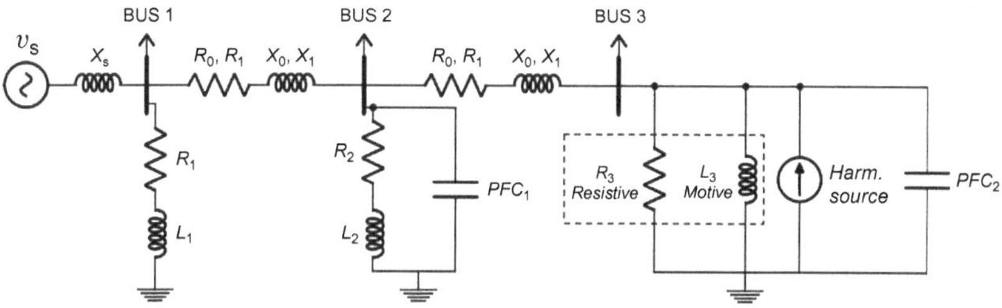

FIGURE 5.18 Test system 2 for harmonic studies.

TABLE 5.4 Technical parameters of the test systems.

Source equivalent	$v_s = 13.8$ kV	$S_{CC} = 30$ MVA		$L_s = 0.0107$ H
	$f = 60$ Hz			$R_s = 0.001$ Ω
Test system 1				
Load	$P = 743$ kW		$Q = 247$ kVAr	
PFC = PFC_2	$Q = -247$ kVAr	$C = 5.4$ μF		$V_{nom} = 13.8$ kV
Injected harmonics (in A)	$I_5 = 0.840$	$I_7 = 0.601$	$I_{11} = 0.382$	$I_{13} = 0.323$
Test system 2				
Bus 1	$P = 1$ MW		PF = 0.9*ind.*	
Bus 2	$P = 2$ MW	PF = 0.9*ind.*		$PFC_1 = 970$ kVAr
Bus 3	Same parameters of test system 1			
Lines (5 km each)	$R_0 = 0.8767$ Ω/km		$R_1 = 0.2112$ Ω/km	
	$X_0 = 1.6847$ Ω/km		$X_1 = 0.2510$ Ω/km	

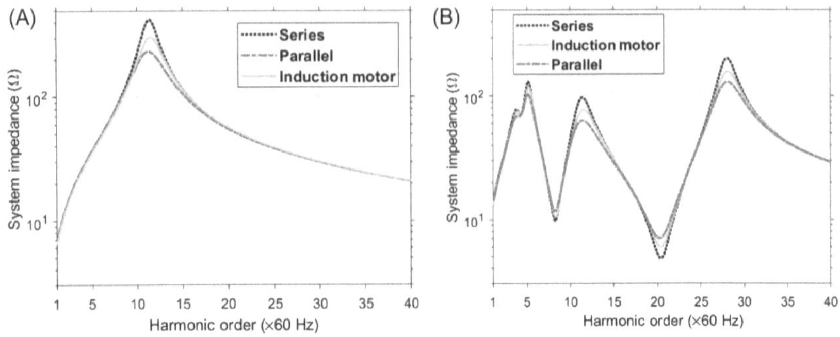

FIGURE 5.19 Frequency scan for (A) test system 1 and (B) test system 2.

Frequency scans on the connection point of the load models were done, and the relation between system impedance and frequency can be seen in Fig. 5.19.

To verify the sensitivity of the aggregate load models for both load type and load composition, three scenarios corresponding to different grid topologies of the test systems are selected, which are stressed as follows:

- Case 1: refers to test system 1;
- Case 2: refers to test system 2, the way it was described;
- Case 3: also refers to system 2, but now PFC_1 is switched off (out of service).

5.4.3.1 Model sensitivity

Model sensitivity consisted of analyzing the aggregate load models composed of 50% of induction motors, and the remaining load is resistive. The frequency scan clarified the differences among the three load models, especially the peak impedance magnitudes, shown in Fig. 5.20. The voltage distortion analyses were done for Cases 2 and 3 only, and the graphics are displayed in Fig. 5.21.

In Fig. 5.20 the first resonant peak moved from 11th harmonic on Case 1 to the 5th harmonic on Case 2. The series model presented the highest impedance magnitude, while the parallel model had the highest damping. The induction motor model was shown to be balanced between the two values of impedance magnitude. The same approach is shown in Fig. 5.21, by analyzing the voltage distortion on the four harmonics of interest (5th, 7th, 11th, and 13th). Switching off PFC_1 led to a profile change on the harmonic voltages, comparing Case 2 with Case 3.

5.4.3.2 Load composition

Load composition comprises the analysis of induction motor load only, under three different mixture of the load—a share of motor loads corresponding to

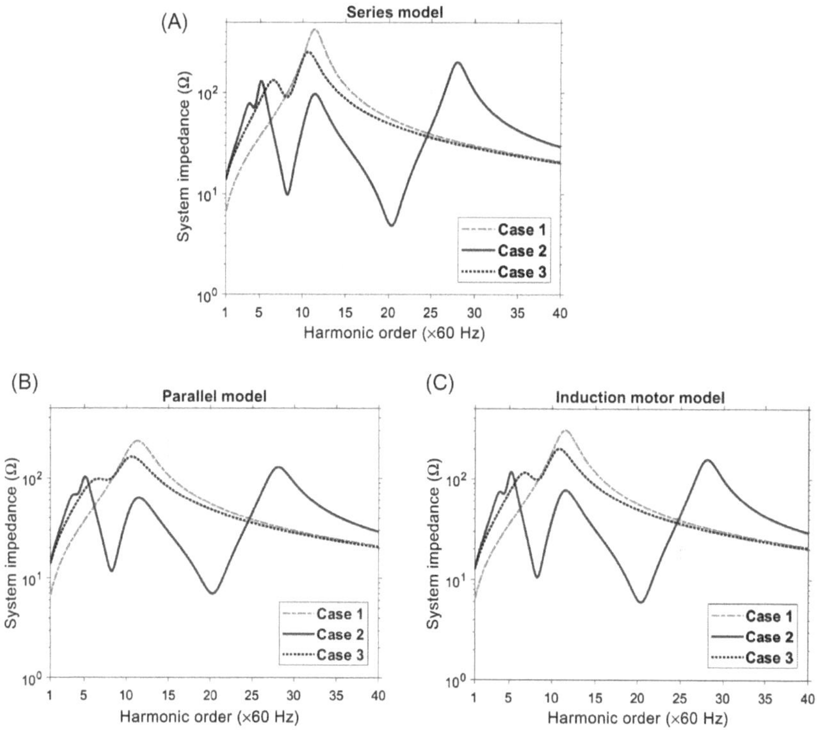

FIGURE 5.20 Frequency response for each load model in each test case.

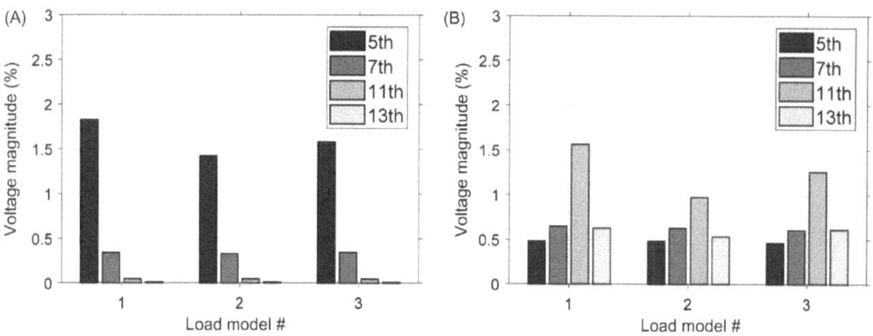

FIGURE 5.21 Voltage distortion in bus 3 for each load model in (A) Case 2 and (B) Case 3.

25%, 75%, or 90% of the aggregated linear load. The three cases were simulated, but in two different conditions:

- PFC kept constant on the default case ($K = 0.5$) and
- PFC following the load variation.

The plots with constant PFC are shown in Figs. 5.22 and 5.23.

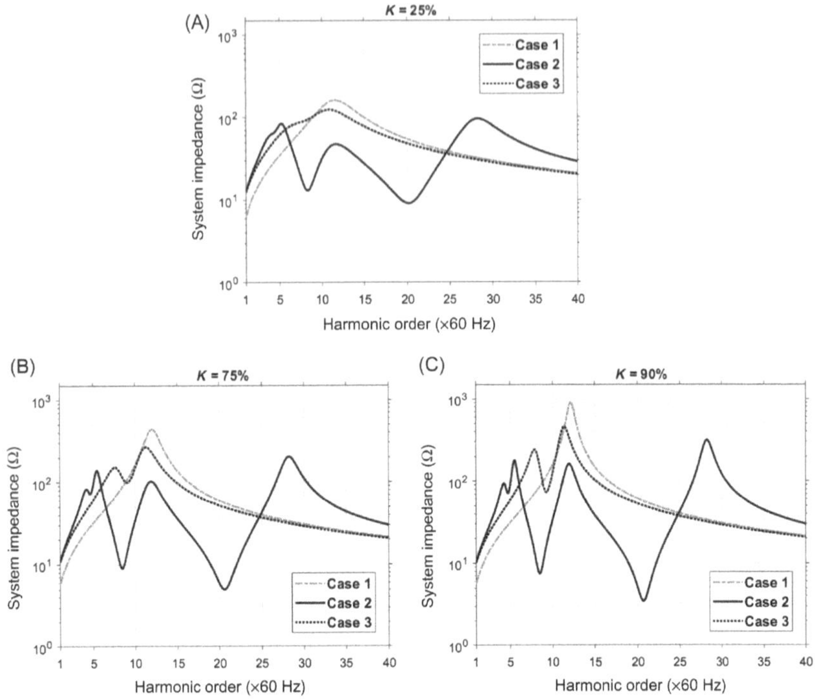

FIGURE 5.22 Frequency response with PFC fixed and load composition of induction motors corresponding to (A) 25%, (B) 75%, and (C) 90% of the total load demand. *PFC*, Power factor correction.

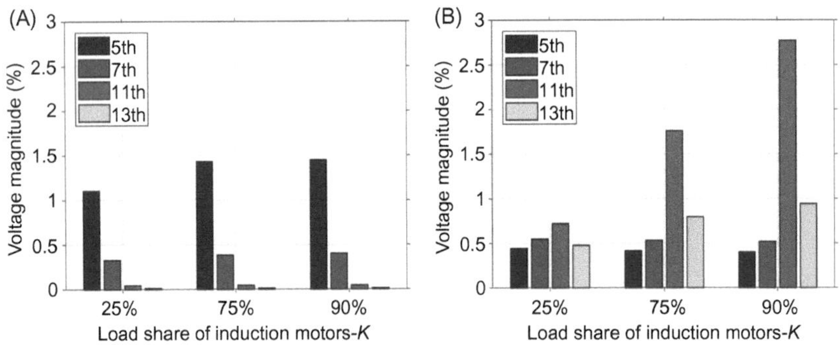

FIGURE 5.23 Voltage distortion versus load composition, fixed PFC, at bus 3 for (A) Case 2 and (B) Case 3. *PFC*, Power factor correction.

When $K = 25\%$, the major share of the aggregate load is still resistive, causing a large damping on impedance magnitudes, especially at resonant points. With the increase of induction motors, the damping declines, as seen in Fig. 5.22. The higher is the share of induction motors, the greater is the

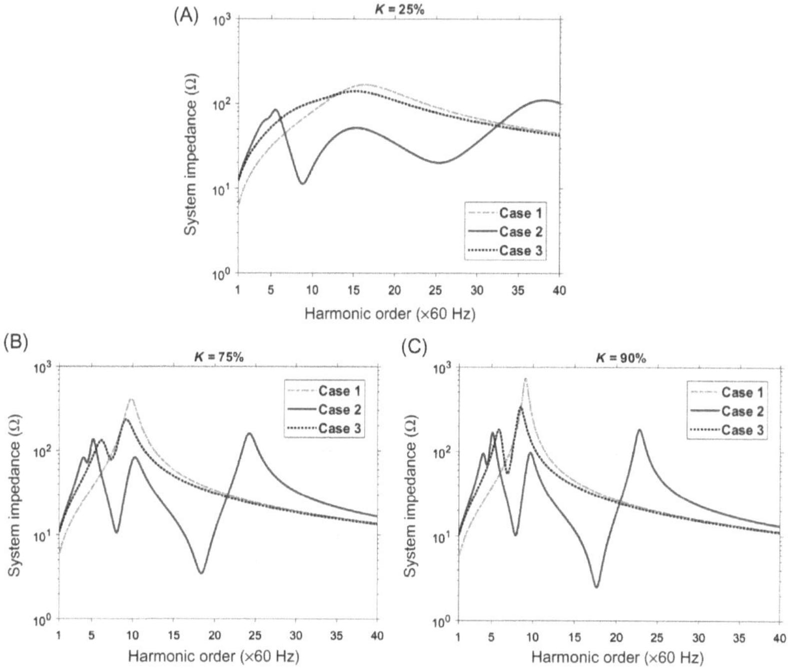

FIGURE 5.24 Frequency response with variable PFC and load composition of induction motors corresponding to (A) 25%, (B) 75%, and (C) 90% of the total load demand. *PFC*, Power factor correction.

voltage distortion, as shown in Fig. 5.23, although Case 3 exhibited greater sensitivity than Case 2.

Now, the results with variable PFC with the load composition are shown in Figs. 5.24 and 5.25. In this condition the capacitor bank is adjusted to match the reactive power consumption and correcting the PF, thus leading to a more realistic situation.

Similar impedance magnitudes are seen in Fig. 5.24, compared to the previous results, but variations occur in resonant frequencies, mostly because of the variable capacitance. The voltage distortion displayed in Fig. 5.25 still increases with the growth of the share of induction motors, but a higher sensitivity appeared on Case 2 rather than Case 3.

5.5 Result analysis

5.5.1 Case 1: Volt/Var and Volt/Watt controls

With the increasing number of PV systems in the distribution system, there are also growing concerns about problems that may arise, such as an overvoltage. The use of the inverters in this work was meant to mitigate the

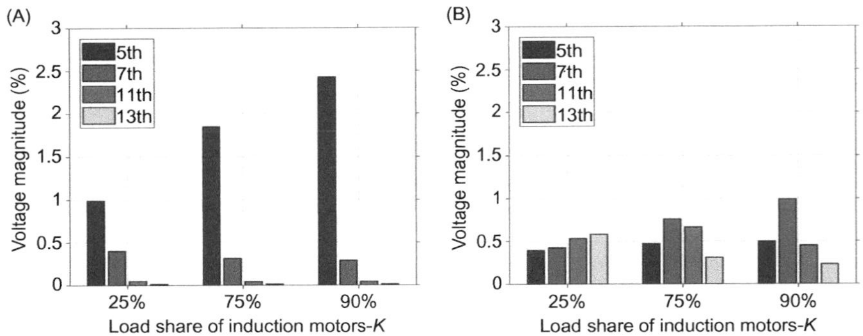

FIGURE 5.25 Voltage distortion versus load composition, PFC variable. Seen from bus 3 for Cases (A) 2 and (B) 3. *PFC*, Power factor correction.

overvoltage condition. The Volt/Var control used by OpenDSS has proven to be an effective alternative as long as care is taken with the inverter rating specifications.

Other methodologies have been presented when the control is not able to regulate the voltage. The use of the Volt/Watt control, despite being able to maintain the voltage in the proper range, caused a considerable reduction on output power of the PV, which in most cases is not desirable. The insertion of multiple Volt/Var controls was not very satisfactory for local voltage regulation and would require appropriate coordination and communication among inverters. In addition, the insertion of energy storage in the PV system proved to be an interesting alternative, regulating the voltage in the generation period of the PV and relieving the system in moments of higher demand.

5.5.2 Case 2: Harmonics

This case study showed how different load models may impact harmonic analyses. Also, load composition has considerable influence over the magnitude and the frequency of the resonant peaks. Capacitors should be modeled as accurately as possible, including filters of PV inverters, smart grid devices, and other electronic-based loads.

The induction motor model was proposed for a low or moderate share of motors in an aggregate linear load. Therefore this model is slightly superior to the series and parallel models in this study case. But remember that there is not a unique model fitting for all frequencies.

Incorrect modeling of linear loads may lead to unrealistic harmonics evaluations. Most harmonic discrepancies of aggregated linear loads occur near parallel resonant points. Furthermore, the background harmonics play an important role in harmonic studies, such as for hosting capacity evaluation on modern power systems.

5.6 Conclusion

This chapter presented methodologies to address some challenges that may occur in power distribution networks with the increasing penetration of DG, using the modeling tools of the open-source software OpenDSS.

Impacts of high penetration levels of DG in distribution systems were presented, including not only the technical aspects but also socioeconomic impacts. Smart functions of power inverters were reviewed, since those devices have a high potential to play a key role in the operation and control of future power systems. A description of the modeling and analyses tools of OpenDSS was presented, including the main aspects for modeling loads, storage units, PV systems, other linear, and nonlinear electrical components.

The first proposed case analyzed was a voltage control method in the IEEE 13-bus using smart inverter functions. Results have shown that Volt/Var control is a good choice for overvoltage mitigation, especially when this control is applied to different buses with DG along the feeder. Alternatives using Volt/Watt control and batteries for energy storage have also shown excellent results for voltage management, although DG power curtailment and investment costs are variables that can interfere with the overall performance.

At last, the second case presented a series of harmonic studies focusing on the impacts of aggregated linear loads modeling on the system impedance and voltages. The results show that different modeling representations have significant differences near the parallel resonant points, and load composition may impact voltage distortion and system resonance. Moreover, capacitor banks and resistive linear loads cause large impacts on harmonic resonances and harmonic damping, respectively, and they should be precisely represented in the simulations. Harmonic studies usually have a minor concern for DG connection; however, with the insertion of new types of loads and behaviors—for example, electrical vehicles, demand response, and electronic-based loads—it is recommended doing harmonic studies for connection of new distributed generators and capacitors for PFC.

Acknowledgment

The authors thank to CNPq, CAPES, FAPEMIG, and INERGE for partially supporting this work. The authors L.C. Ribeiro Jr. and F.L. Vieira thank the scholarship grants received from CAPES [Coordenação de Aperfeiçoamento de Pessoal de Nível Superior Brasil (CAPES) (Finance Code 001)].

References

[1] L.C. Ribeiro Jr., Inversores Inteligentes em Sistemas Fotovoltaicos para Controle Intergrado de Funções utilizando o OpenDSS, (M.Sc. dissertation) (Electrical Eng.), Federal University of Itajuba (UNIFEI), 2018.

[2] M. Yongfei et al., Analysis of the influence of distributed generation access on the operation and management of distribution network, in: 2016 International Conference on Smart City and Systems Engineering (ICSCSE), 2016, pp. 194–196.

[3] M. Mcgranaghan, T. Ortmeyer, D. Crudele, T. Key, J. Smith, P. Barker, Renewable systems interconnection study: advanced grid planning and operations, Sandia Rep. SAND2008-0944 P. Sandia Natl. Lab (2008) 1–123.

[4] E.J. Coster, J.M.A. Myrzik, B. Kruimer, W.L. Kling, Integration issues of distributed generation in distribution grids, Proc. IEEE 99 (1) (2011) 28–39.

[5] R.A. Walling, R. Saint, R.C. Dugan, J. Burke, L.A. Kojovic, Summary of distributed resources impact on power delivery systems, IEEE Trans. Power Deliv. 23 (3) (2008) 1636–1644.

[6] F.T. Dai, Impacts of distributed generation on protection and autoreclosing of distribution networks, in: 10th IET International Conference on Developments in Power System Protection (DPSP 2010), Managing the Change, 2010, pp. 1–5.

[7] S.M. Brahma, A.A. Girgis, Development of adaptive protection scheme for distribution systems with high penetration of distributed generation, IEEE Trans. Power Deliv. 19 (1) (2004) 56–63.

[8] Electric Power Research Institute (EPRI), Common Functions for Smart Inverters: 4th Edition, Tech. Rep., 2016. [Online]. Available from: <https://www.epri.com/#/pages/product/3002008217/>.

[9] M.S. ElNozahy, M.M.A. Salama, Technical impacts of grid-connected photovoltaic systems on electrical networks—a review, J. Renew. Sustain. Energy 5 (3) (2013).

[10] M. Bollen, F. Hassan, Integration of Distributed Generation in the Power System, John Wiley & Sons Ltd, 2011.

[11] P.F. Ribeiro, A.C.Z. de Souza, B.D. Bonatto, Reflections about the philosophy of technology in the emerging smart power systems, in: 2017 Ninth Annual IEEE Green Technologies Conference (GreenTech), 2017, pp. 195–202.

[12] J. Smith, Modeling high-penetration PV for distribution interconnection studies: smart inverter function modeling in OpenDSS, Tech. Rep. Electric Power Research Institute (EPRI), 2013, p. 76.

[13] R.F. Arritt, R.C. Dugan, Matching the IEEE Test Feeder short circuit results, in: Proc. PES T&D 2012, 2012, pp. 1–7.

[14] R.C. Dugan, Reference guide: the open distribution system simulator (OpenDSS), Training Materials, Electric Power Research Institute (EPRI), 2016.

[15] J. Sexauer, New user primer: the open distribution system simulator (OpenDSS), Training Materials, Electric Power Research Institute (EPRI), 2012, p. 38.

[16] D. Montenegro, R.C. Dugan, OpenDSS and OpenDSS-PM open source libraries for NI LabVIEW, in: 2017 IEEE Workshop on Power Electronics and Power Quality Applications (PEPQA), 2017, pp. 1–5.

[17] D. Montenegro, R.C. Dugan, How to speed up your co-simulation using OpenDSS COM interface, in: OpenDSS Discussion Forum [Online]. Available from: <https://docplayer.net/57460449-How-to-speed-up-your-co-simulation-using-opendss-com-interface.html>, 2015.

[18] C. Rocha, P. Radatz, Algoritmo de Fluxo de Potência do OpenDSS, in: Tech. Rep. University of Sao Paulo, 2017, p. 24.

[19] Electric Power Research Institute (EPRI), OpenDSS PVSystem Element Model Version 1,: OpenDSS Documentation, 2011. [Online]. Available from: <https://sourceforge.net/p/electricdss/code/HEAD/tree/trunk/Doc/OpenDSSPVSystem Model.doc>.

[20] Electric Power Research Institute (EPRI), OpenDSS Storage Element and Storage Controller Element, in: OpenDSS Documentation [Online]. Available from: <https://sourceforge.net/p/electricdss/code/HEAD/tree/trunk/Doc/OpenDSSSTORAGE Element.doc>, 2011 (accessed 26.10.18).

[21] Electric Power Research Institute (EPRI), Load modeling in harmonics analysis with OpenDSS, in: OpenDSS Documentation [Online]. Available from: <https://sourceforge.net/p/electricdss/code/HEAD/tree/trunk/Doc/HarmonicsLoadModeling.docx>, 2015 (accessed 26.10.18).

[22] Z. Huang, Y. Cui, W. Xu, Application of modal sensitivity for power system harmonic resonance analysis, IEEE Trans. Power Syst. 22 (1) (2007) 222–231.

[23] L.C. Ribeiro, J.P.O.S. Minami, B.D. Bonatto, P.F. Ribeiro, A.C.Z. de Souza, Voltage control simulations in distribution systems with high penetration of PVs using the OpenDSS, in: 2018 Simposio Brasileiro de Sistemas Eletricos (SBSE), 2018, pp. 1–6.

[24] W. Sunderman, R. Dugan, B. Seal, Advanced inverter controls for distributed resources, in: 22nd International Conference and Exhibition on Electricity Distribution (CIRED 2013), 2013, pp. 1–4.

[25] IEEE PES, Distribution system analysis subcommittee's distribution Test Feeder Working Group. [Online]. Available from: <http://sites.ieee.org/pes-testfeeders/resources/>, 2018 (accessed 27.10.18).

[26] IEEE Standards Coordinating Committee 21, IEEE application guide for IEEE Std 1547, IEEE standard for interconnecting distributed resources with electric power systems, in: IEEE Std 1547.2-2008, 2009.

[27] J. Smith, et al., Power quality aspects of solar power, Tech. Rep, CIGRÉ, 2016, p. 109.

[28] R. González, A. Arguello, G. Valverde, J. Quirós-Tortós, OpenDSS-based distribution network analyzer in open source GIS environment, in: 2016 IEEE PES Transm. Distrib. Conf. Expo. Am. PES T D-LA 2016, 2017, pp. 1–6.

[29] D. Montenegro, R. Dugan, G. Ramos, Harmonics analysis using sequential-time simulation for addressing smart grid challenges, in: 23rd Int. Conf. Electricity Distribution, Lyon, 2015, pp. 15–18.

[30] P. Radatz, N. Kagan, C. Rocha, J. Smith, R.C. Dugan, Assessing maximum DG penetration levels in a real distribution feeder by using OpenDSS, in: Proc. Int. Conf. Harmon. Qual. Power, ICHQP, vol. 2016–Decem, 2016, pp. 71–76.

[31] Task Force on Harmonics Modeling and Simulation, Impact of aggregate linear load modeling on harmonic analysis: a comparison of common practice and analytical models, IEEE Trans. Power Deliv. 18 (2) (2003) 625–630.

[32] P.F. Ribeiro, Investigations of Harmonic Penetration in Transmission Systems, (Ph.D. thesis), Victoria University of Manchester, England, 1985.

[33] F.L. Vieira, P.F. Ribeiro, B.D. Bonatto, T.E.C. Oliveira, Harmonic studies in OpenDSS considering renewable DG and aggregate linear load models, in: 2018 13th IEEE International Conference on Industry Applications (INDUSCON), 2018, pp. 202–207.

Chapter 6

Adaptive estimation and tracking of power quality disturbances with classification for smart grid applications

Papia Ray[1], Harish Kumar Sahoo[2] and Ganesh Kumar Budumuru[1]
[1]*Department of Electrical Engineering, Veer Surendra Sai University of Technology, Burla, India,* [2]*Department of Electronics & Telecommunication Engineering, Veer Surendra Sai University of Technology, Burla, India*

6.1 Introduction

Nowadays, power quality (PQ) has become one of the most important issues for both utilities and consumers and plays a vital role in the electrical power systems. Delivering an uninterrupted, high-quality power supply to the end users is the expectation of electric power systems. The quality of power supply means the capacity of the power system to deliver undistorted voltage, current, and frequency signals. Any deviation manifested in frequency, current, and voltage from the rated values, which results in the failure of consumer equipment, is described as PQ problem [1]. The sources of issues that affected the quality of power are generally rapid increase of sensitive loads, power electronic devices, lightning, switching of capacitor banks, smart transmission system, integration of renewable energy sources, and nonlinear loads. The PQ events are categorized into three types followed by the deviations in the magnitude such as interruption, voltage swell, and voltage sag because of the inequity of heavy or light loads and power systems faults. Sudden transients, such as spikes or impulsive transients because of lightning and capacitor banks switching, come under second category. Finally, steady-state harmonics, such as flickers and notches due to nonlinear load applications as well as power electronic converters [2], come under PQ disturbances. To improve the PQ, we must know the reasons for PQ disturbances and mitigate them as early as possible; otherwise it will disturb the whole system.

Decision Making Applications in Modern Power Systems. DOI: https://doi.org/10.1016/B978-0-12-816445-7.00006-2

An adaptive filter is a system with a linear filter that has a transfer function controlled by variable parameters and a means to adjust those parameters according to an optimization algorithm. The property of an adaptive filter is self-modifying its frequency response to change the behavior in time, allowing the filter to adapt the response to the input signal characteristics change. The adaptive filters have various applications such as echo cancellation in the telephones, signal processing in the radars, navigation systems, biometric signal processing, navigation signals, and communications channel equalization. The main purpose of an adaptive filter in noise cancellation is to eliminate the noise from a signal adaptively to improve the signal-to-noise ratio (SNR). For accurate assessment of PQ events in the presence of noise, tremendous efforts are being made for a long period of time to design effective and robust algorithms. The occurrence of harmonics frequently causes communication interference, resonance of mechanical devices, and melting of magnetic parts of electrical appliances, etc. Nowadays, the detection and removal of harmonics using appropriate harmonic filter and forecasting of PQ disturbances are key research aspects of power engineers. The methods for detection of PQ events have been classified into two types; those are parametric and nonparametric methods. Fourier transform (FT), wavelet transform (WT), *S*-transform (ST), *H*-transform, etc., all come under the nonparametric methods and are restricted by the length of the data. Adaptive filtering is popular parametric estimation technique to track and estimate the PQ events. A common adaptive filter design is based on transversal filter with adaptive weight update mechanism such as least mean square (LMS) adaptive filtering algorithm that has been widely used due to its simplicity and numerical robustness. On the other hand, normalized LMS (NLMS) and recursive least square (RLS) give better convergence properties than LMS. To estimate amplitude, phase, and frequency, an extended Kalman filter (EKF) has been implemented. Further the EKF approach has the advantage that the estimates are computed recursively using one-step prediction [3–5].

In general, several researchers in this area have applied one of the well-known signal-processing techniques to extract the features and complete the classification process by using an artificial intelligence technique as a classifier. The signal-processing techniques give some redundant features that affect the efficiency of the classifiers. Moreover, there is no discussion on how to set the best parameters for the classifiers. Only few researchers have attempted optimization techniques for selecting the suitable feature subset and selection of parameter. In this view, signal-processing techniques for feature extraction and artificial intelligent techniques for the classification are the most important parts of the pattern recognition of PQ disturbances. The feature extraction stage provides a set of statistical data to make the analysis more effective. The set of feature extraction is then used as input for the classification system. In spite of technical advancement in signal-processing techniques, the proper selection of feature extraction is still a challenge.

Thus the optimal feature selection techniques have been proposed to retain the useful features and discard the redundant features.

To extract the features of disturbed signal, there are many feature extraction methods, such as WT [6], FT [7,8], ST [9,10], short-term FT, and Hilbert transform (HT) [11,12], which were implemented, and then extracted features are fed to pattern classifiers, such as artificial neural network (ANN) [13,14], probabilistic neural network (PNN) [15], fuzzy logic [16], and support vector machine (SVM) [17], to classify PQ disturbances. There are some advantages and disadvantages for each and every technique. In the proposed work, empirical mode decomposition (EMD) with HT has been implemented for feature extraction of the PQ disturbances. This is an inventive technique in which the distorted signal is decomposed into number of intrinsic mode functions (IMFs). We can get instantaneous frequencies as well as amplitudes of the signal by applying HT to the IMFs. For classification purpose, ANN and PNN have been developed. Further, for better classification, efficiency SVM has been implemented.

6.2 Methodologies for efficient estimation of power quality disturbances by using adaptive filters

In this section, various methodologies for efficient estimation of PQ events by using adaptive filters are discussed.

6.2.1 Signal model for power quality disturbances and harmonics estimation

Efficient signal-processing architectures are used to design PQ estimation models. State-space modeling in real and complex forms can be implemented using state variables to estimate sag, swell, notch, and harmonic parameters. State-space modeling is quite popular to implement Kalman filtering algorithm for PQ estimation.

6.2.1.1 Signal model for power quality disturbances estimation

PQ disturbances, such as voltage sag, swell, notch, and momentary interruption [18,19], are related to time variation of signal amplitudes and can easily be tracked from estimated state variables. The number of state variable in a state vector depends on the nature of the PQ disturbances. The following mathematical analysis describes the state-space models using three state variables.

y_k is the noisy observed signal generated by a sinusoid z_k in the presence of white Gaussian noise v_k.

$$y_k = z_k + v_k \tag{6.1}$$

where

$$z_k = a_1 \sin\left(k\omega_1 T_s + \phi_1\right) \tag{6.2}$$

where ω_1 is the fundamental of angular frequency, ϕ_1 is the fundamental of phase angle, and a_1 is the fundamental amplitude of the signal.

The observation noise, v_k, is a Gaussian white noise with zero mean and variance, σ_v^2, and the covariance of measured errors is $R_k = E\left[v_k v_k^{*T}\right]$. The sinusoid can be represented by using three complex state variables as

$$x_{k(1)} = e^{j\omega_1 T_s} \tag{6.3}$$

$$x_{k(2)} = a_1 e^{j(k\omega_1 T_s + \phi_1)} \tag{6.4}$$

$$x_{k(3)} = a_1 e^{-j(k\omega_1 T_s + \phi_1)} \tag{6.5}$$

The state-space model can be formulated by using state and measurement equations as given in the following equations:

$$\text{State equation } x_{k+1} = f(x_k) + G\,w_k \tag{6.6}$$

$$\text{Measurement equation } y_k = Hx_k + v_k \tag{6.7}$$

where

$$x_k = \left[x_{k(1)} \quad x_{k(2)} \quad x_{k(3)}\right]^T \tag{6.8}$$

The state transition matrix can be obtained from state equation using Taylor series expansion as

$$F_k = \begin{bmatrix} 1 & 0 & 0 \\ x_{k(2)} & x_{k(1)} & 0 \\ -x_{k(3)}/x_{k(1)}^2 & 0 & 1/x_{k(1)} \end{bmatrix} \tag{6.9}$$

The measurement matrix is given by

$$H_k = \begin{bmatrix} 0 & -0.5i & 0.5i \end{bmatrix} \tag{6.10}$$

Frequency, $\hat{f}_{(k)}$, and amplitude, $\hat{a}(k)$, can be estimated from state variables as shown in the following equations:

$$\hat{f}_{(k)} = \frac{1}{2\pi\Delta T}\left[\mathrm{Im}(\ln(\hat{x}_{k(1)}))\right] \tag{6.11}$$

$$\hat{a}(k) = |\hat{x}_{k(1)}| \tag{6.12}$$

6.2.1.2 Signal model for harmonic estimation

Similar complex state-space model can also be used to estimate harmonic parameters and decaying DC components [20]. If the power signal is considered with fundamental, third, and fifth harmonics and decaying DC component, state-space model can be formulated using nine complex state variables.

$$\left\{\begin{array}{l} x_k(1)=e^{j\omega_1 T_s} \\ x_k(2)=a_1 e^{j(k\omega_1 T_s+\phi_1)} \\ x_k(3)=a_1 e^{-j(k\omega_1 T_s+\phi_1)} \\ x_k(4)=a_3 e^{j(k3\omega_1 T_s+\phi_3)} \\ x_k(5)=a_3 e^{-j(k3\omega_1 T_s+\phi_3)} \\ x_k(6)=a_5 e^{j(k5\omega_1 T_s+\phi_5)} \\ x_k(7)=a_5 e^{-j(k5\omega_1 T_s+\phi_5)} \\ x_k(8)=a_{DC} e^{-\alpha k T_s} \\ x_k(9)=e^{-\alpha k T_s} \end{array}\right\} \tag{6.13}$$

The corresponding state vector can be expressed as

$$x_k=\begin{bmatrix} x_k(1) & x_k(2) & x_k(3) & x_k(4) & x_k(5) & x_k(6) & x_k(7) & x_k(8) & x_k(9) \end{bmatrix}^T \tag{6.14}$$

The state transition matrix, F_k, and measurement matrix, H_k, can be generated by Taylor series expansion neglecting higher order derivative terms.

$$F_k=\begin{pmatrix} 1 & 0 & 0 & 0 & 0 & 0 & 0 & 0 & 0 \\ x_k(2) & x_k(1) & 0 & 0 & 0 & 0 & 0 & 0 & 0 \\ \dfrac{-x_k(3)}{x_k(1)^2} & 0 & \dfrac{1}{x_k(1)} & 0 & 0 & 0 & 0 & 0 & 0 \\ x_k(4) & 0 & 0 & x_k(1) & 0 & 0 & 0 & 0 & 0 \\ \dfrac{-x_k(5)}{x_k(1)^2} & 0 & 0 & 0 & \dfrac{1}{x_k(1)} & 0 & 0 & 0 & 0 \\ x_k(6) & 0 & 0 & 0 & 0 & x_k(1) & 0 & 0 & 0 \\ \dfrac{-x_k(7)}{x_k(1)^2} & 0 & 0 & 0 & 0 & 0 & \dfrac{1}{x_k(1)} & 0 & 0 \\ 0 & 0 & 0 & 0 & 0 & 0 & 0 & x_k(9) & x_k(8) \\ 0 & 0 & 0 & 0 & 0 & 0 & 0 & 0 & e^{-\alpha T_s} \end{pmatrix} \tag{6.15}$$

$$H_k=\begin{bmatrix} 0 & -0.5i & 0.5i & -05i & 0.5i & -0.5i & 0.5i & 1 & 1 \end{bmatrix} \tag{6.16}$$

The amplitudes and phases of the harmonics can be estimated as shown from Eqs. (6.17) to (6.20).

$$\begin{aligned}\hat{a}_1(k) &= |\hat{x}_{k(2)}| \\ \hat{a}_3(k) &= |\hat{x}_{k(4)}| \\ \hat{a}_5(k) &= |\hat{x}_{k(6)}|\end{aligned} \tag{6.17}$$

$$\overset{\Lambda}{\phi_1} = \left(\frac{1}{a_1}\right)\text{imag}\left[\log\left(\frac{x_{k(2)}}{x_{k(1)}^{k}}\right)\right] \tag{6.18}$$

$$\overset{\Lambda}{\phi_3} = \left(\frac{1}{a_3}\right)\text{imag}\left[\log\left(\frac{x_{k(4)}}{x_{k(1)}^{3k}}\right)\right] \tag{6.19}$$

$$\overset{\Lambda}{\phi_5} = \left(\frac{1}{a_5}\right)\text{imag}\left[\log\left(\frac{x_{k(6)}}{x_{k(1)}^{5k}}\right)\right] \tag{6.20}$$

6.2.2 Adaptive filtering algorithms for power quality estimation

The state-space model discussed in the previous section cannot track and estimate the time-varying PQ disturbances if the state variables are not updated recursively. The updated state variables will provide the estimated values of amplitude, frequency, and phase parameters of distorted power signals. The mathematical formulation and weight update equations have been described in this section. The adaptive filtering algorithm starts from a predetermined set of initial conditions, which represents some statistical behavior of the environment. In the case of stationary environment, the algorithm converges to the optimum Wiener solution in some statistical sense after successive iterations. In a nonstationary environment such as PQ disturbances, the algorithm can track time variations in the statistics of the input data.

6.2.2.1 *Least mean square algorithm*

LMS algorithm is simple to implement and is a class of stochastic gradient algorithm. According to LMS algorithm, recursive relation for updating the tap weight vector can be expressed as

$$\hat{w}(n+1) = \hat{w}(n) + \mu u(n)e^*(n) \tag{6.21}$$

In the weight updating expression, the filter output is given by

$$y(n) = \hat{w}(n)u^H(n) \tag{6.22}$$

and estimation error is given by

$$e^*(n) = d^*(n) - y(n) \tag{6.23}$$

The step size parameter, μ, plays a vital role for the convergence of the algorithm.

6.2.2.2 Recursive least square algorithm

RLS filtering algorithm is based on matrix inversion lemma. The rate of convergence of this filter is typically much faster than the LMS algorithm due to the fact that input data is whitened by using the inverse correlation matrix of the data, assumed to be of zero mean. But RLS is computationally more complex than LMS.

A weighting factor is introduced to the definition of $\xi(n)$ as

$$\xi(n) = \sum_{i=1}^{n} \beta(n,i)|e(i)|^2 \tag{6.24}$$

where $e(i)$ is the difference between the desired response, $d(i)$, and the output, $y(i)$

$$e(i) = d(i) - y(i) = d(i) - w^H(n)u(i) \tag{6.25}$$

where $u(i)$ is the tap input vector at time i, defined by

$$u(i) = [u(i), u(i-1), \ldots, u(i-M+1)]^T \tag{6.26}$$

and $w(n)$ is the tap weight vector at time n, defined by

$$w(n) = [\omega_0(n), \omega_1(n), \ldots, \omega_{M-1}(n)] \tag{6.27}$$

The algorithm estimates iteratively by initializing weight vector and estimation covariance to zero.

$$\hat{w}(0) = 0 \ , P(0) = \delta^{-1}I \tag{6.28}$$

and δ is the small positive constant for high SNR and the large positive constant for low SNR.

The recursive formulation of the algorithm can be expressed by Eq. (6.29) as

$$\left\{ \begin{array}{l} \text{for } n = 1, 2, \ldots \\ \pi(n) = P(n-1)u(n) \\ k(n) = \dfrac{\pi(n)}{\lambda + u^H(n)\pi(n)} \\ \xi(n) = d(n) - \hat{w}^H(n-1)u(n) \\ \hat{w}(n) = \hat{w}(n-1) + k(n)\xi^*(n) \\ P(n) = \lambda^{-1}P(n-1) - \lambda^{-1}k(n)u^H(n)P(n-1) \end{array} \right\} \tag{6.29}$$

The $M \times M$ matrix $P(n)$ is referred to as inverse correlation matrix, that is, $P(n) = \Phi^{-1}(n)$, and $M \times 1$ vector $k(n)$ is referred to as the gain vector.

6.2.2.3 Kalman filtering algorithm

The Kalman filter is computationally more efficient as the estimation depends only on one-step predicted value rather than a large set of past values. The Kalman filter is basically used for the estimation of a state vector in a linear model of a dynamical system. But if the model is nonlinear, Kalman filtering can be extended through a linearization procedure. The resulting filter is referred to as the EKF in which estimation process is divided into state prediction and state update.

1. *State prediction*

$$\tilde{x}_{k+1|k} = f(\hat{x}_{k|k}) \tag{6.30}$$

$$P_{k+1|k} = F_k P_{k|k} F_k^* + Q_k \tag{6.31}$$

The symbols $\sim$ and $\wedge$ stand for predicted and estimated values, respectively.

2. *State update*

The state update equation is formulated by using predicted state variable and innovation vector.

$$y_k - H_k \tilde{x}_{k-1|k} \tag{6.32}$$

$$\hat{x}_{k+1|k} = \tilde{x}_{k-1|k} + K_k (y_k - H_k \tilde{x}_{k-1|k}) \tag{6.33}$$

Along with the state vector, the Kalman gain is also updated, which plays a significant role in the improvement of the tracking behavior of the algorithm.

$$K_k = P_{k|k-1} H_k^{*T} [H_k P_{k|k-1} H_k^{*T} + R_k]^{-1} \tag{6.34}$$

By using the update value of the Kalman gain, estimation error covariance can also be updated as per the following equation:

$$P_{k|k} = P_{k|k-1} - K_k H_k P_{k|k-1} \tag{6.35}$$

6.2.3 Sparse model–based adaptive filters

Sparse modeling of adaptive filters is the current research focus due to reduction in computational complexity, which will help to design low complex PQ estimation models for real-time applications. In this section, norm-based sparsity is introduced in standard EKF algorithm.

The inherent sparsity of the filter is exploited by incorporating an ℓ_1 norm penalty into the quadratic cost function. Inclusion of ℓ_1 relaxation to the cost function will help one to obtain the original sparse solution as compared to ℓ_0 and ℓ_2 norms. The modified cost function with ℓ_1 norm penalty can be expressed as

$$J_1(n) = \frac{1}{2} e^2(n) + \delta \| w(n) \|_1 \tag{6.36}$$

where $\left\|\cdot\right\|_1$ denotes the ℓ_1 norm of coefficient vector, and δ is the weight assigned to the penalty term. The cost function is convex, and it is expected that the EKF algorithm converges to optimum value under some constraints. The new state update equation for EKF can be expressed as

$$x(n+1) = x(n) - \rho \operatorname{sgn}(x(n)) + ke(n) \tag{6.37}$$

6.2.4 FPGA implementation of adaptive filters used in power quality estimation

Adaptive PQ estimation models can be designed by MATLAB/SIMULINK by implementing the mathematical equations using suitable blocks, which is shown in Fig. 6.1, but real-time hardware implementation of adaptive filtering–based PQ estimation models is quite difficult due to computational complexity of the model and algorithm. Generally computational complexity and quantization effects degrade the tracking and estimation accuracy of the algorithms. Adaptive filtering–based PQ estimation model can also be designed through Xilinx blockset available in MATLAB/SIMULINK library, which is quite suitable for field programmable gate array (FPGA) hardware platform. ML506 is an example of general purpose evaluation and development platform, and System Generator for DSP is the industry's leading high-level tool for designing high-performance DSP systems. Fig. 6.2 shows the connection of Virtex 5 series board to the laptop. Fig. 6.3 shows the different sections of adaptive filtering–based estimation model designed using Xilinx blockset.

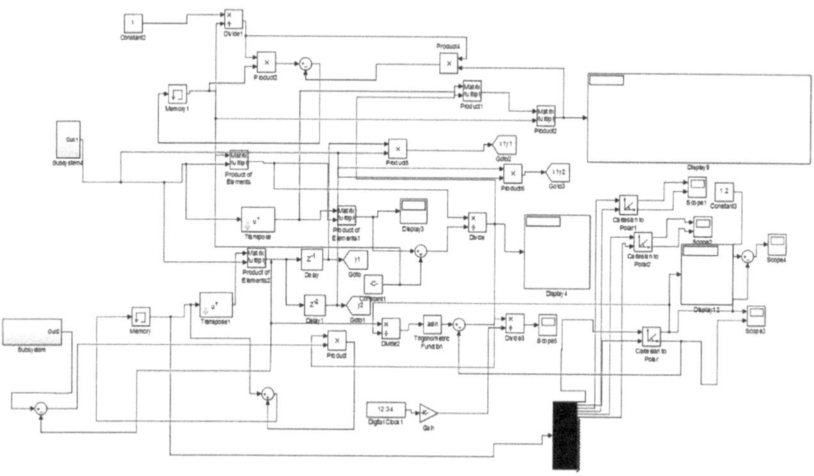

FIGURE 6.1 SIMULINK modeling of adaptive filter.

FIGURE 6.2 FPGA connection using joint test action group (JTAG)cable with the laptop.

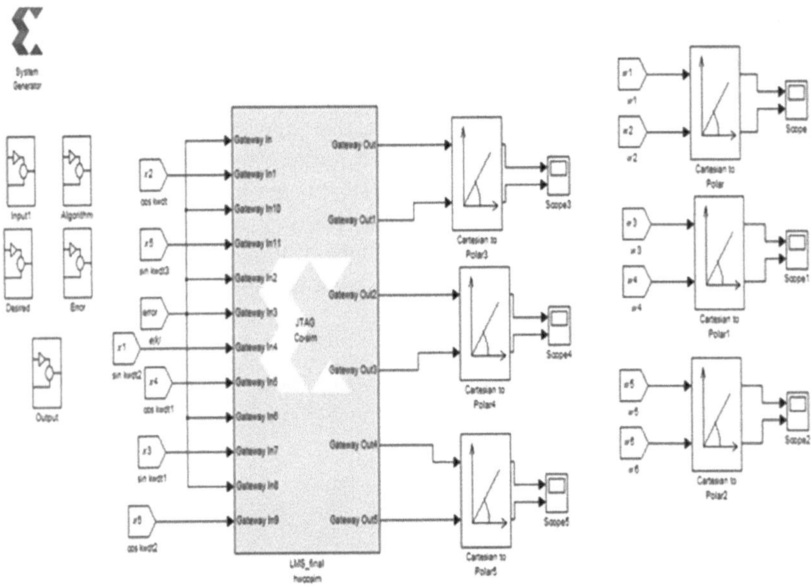

FIGURE 6.3 FPGA modeling using Xilinx blockset.

6.2.5 Simulation results and discussion

Adaptive filters play an important role in designing the PQ estimation models. To judge the tracking and estimation accuracy of different models, simulations have been carried out by using MATLAB/SIMULINK before testing the model in FPGA platform. Simulated comparison results were thoroughly analyzed to gain the knowledge about the adaptive filtering suitable for a specific type of PQ disturbance. The results presented in Fig. 6.4 show a

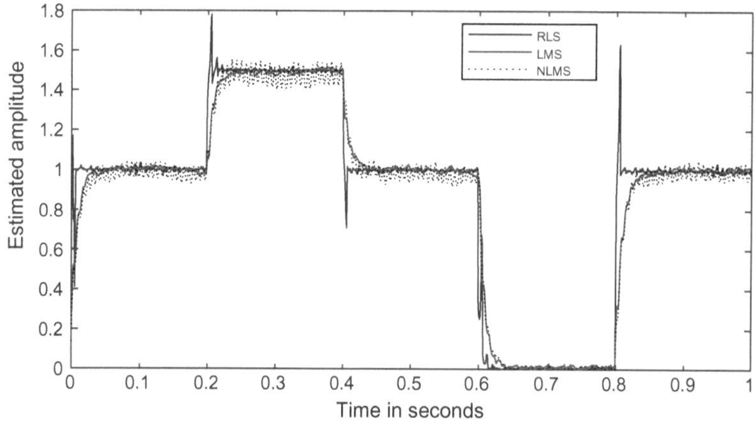

FIGURE 6.4 Estimated amplitude in the presence of swell and momentary interruptions.

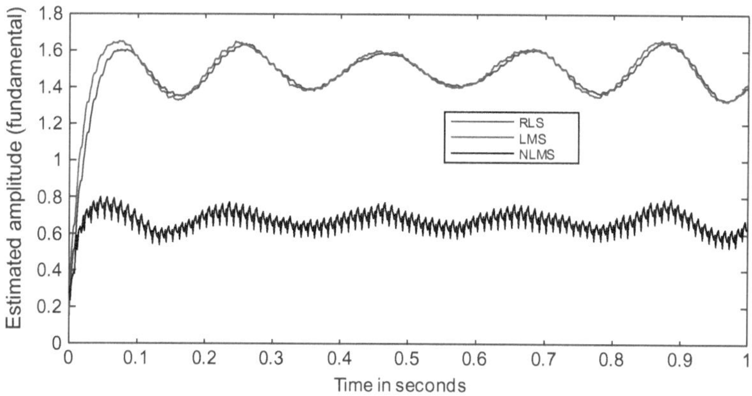

FIGURE 6.5 Estimated amplitude of fundamental harmonic component.

comparison between LMS, NLMS, and RLS algorithms in the presence of swell and momentary interruptions, which clearly indicates that RLS has better estimation accuracy than the other two algorithms. Similarly, Fig. 6.5 describes the comparison results of time-varying fundamental harmonic amplitudes obtained through LMS, NLMS, and RLS-based PQ estimation models.

6.3 Methodologies for feature extraction and classification of power quality disturbances

To extract the feature of PQ events the combination of EMD with HT has been implemented. Thereafter for classification purpose, various pattern recognition techniques, such as artificial neural network, PNN, and SVM, have been applied, which is briefly discussed later.

6.3.1 Empirical mode decomposition

The main idea of EMD technique is to identify the oscillatory modes present in the time scales defined by the interval between local extrema of the composite signal [21,22]. The steps to get IMF from a distorted signal are as follows:

Step 1. Find out local maxima and minima of the signal $S(t)$.
Step 2. Interpolate between maxima to get upper envelope.
Step 3. Interpolate between minima to get lower envelope.
Step 4. Compute the mean of the upper and lower envelope $m(t)$ by

$$m(t) = \frac{\left(e_{upper}(t) + e_{lower}(t)\right)}{2}. \tag{6.38}$$

where $e_{\text{upper}}(t)$ and $e_{\text{lower}}(t)$ is the upper and lower envelopes of the signal $S(t)$.
Step 5. Extract

$$c_1(t) = S(t) - m(t). \tag{6.39}$$

$c_1(t)$ is an IMF if it satisfies two conditions:
Condition 1: The number of local extrema of $c_1(t)$ is equal to or differ from the number of zero crossing of $c_1(t)$ by one.
Condition 2: The average of $c_1(t)$ logically be zero.

If $c_1(t)$ does not fulfill, the above two conditions then repeat the steps from 1 to 4 on $c_1(t)$ instead of $S(t)$.
Step 6. Calculate residue, $r_1(t)$:

$$r_1(t) = S(t) - c_1(t). \tag{6.40}$$

Step 7. If the value of residue, $r_1(t)$, exceeds the threshold error tolerance value then repeat steps from 1 to 7 to obtain the next IMF and new residue.

If n number of IMFs are obtained from an iterative manner, the original signal can be reconstructed as

$$S(t) = \sum_{n} c_i(t) + r(t). \tag{6.41}$$

6.3.2 Hilbert transform

An analytic signal has a real part as well as an imaginary part. Magnitude of the analytic signal gives the magnitude spectrum, and phase angle of the analytic signal gives phase spectrum. From these spectrums, features, such as standard deviation of amplitude, standard deviation of phase, and signal energy, are extracted. For a real-valued signal $a(t)$, the HT is defined by the principal value integral [23].

$$b(t) = \frac{1}{\pi} \int_{-\infty}^{+\infty} \frac{a(t')}{t - t'} dt' \tag{6.42}$$

$$c(t) = a(t) + jb(t) = d(t)\exp(j\theta(t)) \tag{6.43}$$

where $d(t)$ and $\theta(t)$ are, respectively, the amplitude and phase of analytic function whose expressions are as

$$d(t) = \sqrt{a^2(t) + b^2(t)} \tag{6.44}$$

$$\theta(t) = \arctan\left(\frac{b(t)}{a(t)}\right) \tag{6.45}$$

The instantaneous frequency is then defined by $\omega(t) = d\theta(t)/dt$. Both the instantaneous amplitude and the instantaneous frequency are the function of time which can be calculated for every IMF at every time-step.

6.3.3 Artificial neural network

In various power system applications and for PQ event classification purpose, the artificial neural network has been mostly utilized. Data clustering, classification, function approximation, and optimization are the capabilities of ANN technique [24]. The methodologies that are based on ANN have been proved efficient for resolving the problems in real time. The patterns are regularly used based on learning from examples for the classification. For each type of ANN, the learning rules are different until they are able to recognize pattern features from a set of training data, and on the basis of features, it uses to classify the new data. The capabilities of self-tuning and self-learning are the salient features of ANN. Fig. 6.6 shows architecture of ANN. The ANN is flexible, which can be used in real-time applications for the classification of PQ events [25].

6.3.4 Probabilistic neural network classifier

A PNN is a kind of feed-forward neural network, which is suitable for classification and pattern recognition problems [26]. This model is composed of two layers, that is, the radial basis layer and the competitive layer. The operations are organized into a multilayered feed-forward network with four layers, followed by input layer, hidden layer, pattern layer, and output layer, which is shown in Fig. 6.7.

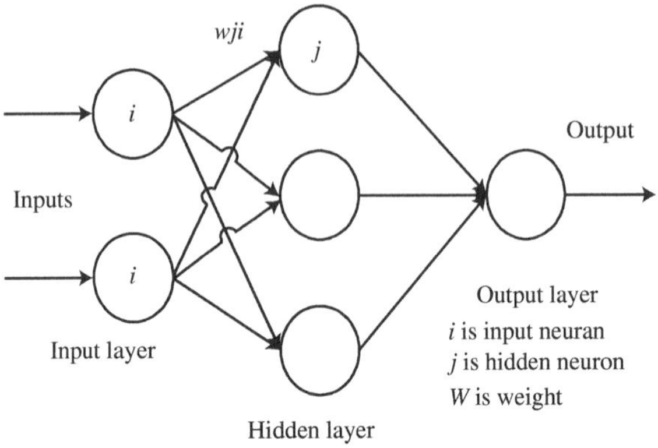

FIGURE 6.6 Feed-forward neural network algorithm structure.

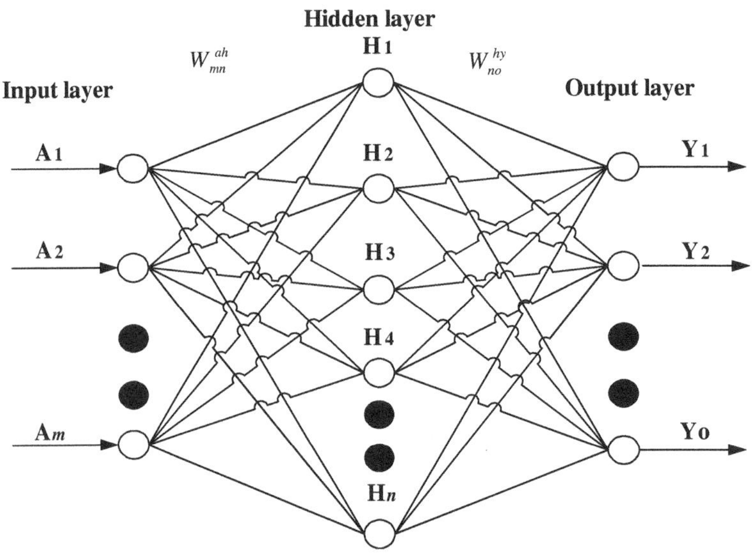

FIGURE 6.7 Architecture of PNN. *PNN*, Probabilistic neural network.

Probabilistic density function is given by

$$f_t(A) = \frac{1}{C_t}\sum_{o=1}^{C_t}\exp\left(-\frac{|A - A_{to}|}{2\lambda^2}\right) \tag{6.46}$$

Applying Eq. (6.46) to the output vector H, the hidden PNN layer becomes

$$H_n = \exp\left(\frac{-\sum_m\left(\left(A_m - W_{mn}^{ah}\right)^2\right)}{2\lambda^2}\right) \tag{6.47}$$

$$\text{net}_o = \left(\frac{1}{N_o}\right)\sum_n W_{no}^{ny} H_n \text{ and, } \text{net}_o = \max_t (\text{net}_t) \tag{6.48}$$

where

m	No. of input layers
n	No. of hidden layers
o	No. of output layers
t	No. of training examples
C	No. of classifications
Λ	Smoothing parameter
A	Input vector
$\lvert A - A_{to} \rvert$	Euclidean distance between the vectors, A and A_{to}
W_{mn}^{ah}	Connection weight between the A and Y layers
W_{no}^{hy}	Connection weight between H and O layers

6.3.5 Support vector machine

Based on the statistical learning theory, an adaptive computational powerful tool called SVM has been implemented by Vapnik for both regression and classification [27,28]. It executes a nonlinear mapping of the input vectors to a high-dimensional feature space, and to determine the generalization ability of the classifier, optimal hyperplane has been implemented. For a given set of training data belonging to different categories of the target variable, training algorithm of SVM fault classifier [29] builds a model that is represented by features in space mapped, so that the features of separate category are divided by a clear gap. Then a hyperplane is defined as the gap in which the categories are separated. To maximize the gap between the categories a radial basis function (RBF) has been implemented in this chapter as kernel parameter, which makes the hyperplane optimal. After that, the features of testing data set are mapped into the same plane that is hyperplane and is validated by the trained SVM model [30]. The main advantages of SVM are prone to overfitting, which does not converge into local minima and sparse and gives a global solution. It is very important to select proper SVM parameters so that high accuracy in the classification of PQ events and good generalization performance can be achieved. For classification purpose, support vector classifier (SVC) has been used in this chapter. For SVM parameters, library of SVM (LIBSVM) [30], and for optimal value of parameters, particle swarm optimization (PSO) technique has been implemented in this chapter. To make the hyperplane optimal, RBF is used as the kernel parameter, which further maximizes the gap between the two categories. Two additional parameters, namely, cost parameter or soft parameter (c) and gamma parameter (g), have been taken from LIBSVM. The soft parameter or cost parameter (c) gives the trade-off between forced, rigid margin, and train errors, and gamma parameter controls the shape and the radius of the hyperplane, and

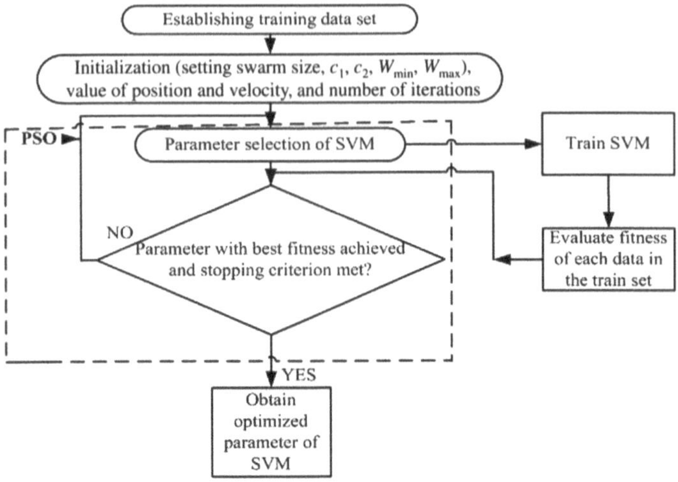

FIGURE 6.8 Flowchart of optimal parameter selection process of SVM by PSO. *PSO*, Particle swarm optimization; *SVM*, support vector machine.

also the number of support vectors is increased by increasing the gamma parameter. To select the best SVM parameter, PSO has been applied here, which is enumerated in Fig. 6.8. In Eq. (6.49), f is the fitness value that is represented mathematically for PSO and is assumed as the mean squared error (MSE) (residual mean square value), which is given as

$$f = \sqrt{\frac{1}{N}\sum_{k=1}^{N}\left[y^{\Lambda}(k) - y(k)\right]^2} \tag{6.49}$$

where $y^{\wedge}(k)$ is the output SVM predictor, $y(k)$ is the test samples, and N is the number of test samples. During training process, the optimal parameters of SVM were selected by PSO.

6.3.6 Power quality event classification

To identify the exact PQ event, it is required to extract the features of disturbed signal such as standard deviation of amplitude, standard deviation of phase, signal energy, mean amplitude, variance, and mean. Among all the extracted features, the standard deviation of amplitude and phase, and signal energy give distinct information about the events. So in this chapter, these three features are considered for the classification of PQ events. In this chapter, the following seven PQ events given are considered for analysis. These signals are generated by using MATLAB/SIMULINK environment by considering a system having two generators on both sides feeding a long transmission line with different abnormal conditions such as symmetrical fault

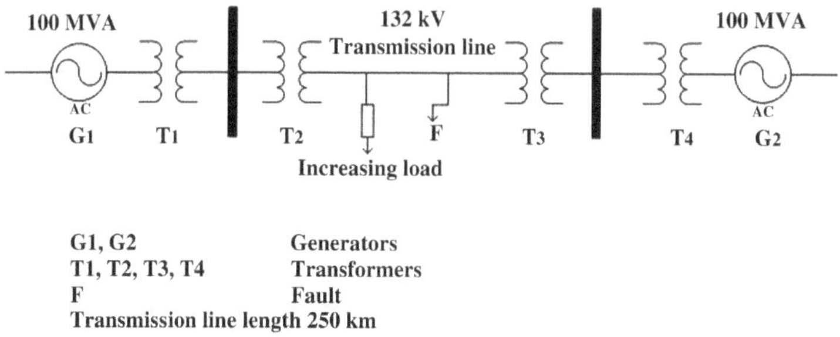

FIGURE 6.9 System model under study.

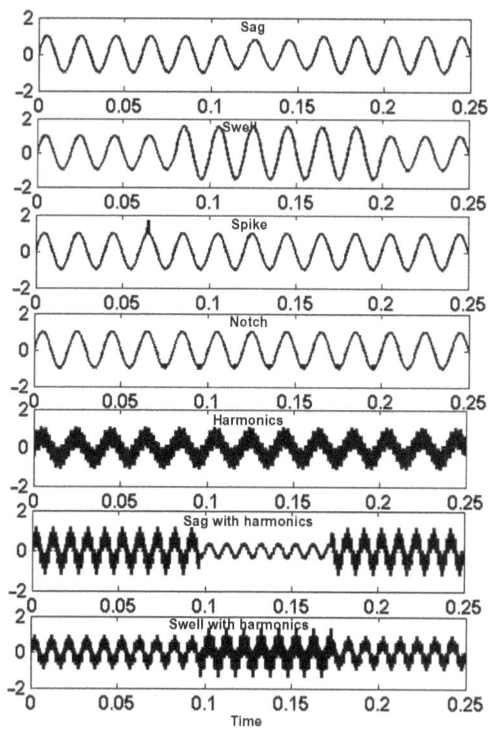

FIGURE 6.10 Different types of PQ event. *PQ*, Power quality.

and sudden loading of large load at different distances, which is shown in Fig. 6.9. One sample of each of the events is shown in Fig. 6.10.

1. Sag (S1)
2. Swell (S2)
3. Sag with harmonic (S3)
4. Swell with harmonic (S4)

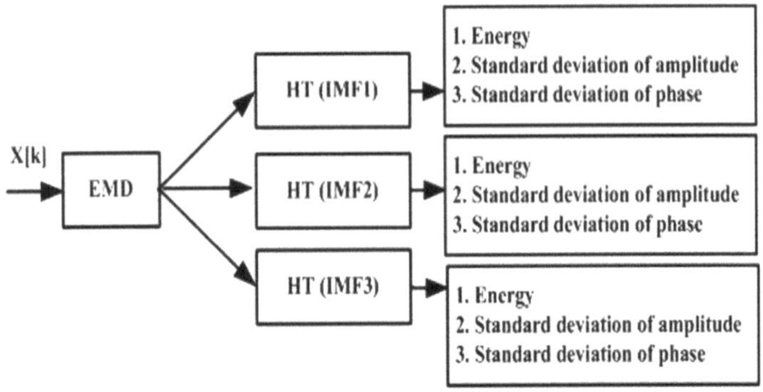

FIGURE 6.11 Flow diagram of feature extraction from disturbed waveform with EMD. *EMD*, Empirical mode decomposition.

5. Spike (S5)
6. Harmonics (S6)
7. Notch (S7)

To extract features from the seven signals mentioned previously, the steps given later are followed. The process of the flow diagram is shown in Fig. 6.11.

Step 1: EMD is applied to the PQ events to get IMFs.
Step 2: After getting IMFs, the first three IMFs are considered for the analysis, as in EMD, most of the signal energy lie in the first three IMFs.
Step 3: Apply HT to the extracted IMFs.
Step 4: Calculate standard deviation of amplitude and phase, energy from the amplitude, and phase spectrum of HT.

After the extraction of signal features, the classification of different PQ events is carried out by using ANN, PNN, and SVM for EMD–HT feature extraction method.

6.3.7 Results and discussion

The classification results of PQ events are discussed in the following section with different schemes.

6.3.7.1 Classification of power quality events by using ANN and PNN

After the extraction of signal features, the classification of different PQ events is carried out by using ANN and PNN. To do classification, for each of the events, 45 cases are considered here. For the training of neural network, 175 samples are considered, which have 25 samples from each seven PQ events (sag, swell, sag with harmonics, swell with harmonics, spike,

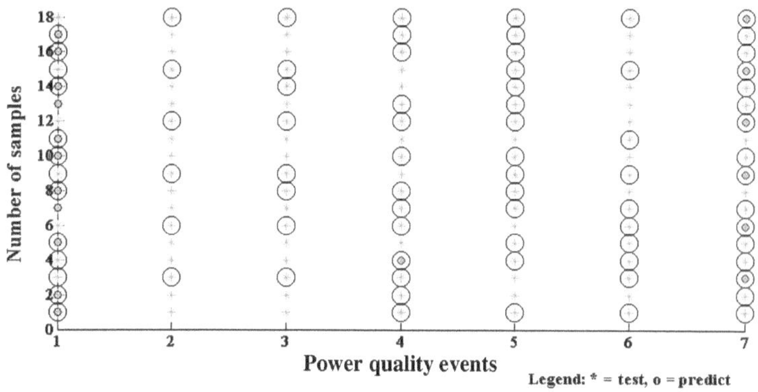

FIGURE 6.12 Plot for PQ events with EMD−HT−ANN. *EMD*, Empirical mode decomposition; *HT*, Hilbert transform; *PQ*, power quality.

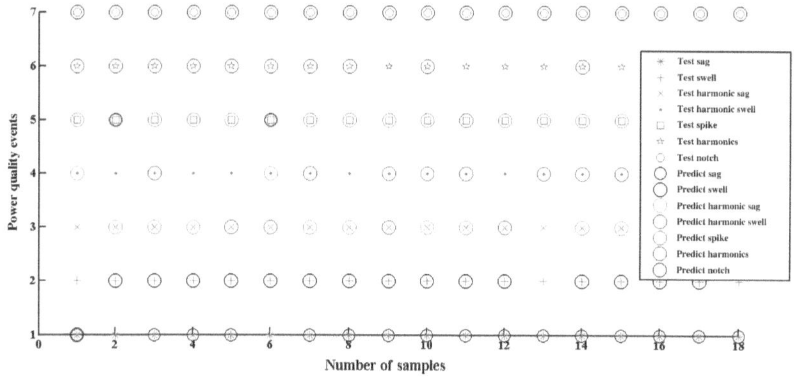

FIGURE 6.13 Plot for PQ events with EMD−HT−PNN. *EMD*, Empirical mode decomposition; *HT*, Hilbert transform; *PNN*, probabilistic neural network; *PQ*, power quality.

harmonics, and notch), and 126 samples are considered for testing, that is, 18 samples from each of PQ events. From the simulation result, classification accuracy obtained is 65.8%, that is, 83 test samples were classified correctly out of 126 test samples by using the EMD−HT−ANN scheme, and the classified samples are marked as round symbol, which is shown in Fig. 6.12. By using the EMD−HT−PNN scheme the classification accuracy obtained is 80.9%, that is, 102 test samples were classified correctly out of 126 test samples, which is shown in Fig. 6.13. A comparative study among EMD−HT−ANN and EMD−HT−PNN has been done. Tables 6.1 and 6.2 conclude that overall efficiency of EMD−HT−ANN is 65.8% and EMD−HT−PNN is 80.9%. The parameters of ANN and PNN are enumerated in the Appendix.

TABLE 6.1 Classification results for empirical mode decomposition–Hilbert transform–ANN.

Sl. no.	Power quality event	Total no. of samples	No. of samples classified correctly	Classification accuracy (%)
1	S1	18	15	83.3
2	S2	18	6	33.3
3	S3	18	8	44.4
4	S4	18	13	72.2
5	S5	18	14	77.7
6	S6	18	10	55.5
7	S7	18	17	94.4
Overall classification accuracy				*65.8*

TABLE 6.2 Classification results for empirical mode decomposition–Hilbert transform–probabilistic neural network.

Sl. no.	Power quality event	Total no. of samples	No. of samples classified correctly	Classification accuracy (%)
1	S1	18	16	88.8
2	S2	18	15	83.3
3	S3	18	15	83.3
4	S4	18	9	50
5	S5	18	18	100
6	S6	18	11	61.1
7	S7	18	18	100
Overall classification accuracy				*80.9*

6.3.7.2 Classification of power quality events using support vector machine

In this section, SVM has been used for fault classification. A detailed discussion on SVM has already been done in Section 6.3.5. In this work, LIBSVM [30] has been referred to for the parameters of SVM. Seven PQ events have

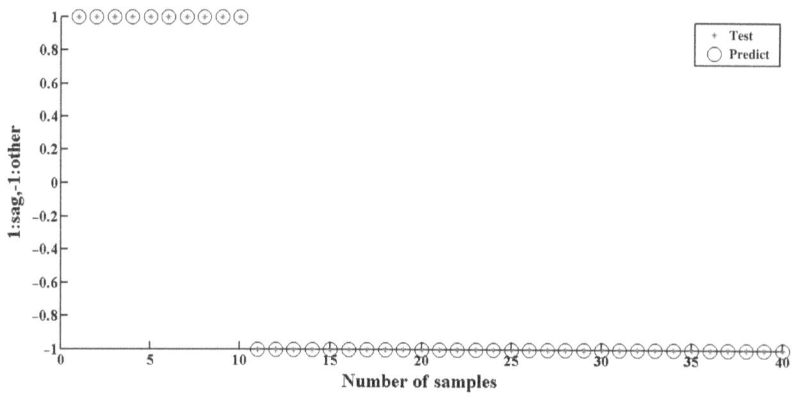

FIGURE 6.14 Sag detection plot.

been taken for classification purpose. The description of the data has already been given, and preprocessing of the data samples has been done with EMD that is described in Section 6.3.6. In order to analyze the classification results, two cases have been carried out. The first case tells about the sag detection, and in the second case, seven PQ events have been taken for classification purpose. The parameters of SVM are optimized by PSO, which is shown in Fig. 6.8 of Section 6.3.5. The parameters of PSO implemented here are given in the Appendix.

Case 1: Training samples taken are 56 (8 samples each of the 7 PQ events), and the testing samples considered are 40 (10 sag samples and 5 samples each of the rest 6 PQ events). The PQ events for the training target matrix have been assigned as 1: sag and −1: other six PQ events. SVM used is a nu-support vector classifier, and the kernel function used is RBF. The other kernel parameters of SVM used are the cost function $(c)=2$ and the gamma parameter $(g)=1$. PSO has been used to obtain the cost and gamma parameter values. The flowchart of SVM parameter optimization with PSO has already been shown in Fig. 6.8 of Section 6.3.5. Simulation result of sag detection is shown in Fig. 6.14. In Fig. 6.14, "predict" is the output of SVC, and "test" denotes the test samples.

It can be observed from Fig. 6.14 that 100% sag detection is obtained with other 6 PQ events, that is, swell, sag with harmonics, swell with harmonics, spike, harmonics, and notch. Fig. 6.15 shows the boundary plot of sag with other PQ events.

The inner circle shows the sag region, whereas the outer circle shows the other PQ event region. The inner circle denotes the number of sag samples, that is, 10, whereas the outer circle denotes 5 samples each for other 6 PQ events $(5\times 6=30)$. It can be depicted from Fig. 6.15 that 100% sag detection is achieved.

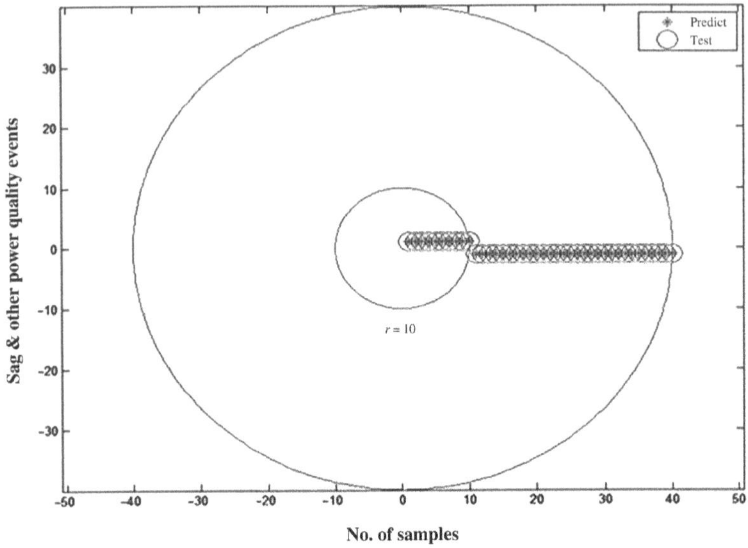

FIGURE 6.15 Boundary plot of SVC output showing sag detection.

Case 2: The number of training samples taken is 175 (25 samples each of the 7 PQ events), and the number of testing samples considered is 126 (18 samples each of the 7 PQ events). The PQ events for the training target matrix have been assigned as 1: sag, 2: swell, 3: sag with harmonics, 4: swell with harmonics, 5: spike, 6: harmonics, and 7: notch. The other kernel parameters of SVM used are the cost function $(c) = 2.5$ and the gamma parameter $(g) = 1.4$. PSO has been used to obtain the cost and gamma parameter values. The flowchart of SVM parameter optimization with PSO has already been shown in Fig. 6.8 of Section 6.3.5. From the simulation result, classification accuracy obtained is 94%, that is, 119 test samples were classified correctly out of 126 test samples. The classification result of seven PQ events is given in Table 6.3.

Simulation result of the detection of seven PQ events is shown in Fig. 6.16. In Fig. 6.16, "predict" is the output of support vector classifier (SVC), and "test" denotes the test samples.

It can be observed from the two cases discussed previously that as we take a more samples for training and testing purpose in SVC, the accuracy decreases. Also it can be noticed that sag event is detected 100% in both the cases. So it can be concluded that the hybrid technique, that is, EMD with SVC is recommended for sag and notch event detection. However, the classification accuracy of other PQ events such as swell, sag with harmonics, swell with harmonics, spikes, and harmonics also gives pretty good classification results. In order to show the superiority of EMD−SVC technique,

TABLE 6.3 Classification accuracy of seven power quality events.

Sl. no.	Power quality event	Total no. of samples	No. of samples classified correctly	Classification accuracy (%)
1	S1	18	18	100
2	S2	18	17	94.4
3	S3	18	16	88.8
4	S4	18	17	94.4
5	S5	18	17	94.4
6	S6	18	16	88.8
7	S7	18	18	100
Overall classification accuracy				*94.4*

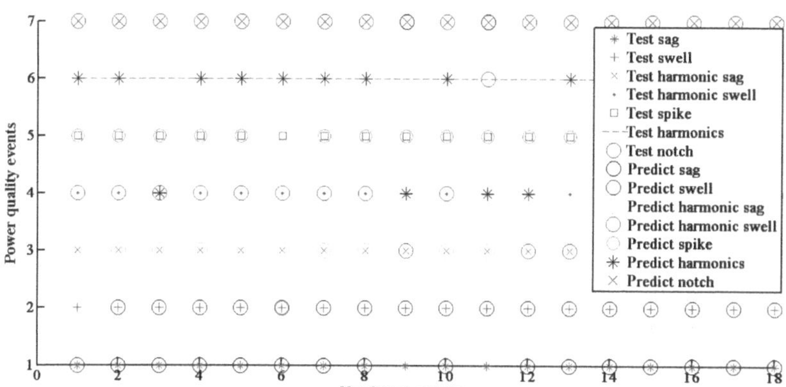

FIGURE 6.16 Power quality event classification with EMD–HT–SVC. *EMD*, Empirical mode decomposition; *HT*, Hilbert transform.

comparison with other conventional techniques such as EMD–ANN and EMD–PNN has been done, which is shown in Table 6.4.

It can be seen from Table 6.4 that EMD–HT–SVC scheme gives better classification accuracy as compared to EMD–HT–ANN and EMD–HT–PNN schemes. In order to validate the proposed scheme for the PQ classification, a comparison with other researcher's scheme is done, which is shown in Table 6.5.

It can be observed in Table 6.5 that the proposed scheme in this chapter gives a better classification accuracy of PQ events as compared to research work by others. However, further work needs to be done in future to enhance the classification accuracy.

TABLE 6.4 Comparison of classification accuracy of power quality events by different techniques.

Sl. no.	Classification accuracy (%)		
Power quality events	EMD–HT–ANN	EMD–HT–PNN	EMD–HT–SVC
S1	63	78	100
S2	60	72	94.4
S3	56	73	88.8
S4	66	76	94.4
S5	63	73	94.4
S6	59	78	88.8
S7	74	82	100
Overall efficiency	*63%*	*76%*	*94%*

EMD, Empirical mode decomposition; *HT*, Hilbert transform; *PNN*, probabilistic neural network.

TABLE 6.5 Comparison scheme.

Sl. no.	Scheme	Classification accuracy of PQ events (%)
1	[14]	93
2	[19]	90
3	[20]	89
4	[21]	90
5	[22]	93
6	*Proposed one (with EMD–HT–SVC)*	*94.4*

EMD, Empirical mode decomposition; *HT*, Hilbert transform; *PQ*, power quality.

6.3.8 Conclusion

The first part of the chapter focuses on the efficient tracking estimation of PQ disturbances by using adaptive filters, and the second part discusses a novel approach for the detection of PQ events. It was concluded from the first part of the chapter that adaptive filters play an important role in designing the PQ estimation models. To judge the tracking and estimation accuracy of different models, simulations have been carried out by using MATLAB/SIMULINK before testing the model in FPGA platform. Simulated comparison results show a comparison between LMS, NLMS, and RLS algorithms in the presence of swell and momentary interruptions, which clearly indicates that RLS has better estimation accuracy than the other two algorithms. The second part of the chapter deals with the classification of seven PQ events, that is, sag, swell, harmonics, sag with harmonics, swell with harmonics, notch, and spikes. The seven PQ event signals (sag, swell, harmonics, sag with harmonics, swell with harmonics, notch, and spikes) are generated by using MATLAB/SIMULINK environment by considering a system having two generators on both sides feeding a long transmission line under different abnormal conditions such as symmetrical fault and sudden loading of large load at different distances. It was concluded from the simulation results of the second part of this chapter that the EMD–HT–SVM technique gives better results (94.4%) as compared to EMD–HT–ANN (65.8%) and EMD–HT–PNN (80.9%) techniques.

Appendix

Parameters of ANN

TABLE A1 Details of the ANN parameters.

Network type	Feed-forward back propagation network
Training function	Levenberg–Marquardt
Size of first hidden layer	20
Size of second hidden layer	05
Train parameter goal	7×10^{-9}
Performance function	MSE
No. of epochs	1000

MSE, Mean squared error.

Parameters of probabilistic neural network

Kernel function used in PNN: RBF

Spread factor (σ) = 0.10.

Parameters of particle swarm optimization

$C_1 = 2$, $C_2 = 2$, particle size = 20, no. of iteration = 300, $W_{min} = 0.2$, $W_{max} = 0.6$. "*W refers to weight*".

References

[1] R.C. Dugan, M.F. Mc Granaghan, H.W. Beaty, Electrical Power Systems Quality, McGraw-Hill, New York, 1996.

[2] S. Khokhar, A.A.M. Zin, A.S. Mokhtar, N.A.M. Ismail, MATLAB/Simulink based modeling and simulation of power quality disturbances, in: Proceedings of IEEE Conference on Energy Conversion (CENCON), 2014, IEEE, 2014, pp. 445–450.

[3] C.-H. Huang, C.-H. Lee, K.-J. Shih, Y.-J. Wang, Frequency estimation of distorted power system signals using a robust algorithm, IEEE Trans. Power Syst. 23 (1) (2008) 41–51.

[4] S.J. Julier, J.K. Uhlmann, H.F. Durrant-Whyte, A new method for nonlinear transformation of means and covariances in filters and estimators, IEEE Trans. Autom. Control 45 (3) (2000) 477–482.

[5] P.K. Dash, R.K. Panda, G. Panda, An extended complex Kalman filter for frequency measurement of distorted signals, IEEE Trans. Instrum. Meas. 49 (4) (2000) 746–753.

[6] S. Santoso, E.J. Powers, W.M. Grady, P. Hofmann, Power quality assessment via Wavelet transform analysis, IEEE Trans. Power Deliv. 11 (2) (1996) 924–930.

[7] Y.H. Gu, M.H.J. Bollen, Time–frequency and time-scale domain analysis of voltage disturbances, IEEE Trans. Power Deliv. 15 (4) (2000) 1279–1284.

[8] P.S. Wright, Short-time Fourier transforms and Wigner–Ville distributions applied to the calibration of power frequency harmonic analyzers, IEEE Trans. Instrum. Meas. 48 (2) (1999) 475–478.

[9] P.K. Dash, B.K. Panigrahi, G. Panda, Power quality analysis using *S*-transform, IEEE Trans. Power Deliv. 18 (2) (2003) 406–411.

[10] W.C. Lee, P.K. Dash, S-transform-based intelligent system for classification of power quality disturbance signals, IEEE Trans. Ind. Electron. 50 (4) (2003) 800–805.

[11] N. Senoroy, S. Siddharth, P.F. Ribeiro, An improved Hilbert-Huang method for analysis of time-varying waveforms in power quality, IEEE Trans. Power Syst. 22 (4) (2007).

[12] Y. Huang, Y. Liu, Z. Hong, Detection and location of power quality disturbances based on mathematical morphology and Hilbert-Huang transform, in: 9th IEEE International Conference on Electronic Measurement and Instruments, Beijing, China, August 2009.

[13] M.K. Saini, R. Kapoor, Classification of power quality events—a review, Electr. Power Syst. Res. 43 (2012) 11–19.

[14] A.K. Ghosh, D.L. Lubkeman, The classification of power system disturbance waveforms using a neural network approach, IEEE Trans. Power Deliv. 10 (1) (1995) 109–115.

[15] C.-C. Liao, Enhanced RBF network for recognizing noise-riding power quality events, IEEE Trans. Instrum. Meas. 59 (6) (2010) 1550–1561.

[16] Y. Liao, J.-B. Lee, A fuzzy-expert system for classifying power quality disturbances, Electr. Power Energy Syst. 26 (2004) 199–205.

[17] B. Biswal, M.K. Biswal, P.K. Dash, S. Mishra, Power quality event characterization using support vector machine and optimization using advanced immune algorithm, Neuro Comput. 103 (2013) 75–86.

[18] H.K. Sahoo, P.K. Dash, N.P. Rath, Frequency estimation of distorted non-stationary signals using complex H∞ filter, Int. J. Electron. Commun. 66 (2012) 267–274.

[19] H.K. Sahoo, P.K. Dash, Robust estimation of power quality disturbances using unscented H∞ filter, Int. J. Electr. Power Energy Syst. 73 (2015) 438–447.

[20] H.K. Sahoo, P.K. Dash, N.P. Rath, B.N. Sahu, Harmonic estimation in a power system using hybrid H∞-adaline algorithm, in: IEEE International Conference (TENCON 2009), Singapore, 23–26 Jan. 2009, pp. 1–6.

[21] S. Shukla, S. Mishra, B. Singh, Empirical-mode decomposition with Hilbert transform for power quality assessment, IEEE Trans. Power Deliv. 24 (4) (2009) 2159–2165.

[22] S. Shukla, S. Mishra, B. Singh, Power quality event classification under noisy conditions using EMD-based de-noising techniques, IEEE Trans. Ind. Inform. 10 (2) (2014) 1044–1054.

[23] N.E. Huang, et al., The empirical mode composition and the Hilbert spectrum for nonlinear and non-stationary time series analysis, Proc. R. Soc. Lond. A 454 (1998) 903–995.

[24] K. Ghosh Atish, L. Lubkeman David, The classification of power system disturbance waveforms using a neural network approach, IEEE Trans. Power Deliv. 10 (1) (1995) 109–115.

[25] M.K. Saini, R. Kapoor, Classification of power quality events—a review, Electr. Power Syst. Res. 43 (2012) 11–19.

[26] S. Misra, C.N. Bhende, B.K. Panigrahi, Detection and classification of power quality disturbances using s-transform and probabilistic neural network, IEEE Trans. Power Deliv. 23 (1) (2008) 280–287.

[27] V. Vapnik, The Nature of Statistical Learning Theory, Springer-Verlag, New York, 1995.

[28] Khasnobish A., Bhattacharya S., Konar A., Tibarewala D.N., Nagar A.K. A two-fold classification for composite decision about localized arm movement from EEG by SVM and QDA techniques, in: IEEE International Joint Conference on Neural Networks, San Jose, CA, 2011.

[29] U.B. Parikh, B. Das, R. Maheshwari, Fault classification technique for series compensated transmission line using support vector machine, Int. J. Electr. Power Energy Syst. 32 (6) (2010) 629–636.

[30] LIBSVM—a library for support vector machines. Available from: <http://www.csie.ntu>.

Chapter 7

Role of microphasor measurement unit for decision making based on enhanced situational awareness of a modern distribution system

Soham Dutta[1], Pradip Kumar Sadhu[1], Maddikara Jaya Bharata Reddy[2] and Dusmanta Kumar Mohanta[3]

[1]Department of Electrical Engineering, Indian Institute of Technology (Indian School of Mines), Dhanbad, Jharkhand, India, [2]Department of Electrical and Electronics Engineering, National Institute of Technology, Tiruchirapalli, Tamil Nadu, India, [3]Department of Electrical and Electronics Engineering, Birla Institute of Technology, Mesra, Ranchi, Jharkhand, India

7.1 Introduction

Phasor measurement unit (PMU) has been a great success in power transmission system. It enhanced the wide-area situational awareness of the transmission system by providing accurate phasor measurements [1]. The furnished PMU data is time stamped with an accuracy of a microsecond. Today, most countries have thousands of PMU deployed across the transmission grid to obtain a wide dynamic snapshot of the transmission system.

Earlier, the traditional distribution system used to follow a radial topology. The power used to flow from high-voltage to low-voltage grid, and this flow was always unidirectional. Hence, the design considerations of the distributions system were confined to peak loads and fault current level. It was not necessary to continually observe the distribution system operation. The rapid development in the renewable energy sector contributed to the advancement of renewable distributed generation (DG) [2]. The introduction of deregulation market and net energy metering concept further motivated independent power production [3–5]. Therefore several renewable as well as nonrenewable DG started to evolve in the distribution system. Due to this, the distribution system saw a paradigm change. Instead of being passive, it became

Decision Making Applications in Modern Power Systems. DOI: https://doi.org/10.1016/B978-0-12-816445-7.00007-4

active with bidirectional power flow. This introduced several challenges, including unintentional islanding, voltage volt-ampere reactive (VAR) control, and adaptive protection [6,7], in the distribution system, which needed to be continuously observed. Multiple DG connection in a feeder introduced uncertainty, vulnerability, variability, and complexity. Hence, a more refined measurement for smooth operation of distribution grid was needed.

Motivated from the accomplishment of the PMUs in making the transmission grid completely observable, the concept of microphasor measurement unit (μPMU) was materialized to provide new, efficient, and reliable strategies for the proper management of the distribution system [8]. A μPMU is a device that provides synchronized voltage, current, and phase angles with ultraprecision [9]. The first software and hardware of the μPMU were developed in a collaborative project among Power Sensors Limited, the University of California, and Lawrence Berkeley National Lab (LBNL) in 2013. The project was entirely sponsored by the US Department of Energy ARPA-E with the aim of developing a tool for enhanced observability, understandability, and manageability of the distribution grid [10,11]. Such μPMUs were installed in several electrical utilities distribution system termed as μPMU network (μPnet) to make it more suitable for practical applications. Fig. 7.1 shows such a network. The μPMUs are installed at various locations of a distribution feeder such as substation and feeder end [10]. Each μPMU transfers data to a μPnet where it is equated against other μPMU data to reach a decision. Berkeley Tree Database (BTrDB) was created to store a huge amount of data [10]. Fig. 7.2 demonstrates the abilities of the μPMU that is approximately shown in a logarithmic scale. Since then, several other projects have started digging the potential applications of the μPMU.

7.2 Need of microphasor measurement unit in modern distribution system

Earlier, there was little monitoring of distribution system as compared to transmission system. The distribution system possessed a passive nature, that is, power flow was unidirectional (from high-voltage to low-voltage grid). Hence, monitoring of distribution system was not paid much attention. However, in modern distribution system, with its active nature, that is, bidirectional power flow, it becomes necessary for the utility to monitor the distribution system due to the following reasons [12,13]:

1. *Integration of DGs in distribution system*: Due to the advancement of technology in green energy, a large number of DGs are now being connected to the distribution system. There are several topologies to connect the DGs to the distribution systems. Some connections (mostly hydro and wind) are directly connected at bus level, that is, to the medium voltage side [14,15]. The distribution companies are responsible for these

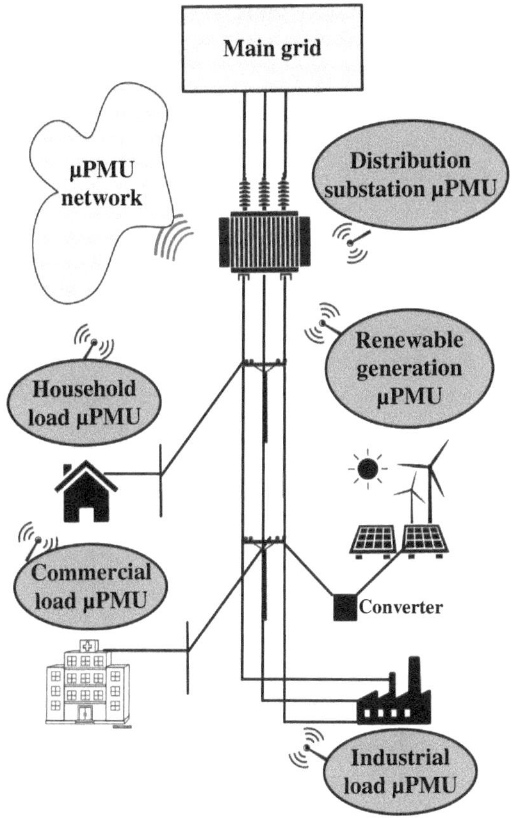

FIGURE 7.1 μPnet arrangement. *μPnet*, Microphasor measurement unit network.

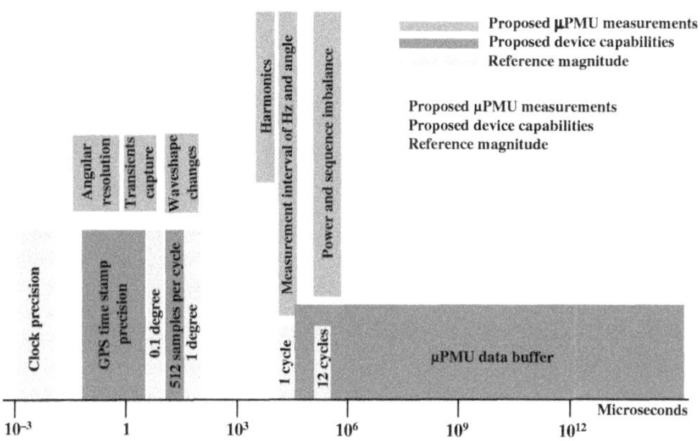

FIGURE 7.2 Abilities of μPMU in a logarithmic timescale. *μPMU*, Microphasor measurement unit.

connections. Other connections (mostly solar) are connected at the consumer level, that is, to the low voltage side. The consumers are responsible for these connections [4–5].

2. Such active sources include solar photovoltaics, fuel battery, etc. These have introduced certain challenges. Unidirectional power flow has been replaced by bidirectional power flow, which affects the traditional protection practices, voltage, and frequency control. The intermittent nature of the green energy sources also introduces harmonics and transients. All these need to be addressed by the utility through an appropriate device.
3. *High power reliability*: Due to deregulation, the utilities are striving hard to increase the power reliability. With the help of suitable monitoring devices in the distribution system, the reliability of power supply can be enhanced such as restoring power supply rapidly after a power failure.
4. *Change of loads nature*: Nowadays, most of the motor-based equipment as well as other equipment use power electronic devices. This makes the loads less affected by voltage and frequency and, in turn, injects harmonics. The power quality thus gets affected. Thus a suitable device to monitor and control these harmonics is required.
5. *Increasing geographical sprawl*: The geographical extent of modern distribution system is increasing at a rapid pace. Thus the area for measuring, monitoring, and protection of the distribution system is now not constrained to a small area. Hence, suitable devices to handle this vast extent are needed.
6. *Cost*: Due to all the above reasons, it becomes necessary to install a large number of measuring and protection devices in the distribution system. Installing such devices for individual protection purpose is uneconomical. Therefore devices that will make distribution system completely observable by installing them in a few number at an economical cost are needed.

Besides all the above reasons, the μPMUs have to face more challenges than the PMU in the transmission system due to the following inherent characteristics of the distribution system.

1. *Low X/R ratio*: The distribution system is inherently resistive. Hence, real and reactive power flow equations cannot be decoupled, and standard power flow approximate equation that is used for transmission system as in Eq. (7.1) becomes invalid, where V_a and V_b are the voltages of two points, and X is the inductance between two points. The operating states thus cannot be derived from these equations. Hence, a separate set of equations is needed to define the sensitivity of reactive and active power to θ_{ab} and V [12].

$$P_{ab} \approx \frac{|V_a||V_b|}{X} sin\theta_{ab} \tag{7.1}$$

2. *Small line length*: The distances in distribution system are smaller than transmission system. Hence, both phase angle as well as magnitude difference of electrical quantities between nodes are much smaller leading to small power flows. The phase angle difference in a distribution system is typically in the range of tenths to hundredths of a degree unlike transmission system where the difference is in the range of degrees. Moreover, the difference is much smaller than nonrandom measurement errors and measurement noise. This calls for higher precision measurement in a μPMU.
3. *Noise corruption*: The signals in a distribution system are heavily corrupted with noise. This noise occurs due to the proximity of numerous devices connected closely in the distribution system. These devices include loads, transformer, capacitor banks, and switchgear that introduce harmonics and transients. Thus the signals obtained in the distribution system need to be accurately denoised through appropriate methods.

7.3 Synchrophasor technology

Synchrophasor technology refers to calculation of phasor from measured data samples using a standard common time reference. This technology was first introduced in PMU [16]. The PMUs used it to measure voltage, current, etc. at different locations of the transmission system but with a common time reference.

To understand the underlying philosophy of phasor generation, consider an ideal sinusoidal signal as in Eq. (7.2), where ω is the frequency of the signal (in radians per second), θ is the phase angle (in rad), and X_{max} is the maximum magnitude of the signal [16].

$$x(t) = X_{max} cos(\omega t + \theta) \tag{7.2}$$

Expressing the cosine term in exponent form, Eq. (7.2) can be written as follows:

$$x(t) = \left\{Re(X_{max} e^{i(\omega t+\theta)})\right\} = Re[\{e^{i\omega t}\} X_{max} e^{i\theta}] \tag{7.3}$$

Ignoring the term $e^{i(\omega t)}$, Eq. (7.3) can be represented by a complex number X_p as in the following equation, X_p being the phasor representation:

$$X_p = \left(\frac{X_{max}}{\sqrt{2}}\right) e^{i\theta} = X_{rms}[cos\theta + isin\theta] \tag{7.4}$$

An ideal sinusoid signal along with its phasor representation is depicted in Fig. 7.3A and B, respectively [16]. The phase angle of the phasor depends on the instant at which $t = 0$. It is the angle between the axis $t = 0$ and the instant at which the signal reaches its maximum. The length of the phasor is

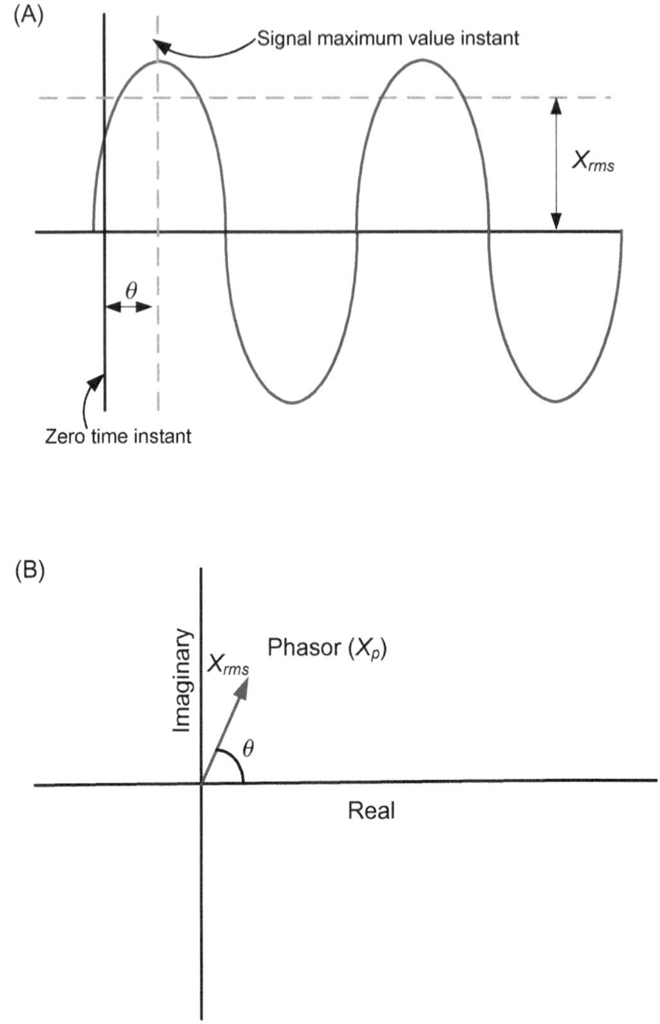

FIGURE 7.3 (A) An ideal sinusoid signal and (B) phasor representation an ideal sinusoid signal.

the *rms* value of the sinusoid signal. In power system, most of the signals are corrupted with noise and other frequencies. In such cases, only the principal frequency is considered for phasor representation. The extraction of this frequency is done with Fourier transform. For discrete data, discrete Fourier transform or fast Fourier transform is used [16].

For synchrophasor concept, consider two sinusoid signals of different magnitude and angle as shown in Fig. 7.4A. The equivalent phasor representation of the two signals is represented in Fig. 7.4B. It can be seen that at a

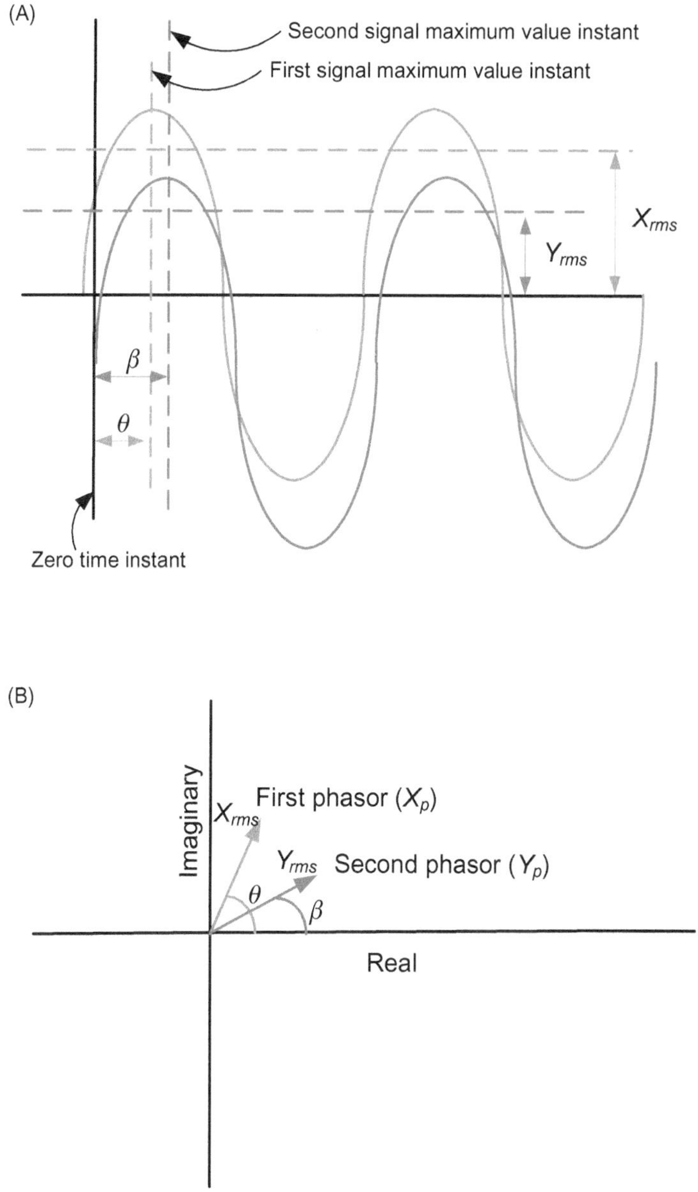

FIGURE 7.4 (A) Two sinusoids signal and (B) synchrophasor technology for two sinusoids signal.

same time reference, two phasors are obtained. Therefore the phase difference between the two signals along with magnitude information can be accurately obtained. This can be extended to numerous signals to obtain synchronized measurements. Thus the synchrophasor technology can be exploited in the operation of μPMU to obtain various measurements from different locations in a distribution system to obtain phasors based on a common time reference.

7.4 Principal components of a basic microphasor measurement unit

There are multiple types of μPMU available in the market produced by different companies. Every manufacturer has different configurations. Hence, it is not possible to show the configurations individually. However, a basic μPMU incorporating all the common and essential features can be discussed.

Fig. 7.5 shows the principal components of a basic μPMU. All the μPMUs primarily incorporate a current transformer/potential transformer (CT/PT) module, a low-pass filter module, analog-to-digital (A/D) converter module, CPU module, GPS module, communication interface module, and a power module. The analog inputs comprise the current and voltages of the measuring circuit obtained from the secondary side of the CT and voltage transformer, respectively. All the three-phase current and voltage are fed to the μPMU in the case of a three-phase line. Similarly, for two-phase line, both the phase current and voltage are used in μPMU. Generally, μPMUs are not used frequently in single phase line.

In the CT/PT module (Md1), after acquiring the current and voltage signals, shunts transformers scale down the signals in a range suitable to be fed to A/D converter. This range is generally between −10 and +10 V. Before feeding the signal to A/D converter, high-frequency noises are discarded with the low-pass filter module (Md2). The cutoff frequency of such filters are typically less than half the sampling frequency in order to maintain the

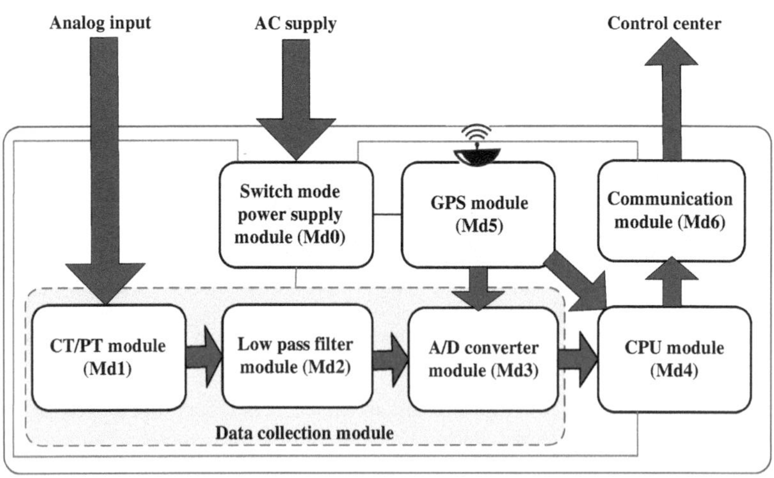

FIGURE 7.5 Principal components of a basic μPMU. *μPMU*, Microphasor measurement unit.

Nyquist criteria. The filters are designed such that all the analog signals receive same phase shift and attenuation. This helps in maintaining both constant phase angle as well as magnitude difference between various signals. The signals are then sampled at a suitable sampling frequency in the A/D (Md3) converter. The sampling frequency varies from application to application. The CT/PT module, low-pass filter module, and the A/D module constitute the data collection module of μPMU.

The CPU module incorporates a microprocessor that estimates voltage and current phasors from the signals obtained from A/D module. Depending on the application, apart from voltage and current phasors, several other parameters can be estimated. Most commonly, frequency and rate of change of frequency (ROCOF) is estimated for most of the applications. With the help of the GPS module (Md5), the parameters are time stamped with reference to universal coordinated time (UTC). The GPS system is shown in Fig. 7.6. It consists of several satellites arranged in six orbital planes having an angular difference of 60 degrees. The GPS satellites provide the GPS time (different from Greenwich Mean Time). The GPS receivers correct the GPS time and provide UTC time required by μPMU. Hence, the output of a μPMU essentially consists of time-stamped voltage and current phasor (i.e., phase angle and magnitude), frequency, and ROCOF.

The communication interface module (Md6) consists of suitable modems required to transfer the time-stamped μPMU outputs to higher level in the power system hierarchy. Communication of data is an important aspect of μPMU. Two factors are taken into account while considering the communication channel of μPMU for a particular application—channel capacity and latency. Channel capacity is the amount of data (in kilobits or megabits) that

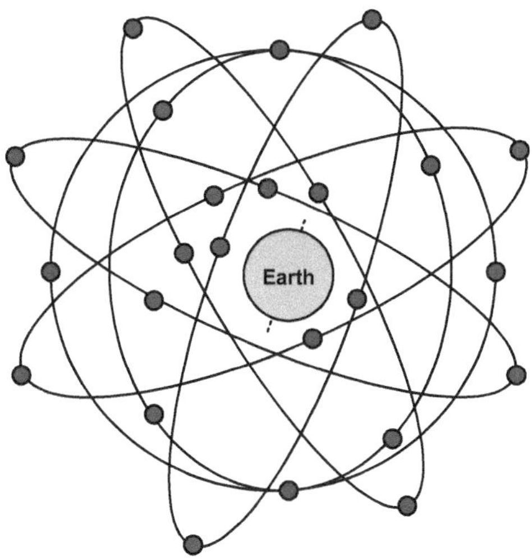

FIGURE 7.6 GPS satellite system.

a channel can sustain in 1 second. Latency is the time elapsed between time of data creation and time of data available at the receiving unit or location. Real-time μPMU applications need a low-latency communication channel. Commonly used communication channels are switched telephone circuits (high latency), power line carrier (moderate latency), and fiber optics (extremely low latency). The most advanced communication network followed nowadays is 4G and 5G. The power needed for the operation of the modules is supplied by the switch mode power supply module (Md0).

7.5 Decision application of microphasor measurement unit in modern distribution system

A plethora of decision applications of μPMU in modern distribution system is possible [12,13,17,18]. The decision application can be broadly categorized into noninstantaneous and instantaneous decisions. Noninstantaneous decisions include nonreal-time decisions such as decisions regarding planning, modeling, and equipment health, which help the utility to understand the past and present conditions of the distribution system. Instantaneous decisions include more or less real-time decisions such as topology reconfiguration, island detection, and fault detection. The important applications are explained in detail as follows:

1. *Island*: Islanding in distribution system is defined as a condition when a part of the distribution grid is electrically isolated from the main utility grid but still gets power from the DGs connected in that part [7]. Fig. 7.7 shows the island concept. Islanded microgrid areas 1, 2, 3, and 4 are formed when circuit breakers 1, 2, 3, and 4 open, respectively. Islanding may be intentional or nonintentional. While intentional islanding does not pose any problem to the isolated grid as it is planned, unintentional islanding is unplanned (caused by faults, human error, lightning, etc.) and thus poses adverse problems in the isolated grid. The significant problems of unintentional island and how μPMUs can resolve those decision problems are listed below:
    a. *Information to DG and utility about island*: When island occurs, both the DG and the utility should be made aware about it to prevent adverse cascading condition. The phase angle measured by μPMUs of the main grid and the islanded area can be compared to detect island and broadcast the islanded information, creating a wide-area situational awareness [19].
    b. *Active and reactive power balance*: An islanded grid is a miniature form of the main grid that needs for a balance between generated power and consumed power to maintain voltage and frequency to a nominal value. However, due to the renewable energy sources, it becomes difficult. While PV sources have zero inertia and react to

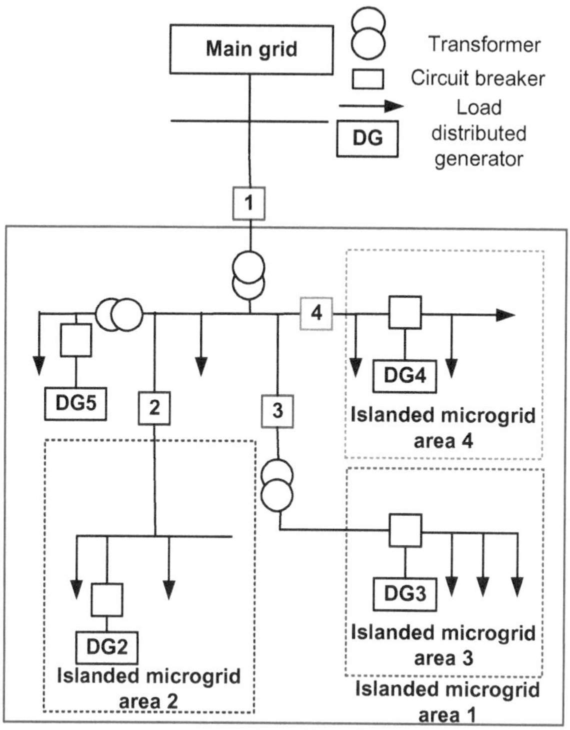

FIGURE 7.7 Island concept.

changes quickly, hydro generators have inertia and therefore react to changes slowly. Therefore the outputs of different generators need to be coordinated [20]. As μPMU measures frequency quickly and precisely, control measures such as power injections from generators or shedding of noncritical loads can be easily and accurately implemented in the islanded area to maintain power balance [19].

c. *Resynchronization*: For resynchronizing the islanded grid with the main grid at the point of common coupling, frequency and phase angle need to be matched. The frequency and phase angle information of the μPMU eliminate the need of resynchronizing devices colocated with the breakers [21].

d. *Safety*: In islanded condition the utility maintenance workers are unaware of the fact that the downed lines are still energized in the islanded area. This can prove detrimental to the utility workers. The decision of the μPMU whether island has occurred or not and its subsequent decision result to the utility can prevent such casualty [19,22].

2. *Fault identification and fault location*: For high-impedance faults in distribution system, the fault current magnitude is much smaller in magnitude and close to the magnitude of load current. Therefore the protective arrangements are unable to function properly in high-impedance faults. Moreover, the quality of the data available in the distribution system is insufficient to support any appropriate fault detection algorithm. Besides identifying the fault, knowing the location of the fault is also necessary. Common method of detection of faults involves manually examining the faulty line section that consumes a lot of manpower as well as time. It becomes a costly process for underground cables.

   With the aid of μPMU, faults can be identified accurately. From the phasors recorded by various μPMUs in the circuit, the impedance between various points can be determined through different algorithms. For example, the impedance between a μPMU and fault point can be computed with the third harmonic current and voltage phasors as per Eq. (7.5), where f and j denote the fault location and μPMU location, respectively [23].

$$Z_{jf} = \frac{V_j - V_f}{I_j} \tag{7.5}$$

   So the detection methodology does not depend only on the magnitude of current, and hence no chance of confusion. Faults can be identified almost instantaneously when the impedance of a particular line section becomes extremely high. Since impedance is calculated, the fault location can be automatically identified by taking into account the impedance between the faulted point and μPMU on its either side [24]. There are some other complex algorithms that use μPMU measurements to detect faults and its location in distribution system [25]. In general the algorithm followed for fault detection is shown in Fig. 7.8 for n number of μPMU considered.

   Thus μPMU can drastically reduce the time for fault decisions and also reduce manual labor and cost for the determination of fault location. Moreover, as the fault information is communicated to nearby substations of the faulted area, the substations become aware and prepared for post-fault effects on the distribution system. This increases the reliability of power in the distribution system. Thus the μPMU can accurately decide fault situations and increase the situational awareness during fault conditions.
3. *Reconfiguration*: Distribution network reconfiguration is a practice of modifying the topology of the distribution feeders by changing the closed/open status of tie and sectionalizing switches [26]. Fig. 7.9 shows some simple reconfiguration possibilities in a typical distribution system. The solid line shows the normal configuration of the distribution system, while the dotted line shows reconfigured network with the help of tie

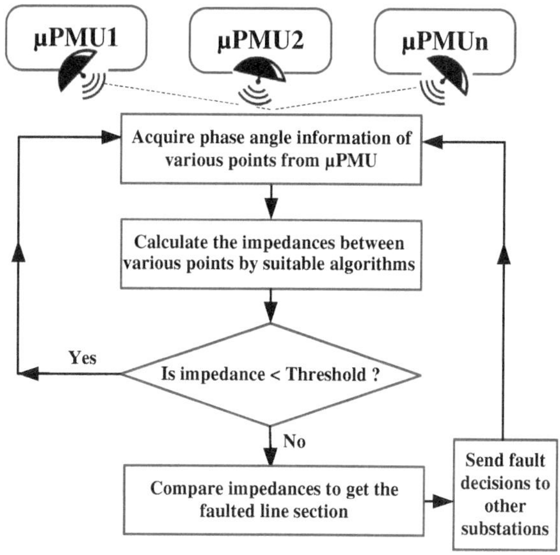

FIGURE 7.8 Flowchart of the proposed algorithm.

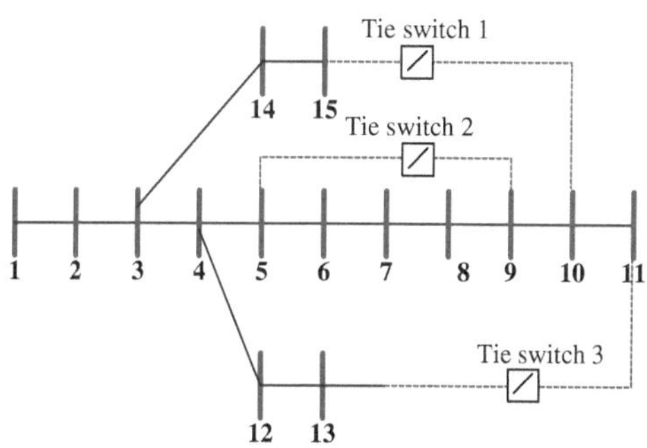

FIGURE 7.9 Reconfiguration possibilities in a typical distribution system.

switches. It is generally done to transfer loads from heavily stressed feeders to less stressed feeders to avoid overloading failures. It is also done to manage power restoration after a fault scenario. It is a multiobjective as well as complex procedure, and the algorithms should be quick enough to handle several load flow calculations making the decisions in a timely manner.

Reconfiguration by manual labor is a time-consuming task and technically infeasible. An automated reconfiguration procedure will need an array of sensors to acquire the magnitude data of voltages. Even if the magnitude data on each side of breaker is acquired by suitable data acquisition systems, there is a need for phase angle information. If the breakers are closed at identical *rms* voltages on its either side but with a large phase angle difference (rare case in a traditional distribution system when feeders are fed from the same sources but very common case in modern distribution system when feeders are fed from different sources), a large circulating current can flow leading to extreme damages. Thus complete information of data including phase as well as magnitude is required for proper reconfiguration of distribution feeders.

The μPMU can support reconfiguration processes appropriately by providing complete voltage information, especially the voltage angle. This will enable an automated system to take quick decision about reconfiguration. Such rapid decisions will also decrease the outage times, thus increasing the power reliability.

4. *Planning*: In the passive distribution system, planning was simple. However, in active distribution system, a more careful planning is required. In the modern system, power flows bidirectionally. In addition to this, modern distribution system also possesses huge automation system and complex loads. With increase in the span of distribution system, these issues are likely to get escalated. Thus the errors introduced in the planning stage may pose a great problem to the orderly operation of the distribution system. The accuracy of the planning models depends on the measurement accuracy and several other features such as load value, impedance, and topology. Out of these, load value estimation and feeder impedance are the two major areas of introduction of error. Moreover, these parameters continually change with time. Therefore they should be updated on a regular basis. Thus the inaccurate distribution modeling is due to the absence of proper observability of the system and insufficient measured data for calibrating or validating the performance of a model. This insufficient data, in turn, imposes a limit on DG connection to a particular feeder [27]. Without knowing the DG characteristics, that is, how the DG will behave during various grid connections, planners are unable to produce a safe and reliable planning.

   The μPMU with its high-speed and accurate data measurement can make strategic decisions in modern distribution system planning and operational models. It provides real impedance value unlike calculated values of the traditional approaches. This provides a new level of detail and significantly improves the accuracy of DG interconnection planning. The μPMU measurements improve model validation contributing to correct voltage profile estimation and reverse power flows. Moreover, with time as network conditions evolve, the μPMU information can empower the planners to adapt to the changes by fine-tuning their models.

5. *Cyberattack*: Cyberattacks have become prevalent nowadays in distribution power system where attackers hack the information and modify it in order to disrupt the operation of the system [28]. The μPMU can decide whether the information are hacked or not. It can be easily identified by

comparing the SCADA data consistency with μPMU data that uses a separate independent cyber-network. Cyberattacks can only go undetected if both the networks are hacked in exact coordination, which is an extremely difficult task. The μPMU can also decide cyberattack instants by identifying unanticipated operations or change in topology through its synchronized measurements.

6. *Fault-induced delayed voltage recovery (FIDVR) detection*: The modern distribution system has numerous number of induction motors. When there is low voltage due to a fault in the grid, the induction motors get stalled. Just after the fault is cleared, these stalled motors consume a large current, which does not allow the distribution voltage recover in a timely manner. Hence, the motors get tripped, and the voltage finally comes to a nominal value. This is known as FIDVR [29]. The excessive current flow in the motor can reduce the life of the motors. FIDVR causes voltage sag for several of seconds, and when the induction motors trip, it causes a sudden over voltages due to the capacitors that are used to boost up the voltage in the FIDVR condition. This sudden overvoltage affects all the consumers. Hence, FIDVR mitigation is a major issue.

Through its rapid data monitoring capability and suitable time sampling, μPMU can come to aid in detecting and diagnosing FIDVR. Although installing μPMU just for FIDVR may not justify its cost, FIDVR diagnosis can be incorporated as an additional function of the μPMU.

The μPMU requirements for the above applications in terms of resolution, accuracy, latency, and continuity are tabulated in Table 7.1 [12]. However, the applications of μPMU cannot be confined to the above applications only. As the modern distribution is continuously evolving, various applications of μPMU are emerging, which is increasing the foothold of μPMU in the distribution system. Some other applications involve control of voltage in the power system (other than the islanded system), health monitoring of various distribution system equipment, oscillation detection, power quality monitoring, etc. The extent of application of μPMU in making various distributions for situation awareness will get more justified with increase in time.

7.6 Open microphasor measurement unit data for research study

To advance the research of μPMU in distribution system application, a data set of μPMU data is provided freely by LBNL [30,31]. The μPMU has a sampling rate of 512 samples per cycle. Twelve streams of 120 Hz precise values with a GPS time accuracy up to 100 ns are being provided by each μPMU. The data consists of three-phase current and voltage magnitude along with the phase angle acquired by three μPMUs at LBNL from different buses in distribution system. The data provided is for the time frame between

TABLE 7.1 Microphasor measurement unit (μPMU) requirements for different application.

μPMU application	Resolution	Accuracy	Latency	Continuity
Island	Less than one cycle	Phase angle accuracy up to 0.01 degree	Subsecond latency	Continuous
Fault identification and location	Less than one cycle	Phasor absolute accuracy is a limiting factor	Order of 1 s	Continuous
Reconfiguration	1 s	99%	Order of 1 s	Continuous
Planning	Less than one cycle	0.0001 p.u.	No particular latency	No particular continuity
Cyberattack	Less than one cycle	99.50%	Order of 1 s	Continuous
FIDVR	Less than one cycle	99.50%	Order of 1 s	Continuous

FIDVR, Fault-induced delayed voltage recovery.

October and December for the year 2015. The data can be accessed graphically through the website https://archive.upmu.org/ with username as LBNL team and password as chocolateeclair. The data in .csv format can be downloaded from the website http://powerdata.lbl.gov/ [30,31].

The μPMU uses Ethernet and 4G LTE cellular network to transmit data to the BTrDB server, where the data are stored. Researchers can inquire about this data from the server and also plot the data in various time resolutions. It also provides features such as locating maximum, minimum, or average value within a time frame and also enables the users to identify the instants when the data exceeds or is less than a threshold values. Thus it provides the researchers a real-time view of the various events in the distribution system.

7.7 Conclusion

The PMUs have already ascertained its applicability in the stomping ground of power transmission system. With the spread of the DGs at full length in the power distribution grid and hence depleting it from its quasistationary nature, μPMUs will become obligatory for the measurement, decision process as well as protective part of the distribution system.

As voltage and current magnitude and phase angle measurements are the most important aspects for safe, reliable, and automated distribution system, μPMUs would be indispensable in modern distribution system. Initially, the cost of installing such μPMUs would seem to be costly in comparison with several standalone measurement systems in the case of a small and simple distribution system. However, with the increased complexity of the modern smart distributions system, as more functions will be integrated to these μPMUs, the implementation cost would become much economical than manned operation or other individual monitoring systems. The application of μPMUs in the distribution system is still in a native stage. Various research and projects are still going on to make these more instrumental and usable for the wider distribution area. As the μPMUs will get incorporated in the system, their potential applications will become broader, clearer, and robust. The deployment of such μPMUs will guarantee to enhance the decision-making based on enhanced situational awareness of modern distribution system.

References

[1] J. De La Ree, V. Centeno, J.S. Thorp, A.G. Phadke, Synchronized phasor measurement applications in power systems, IEEE Trans. Smart Grid 1 (1) (2010) 20–27.

[2] H. Lund, Renewable energy strategies for sustainable development, Energy 32 (6) (2007) 912–919.

[3] K.S. Parmar, S. Majhi, D.P. Kothari, LFC of an interconnected power system with multi-source power generation in deregulated power environment, Int. J. Electr. Power Energy Syst. 57 (2014) 277–286.

[4] S. Dutta, D. Ghosh, D.K. Mohanta, Optimum solar panel rating for net energy metering environment, International Conference on Electrical, Electronics, and Optimization Techniques (ICEEOT), IEEE, March 2016, pp. 2900–2904.

[5] S. Dutta, D. Ghosh, D.K. Mohanta, Location biased nature of net energy metering, 2016 International Conference on Computation of Power, Energy Information and Communication (ICCPEIC), IEEE, April 2016, pp. 350–355.

[6] Y. Liu, P. Zhang, X. Qiu, Optimal volt/var control in distribution systems, Int. J. Electr. Power Energy Syst. 24 (4) (2002) 271–276.

[7] S. Dutta, P.K. Sadhu, M.J.B. Reddy, D.K. Mohanta, Shifting of research trends in islanding detection method—a comprehensive survey, Prot. Control Mod. Power Syst. 3 (1) (2018) 1.

[8] J. Liu, J. Tang, F. Ponci, A. Monti, C. Muscas, P.A. Pegoraro, Trade-offs in PMU deployment for state estimation in active distribution grids, IEEE Trans. Smart Grid 3 (2) (2012) 915–924.

[9] B. Pinte, M. Quinlan, K. Reinhard, Low voltage micro-phasor measurement unit (μPMU), 2015 IEEE Power and Energy Conference at Illinois (PECI), IEEE, February 2015, pp. 1–4.

[10] A. von Meier, D. Culler, A. McEachen, R. Arghandeh, Micro-synchrophasors for distribution systems, in: IEEE PES Innovative Smart Grid Technologies Conference (ISGT), 2014, pp. 1–5.

[11] PSL (Ed.), PQube Specifications, PSL, 2015. Available from: <https://www.powerstandards.com/product/pqube-3e/specifications/>.

[12] A. von Meier, E. Stewart, A. McEachern, M. Andersen, L. Mehrmanesh, Precision micro-synchrophasors for distribution systems: a summary of applications, IEEE Trans. Smart Grid 8 (6) (2017) 2926–2936.

[13] J.H. Eto, E. Stewart, T. Smith, M. Buckner, H. Kirkham, F. Tuffner, et al., Scoping study on research and development priorities for distribution-system phasor measurement units, 2015. Available from: <https://prod-ng.sandia.gov/techlib-noauth/access-control.cgi/2016/163546r.pdf>.

[14] A.K. Pandey, S. Dutta, P.K. Sadhu, M.J.B. Reddy, D.K. Mohanta, Islanding detection employing net energy meter signals in a DFIG wind turbine based microgrid, 2019 International Conference on Computer, Electrical & Communication Engineering (ICCECE), IEEE, January 2019 [Accepted].

[15] S. Verma, S. Dutta, P.K. Sadhu, M.J.B. Reddy, D.K. Mohanta, Islanding detection using bi-directional energy meter in a DFIG based active distribution network, 2019 International Conference on Computer, Electrical & Communication Engineering (ICCECE), IEEE, January 2019 [Accepted].

[16] A.G. Phadke, J.S. Thorp, Synchronized Phasor Measurements and Their Applications, vol. 1, Springer, New York, 2008.

[17] E.O. Schweitzer, D. Whitehead, G. Zweigle, K.G. Ravikumar, G. Rzepka, Synchrophasor-based power system protection and control applications, Modern Electric Power Systems (MEPS), 2010 Proceedings of the International Symposium, IEEE, September 2010, pp. 1–10.

[18] M. Wache, D.C. Murray, Application of synchrophasor measurements for distribution networks, 2011 IEEE Power and Energy Society General Meeting, IEEE, July 2011, pp. 1–4.

[19] S. Dutta, P.K. Sadhu, M.J.B. Reddy, D.K. Mohanta, Smart inadvertent islanding detection employing p-type μPMU for an active distribution network, IET Gener. Transm. Distrib. 12 (20) (2018) 4615–4625.

[20] A. Eshraghi, M. Motalleb, E. Reihani, R. Ghorbani, Frequency regulation in islanded microgrid using demand response, 2017 North American Power Symposium (NAPS), IEEE, September 2017, pp. 1–6.

[21] S. Skok, K. Frlan, K. Ugarkovic, Detection and protection of distributed generation from island operation by using PMUs, Energy Procedia 141 (2017) 438–442.

[22] A. Borghetti, C.A. Nucci, M. Paolone, G. Ciappi, A. Solari, Synchronized phasors monitoring during the islanding maneuver of an active distribution network, IEEE Trans. Smart Grid 2 (1) (2011) 82–91.

[23] M. Farajollahi, A. Shahsavari, H. Mohsenian-Rad, Location identification of high impedance faults using synchronized harmonic phasors, 2017 IEEE Power & Energy Society Innovative Smart Grid Technologies Conference (ISGT), IEEE, April 2017, pp. 1–5.

[24] J. Lee, Automatic fault location on distribution networks using synchronized voltage phasor measurement units, ASME 2014 Power Conference, American Society of Mechanical Engineers, July 2014, pp. V002T14A008, 8 pages.

[25] M. Farajollahi, A. Shahsavari, E. Stewart, H. Mohsenian-Rad, Locating the source of events in power distribution systems using micro-PMU data, IEEE Trans. Power Syst. 33 (6) (2018) 6343–6354.

[26] R. Rajaram, K.S. Kumar, N. Rajasekar, Power system reconfiguration in a radial distribution network for reducing losses and to improve voltage profile using modified plant growth simulation algorithm with distributed generation (DG), Energy Rep. 1 (2015) 116–122.

[27] M.W. Davis, R. Broadwater, J. Hambrick, Modeling and Verification of Distributed Generation Equipment and Voltage Regulation Equipment for Unbalanced Distribution Power Systems, National Renewable Energy Laboratory, Golden, CO, 2007. NREL/SR-581-41885.

[28] W. Wang, Z. Lu, Cyber security in the smart grid: survey and challenges, Comput. Networks 57 (5) (2013) 1344–1371.

[29] North American Electric Reliability Corporation (NERC), A Technical Reference Paper Fault-Induced Delayed Voltage Recovery, NERC, Princeton, NJ, 2009. Available from: <http://www.nerc.com/docs/pc/tis/FIDVR_Tech_Ref%20V1-2_PC_Approved.pdf>.

[30] E. Stewart, A. Liao, C. Roberts, Open μPMU: a real world reference distribution micro-phasor measurement unit data set for research and application development. LBNL Technical Report 1006408, October 2016.

[31] S. Peisert, R. Gentz, J. Boverhof, C. McParland, S. Engle, A. Elbashandy, et al., LBNL open power data, LBNL Technical Report, May 2017, doi:10.21990/C21599.

Chapter 8

Effects of electrical infrastructures in grid with high penetration of renewable sources

Yuri R. Rodrigues[1], Antonio Carlos Zambroni de Souza[2] and Paulo Fernando Ribeiro[2]
[1]*School of Engineering, The University of British Columbia, Kelowna, BC, Canada,* [2]*Institute of Electrical System and Energy, Federal University of Itajuba, UNIFEI, Itajubá, Brazil*

Nomenclature

β_n	maximum loadability of each node
$\beta 0$	initial condition
$\Delta\lambda$	increase parameter
I_n	nodal current
$I_{n_{max}}$	maximum nodal current
J	Jacobian
m	electric vehicle position
M	total number of electric vehicles on the node
n	node number
N	total number of nodes
$P_{\text{available Node}}$	maximum load increase available on the node
P_{demand}	load demand
$P_{\text{max technical}}$	maximum loadability on the node due to technical restrictions
$P_{n_{max}}$	maximum nodal power
P_{storable}	maximum power storable on the node
$P_{\text{storable Network}}$	maximum power storable on the network
P_{supply}	power required to be supplied by flexible resources
P_G	actual generation of each flexible resource
$P_{G\,\text{available Node}}$	available nodal generation due to flexible resource

Decision Making Applications in Modern Power Systems. DOI: https://doi.org/10.1016/B978-0-12-816445-7.00008-6

$P_{G\max}$	maximum generation of each flexible resource
$P_{G\ \max_{\text{Node}}}$	maximum nodal generation due to flexible resources
$P_{G\ \max_{\text{Network}}}$	maximum network generation due to flexible resources
V_n	nodal voltage
$V_{n_{\min}}$	minimum nodal voltage
VT	tangent vector

8.1 Introduction

New social and environmental requirements for more efficient and cleaner, while still profitable, energy systems have positively driven the development of new technologies for an electric power system. Among them, distributed synchronous and renewable generators [distributed generators (DGs)], flexible loads, which can also become flexible resources, such as electric vehicles (EVs) and storage units, have led to a major transformation of the distribution systems that now have a special branch of assets with high capacity to support the network at various levels. In turn, novel and innovative planning approaches are required to make the most of these new available resources.

Traditionally, distribution system planning would only consider a single directional injection flow from the connection between the main grid and substation into the distribution network branches. However, the insertion of distributed sources requires the consideration of multidirectional flowing possibilities, which calls for the development of new tools and procedures to these systems planning, operation, and optimization [1]. Also, due to the significant impact and importance associated to DGs, several studies are available in the literature as optimal placement [2], sitting and sizing [3], operation [4,5], control [6], and stability [7]. Furthermore, the association of EVs and storage units is creating a new perspective on distribution system planning and operation. These elements may be defined as flexible loads since their demand requirements can be adjusted over a period as well as flexible resource that is able to supply power to the network through their stored energy [8]. These features enable a full range of applications and possibilities, such as demand response, peak shaving, ancillary services, and many other features, that are very important and costly for power systems operation. Furthermore, the benefits provided by these units flexible operation may offer significant deferral of investments in many levels.

In the literature, several works addressing the aspects necessary for the application of these elements are available. Ref. [6] presents solutions for the control and management of hybrid AC/DC microgrids with DG penetration. In [9,10], controlled charging processes for EVs are proposed; however, no consideration of its operation as flexible source is presented. Instead, this topic is studied in [11] for balancing wind power and load fluctuations.

Further studies include demand-side management and demand response [12], grid support [13], optimal power flow [14], integrating of flexible distribution and storages [15], as well as several control strategies for microgrids islanded operation [5,16–22]. Nevertheless, these works are mainly focused on individual requirements for each of these technologies operation, whereas these resource applications are either limited to grid support, in the cases where the microgrid is operating connected to the bulk system, or for the preservation of islanded operation. In this perspective, there is still an opening for works developing integrated applications of these resources for both connected and islanded operative scenarios, as well as identifying the effects that these interactions may imply to these resources operation and grid support.

In this perspective, this chapter proposes a smart coordinated approach considering multiple applications of these elements and how they may interact with each other to provide even better solutions and novel applications to the grid during both connected and islanded modes. These work contributions occur in many levels of distribution system planning and provide a significant capacity of investment deferral and enhancement of grid operability. First, a controlled charging process of EVs and storage units is presented to avoid possible overcurrents in transformers and transmission lines, as well as undervoltages at the nodes. Following that, a second contribution is associated to the prevention of possible outages in distribution networks capable of operating in islanded mode. For this, local controls are associated with these units being their responsibility to perform phase balancing and control actions to sustain the network operation within satisfactory limits. This provides a significant increase in quality indexes, leading to better social welfare and reduction of economic losses. Also, a third contribution is obtained from flexible loads capacity to work as generators, being this aspect employed to increase the service capacity of the available infrastructure, providing significant investment deferral in new infrastructures as local generators and transmission lines, aiding for peak shaving and mitigation of transmission congestion.

The following integrated contributions from the smart coordinated operation of renewables and flexible resources can be highlighted:

1. Coordinated operation of DGs, renewables, and flexible resources
2. Flexible resources applied to power system assistance, for example, peak shaving and mitigation of transmission congestion
3. Islanded operation with coordination between local generation and flexible resources
4. Improved planning of distribution system
5. Deferral of investments and increase in social welfare

The chapter is organized as follows: Section 8.2 presents the fundaments for a coordinated operation of local generation and flexible resources.

Section 8.3 depicts the application of flexible resources to assist distribution system operation. Section 8.4 discusses the technical challenges of islanded microgrid operation. Section 8.5 develops the proposed smart coordination methodology for the improvement of distribution system planning. Section 8.6 validates the proposed methodology with a holistic operative scenario with multiples events. Finally, Section 8.7 draws the final remarks and conclusions.

8.2 Coordinated operation of local generation and flexible resources

The coordination of local generation and flexible resources, for example, EVs and storages, can provide significant enhancement of distribution systems service capacity and operational conditions. These elements must be combined in such a way that when the network requires assistance, it is possible to use the stored energy to provide the necessary power back to the grid. However, before taking action on the use of these powerful resources, an effective controlled charging methodology is necessary to avoid the bad influences that the insertion of these elements can generate [23]. As highlighted in previous works, the operation of these units without control of their charging process can lead to a significant violation of the system operative limits possibly leading to overcurrents in transformers, transmission lines and undervoltages at nodes [9,24]. Because of this, controlled charging processes must be addressed.

These controllers will be responsible for sustaining the system operation within acceptable limits with the insertion of these new elements; such control can defer investment in new infrastructures and equipment renewal to settle these new units. In the literature, several approaches to address the controlled charging processes of EVs and energy storages are presented [9,10,24–26]. In this work a method based on the maximum loadability of each system node is employed [26]. This is accomplished with the help of the continuation method [27], being the maximum available power in the node defined as the maximum increase in demand, Eq. (8.1), which does not break the established operational and technical limits, as described in Eqs. (8.2)–(8.5).

$$\beta_n = \beta 0_n (1 + \Delta\lambda) \tag{8.1}$$

$$\beta_n \leq P_{n_{\max}} \tag{8.2}$$

$$I_n \leq I_{n_{\max}} \tag{8.3}$$

$$V_n \geq V_{n_{\min}} \tag{8.4}$$

$$\lambda < \mathrm{VT}^T \cdot J \cdot \mathrm{VT} \tag{8.5}$$

These limits can either be related to equipment and infrastructures technical limits as conductors, transformers, etc., Eqs. (8.2) and (8.3), or operative limits as undervoltage or system stability, (8.4) and (8.5).

Once determined the maximum loadability, the maximum storage capacity of each node can be determined by Eq. (8.6). The actual usable amount of the storage capacity after considering technical and operative limits is described in Eq. (8.7), while the total available storage capacity of the network is obtained by Eq. (8.8). This process must be repeated for each integration interval, since variations in demand and charging of EVs and storage elements can lead to different network storage capacities.

$$p_{\text{available}_{\text{Node}}}(n) = \beta_n(n) - p_{\text{demand}}(n) \tag{8.6}$$

$$P_{\text{storable}} = \begin{cases} P_{\substack{\text{available},\\ \text{Node}}} & \text{if} \quad P_{\substack{\text{available}\\ \text{Node}}} < P_{\substack{\text{max}\\ \text{technical}}} \\ P_{\substack{\text{max}\\ \text{technical}}}, & \text{if} \quad P_{\substack{\text{available}\\ \text{Node}}} \geq P_{\substack{\text{max}\\ \text{technical}}} \end{cases} \tag{8.7}$$

$$p_{\text{storable}_{\text{Network}}} = \sum_{n=1}^{N} p_{\text{storable}}(n) \tag{8.8}$$

With this knowledge, it is possible to efficiently coordinate the operation of renewables with flexible loads determining the amount of surplus generation possible to be stored without compromising the network operating conditions. It is worth mentioning that agreements between EV owners, charging stations, and utilities must be performed to flexibilize these units operation either as flexible load or resources. The use of EVs will depend on owners' acceptance to provide some energy to the grid in return of economic compensation; otherwise, they will remain as typical loads connected to the grid. Several devices focused on enabling these features and depicting the necessary infrastructure are in development [28,29].

8.3 Flexible resources applied to distribution network assistance

The flexible operation capacity of EVs and storage units enable their use for management actions as load shaving and reduction of transmission congestion. These actions can significantly increase the grid service capacity and defer investments in new generating units and/or transmission infrastructures. In this sense, during peak load, these flexible units are triggered into generator mode, assisting the grid with the energy stored during the moments of a surplus generation, which provides additional benefits as losses reduction once the current flow is no longer concentrated from the main grid to the distribution system branches.

The impact of this strategy will depend on how close to the operational limit are the transmission lines and which is the percent capacity of generation by the flexible resources in the face of the system peak load. In this sense, different possible scenarios must be performed during the planning stage to guarantee that these actions will be able to be fulfilled.

To efficiently accomplish these goals, some aspects must be assessed at different network levels [26] consisting of three main planes: resource, node, and network-wide. The first level is featured at the resources connection, it is typically related to the current status of the unit and its connection to the grids, including aspects, such as generation availability or unavailability, maximum generation capacity, actual state of charge (SOC), maximum energy storage capacity, electrical connection to the grid (single or multiphase), and other features related to the respective unit. This last feature is of uttermost importance since distribution system presents significant unbalance between phases. The use of these units can be an efficient solution to provide the system phase balance, especially in situations where the distribution network is operating islanded from the main grid.

The second level regards to the aggregation of flexible resources available for the contribution at the nodal level. In this level the maximum contribution previously obtained is associated to other limiting factors that take place at the nodal level such as conductors and transformers current levels. The third level is the final barrier, it considers the available net contribution obtained in the previous levels with the limits featured in a network-wide perspective, like the ones imposed by central controllers.

A flowchart depicting the main restricting aspects of flexible resources contribution is shown in Fig. 8.1.

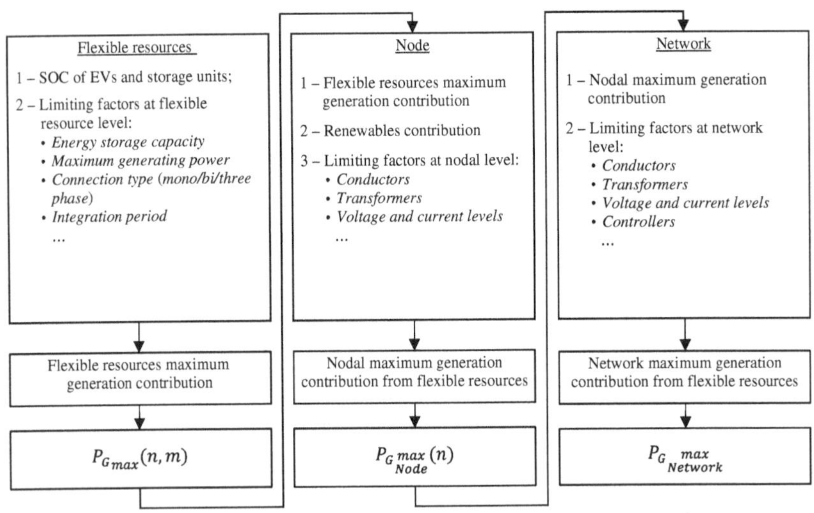

FIGURE 8.1 Flowchart main possible restricting aspects for flexible resources generating contribution.

In this sense the capacity of these units to assist in any process, be it peak shaving, congestion reduction, islanded operation supply, etc., is limited by the maximum network generation contribution, as shown in the following equation.

$$P_{\text{supply}} \leq P_{G\,\underset{\text{Network}}{\text{max}}} \tag{8.9}$$

Once defined the respective management action to be performed, the contribution of each EV and storage element must be determined. In this work a linear contribution seeking equity in the participation of each flexible resource based on the available stored energy and the total storage capacity is employed. The formulation is described in the following equations.

$$P_{G\underset{\text{Node}}{\text{available}}}(n) = \sum_{m=1}^{M(n)} P_{G\text{max}}(n,m) \tag{8.10}$$

$$P_G(n,m) = \frac{P_{\text{supply}}}{P_{G\,\underset{\text{Network}}{\text{max}}}} \cdot P_{G\,\underset{\text{Node}}{\text{max}}}(n) \cdot \frac{P_{G\text{max}}(n,m)}{P_{G\underset{\text{Node}}{\text{available}}}(n)} \tag{8.11}$$

The first term, $P_{\text{supply}}/P_{G\,\underset{\text{Network}}{\text{max}}}$, determines the percentage of the network generating capacity provided by flexible resources necessary to addresses the demand–supply requirement, be this peak shaving, islanded operation support, transmission congestion, ancillary service, or any other application. The second term, $P_{G\,\underset{\text{Node}}{\text{max}}}(n)$, associates the maximum power contribution available in each distribution system node. At last, the third parcel, $P_{G\text{max}}(n,m)/P_{G\underset{\text{Node}}{\text{available}}}(n)$, is responsible to determine the power share to be assumed by each storage element connected to the respective node. This process must be executed for all system nodes in order to define each component supplying share.

8.4 Islanded microgrids operation

The development of distribution systems into microgrids can allow these networks to operate isolated from the bulk system. In this perspective, possible outages due to failures at the main grid can be avoided if the microgrid has local generation and control capacity to sustain the system operation. This feature provides a significant increase in quality indexes, leading to better social welfare and reduction of economic losses.

For this to become technically feasible, local controls must be associated to the local DGs and flexible resources, being their responsibility to perform the control strategies that were previously held by the main grid as phase balancing, primary and secondary controls.

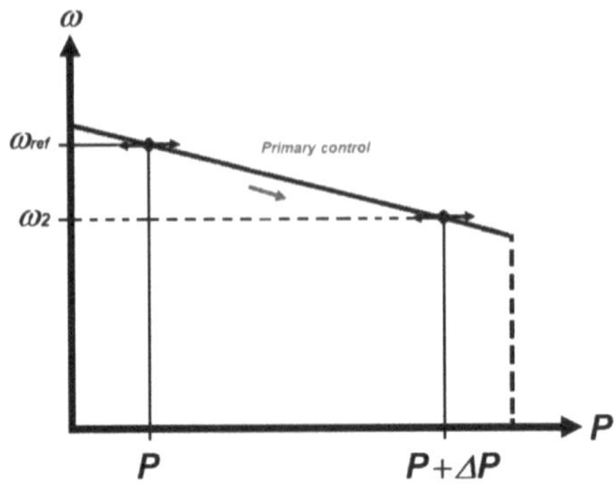

FIGURE 8.2 Primary droop control.

8.4.1 Primary control

A general practice is the application of droop control to perform primary regulation [16]. This approach is very efficient due to the naturally distributed characteristic and no requirements for communication. However, these control actions may sometimes be not sufficient to maintain the network operation within satisfactory levels, especially in scenarios where the microgrid is subject to significant demand variations.

A demonstration of primary droop control response is described in Fig. 8.2. A perturbation in the system caused by an increase in demand requires the adjustment of the generating units' contribution to attend this additional supplying requirement. The primary control adjusts the generators to meet this new operating condition. However, this adjustment is accomplished with an inversely proportional decrease in frequency related to the machines' droop coefficients [17,18].

8.4.2 Secondary control

Secondary control strategies are the ones responsible for adjusting the system operation back to the reference level after the primary control response to systems perturbation. These controls are typically associated with a small group of predefined generating units and usually require some sort of communication between the generating units [22]. Nonetheless, recent works have proposed new methodologies without communication requirements [20,21].

As described in Fig. 8.3, after the system is subjected to a disturbance, the primary control responses adjust the generators to supply the new demand requirements. However, it comes with a frequency offset. This frequency deviation is corrected by secondary control strategies, returning the system to the operation as close as possible to the reference level.

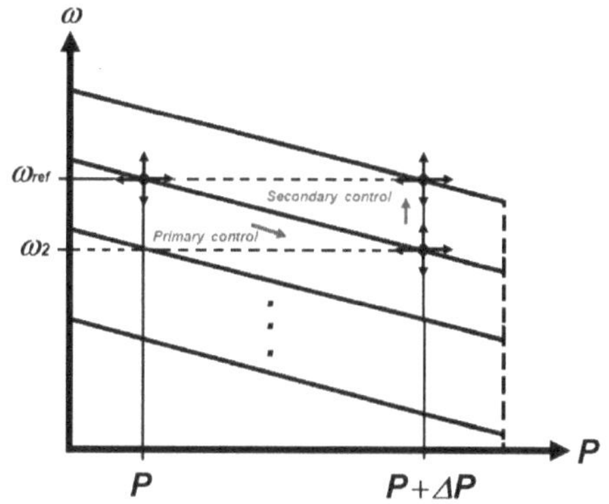

FIGURE 8.3 Secondary control.

8.5 Smart coordinated methodology

The coordinate operation of DGs, renewable resources, EVs, and storages is a great challenge to be fulfilled during the planning stage. The available resources must be addressed and different operative scenarios considering the uncertainties associated with prediction factors stressed. These actions are necessary to deliver the most broad and precise operational procedure to the systems' operators, as well as to identify whether new infrastructures or equipment renewal are required.

The proposed intelligent coordinate operation gives distribution systems the ability to take full advantage of the possibilities enabled by the new associated features. Because of this, first, a controlled charging methodology for flexible loads is performed. This ensures that the insertion of these new units will not violate the system limits. Following that, local control strategies of primary and secondary levels are associated to the generating units that can flexibilize their generation contribution, enabling the distribution network operating as an islanded microgrid capable of islanding. This includes EVs and storage units. Also, grid support features are enabled, so that the flexible resources can provide their stored energy to the grid during moments that require extra generation, performing actions as islanded supplying, phase balancing, peak shaving, and reduction of transmission congestion.

To validate the proposed integrated operation a benchmark test system with real data and different elements for distribution system analysis featuring unbalance between phases, voltage regulators, renewable sources, DGs, massive EV penetration, storage units, and islanding capacity is considered. This system is based on the IEEE 34 bus [30], being the considerations about single-phase and three-phase nodes presented in [9] adopted in this work.

The diagram of the test system is shown in Fig. 8.4. The data of the flexible loads are shown in Table 8.1 while distributed and renewable sources are depicted in Table 8.2.

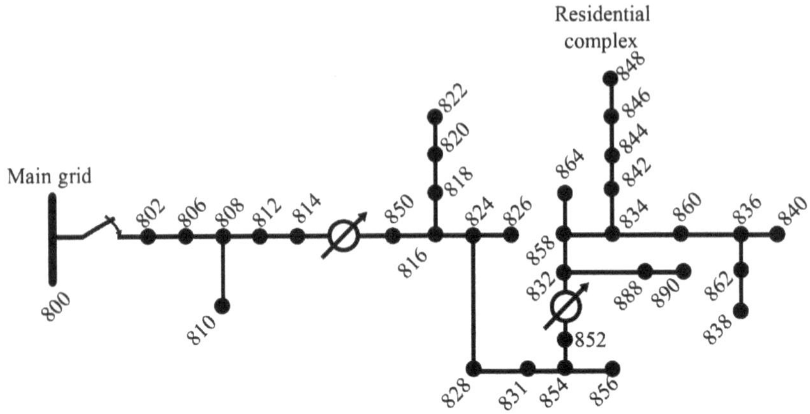

FIGURE 8.4 IEEE 34 bus [30].

TABLE 8.1 Flexible loads.

Model	Battery type	Energy (kW h)	DOD (%)
EV model 1	NiMH	27.4	80
EV model 2	Li-ion	5.33	60
EV model 3	Li-ion	16	65
EV model 4	Li-ion	24	65
EV model 5	Li-ion	16	65
Stationary battery	Lit-ion	1000	88

DOD, Depth of discharge; *EV*, electric vehicle.

To stress the possibilities enabled by the coordinated operation of renewables and flexible resources, an operative scenario considering multiple events in a day period was employed, so that the interactions and effects between the different applications can be evaluated. Also, due to the uncertainties associated with the expected penetration of EVs, several scenarios considering different levels of EV penetration have been assessed, to the point where the capacity of the flexible resources becomes insufficient to provide the necessary assistance to the network. In this sense, additional storage capacity is associated to allow systems' requirements to be met. This type of analysis is fundamental during the planning stage, so that possible future outcomes due to forecast errors associated to uncontrollable factors, as the EVs market growth, can be assessed in advance.

TABLE 8.2 Distributed and renewable generation.

Source	Generation type	Location (bus)	Nominal power (kW)	Operational power (kW)
Hydro	Controllable	800	2500	2000
Thermal	Controllable	856	1000	700
Wind	Noncontrollable	822	3 × 225	Variable
Wind	Noncontrollable	826	3 × 450	Variable
Wind	Noncontrollable	838	3 × 450	Variable
Solar	Noncontrollable	812	3 × 500	Variable
Solar	Noncontrollable	844	3 × 500	Variable

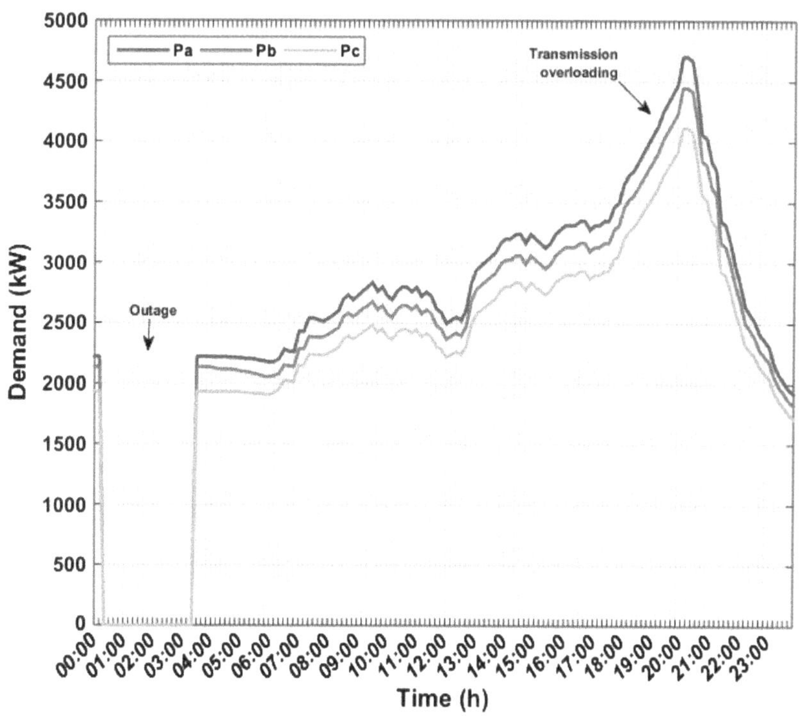

FIGURE 8.5 Distribution system demand profile.

8.6 Results

The results were obtained for a day period analysis, and two very interesting characteristics were associated with this planning study. First, a failure in the bulk system at 0020 hours led to an outage at the distribution grid that lasted until 0320 hours. The second feature is due to the predicated demand that will possibly cause transmission congestion during peak hours in the branch connecting the main grid with the distribution system substation at node 800. The transmission limit per phase at this branch is 2500 kW.

The analyzed demand and featured events are described in Fig. 8.5. The generation contribution is depicted in Fig. 8.6. As one can notice, local generation supplies a significant share of the system demand; however, due to the considerable peak load, a supplying contribution higher than the actual transmission capacity is required from the main grid. This scenario would entail in the expansion of the transmission infrastructure or new investment in local generation capacity.

A detailed three-phase representation of the system generation is shown in Fig. 8.7. It is possible to observe that the analyzed environment is an unbalanced distribution network with *phase a* as the highest loaded phase. This unbalance in the grid requires special actions to be taken. Still, while the distribution system is operating connected to the main grid, phase balance as well as primary and secondary controls are from the bulk system responsibility, being the local generation provided as a balanced contribution.

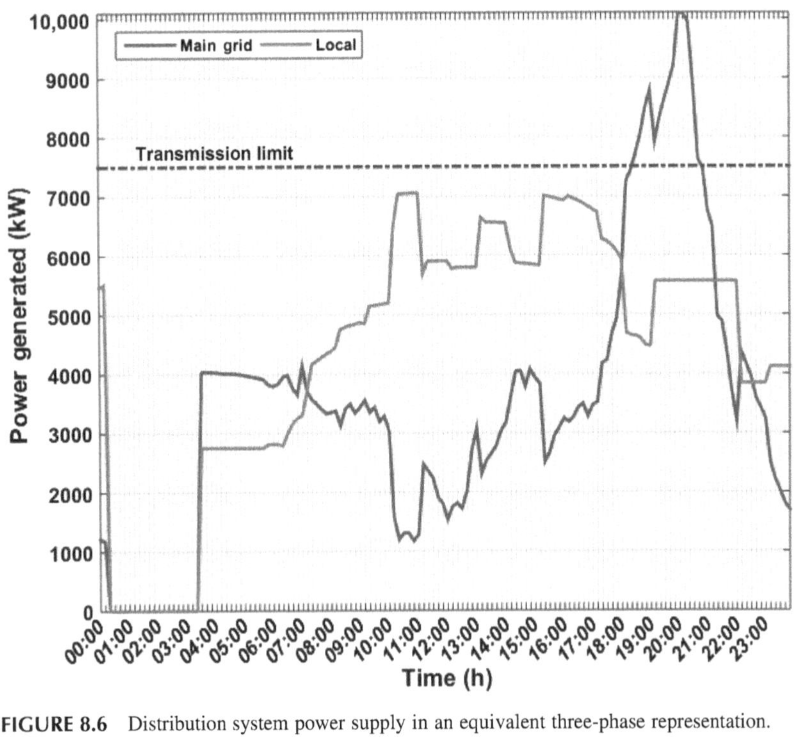

FIGURE 8.6 Distribution system power supply in an equivalent three-phase representation.

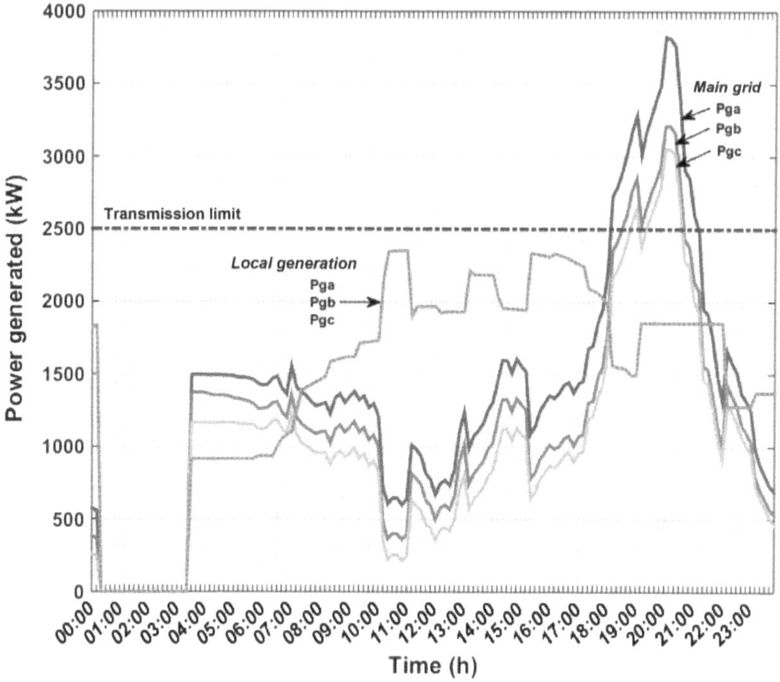

FIGURE 8.7 Distribution system power supply in a per phase representation.

Nevertheless, if the proposed smart coordinated operation is employed, the analyzed distribution system becomes an actual microgrid environment with local control capacity and ability of taking advantage of the opportunities created by the available flexible resources. In this perspective the previously observed operative reality can be drastically improved in multiples aspects, such as for deferring investment in the expansion of the transmission infrastructure or increasing the local generation capacity.

The new operative reality can be assessed through the new demand profile of the distribution system described in Fig. 8.8. The measures required to perform these actions are observable by the microgrids' supplying patterns shown in Fig. 8.9.

First, the previously observed outage during dawn can be entirely eliminated significantly improving the service quality and social welfare. This is possible since the local generation, with the help of flexible resources, is capable of supplying the microgrid demand and performs the necessary controls to sustain the microgrid operation within acceptable limits.

This perspective is stressed in Fig. 8.10 presenting the frequency variation of the microgrid during the complete analyzed period. It is possible to observe that during islanded operation, the system controls are assumed by

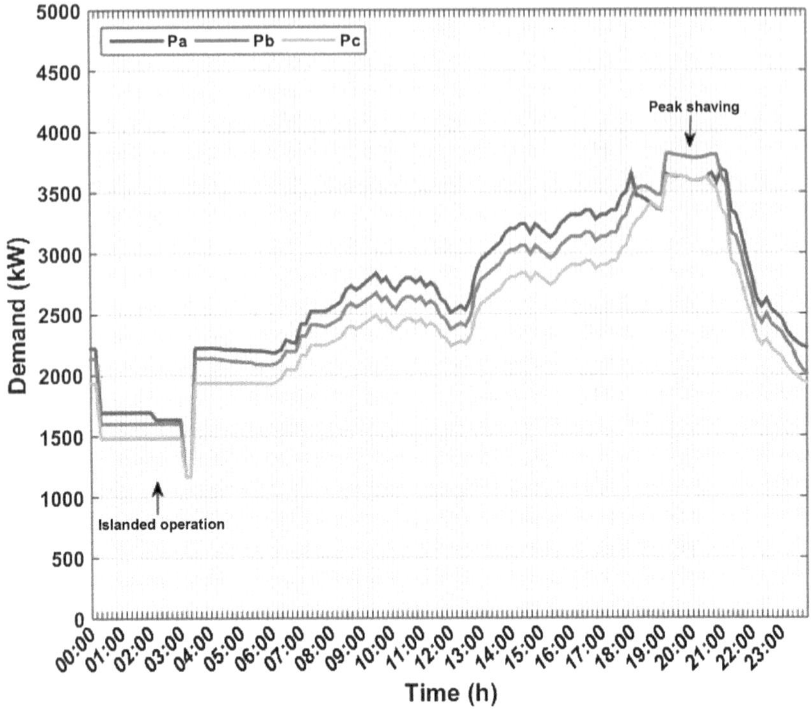

FIGURE 8.8 Microgrid demand profile.

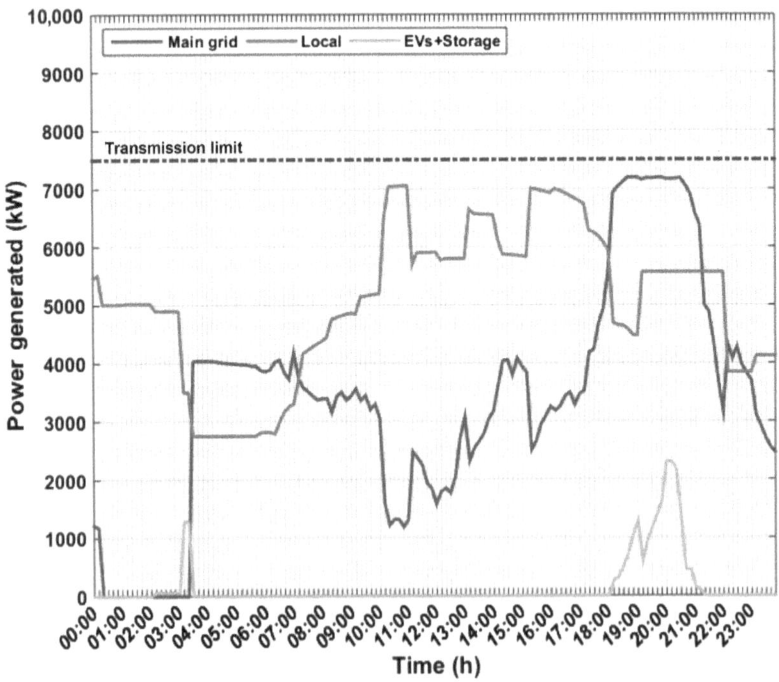

FIGURE 8.9 Microgrid power supply in an equivalent three-phase representation.

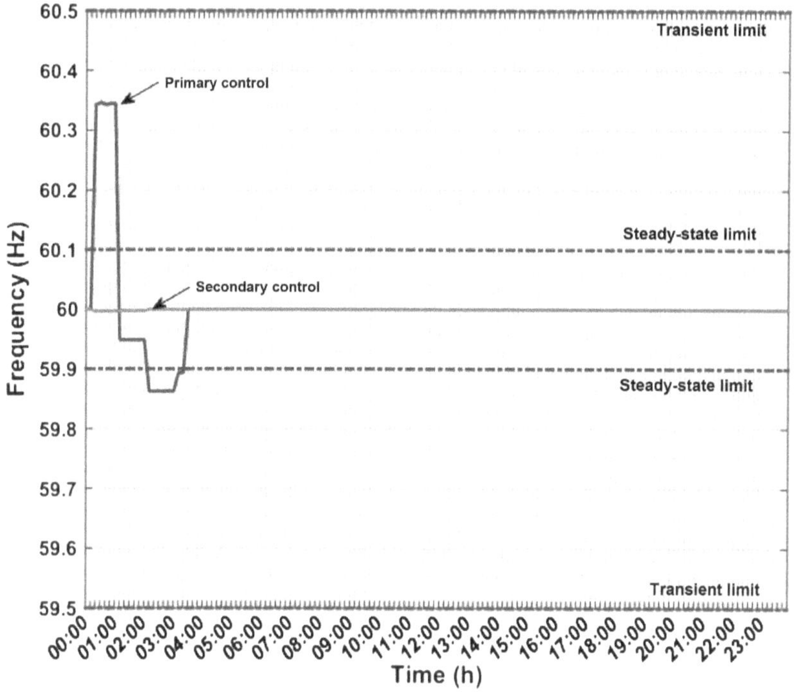

FIGURE 8.10 Microgrid frequency variation.

the microgrid. After the primary control response the distribution system starts to operate in the transient region, 60.5–60.1/59.9–59.5 Hz [31]. The operation in these levels is allowed for 30 seconds, being necessary its correction to steady-state limits, 60.1–59.9 Hz [31]. In this sense, secondary control strategies are required to correct the frequency deviation and guarantee satisfactory operating condition within limits established by the system operator. A detailed representation of the microgrid frequency variation during islanded operation is shown in Table 8.3. It should be noted that the presented values refer to the frequency levels after the transient response of the controllers.

Furthermore, the three-phase representation of microgrids' supplying is described in Fig. 8.11. It is possible to observe that when operating in islanded mode, local controllers must provide the required unbalance contribution. This process is divided into two scenarios; first, if the distributed and renewable generators are capable of providing the necessary supply to the grid demand, these units are responsible to perform primary and secondary controls, as well as phase balancing. Still, as the period of islanding evolves, distributed and renewable generations are not sufficient to supply the demand of the microgrid. In this sense, flexible resources assistance is requested to provide the remaining demand supply. During these moments,

TABLE 8.3 Microgrid frequency variation in islanded mode.

Control response	Frequency (Hz)							
Primary	60.34	60.34	60.34	59.94	59.94	59.86	59.86	59.89
Secondary	59.99	59.99	59.99	59.99	59.99	60.00	60.00	60.00
Time (h)	00:20	00:30	01:00	01:30	02:00	02:30	03:00	03:20

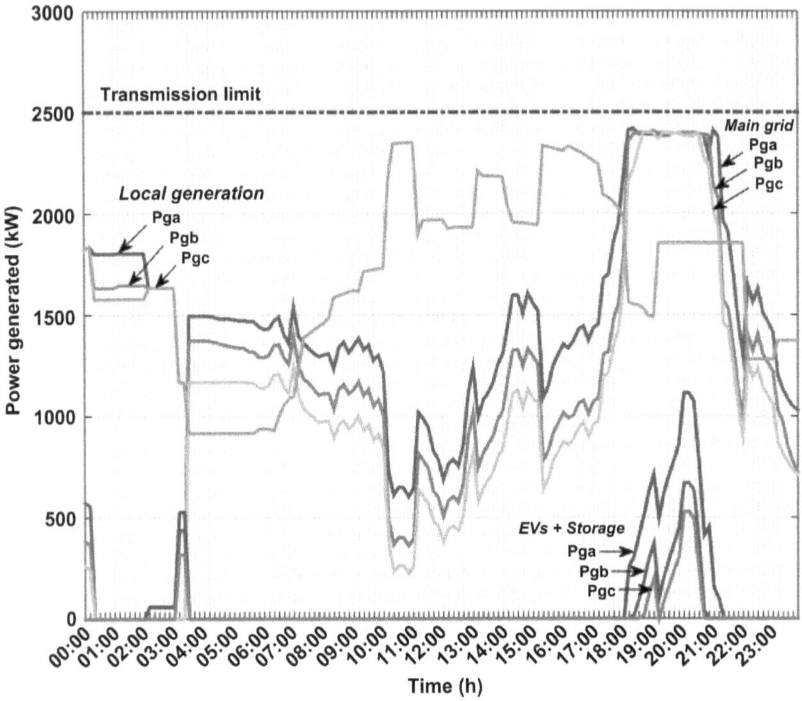

FIGURE 8.11 Microgrid power supply in a per phase representation.

flexible resources assume the phase balancing responsibility, allowing a balanced contribution by the local generation.

The second critical aspect while planning this system is due to the possible transmission congestion by the main grid supplying at peak hours. However, with the conversion of EVs and storage units from flexible loads into flexible resources, this problematic contingency can be completely overcome. These units can provide the power required by the grid based on their stored energy, being responsible to provide the surplus power that the transmission line cannot conduct due technical restrictions, plus a security margin at the discretion of the system operator. In addition, due to the distribution of EVs in different phases, their contribution can be unbalanced, allowing a balanced contribution from the bulk system, being a very effective mode of providing phase balance for the microgrid.

These features are well characterized in Fig. 8.11. This action significantly increases the microgrid service capacity and provides a substantial deferral of investments in new infrastructures.

Nevertheless, once at the planning stage, the number of EVs in operation is based on expectations which can be associated with significant uncertainties. To guarantee the broadest and as precise as possible operational procedure to the system operators identifying whether new infrastructures or equipment renewal will be necessary, several simulating scenarios considering different levels of EV penetration were performed so that the microgrid demand will be safely supplied in the future. This procedure is fundamental to provide alternative schemes allowing the accomplishment of the system requirements even if the predicted EV penetration is well below expectations.

These scenarios are depicted in Table 8.4, and the results are described in Fig. 8.12.

Fig. 8.12 also highlights the importance of controlled charging. During islanded operation, flexible loads were operated as flexible resources and assisted in the grid supply. So, after reconnection to the main grid, EVs and storage units require additional supply to recharge the stored energy provided to the grid. Still, due to the controlled charging, it is possible to notice by comparing Figs. 8.5 and 8.8 that no sharp variation in demand was added due to these units after reconnection to the main grid or after peak shaving, being the additional demand request proportional to the levels of EV

TABLE 8.4 Scenarios of electric vehicle penetration and batteries.

Case	Number of EVs				Stationary batteries banks
	One-phase node	Three-phase node	Residential complex	Total	
110%	17	50	66	1297	10
100%	15	45	60	1176	10
90%	13	40	54	1035	10
70%	10	30	42	781	10
50%	7	23	30	596	10
50% + Extra storage	7	23	30	596	25

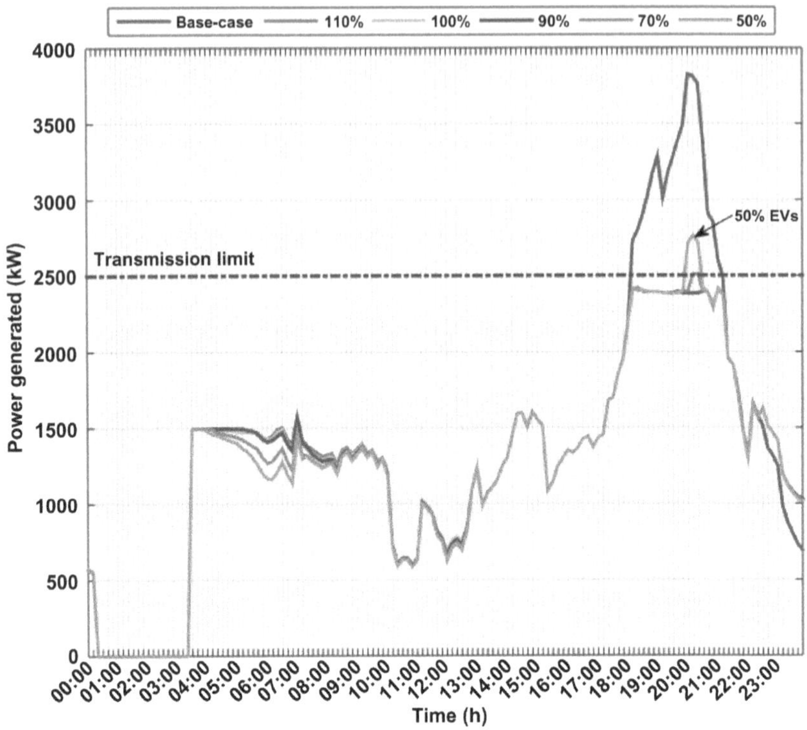

FIGURE 8.12 Main grid generation for microgrid supply—critical phase.

penetration. From Fig. 8.12, it is possible to observe that this increase in demand is proportional to the penetration of EVs after reconnection, since the demand required by these units charging does not surpass the maximum allowed limit. Yet, after performing the peak shaving, these units SOC were significantly depleted in all scenarios of EV penetration (50%−110%). And even though the different scenarios of EV penetration would lead to unlike charging power requests, the actual increase in demand faced by the grid is the same for all scenarios. This is due to the controlled charging process that limits the maximum demand associated by these units, in order to sustain the system operation within satisfactory limits. This is clear in Fig. 8.12, as the charging power required by these units starting at 2220 hours is the same regardless of the percentile of EV penetration.

From Fig. 8.12, it can be noted that the peak shaving methodology to avoid transmission congestion is satisfactorily performed up to a penetration rate of 70% from the EV base case (100%). However, for rates below this level, as observed in the critical case of 50% of EV penetration, the microgrid would not be able to avoid transmission congestion due to the main grid supply. This result is of paramount importance as it raises concern

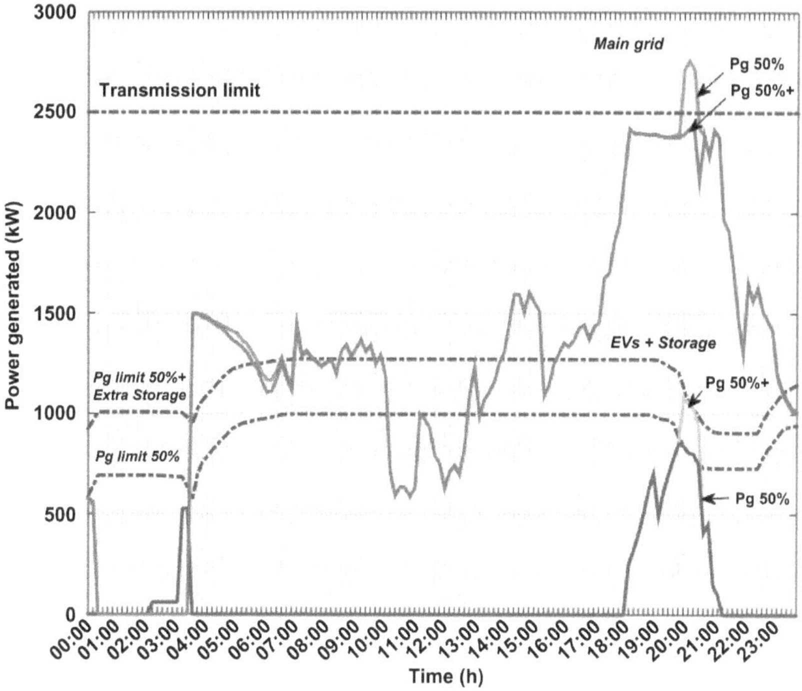

FIGURE 8.13 Analysis of extra storage influence on peak shaving—critical phase.

about the need to provide alternative schemes in cases where the EV penetration is well below expectations. In this sense an iterative process was developed to determine the necessary volume of extra storage required by the critical case, 50% of EV penetration, to ensure that system requirements are met.

The results of this analysis are shown in Fig. 8.13. The associated additional storage capacity provides the necessary increases in the power generation limit of flexible resources to avoid transmission congestion. It is worth mentioning that these limits are defined by a combination of factors as described in Fig. 8.1.

8.7 Conclusion

The coordination of distributed generation, renewables, and flexible resources enables a completely new level of applications to distribution systems. In this perspective the smart management of these elements during the planning stage, followed by a comprehensive assessment of predicted values that may become well below expectation, as is the case for EV penetration, can provide a holistic and improved distribution system planning.

As presented by the results, the proposed methodology can provide multiple benefits to the grid even with events in the same day period. Between the implemented features, the controlled charging process of flexible loads ensured the grid operation within satisfactory limits. Further, the coordination of local generation and flexible resources with local controls provided the distribution system islanding capacity, avoiding outages due to failures in the bulk system. As well, the use of flexible resources for grid assistance applications fulfilled upcoming necessities for peak shaving and mitigation of transmission lines congestion.

Taking advantage of these features gives a significant improvement of the distribution system service capacity, great deferral of investments, enhancement of quality indexes, and better social welfare. Finally, it is possible to conclude that this chapter provided a coordinated operation of DGs, renewables, and flexible resources, allowing the distribution system islanded operation and the application of flexible resources to the grid assistance. This provides a better overall planning of the distribution system due to the smart coordination and multiple applications of the available resources.

Acknowledgments

This work was partially supported by CAPES, CNPq, FAPEMIG, and INERGE. The author Yuri R. Rodrigues especially thanks CAPES Notice No. 18/2016 of the Full Doctoral Program Abroad/Process No. 88881.128399/2016-01.

References

[1] A. Keane, L.F. Ochoa, C.L.T. Borges, G.W. Ault, A.D. Alarcon-Rodriguez, R.A.F. Currie, et al., State-of-the-art techniques and challenges ahead for distributed generation planning and optimization, IEEE Trans. Power Syst. 28 (2013) 1493–1502.

[2] P.S. Georgilakis, N.D. Hatziargyriou, Optimal distributed generation placement in power distribution networks: models, methods, and future research, IEEE Trans. Power Syst. 28 (2013) 3420–3428.

[3] G. Celli, E. Ghiani, S. Mocci, F. Pilo, A multiobjective evolutionary algorithm for the sizing and siting of distributed generation, IEEE Trans. Power Syst. 20 (2005) 750–757.

[4] P.H. Nguyen, W.L. Kling, I.G. Kamphuis, P.F. Ribeiro, Integration of agent-based functions to facilitate operation of Smart Distribution Networks, Innovative Smart Grid Technologies (ISGT Europe) (2011) 1–5.

[5] J.M. Guerrero, J.C. Vasquez, J. Matas, L.G. de Vicuna, M. Castilla, Hierarchical control of droop-controlled AC and DC microgrids—a general approach toward standardization, IEEE Trans. Ind. Electron. 58 (2011) 158–172.

[6] B.N. Eghtedarpour, E. Farjah, Power control and management in a hybrid AC/DC microgrid, IEEE Trans. Smart Grid 28 (2014) 1494–1505.

[7] A.B. Almeida, E.V. De Lorenci, R.C. Leme, A.C. Zambroni De Souza, B.I.L. Lopes, K. Lo, Probabilistic voltage stability assessment considering renewable sources with the help of the PV and QV curves, IET Renew. Power Gener. 7 (2013) 521–530.

[8] M.S. Whittingham, History, evolution, and future status of energy storage, Proc. IEEE 100 (2012) 518–1534.

[9] D.Q. Oliveira, A.C. Zambroni de Souza, L.F.N. Delboni, Optimal plug-in hybrid electric vehicles recharge in distribution power systems, Electr. Power Syst. Res. 98 (2013) 77–785.

[10] C. Jin, X. Sheng, P. Ghosh, Optimized electric vehicle charging with intermittent renewable energy sources, IEEE J. Select. Top. Signal Process. 8 (2014) 1063–1072.

[11] H. Yang, H. Pan, F. Luo, J. Qiu, Y. Deng, M. Lai, et al., Operational planning of electric vehicles for balancing wind power and load fluctuations in a microgrid, IEEE Trans. Sustain. Energy 8 (2017) 592–604.

[12] I. Lampropoulos, W.L. Kling, P.F. Ribeiro, J. van den Berg, History of demand side management and classification of demand response control schemes, in: IEEE Power and Energy Society General Meeting (PES), 2013, pp. 1–5.

[13] P. Kadurek, J.F.G. Cobben, W.L. Kling, P.F. Ribeiro, Aiding power system support by means of voltage control with intelligent distribution substation, IEEE Trans. Smart Grid 5 (1) (2014) 84–91.

[14] P.H. Nguyen, W.L. Kling, P.F. Ribeiro, Smart power router: a flexible agent-based converter interface in active distribution networks, IEEE Trans. Smart Grid 2 (2011) 487–495.

[15] M.S. Illindala, H.J. Khasawneh, A.A. Renjit, Flexible distribution of energy and storage resources: integrating these resources into a microgrid, IEEE Ind. Appl. Mag. 21 (2015) 32–42.

[16] C.A. Cañizares, R. Palma-Pehnke, Trends in microgrid control, IEEE Trans. Smart Grid 5 (4) (2014) 1905–1919.

[17] Y.R. Rodrigues, M.F.Z. Souza, A.C. Zambroni de Souza, B.I.L. Lopes, Unbalanced load flow for microgrids considering droop method, in: IEEE PES General Meeting, 2016.

[18] M.R. Monteiro, Y.R. Rodrigues, J.P.O.S. Minami, A.C. Zambroni de Souza, P.F. Ribeiro, L. Wang, et al., Unbalanced frequency dependent load flow for microgrids, in: IEEE PES General Meeting, 2018.

[19] J.A.P. Lopes, S. Member, C.L. Moreira, A.G. Madureira, Defining control strategies for microgrids islanded operation, IEEE Trans. Power Syst. 21 (2) (2006) 916–924.

[20] J.M. Rey, P. Martí, M. Velasco, J. Miret, M. Castilla, Secondary switched control with no communications for islanded microgrids, IEEE Trans. Ind. Electron. 64 (2017) 8534–8545.

[21] Y.R. Rodrigues, M.R. Monteiro, A.C. Zambroni de Souza, P.F. Ribeiro, L. Wang, et al., Adaptative secondary control for energy storage in island microgrids, in: IEEE PES General Meeting, 2018.

[22] J.M. Guerrero, J.C. Vasquez, J. Matas, L.G. de Vicuna, M. Castilla, Hierarchical control of droop-controlled AC and DC microgrids—a general approach toward standardization, IEEE Trans. Ind. Electron. 58 (2011) 158–172.

[23] K. Qian, C. Zhou, M. Allan, Y. Yuan, Modeling of load demand due to EV battery charging in distribution systems, IEEE Trans. Power Syst. 26 (2011) 802–810.

[24] N. Chen, C.W. Tan, T.Q.S. Quek, Electric vehicle charging in smart grid: optimality and valley-filling algorithms, IEEE J. Select. Top. Signal Process. 8 (2014) 1073–1083.

[25] E. Sortomme, M.M. Hindi, S.D. James MacPherson, S.S. Venkata, Coordinated charging of plug-in hybrid electric vehicles to minimize distribution system losses, IEEE Trans. Smart Grid 2 (2011) 198–205.

[26] Y.R. Rodrigues, A.C. Zambroni de Souza, P.F. Ribeiro, An inclusive methodology for plug-in electrical vehicle operation with G2V and V2G in smart microgrid environments, Int. J. Electr. Power Energy Syst. 102 (2018) 312–323.

[27] F.W. Mohn, A.C.Z. de Souza, Tracing PV and QV curves with the help of a CRIC continuation method, IEEE Trans. Power Syst. 21 (2006) 1115–1122.

[28] Gadh R., Chattophadhyay A., Chung C.-Y., Chu P., Prabhu B., Sheikh O., et al., Intelligent Electric Vehicle Charging System, WO2013019989 (A2), 2013.

[29] Miranda, P.E.V., Carreira E.S., Smart energy management systems for electric and hybrid electric vehicles with bidirectional connection, smart energy management system for an energy generator, method for managing energy in a smart energy management system and method for controlling the operation of an energy generator, US2017072804 (A1), 2017.

[30] IEEE PES Distribution System Analysis Subcommittee's Distribution Test Feeder Working Group, 2010. Available from: <http://www.ewh.ieee.org/soc/pes/dsacom/testfeeders/index.html>.

[31] National Agency of Electric Energy – ANEEL Procedures of Distribution of Electric Energy in the National Electrical System – PRODIST Module 8 – Energy Quality, National Electric Energy Agency (ANEEL) Std., Jan. 2017.

Chapter 9

Distributed generation in deregulated energy markets and probabilistic hosting capacity decision-making challenges

Sherif M. Ismael[1], Shady H.E. Abdel Aleem[2], Almoataz Y. Abdelaziz[3] and Ahmed F. Zobaa[4]

[1]Electrical Engineering Division, Engineering for the Petroleum and Process Industries (ENPPI), Cairo, Egypt, [2]Mathematical, Physical and Engineering Sciences Department, 15th of May Higher Institute of Engineering, Cairo, Egypt, [3]Electric Power and Machines Department, Faculty of Engineering, Ain Shams University, Cairo, Egypt, [4]College of Engineering, Design and Physical Sciences, Brunel University London, Uxbridge, United Kingdom

9.1 Introduction

Currently, the collaboration of renewable energy–based distributed generation (DG) with conventional fossil fuel–based power-generation units has led to the development of new practices in energy production associated with various challenges. However, the proposed integration of RES units with their power electronic interfaces (predominantly wind and solar energy systems) into electrical networks and deregulated energy markets has brought its own problems and restrictions [1–3]. For example, if these DG units are not properly allocated, various operational problems in the distribution systems may occur, such as overvoltage risks, overloading of electrical equipment, increased line losses, protection system maloperations, and power quality problems [4]. These problems arise when the DG penetration levels surpass the maximum allowable penetration level, the so-called system's hosting capacity (HC) limit. The evaluation of the HC enables the investor to quantify the impact of his DG units on the performance of the power system by using a set of performance indices (PIs). The selection of these indices hinges on many aspects, such as voltage profile of the system, thermal capability of the feeders, and harmonic distortion levels. Today's power systems are becoming more complex, expandable, dynamic, and nonpredictable;

Decision Making Applications in Modern Power Systems. DOI: https://doi.org/10.1016/B978-0-12-816445-7.00009-8

therefore accurate to-date HC analysis studies are necessary to face the significant challenges resulting from the massive integration of DG into modern electrical power systems. In deregulated energy markets a conflict of interest took place between the DG investors and network operators, as the DG investors aim to install more DG units, while distribution system operators (DSOs) are worried about these pushy DG integration requests and the relevant adverse consequences on the performance of their existing electrical networks. Therefore the HC plays a vital role as a fair and transparent solution that assists decision makers in the assessment of RES integration requests and to decide which integration request should be accepted or rejected [1]. Accordingly, an appropriate HC enhancement technique (case-dependent) should be investigated to allow more DG integration while complying with the system requirements to ensure safe and sustainable operation, as shown in Fig. 9.1.

In any electrical network the network operator is obliged to take frequent decisions to ensure safe and reliable operation of his network. Accordingly, sufficient data are required to be available that, in turn, allows for taking the correct decision in a time-effective framework. Practically, uncertainty, which is simply defined as the absence of confirmed data, exists in many electrical fields. The uncertainty management in electrical power systems has been one of the main concerns of the electrical decision makers, such as DSO and DG investors [5]. Uncertainty in the HC calculations took place due to numerous issues, such as uncertain DG locations and size, the variable nature of the output powers from renewable energy resources, such as wind and photovoltaic (PV) resources,

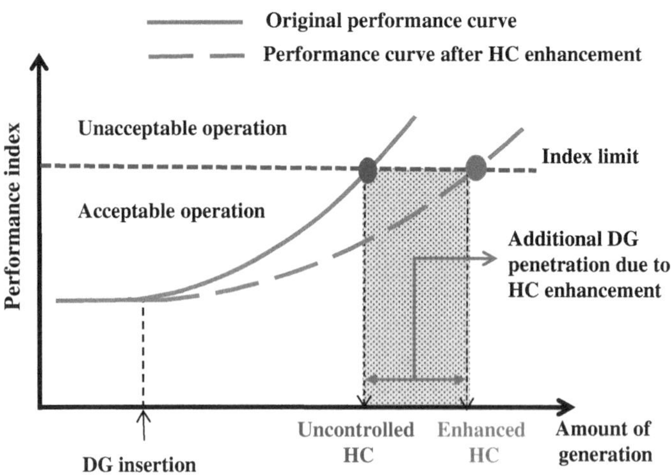

FIGURE 9.1 The concept of HC assessment and benefit from its enhancement. *HC*, Hosting capacity.

and fluctuations of load profiles along the day. Accordingly, the HC calculations should not be a deterministic problem with no randomness involved. However, it should be solved as a probabilistic problem, where uncertainty levels are considered.

To manage these various uncertainties, decision-making techniques have been examined in the presence of uncertainties in many works. Soroudi et al. [5] presented a comprehensive overview of the various uncertainty handling approaches in energy systems. The authors categorized the uncertainty of electrical parameters into two sections as follows:

1. uncertainty in technical parameters, such as load changes, DG output fluctuations, and generation outages and
2. uncertainty in economic parameters, such as the variations in inflation rates, unemployment rate, and gross domestic product.

Afterward, a comprehensive review of the various decision-making techniques was presented and mainly categorized into six categories as follows:

1. Probabilistic approach: where the input parameters of the problem under study are random data with an identified probability distribution function. The commonly known probabilistic approaches or uncertainty handling are Monte Carlo simulation (MCS), point estimate method, and scenario-based approach.
2. Possibilistic approach: depends on the fuzzy sets where the input parameters are represented using a membership function.
3. Information gap decision theory: where the input parameters are grouped into two groups to represent the known parameters and the parameters that are essential to be known.
4. Hybrid possibilistic–probabilistic: where the input parameters are a mix of both possibilistic and probabilistic approaches.
5. Robust optimization: in this approach, the achieved decisions are taken based on solving an optimization problem considering the worst-case scenario of a given uncertain data set.
6. Interval analysis: where the input parameters are assumed to be taken from a known interval.

All the above-listed decision-making techniques were proposed to assist the decision maker in evaluating the consequences of the different aspects of his problem in the presence of uncertain input parameters.

The multicriteria decision-making (MCDM) techniques in the energy planning sector were overviewed by Pohekar et al. [6]; it was concluded that the analytical hierarchy process is the most common and efficient technique for handling multicriteria problems. Wimmler et al. [7] overviewed the various MCDM techniques applied to the selection of the optimal renewable energy resources and storage technologies in the islands.

TABLE 9.1 A comprehensive survey of the decision-making techniques and its applications in the hosting capacity (HC) studies.

Reference	Year	Decision-making technique	Test system	Main findings
[9]	2006	MCS	IEEE 30-bus	Various uncertainties were considered such as DG location and penetration level
[10]	2011	MCS	• 9-bus • 547-bus real system	• Multiobjective DG planning is performed using the NSGA-II technique • Two objectives were investigated, namely, reducing the total cost and minimizing the operational threats that may arise due to violating the predefined constraints
[11]	2010	PEM	Modified IEEE 13-bus system	• A probabilistic load flow was used to handle the uncertainties, such as daily load variations, DG unit fluctuating output, and operational variations in the network configuration • It was concluded using the PEM method led to competitive results compared to other techniques. In addition, the confidence in the estimation of the node voltages has been improved, thus allowing a more confident decision to be taken by the network operators
[12]	2011	MCS	• IEEE 37-bus	• The authors considered various uncertain parameters, such as the stochastic power output of electric vehicles, fluctuating DG output power and integration location, varying fuel prices, and uncertain load growth
[13]	2016	Scenario-based approach	• 69-bus • 152-bus	• A probabilistic multiobjective optimization framework was presented to maximize the HC while complying with the preset technical and economic constraints • Multiobjective DG planning is performed using NSGA-II technique

				• The obtained results showed the effectiveness of the proposed stochastic in handling the various DG uncertainties using a set of predefined scenarios
[14]	2011	Scenario-based approach	• 33-bus	• The uncertainty management in active distribution networks is handled using a scenario-based approach • The authors concluded that considering the uncertainties during the optimization problem lead to confident and realistic solutions
[15]	2011	Possibilistic	• 9-bus • 547-bus	• A fuzzy tool is proposed to analyze the effect of investment and operation of DG units on active losses and network performance under load fluctuations • It was concluded that the proposed model is effective as an analytical tool for network operators that assists them in the decision-making process for network reinforcement or reconfiguration in the presence of DG units and load uncertainties
[16]	2017	Probabilistic	• 8-feeders network	• A three-step stochastic HC evaluation approach is presented
[17]	2017	Probabilistic	• 11-bus • 45-bus	• A risk assessment tool is proposed to calculate the HC of active distribution networks considering several DG uncertainties and lack of confirmed data in the modeling of system problem
[18]	2011	Hybrid probabilistic–possibilistic	• 547-bus	• A hybrid possibilistic–probabilistic was proposed as a tool for analyzing the effect of uncertainty in DG output power on active losses of the networks

DG, Distributed generation; *HC*, hosting capacity; *MCS*, Monte Carlo simulation; *NSGA*, nondominated sorting agent; *PEM*, point estimate method.

In this chapter the HC is presented as a decision-making tool that provides a fair and clear approach to decide which integration request should be accepted or rejected. Various HC PIs, such as overvoltage, thermal overloading, power quality, and protection problems, are explored, and their impacts on deciding the optimum HC enhancement technique assessment are discussed. Finally, the benefits of utilizing the HC approach for both DG investors and DSOs are presented. The application of decision-making techniques in HC assessment and enhancement is examined on a real electrical Egyptian distribution system (EDS). The achieved results revealed that the HC should not be evaluated from a deterministic perspective only, in which no randomness is involved. Instead, it should be viewed as a stochastic problem, whereby real uncertainty levels are considered.

Organization of the remaining chapter is as follows: Section 9.2 presents the decision-making techniques and its applications in the HC studies. Section 9.3 presents an overview of the HC assessment under uncertainty of renewable energy resources. Section 9.4 presents the simulation results and discussion of results. Finally, the conclusion is given in Section 9.5.

9.2 Decision-making techniques and its applications in hosting capacity studies

The presence of uncertain input parameters complicates the decision-making problem. In electricity market the uncertainty in economic, technical, and operational aspects is considered noticeable burden for the DSOs and planners as they need to take harsh decisions in a time-effective manner. Accordingly, to assist network operators and planners, many decision-making techniques were investigated in the literature to examine the network performance in the presence of uncertain parameters, such as the intermittent nature of the DG output power and the location and the daily variations in loading profiles [8]. One can notice that there are almost no published materials that overview the decision-making techniques applied to the HC assessment and DGs allocation problems. To redress this gap, this chapter provides a comprehensive overview of the decision-making techniques applied to the HC assessment as presented in Table 9.1.

9.3 Hosting capacity assessment under uncertainty of renewable energy resources

Recently, it was concluded that deterministic HC studies that ignore the uncertainty of electrical parameters result optimistically that cause a noticeable underestimation to the HC levels achieved from probabilistic assessments [19]. The probabilistic HC assessment is performed to consider the uncertainty in many aspects such as the following:

TABLE 9.2 Comparison between the deterministic and probabilistic hosting capacity (HC) calculation methods.

Method	Deterministic	Probabilistic
Description	• In a deterministic problem the input parameters have specific values that are being processed in a few equations resulting in a set of outputs. The output results have the same values regardless of the number of iterations • A framework is introduced through stepwise increase of the DG penetration at a selected bus, then checking the performance index violations at each iteration till finding the HC of that bus • Then the next bus is examined and the same previous approach is repeated until all the system buses are examined	• Usually used to solve real, complex, and nonlinear problems • In a probabilistic problem the input parameters are not fixed; however, a certain degree of uncertainty is found in representing the inputs • Accordingly, probability distribution functions can be presented to represent the input parameters of a probabilistic problem • Stochastic approaches, such as Monte Carlo, are used to determine how random variation, lack of knowledge, or error affects the performance, of the system that is being studied • In HC studies, Monte Carlo approach is usually used to simulate the uncertainties in DG integration location, rating, and load variations
Complexity	Complex	Case-dependent, according to considered accuracy level
Accuracy	High	Intermediate
Computational time	Large	Intermediate, but it may take large time according to the number of uncertainty simulations required
Achieved HC levels	Conservative HC levels (underestimated)	Practical HC levels (higher HC levels than deterministic)

DG, Distributed generation.

1. the difficulties associated with retrieving established real-time system parameters when calculating the HC analysis studies;
2. the variable nature of the DG output power due to the fluctuations of the environmental conditions;

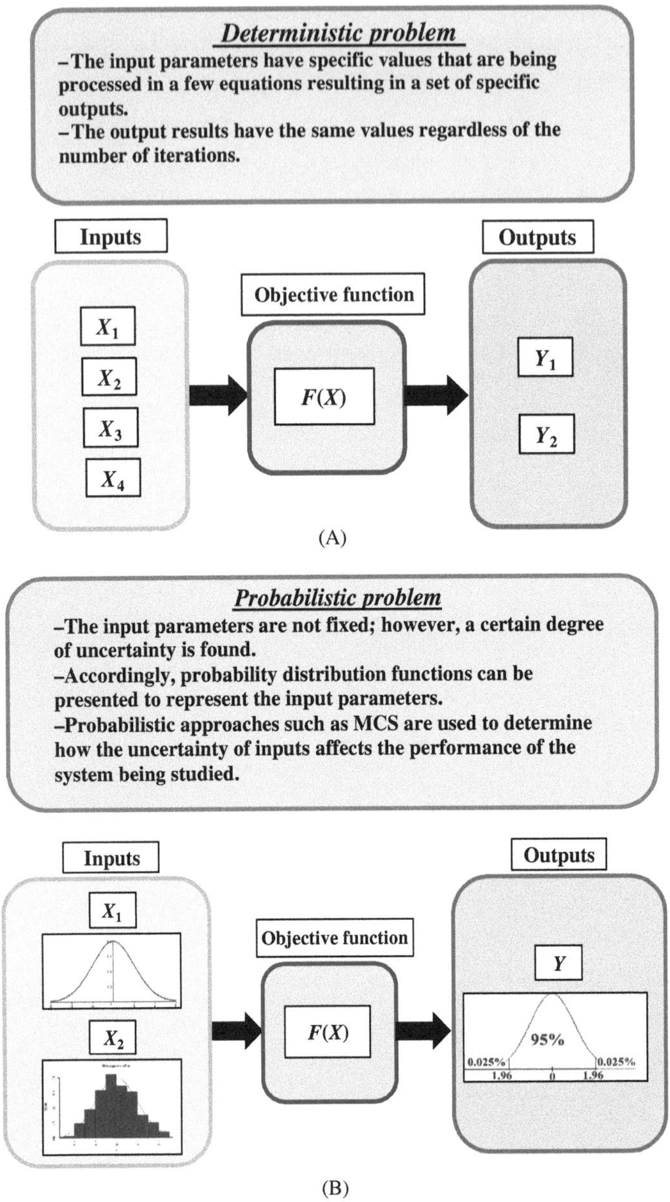

FIGURE 9.2 Deterministic and probabilistic problem formulation. (A) Deterministic problem and (B) probabilistic problem.

3. the uncertainty of DG integration location;
4. the uncertainty of DG unit rating;
5. the daily variations in loading profiles;

6. the variations in network operational topology; and
7. the uncertainty in the economic factors, such as the variations in inflation rates, unemployment rate, and gross domestic product.

Accordingly, the HC of an electrical system will not be a unique level, but a range of levels is obtained according to the uncertainty degree examined.

The MCS approach is considered as one of the most efficient techniques to handle the uncertainty of electrical parameters, such as DG generation fluctuations and load profile changes [16,20].

The Electric Power Research Institute highlighted that the uncertainty in the evaluation of HC levels took place due to the unpredictable DG integration location, variety of unit ratings, the variable nature of DG output power due to climate fluctuations, alteration of load demand, and the absence of accurate system parameters when establishing the HC analysis calculations [21].

A comparison between the deterministic and probabilistic HC calculation methods is presented in Table 9.2 and Fig. 9.2, respectively.

As previously described, the HC assessment should not be handled as a deterministic problem. Instead, it should be solved as a probabilistic problem, whereby appropriate accuracy and uncertainty levels are considered. Throughout the literature, it was concluded that using deterministic solutions based on worst-case scenarios leads to a considerable underestimation of the HC levels. The concept of uncertainty management in HC analysis is illustrated in a simplified manner in Fig. 9.3, while considering the bus voltage as a sample PI for the HC calculation.

From Fig. 9.3, it can be observed that various HC levels exist in the presence of uncertain parameters of the system. In addition, it is clear that a pessimistic HC level (HCU) can be achieved when the system planner considers the upper uncertainty level. On the other hand, an optimistic HC level (HCL) can be obtained when the lower uncertainty level is considered. Furthermore, it is obvious that using high percentiles of the studied index (such as the 95th percentile) leads to a realistic HC level (HC95).

9.4 Overview of related applications

The various uncertainties associated with the massive integration of renewable energy resources into existing electrical distribution networks have been examined by many works. This section provides a comprehensive overview of some relevant researches, which sheds light on the impact of uncertainty handling techniques on the HC assessment and relevant decision-making tools.

In Ref. [22], Lennerhag et al. studied the role of using actual network measurements in the assessment of the HC. The authors concluded that using actual network measurements leads to higher levels of the network's HC, because the dependence on the deterministic calculations only causes a

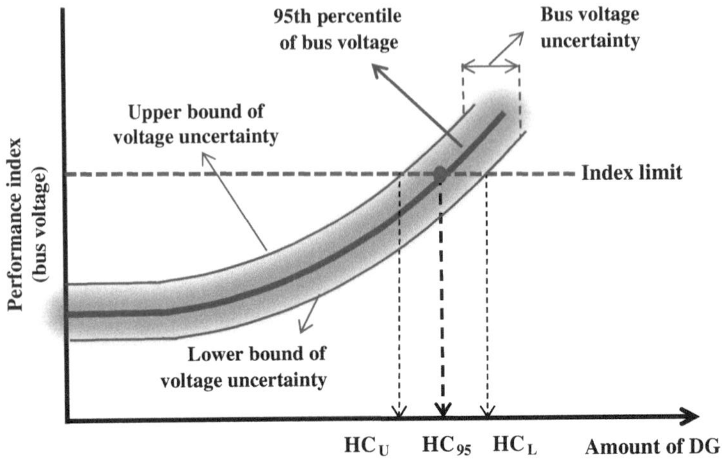

FIGURE 9.3 The concept of uncertainty management in HC analysis.
- HCU: represents the HC value using the upper uncertainty level (*Pessimistic*).
- HC95: represents the HC value using the 95th percentile of bus voltage (*Realistic*).
- HCL: represents the HC value using the lower uncertainty level (*Optimistic*). *HC*, Hosting capacity.

noticeable underestimation of the HC. Furthermore, it was highlighted that the HC is highly dependent on the considered PI and its relevant limit. The authors concluded that the transition from a deterministic value to a high-percentile of the investigated PI may lead to a considerable enhancement in the HC results, without putting unnecessary barriers against additional DG integration and investment. Actual measurements were established in multiple sites in the existing distribution network in Sweden. Five methods for HC calculations were examined starting from conventional estimations to the use of actual measurements. Fig. 9.4 summarizes the reported results of the used methodologies at three different network buses.

From Fig. 9.4 it can be observed that a remarkable variance noticed in the HC levels depends on the methodology used. In addition, it is shown that using the actual measurements (method 5) leads to the highest HC levels.

Furthermore, the authors concluded that using the high percentiles of the examined PI results in higher HC levels compared to the conventional 100% values of the PI, which is commonly known as the "worst-case" value of the PI. Fig. 9.5 presents the impact of using high percentiles of the examined PI on the HC achieved levels.

It can be concluded from Fig. 9.5 that the HC levels achieved by using high percentiles (95th and 99th percentiles) are higher than the HC levels achieved by using the conservative 100% values of the examined PI.

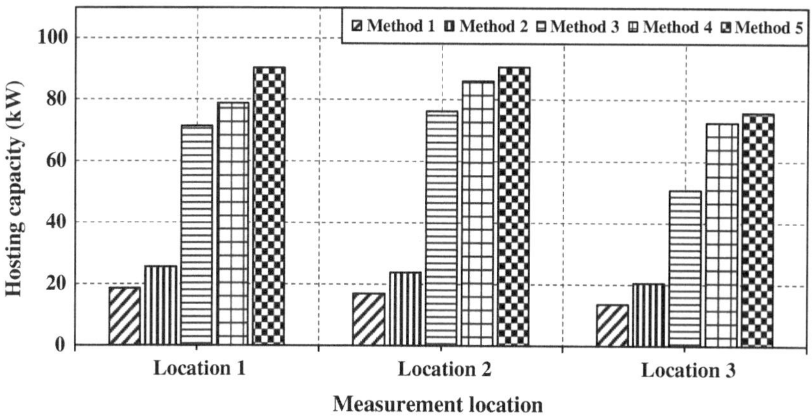

FIGURE 9.4 HC assessment using five different methodologies at three locations. *HC*, Hosting capacity.

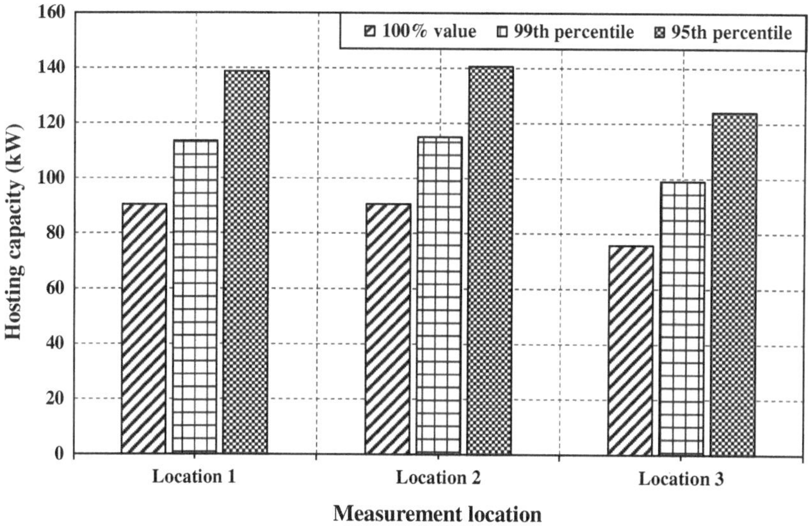

FIGURE 9.5 The impact of using high percentiles of the examined PI on the HC achieved levels. *HC*, Hosting capacity; *PI*, performance index.

Abad et al. [19] investigated the HC assessment considering various uncertain parameters, namely, the uncertainty of the DG data resolution, the DG size, and the DG integration. First, the data resolution effect was examined by considering two figures 30- and 60-minute data resolution. In the 30-minute approach the load profile and the DG output are recorded each for half an hour. The authors concluded that increasing data resolution from

60 to 30 minutes reduces the overall system's HC. Afterward, the uncertainty in DG size and location was investigated. From the DG size perspective, it was found that using a deterministic size of the studied DG instead of using relevant probability distribution curve for the size of the DG unit leads to a considerable underestimation of the system's HC. From the DG location perspective, the authors highlighted that considering a conservative approach for allocating the DG units at the farthest buses leads to a noticeable reduction in the system's HC. Thus it is highly advisable to properly allocate the DG units considering appropriate uniform distribution functions to allow for higher HC levels.

9.5 Problem formulation

The problem of optimal DG sizing and sitting is solved by both deterministic and probabilistic approaches considering the HC maximization as the problem objective function (OF). The problem OF and constraints are described in the following subsections [23].

9.5.1 Objective function

In this work the maximization of the injected active power from connecting DG units is considered as a problem objective that represents the HC maximum allowable level. The HC is expressed as given in Eq. (9.1), where *PDG* represents the injected active power from the DG unit and $S_{\text{load}}^{\text{rated}}$ is the apparent power of the load.

$$\text{HC}(\%) = \frac{P_{\text{DG}}}{S_{\text{load}}^{\text{rated}}} \times 100 \tag{9.1}$$

Thus the OF can be formulated as

$$\text{OF} = \text{Maximize HC} = \text{HC}(\text{DG size}, \text{DG location}) \tag{9.2}$$

where DG size and DG location represent the optimal size and location achieved so far to satisfy the OF while complying with the preset constraints.

9.5.2 Constraints

In this study, five constraints are considered as follows:

1. *Bus voltage constraint*

   The bus rms voltage (V_L^{rms}) should be kept within its specified minimum and maximum limit, thus

$$V_{\min}^{\text{rms}} \le V_L^{\text{rms}} \le V_{\max}^{\text{rms}} \tag{9.3}$$

where $V_{\min}^{\text{rms}}$ and $V_{\max}^{\text{rms}}$ are the minimum and maximum specified rms voltages at load bus, respectively. The minimum and maximum rms bus voltage limits are considered as 0.95 and 1.05 p.u., respectively.

2. *Line capacity constraint*

   The current flow in network's branches is constrained by its thermal limit, as expressed in Eq. (9.6):

$$I_L^{\text{rms}} \leq I_{L,\max}^{\text{rms}} \tag{9.4}$$

   where I_L^{rms} is the total current flowing in the branch and $I_{L,\max}^{\text{rms}}$ is the maximum current carrying capacity (ampacity) of the branch.

3. *DG capacity constraint*

   The total active power produced by the DG unit is recommended to be bounded by the total connected load capacity to avoid excessive reverse power flows. In this study, 100% penetration level is considered as an upper limit for the total DG size [23].

4. *Power factor constraint*

   Power factor (PF) should be kept in its satisfactory range to ensure an optimum energy transfer, thus PF is constrained as follows:

$$\text{PF}_{\min} \leq \text{PF} \leq \text{PF}_{\max} \tag{9.5}$$

   where $\text{PF}_{\min}$ and $\text{PF}_{\max}$ are the minimum and maximum allowed PF limits which are considered as 0.90 lagging and unity, respectively.

5. *Reverse power flow constraint*

To investigate the HC maximization, two approaches are considered: first, inhibiting any reverse power flows in the network branches, and second, allowing the reverse power flows while updating the network protection settings and configurations to allow for these bidirectional power flows.

$$P_{\text{Line}}^J \geq \text{nul} \quad \forall (J \in b) \tag{9.6}$$

where P_{Line} is the net active power flow in branch J, with a network of a total number of branches b.

This constraint will be activated in the first approach (inhibiting reverse power flows), while it will be deactivated in the second approach (allowing reverse power flows).

9.5.3 Load model

From a realistic perspective, the load active and reactive at each bus does not have fixed values along the day. However, a daily load profile is considered to reflect the actual load variations during the day and night. A typical daily load profile relative to rated values is considered in this work as shown in Fig. 9.6 [24].

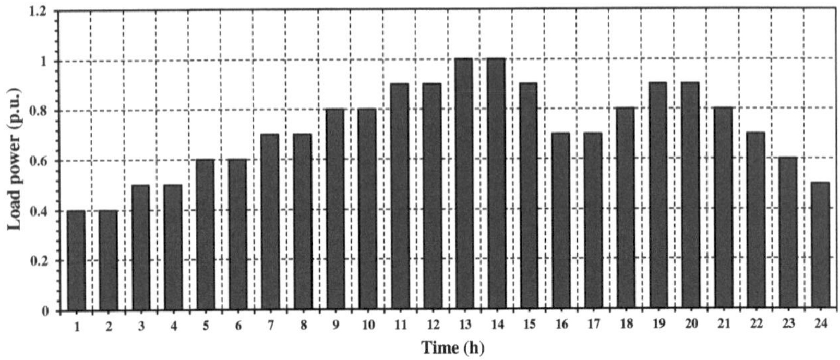

FIGURE 9.6 Typical daily load profile.

9.5.4 Distributed generation unit models

This section presents the models of the examined DG types in this chapter. Two DG types were examined, namely, PV units and wind turbine (WT) generators [24].

1. *PV units*

   The output power of a PV module depends on the site ambient temperature and the solar irradiance in addition to the installation configuration and the PV module characteristics.

   Thus the PV output power at actual site conditions (P^{PV}_{Site}) can be calculated using the following equation [24]:

$$(P^{PV}_{Site}) = P_{STC} \times \frac{I_{Sur}}{1000} \times [1 + \alpha(T_{cell} - 25)] \tag{9.7}$$

   where P_{STC} is the PV module output power at standard test condition in watts. I_{Sur} is the solar irradiance on the surface of the PV module (W/m^2). α is the module temperature coefficient (°C^{-1}) and T_{cell} is the PV module temperature (°C).

   The PV module temperature can be obtained based on the multiple factors, namely, the solar irradiance, ambient temperature, and the module's nominal operating cell temperature (T_{NOCT}), thus [24]

$$T_{cell} = T_{amb} + \frac{I_{sur}}{800} \times (T_{NOCT} - 20) \tag{9.8}$$

   where T_{amb} is the site ambient temperature (°C).

   The site ambient temperature and solar irradiance data used to calculate the PV output power are detailed in Ref. [24]. The utilized solar module data extracted from Ref. [25] are presented in Table 9.3 as follows:

TABLE 9.3 The utilized solar module parameters.

PV module parameter	Value
P_{STC} (W_p)	250
α (°C^{-1})	−0.0045
T_{NOCT} (°C)	46

PV, Photovoltaic.

2. *WT units*

The output power of a WT generator is mainly controlled by two main parameters: first, the wind speed and the considered power/speed characteristics of the WT. The power/speed characteristics of a WT depend on multiple design wind speeds, mainly, the cut-in wind speed (ν_i), the cut-out wind speed (ν_o), and the nominal wind speed (ν_{nom}). Thus the actual output power of a WT (P^{Out}) in percent of its nominal output (P_{nom}) can be calculated as follows [24]:

$$P_{\text{WT}}^{\text{Out}} = \begin{cases} 0 & ,\forall \nu \le \nu_i \\ \dfrac{\nu^2 - \nu_i^2}{\nu_{\text{nom}}^2 - \nu_i^2} & ,\forall \nu_i < \nu < \nu_{\text{nom}} \\ p_{\text{nom}} & ,\forall \nu_{\text{nom}} < \nu \le \nu_o \\ 0 & ,\forall \nu > \nu_o \end{cases} \tag{9.9}$$

The WT design parameters are detailed in Ref. [24]. The utilized WT data extracted from Ref. [26] are presented in Table 9.4.

9.5.5 Deterministic hosting capacity approach

In the deterministic HC approach the load demand, the DG output power, and DG integration location are specified as fixed values. Two schemes are studied: three DG and four DG integration schemes.

9.5.6 Probabilistic hosting capacity approach

In the probabilistic HC approach the various types of uncertainties are considered, mainly the load profile, the PV output power, and the WT output power uncertainty. Thus probabilistic distributions, such as normal probability density function (PDF), are usually utilized to represent the variations of these uncertain parameters. This normal PDF is defined by the expected mean value (μ) and the standard deviation (σ) as follows [27]:

TABLE 9.4 The utilized wind turbine parameters.

WT parameter	Value
P_{nom} (kW)	600
ν_i (m/s)	4
ν_{nom} (m/s)	16
ν_o (m/s)	20

WT, Wind turbine.

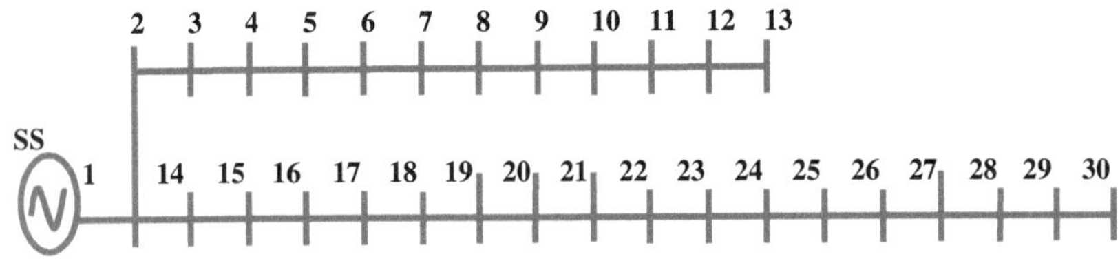

FIGURE 9.7 Arrangement of the real Egyptian distribution system.

$$Y(x) = \frac{1}{\sigma \times 2\pi} \times \exp\left[\frac{-(x-\mu)^{2^2}}{2(\sigma)^2}\right] \tag{9.10}$$

If the limits of the input random variable x are $x_{\min}$ and $x_{\max}$ where $x_{\min} \le x \le x_{\max}$, then the expected mean value (μ), for a 95% confidence level that the random variable x is present within its limits, is calculated as follows [27]:

$$\mu = \frac{0.5 \times (x_{\max} - x_{\min})}{1.96} \tag{9.11}$$

9.6 Case study

In order to present the effect of uncertainty management in HC studies, both deterministic and probabilistic HC assessments are performed on a real EDS. The EDS shown in Fig. 9.7 represents a real system in the East Delta region located in the north of Egypt. The system's nominal voltage and apparent power are 11 kV and 27,221 MVA, respectively. The slack bus voltage (bus 1) is 1 p.u. The total active and reactive load demand of the EDS are 22.4 MW and 14.16 MVAr, respectively [23]. The HC assessment was carried out using MATLAB platform. The line and load data for this system are obtained from Ref. [28]. According to the original system data, the original PF of the base system is too low (0.8457), which violates the applied constraints detailed in Section 9.5.2. Accordingly, PF correction capacitors were considered to enhance the base system PF to reach 0.9229 using four capacitor banks located at buses 9, 18, 21, and 25, respectively, and each of them has a rating of 1.2 MVAr as per the optimal results achieved in Ref. [28]. HC enhancement of the EDS is investigated as a constrained optimization problem using particle swarm optimization (PSO) technique in both deterministic and probabilistic assessments. The widely used PSO algorithm [29] is used to solve the optimization problem presented in this work. PSO method is a heuristic-based method, which is applied effectively in various electrical engineering problems. The parameters of PSO are set to a population size of 10 individuals, cognitive and social acceleration coefficients $c1$ and $c2$ set to 2, minimum and maximum inertia weights set to 0.1 and 1.1, respectively [23], and the maximum number of iterations is set to 100.

9.6.1 Deterministic hosting capacity results

In this chapter, two DG integration schemes have been examined; three DG and four DG schemes. Other arrangements could be explored and the optimal configuration should consider technical, economic, and network operator requirements. Results are reported at two selective durations, at the 14th hour (100% loading) and at the 5th hour (60% loading), according to the daily load profile in Fig. 9.6. The deterministic HC approach simulations are performed under two main conditions:

1. Reverse power flow in network branches is not allowed (no back-feed).
2. Back-feed is allowed, under operators' control and supervision, while updating the network protection settings and configurations to allow for these bidirectional power flows.

In addition, simulations were developed at two loading times, namely, full loading (at the 14th hour) and light loading (at the 5th hour) as shown in the daily load profile in Fig. 9.6. The obtained results of the deterministic HC approach for both conditions (1) and (2) are presented in Tables 9.5 and 9.6, respectively.

From the achieved results in Tables 9.5 and 9.6, it can be noticed that the probabilistic HC levels are higher than the deterministic HC results. In other words, ignoring the uncertainty of electrical parameters results optimistically that cause a noticeable underestimation to the HC levels achieved from probabilistic assessments. Besides, it can be noticed that the HC of a network is not a fixed value along the day. However, as emphasized by the results of Tables 9.5 and 9.6, it is concluded that the network's HC level at full loading ($H = 14$) duration is higher than the HC level during light loading durations ($H = 5$). In addition, it is found that higher HC levels can be obtained by the proper utilization of both active and reactive powers injected by the intelligent DG units, such as PV units controlled by smart inverters. As concluded

TABLE 9.5 Deterministic approach results considering no back-feed power.

	Uncompensated system	4 DG								3 DG					
		$H=14$				$H=5$				$H=14$			$H=5$		
DG location		15	4	3	23	5	16	25	10	14	2	16	18	16	3
DG active power (MW)		3.73	5.71	1.84	0.97	0.93	4.84	0.93	0.65	5.73	3.48	3.08	0.93	4.84	0.93
DG reactive power (MVAr)		1.79	2.74	0.88	0.47	0.45	2.33	0.45	0.31	2.75	1.67	1.47	0.45	2.33	0.45
DG operating PF		0.9				0.9				0.9			0.9		
Active power losses (kW)	805.729	196.7565				274.6019				248.74			316.572		
Active loss reduction (%)		75.6				65.9188				69.128			60.710		
Reactive losses (kVAr)	361.1824	82.39				125.935				101.31			143.604		
Reactive loss reduction (%)		77.18				65.13				71.95			60.241		
V_{min} (p.u.)	0.9463	0.9739				0.96958				0.9694			0.966		
Overall PF	0.8457	0.9459				0.9064				0.9460			90.627		
Hosting capacity (MW)		12.257				7.3566				12.29			6.705		

DG, Distributed generation; *PF*, power factor.

TABLE 9.6 Deterministic approach results with back-feed power allowance.

	Uncompensated system	*4 DG*								*3 DG*					
		$H = 14$				$H = 5$				$H = 14$			$H = 5$		
DG location		11	6	16	19	16	18	7	23	20	25	6	13	19	24
DG active power (MW)		1.02	5.31	0.99	5.31	1.02	5.3	1.02	1.02	6.08	3.14	5.32	1.02	5.32	1.07
DG reactive power (MVAr)		0.48	2.55	0.47	2.52	0.49	2.55	0.49	0.49	2.92	1.51	2.55	0.49	5.56	0.49
DG operating PF		0.9				0.9				0.9			0.9		
Active power losses (kW)	805.729	141.89				193.4798				87.159			222.8126		
Active loss reduction (%)		82.39				75.987				89.183			72.3465		
Reactive losses (KVAr)	361.1824	62.15				91.4341				38.682			105.458		
Reactive loss reduction (%)		82.79				74.6848				89.290			70.802		
V_{min} (p.u.)	0.9463	0.97652				0.97902				0.991			0.97798		
Overall PF	0.8457	0.9426				0.9345				91.077			0.9324		
Hosting capacity (MW)		12.6346				8.3994				14.535			7.3756		

DG, Distributed generation; *PF* power factor.

TABLE 9.7 Probabilistic approach results with back-feed power allowance.

	4 DG								*3 DG*					
	No back-feed				With back-feed				No back-feed			With back-feed		
DG location	17	3	21	18	18	21	2	20	6	14	16	9	8	3
DG active power (MW)	1.01	5.3	1	5.3	2.95	5.56	2.31	2.38	2.37	4.53	5.73	3.13	1.87	7.78
DG reactive power (MVAr)	0.48	2.54	4.8	2.55	1.42	2.67	1.11	1.15	1.14	2.16	2.76	1.56	0.89	3.73
DG operating PF	0.9				0.9				0.9			0.9		
Active power losses (kW)	138.83				156.036				178.95			404.95		
Active loss reduction (%)	82.77				80.6341				77.79			49.74		
Reactive losses (kVAr)	61.048				65.533				75.617			157.87		
Reactive loss reduction (%)	83.097				81.86				79.064			56.288		
V_{min} (p.u.)	0.9795				0.98255				0.9723			0.95988		
Overall PF	0.9427				0.95				0.9428			0.9432		
Hosting capacity (MW)	12.6320				13.2197				12.6634			12.7902		

DG, Distributed generation; *PF* power factor.

in Ref. [23], DG units operating at 0.9 lagging PF help to achieve optimum power loss reduction and hence higher HC levels. In this case, the DG unit acts as a multirole player because the injected active power from DGs supplies the load active power demands and compensates the system active power losses, while the injected DG reactive power helps in improving the voltage profile of the system, relieves the overvoltages arisen from the high DG penetration, and increases the overall PF.

9.6.2 Probabilistic hosting capacity results

In the probabilistic HC approach, an MCS is utilized to develop a large number of probabilities of load active and reactive power, PV output power, and WT output power. For each probability, the 95th percentile of the considered parameter is calculated. In this work, each distribution was obtained from MCS using 1000 samples [30]. An uncertainty tolerance level of 1% was proposed for load power variations, PV output power, and WT output power.

The optimization problem in the presence of uncertain input parameters is performed in the following sequence:

1. The input parameters of the problem are defined in addition to their relevant uncertainty tolerance.
2. The 95th percentiles of the studied uncertain parameters are calculated using MCS for the specified sample.
3. Then, the problem of optimal DG sizing and sitting is activated with a preset objective.
4. To maximize the systems' HC.
5. The OF (HC level) is evaluated, and the compliance with the problem constraints is examined.
6. Steps 2–4 are repeated until reaching the maximum number of iterations.
7. The best value of the achieved objectives (highest obtained HC level) is reported.

From the achieved results in Tables 9.5, 9.6 and 9.7, it can be noticed that the probabilistic HC levels are higher than the deterministic HC results. In other words, ignoring the uncertainty of electrical parameters, results in optimistic that cause a noticeable underestimation to the HC levels achieved from probabilistic assessments. The results obtained of the probabilistic HC approach, with and without back-feed conditions, are presented in Table 9.7.

9.7 Conclusion

Uncertainty in the HC calculations took place due to numerous issues, such as uncertain DG locations and size, the variable nature of the output powers from renewable energy resources, such as wind and PV resources, and fluctuations of load profiles along the day. Accordingly, the HC calculations should not be a deterministic problem with no randomness involved. However, it should be solved as a probabilistic problem, where uncertainty levels are considered. In this chapter the HC is introduced as a decision-making tool that provides a fair and transparent approach to decide when new DG integration requests should be rejected or accepted. A comprehensive overview is presented for the various decision-making techniques and its applications in HC studies. The application of decision-making techniques in HC assessment and enhancement is examined on a real electrical EDS. The achieved results revealed that the HC should not be evaluated from deterministic perspective only in which no randomness is involved. Instead, it should be viewed as a probabilistic problem, depending on the examined uncertainty levels. In addition, it can be noticed that the probabilistic HC levels are higher than the deterministic HC results. In other words, ignoring the uncertainty of electrical parameters results optimistically that cause a noticeable underestimation to the HC levels achieved from probabilistic assessments. In addition, it is found that higher HC levels can be obtained by the proper utilization of both active and reactive powers injected by the intelligent DG units, such as PV units controlled by smart inverters. Finally, further precautions should be taken by network operators to control the excessive reverse power flows and its associated problems resulting from high DG penetration levels.

References

[1] S.M. Ismael, S.H.E. Abdel Aleem, A.Y. Abdelaziz, A.F. Zobaa, State-of-the-art of hosting capacity in modern power systems with distributed generation, Renew. Energy 130 (2019) 1002–1020. Available from: https://doi.org/10.1016/j.renene.2018.07.008.

[2] F. Ding, B. Mather, P. Gotseff, Technologies to increase PV hosting capacity in distribution feeders, IEEE Power Energy Soc. Gen. Meet. 2016 (2016). Available from: https://doi.org/10.1109/PESGM.2016.7741575.

[3] S.M. Ismael, S.H.E. Abdel Aleem, A.Y. Abdelaziz, A.F. Zobaa, "Practical considerations for optimal conductor reinforcement and hosting capacity enhancement in radial distribution systems,", IEEE Access 6 (2018) 27268–27277. Available from: https://doi.org/10.1109/ACCESS.2018.2835165.

[4] G.S. Elbasuony, S.H.E. Abdel Aleem, A.M. Ibrahim, A.M. Sharaf, A unified index for power quality evaluation in distributed generation systems, Energy 149 (2018) 607–622.

[5] A. Soroudi, T. Amraee, Decision making under uncertainty in energy systems: state of the art, Renew. Sustain. Energy Rev. 28 (2013) 376–384. Available from: https://doi.org/10.1016/j.rser.2013.08.039.

[6] S.D. Pohekar, M. Ramachandran, Application of multi-criteria decision making to sustainable energy planning—a review, Renew. Sustain. Energy Rev. 8 (2004) 365–381. Available from: https://doi.org/10.1016/j.rser.2003.12.007.

[7] C. Wimmler, G. Hejazi, E. de Oliveira Fernandes, C. Moreira, S. Connors, Multi-criteria decision support methods for renewable energy systems on islands, J. Clean Energy Technol. 3 (3) (2015) 185–195. Available from: https://doi.org/10.7763/jocet.2015.v3.193.

[8] A.J. Conejo, M. Carrión, J.M. Morales, Decision-Making Under Uncertainty in Electricity Markets, first ed., Springer, New York, 2010.

[9] W. El-Khattam, Y.G. Hegazy, M.M.A. Salama, Investigating distributed generation systems performance using Monte Carlo simulation, IEEE Trans Power Syst. 21 (2) (2006) 524–532.

[10] A. Soroudi, R. Caire, N. Hadjsaid, M. Ehsan, "Probabilistic dynamic multi-objective model for renewable and non-renewable distributed generation planning,", IET Gener. Transm. Distrib. 5 (11) (2011) 1173–1182. Available from: https://doi.org/10.1049/iet-gtd.2011.0173.

[11] C. Su, Stochastic evaluation of voltages in distribution networks with distributed generation using detailed distribution operation models, IEEE Trans. Power Syst. 25 (2) (2010) 786–795. Available from: https://doi.org/10.1109/TPWRS.2009.2034968.

[12] Z. Liu, F. Wen, G. Ledwich, Optimal siting and sizing of distributed generators in distribution systems considering uncertainties, IEEE Trans. Power Delivery 26 (4) (2011) 2541–2551. Available from: https://doi.org/10.1109/TPWRD.2011.2165972.

[13] A. Rabiee, S.M. Mohseni-Bonab, Maximizing hosting capacity of renewable energy sources in distribution networks: a multi-objective and scenario-based approach, Energy (2016). Available from: https://doi.org/10.1016/j.energy.2016.11.095.

[14] V.F. Martins, C.L.T. Borges, Active distribution network integrated planning incorporating distributed generation and load response uncertainties, IEEE Trans. Power Syst. 26 (4) (2011) 2164–2172. Available from: https://doi.org/10.1109/TPWRS.2011.2122347.

[15] A. Soroudi, M. Ehsan, R. Caire, N. Hadjsaid, Possibilistic evaluation of distributed generations impacts on distribution networks, IEEE Trans. Power Syst. 26 (4) (2011) 2293–2301. Available from: https://doi.org/10.1109/TPWRS.2011.2116810.

[16] J. Le Baut, P. Zehetbauer, S. Kadam, B. Bletterie, N. Hatziargyriou, J. Smith, et al., Probabilistic evaluation of the hosting capacity in distribution networks, in: IEEE PES Innovative Smart Grid Technologies European Conference, 2017. Available from: https://doi:10.1109/ISGTEurope.2016.7856213.

[17] H. Al-Saadi, R. Zivanovic, S.F. Al-Sarawi, Probabilistic hosting capacity for active distribution, Networks, IEEE Trans. Ind. Informatics 13 (2017) 2519–2532. Available from: https://doi.org/10.1109/TII.2017.2698505.

[18] A. Soroudi, M. Ehsan, A possibilistic–probabilistic tool for evaluating the impact of stochastic renewable and controllable power generation on energy losses in distribution networks—a case study, Renew. Sustain. Energy Rev. 15 (1) (2011) 794–800. Available from: https://doi.org/10.1016/j.rser.2010.09.035.

[19] M.S. Abad, J. Ma, D. Zhang, A.S. Ahmadyar, H. Marzooghi, Sensitivity of hosting capacity to data resolution and uncertainty modeling, in: Australasian Universities Power Engineering Conference (AUPEC 2018), Auckland, 2018.

[20] H. Al-Saadi, R. Zivanovic, S.F. Al-Sarawi, Probabilistic hosting capacity for active distribution networks, IEEE Trans. Ind. Informatics 13 (2017) 2519–2532. Available from: https://doi.org/10.1109/TII.2017.2698505.

[21] M. Rylander, J. Smith, Stochastic analysis to determine feeder hosting capacity for distributed solar PV, in: EPRI Technical Update No. 1026640, 2012, pp. 1–50.

[22] O. Lennerhag, S. Ackeby, M.H.J. Bollen, G. Foskolos, T. Gafurov, Using measurements to increase the accuracy of hosting capacity calculations, in: 24th International Conference on Electricity Distribution (CIRED), Glasgow, UK, 12–15 June 2017.

[23] S.M. Ismael, S.H.E.A. Aleem, A.Y. Abdelaziz, Optimal sizing and placement of distributed generation in Egyptian radial distribution systems using crow search algorithm,

in: 2018 International Conference on Innovative Trends in Computer Engineering (ITCE), Aswan, 2018, pp. 332–337. Available from: https://doi:10.1109/ITCE.2018.8316646.

[24] R. Jordan Metaheuristic Optimization in Power Engineering, The Institution of Engineering and Technology (IET), London, United Kingdom. Available from: https://doi.org/10.1049/PBPO131E

[25] Solar module "sunmodule" data [Online]: <http://www.siliconsolar.com/documents/data-sheet-solarworld-250w-mono-solar-panel.pdf>.

[26] Wind turbine data "Vestas" [Online]: <https://en.wind-turbine-models.com/turbines/272-vestas-v-44>.

[27] P. Abril, Passive filters' placement considering parameters' variations, Int. Trans. Electr. Energy Syst. (2018) e2727. Available from: https://doi.org/10.1002/etep.2727.

[28] A.A. Abou El-Ela, R.A. El-Sehiemy, A. Kinawy, M.T. Mouwafi, Optimal capacitor placement in distribution systems for power loss reduction and voltage profile improvement, IET Gener. Transm. Distrib. 10 (5) (2016) 1209–1221.

[29] J. Kennedy, R. Eberhart, Particle swarm optimization, in: Neural Networks, 1995. Proceedings of the IEEE International Conference, vol. 4, 1995, pp. 1942–1948.

[30] S. M. Ismael, S. H. E. Aleem and A.Y. Abdelaziz, and A. F. Zobaa, Probabilistic Hosting Capacity Enhancement in Non-Sinusoidal Power Distribution Systems Using a Hybrid PSOGSA Optimization Algorithm, Energies, 12 (6), (2019) 1018, Available from: https://doi:10.3390/en12061018.

Further reading

M.H.J. Bollen, Overvoltages due to wind power-hosting capacity, deterministic and statistical approaches, Electr. Power Qual. Util. Mag. 3 (2) (2018) 2–15.

Chapter 10

Particle swarm optimization applied to reactive power dispatch considering renewable generation

Maíra R. Monteiro[1,2], Yuri R. Rodrigues[2],
Antonio Carlos Zambroni de Souza[1] and Paulo Fernando Ribeiro[1]
[1]*Institute of Electrical System and Energy, Federal University of Itajuba, UNIFEI, Itajubá, Brazil,* [2]*School of Engineering, The University of British Columbia, Kelowna, BC, Canada*

10.1 Introduction

A voltage instability event can be observed when progressive and uncontrollable voltage drops occur after the electrical system was subjected to a disturbance. This condition arises mainly because of the inability to supply reactive power in the area of disturbance, instead requiring large packs of reactive power transmission. This exerts great impact in the systems' active power loss, especially for long lines subjected to heavy load, which should be minimized.

The transmission system losses are mainly influenced by physical and geographical distribution of its reactive power sources, transmission line characteristics, loading condition, and operating voltage of the grid [1]. When the electric power system is operating within recommended parameters, the active power losses can be reduced by means of suitable adjustments of reactive power generating sources. This problem can be solved employing reactive power dispatch control in generators, considering their physical position and system configuration since the reactive power has local characteristics.

However, before performing reactive power dispatch, one should evaluate if the system is operating close to the point of voltage collapse. In this condition the system presents large electric losses and a low voltage profile as shown in [2]; however, control actions aimed at reducing losses can have

Decision Making Applications in Modern Power Systems. DOI: https://doi.org/10.1016/B978-0-12-816445-7.00010-4

adverse effects and lead to an insignificant load margin increase. This scenario is improved when losses-reduction actions are performed in the critical regions.

In the literature, there are different approaches for the reactive power dispatch using particle swarm optimization (PSO). A comprehensive learning PSO for reactive power dispatch is developed in [3]. A combination of PSO and a feasible solution search is applied in wind farms by [4], whereas in [5] the PSO is integrated into multiagent system. The reactive power compensation of radial distribution systems–based multiobjective planning algorithm using PSO is developed in [6], while in [7] a multiobjective optimization problem is formulated to relieve the overvoltage caused by large *PV* penetration and to minimize total line loss. Refs. [8–10] proposed the loss reduction by shunt compensation employing different optimization techniques, the first being solved by PSO and the second by primal-dual method of interior points. In [11] the PSO was employed in training artificial neural networks.

The current chapter presents a reactive power dispatch technique based on the PSO with the objective function of minimizing electrical losses, associated with a dispatched generator selection performed by means of the tangent vector sensitivity analysis. In this way the contribution of this study consists of the union of two methods to solve this problem considering the intermittency of renewable sources. First, the generators are identified using the tangent vector and then supplied to the PSO to perform the dispatch in order to reduce electrical losses. The PSO is employed in this study due to its applications for reactive power control. Moreover, for this study, voltage collapse indices, *PV* and *QV* curves, are applied facilitating the evaluation of the operating conditions related to the voltage safety margin, aiding in the system planning. Using the *PV* curve, it is possible to quantify how close the system is from voltage collapse, while the *QV* curve indicates how far each bus is from the instability region.

The case studies are performed using the IEEE 118-bus–modified system considering penetration of renewable energy sources (RES), wind and solar, and a variable demand profile. The analyses are divided into scenarios with the presence and absence of renewable sources considering different periods of the day. Furthermore, the influence on the active and reactive load margins is also analyzed given that the system configurations change throughout the day. Results are analyzed and discussed seeking to assist the process of decision-making for voltage stability margin at planning. Further, a comparison can be made for different system settings through the dispatch benefit, which represents the system operational gain for additional increments in the generation dispatch.

The chapter structure is defined as follows: first, the voltage collapse index determination is presented in Section 10.2. Section 10.3 discusses the active power losses. Section 10.4 introduces the proposed identification of candidate buses for renewable generation allocation. Section 10.5 showcases the tangent-vector-based generators identification for reactive power dispatch. Section 10.4 presents the PSO technique. Section 10.6 depicts the overall methodology for PSO application to reactive power dispatch considering tangent-vector-based generation selection and renewable generation penetration. Section 10.8 presents the obtained results for the IEEE 118-bus system. At last, Section 10.9 draws the major conclusions of the chapter and final remarks.

10.2 Voltage collapse indexes

Voltage collapse indexes are used to determine the system distance to voltage stability problems. These factors are critical in power system planning and operation analysis, as they can predict operational conditions that should be avoided.

10.2.1 Tangent vector

The tangent vector indicates how the state variables (θ_g, θ_l, and V_l) behave with the variation of a system parameter, such as load increase. This vector identifies the most susceptible buses to voltage collapse for a given operating point, thus being an effective tool for identifying critical buses and preventing the saddle-node bifurcation point [12].

$$TV = \begin{bmatrix} \frac{\Delta\theta_g}{\Delta\lambda} \\ \frac{\Delta\theta_l}{\Delta\lambda} \\ \frac{\Delta V_l}{\Delta\lambda} \end{bmatrix} = [J]^{-1} \cdot \begin{bmatrix} P_{g0} \\ P_{l0} \\ Q_{l0} \end{bmatrix} \quad (10.1)$$

where P_{g0}, P_{l0}, and Q_{l0} are the active power generation, and active and reactive power demands, respectively, λ is the increased parameter, and J is the converged Jacobian matrix of the load flow.

10.2.2 *PV* curve

One of the most widely used methods for static stability analysis is based on the *PV* curve behavior. The maximum point, P_{max}, represents the maximum loading that the system can support without voltage stability loss, as shown in Fig. 10.1. This point corresponds to the load margin, which is defined as the distance from an initial operating point of interest to the saddle-node bifurcation point, that is, voltage collapse condition.

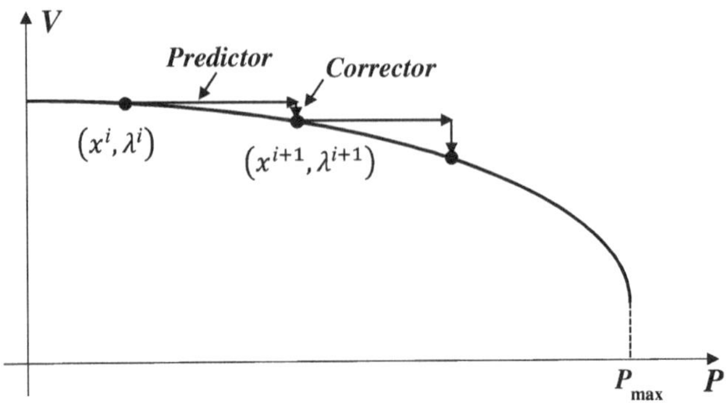

FIGURE 10.1 *PV* curve.

This curve is obtained by the continuation method. It successively increases the system active power generation and active and reactive load demands as per (10.2) and (10.3), up to the point of voltage collapse. To speed up the process, an adaptive predictor step can be adopted considering the inverse of Euclidean norm of the tangent vector (10.4). The corrector step is performed by the load flow. Since the interest of this analysis is restricted to the *PV* curve stable part, a stopping criterion depicted in (10.5) is associated. It anticipates the dominant eigenvalue of the Jacobian matrix, which tends to zero at saddle-node bifurcation point [13,14].

$$P = P_0 \cdot (1 + \Delta\lambda) \tag{10.2}$$

$$Q = Q_0 \cdot (1 + \Delta\lambda) \tag{10.3}$$

$$\Delta\lambda = \frac{k}{\|VT\|} \tag{10.4}$$

$$I_c = TV^T \cdot J \cdot TV \tag{10.5}$$

where k determines the process speed:

$$\begin{cases} k = 1, & \text{regular} \\ k > 1, & \text{accelerated} \\ k < 1, & \text{decelerated} \end{cases}$$

10.2.3 *QV* curve

Another method for static voltage stability analysis is the *QV* curve. This representation indicates the reactive load range of a given system bus, as shown in Fig. 10.2. This is achieved by making the interest bus into a *PV* bus (if it was not) with open reactive power limits and varying its terminal voltage [13,14].

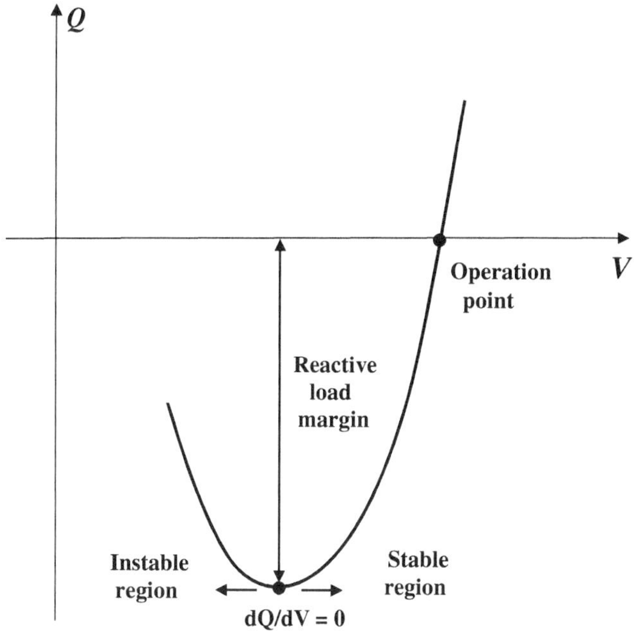

FIGURE 10.2 QV curve.

10.3 Active power losses

The active power losses in branches $i-j$ are given by the differences between the power sent by bus i and received at bus j. Since the power that arrives at bus j is given by the negative value of the power that leaves the bus j, the total system active power losses can be calculated by

$$P_{\text{system}} = \sum_{k=1}^{n} V_{ik} V_{jk} G_k \left[\cos\left(\delta_{(ij)k}\right) + \cos\left(\delta_{(ji)k}\right)\right] - G_k\left(V_{ik}^2 + V_{jk}^2\right) \quad (10.6)$$

where n is the total number of transmission lines, k, V_{*k}, and $\delta_{(*)k}$ are the voltage levels and phase angles at the edges of the transmission line k, and G_k is the conductance of the transmission line k.

10.4 Identification of candidate buses for renewable generation allocation

The allocation of new generating units, in especial variable renewable generation, is bounded by several technical, economical, and political issues. In this section, two methods to identify candidate buses for RES allocation considering only technical criteria are developed. The first method considers the critical buses under the voltage stability perspective, while the second addresses the buses sensitivity to losses.

10.4.1 RES allocation by voltage stability criteria

In this method the critical buses are determined from the ranking of the tangent vector calculated in (10.1), where the critical buses are associated with the ones with the highest absolute value of tangent vector [12]. The ranked candidate buses are given by (10.7).

$$TV_{\text{rank}} = \text{rank}\left(\left\| \begin{bmatrix} \dfrac{\Delta\theta_g}{\Delta\lambda} \\ \dfrac{\Delta\theta_l}{\Delta\lambda} \\ \dfrac{\Delta V_l}{\Delta\lambda} \end{bmatrix} \right\|\right) \tag{10.7}$$

10.4.2 RES allocation by loss sensitivity criteria

This method seeks to identify the buses where the RES allocation would lead to a greater impact in loss reduction. Deriving (10.6) as a function of λ, the sensitivity to losses is obtained in (10.8) [15]. Similarly, the rank of buses' loss sensitivity index leads to the candidate ones where RES placement would have a greater impact in loss reduction.

$$\frac{dP_{\text{syst}}}{d\lambda} = \sum_{k=1}^{n1}\left(\frac{dV_{ik}}{d\lambda}V_{jk} + \frac{dV_{jk}}{d\lambda}V_{ik}\right)A + V_{ik}V_{jk}\frac{dA}{d\lambda} - 2G_k\left(\frac{dV_{ik}}{d\lambda}V_{ik} + \frac{dV_{jk}}{d\lambda}V_{jk}\right) \tag{10.8}$$

where

$$A = G_k\left[\cos\left(\delta_{(ij)k}\right) + \cos\left(\delta_{(ji)k}\right)\right] \tag{10.9}$$

$$\frac{dA}{d\lambda} = G_k\left[2\sin\left(\delta_{(ji)k}\right)\left(\frac{d\delta_{(i)k}}{d\lambda} - \frac{d\delta_{(j)k}}{d\lambda}\right)\right] \tag{10.10}$$

Note that all derivatives are components of the tangent vector.

10.5 Identification of generators for reactive power dispatch using particle swarm optimization

The proposed method for the identification of dispatchable generators for reactive power dispatch uses the tangent vector capacity to effectively identify the system critical buses [12]. Given that the tangent vector indicates the sensitivity of state variables to a desired system parameter, here a method is proposed to determine the generators responsible to perform the reactive dispatch, based on the reactive power variation with the voltage. The modified tangent vector is computed according to the following equation:

$$TV_{\text{mod}} = [J_{\text{mod}}]^{-1} \cdot \begin{bmatrix} 0 \\ \vdots \\ 0.1 \\ \vdots \\ 0 \end{bmatrix} \tag{10.11}$$

where J_{mod} is the calculated Jacobian matrix from the converged power flow with the insertion of reactive power equations of generators that were selected for the dispatch. For the sake of this, the PV bus in the analysis is considered to be of PQ type.

The vector on the right side of (10.11) contains all elements equal to zero except for the equivalent of the reactive power equation of generator under analysis. The value 0.1 is arbitrary and can be any one, since the participation factor of each generator will not change once the system analysis method is linearized.

10.6 Particle swarm optimization for reactive power dispatch

The PSO is an optimization method based on particle swarm and inspired by the sociobiological behavior of birds group [16,17]. For the problem formulation an initial population with size s is determined, where each particle i represents a candidate solution. It initializes randomly for each particle at a current position, x_i, and assigns a current velocity, v_i. Besides, x_i is also the best individual position achieved by particle y_i. From this initial data, the iterative process is performed as follows:

Step 1: For each particle i in s:

1. Calculate the objective function, f, which in this case corresponds to the electrical losses minimization, which is formulated in (10.6).
2. Determine y_i (10.12):

$$y_i(t+1) = \begin{cases} y_i(t) \text{ if } f(y_i(t)) \leq f(x_i(t+1)) \\ x_i(t+1) \text{ if } f(y_i(t)) > f(x_i(t+1)) \end{cases} \tag{10.12}$$

Step 2: Evaluation indicators:

There are two different types of evaluation indicators for improving the position in relation to the desired objective, called gbest (best overall) and lbest (best local) models. These models differ in the manner that they interact between a determined particle and its particles set. In this study the gbest model is considered; thus the particle with the best suitability of entire population, $\hat{y}$, is determined by the following equation, being $i \in 1, \ldots, s$.

$$\hat{y}(t) \in \{y_o(t), y_1(t), \ldots, y_s(t) | f(\hat{y}(t)) = \min [f(y_o(t)), f(y_1(t)), \ldots, f(y_s(t))]\} \tag{10.13}$$

Step 3: For each particle i in s,

1. Update the velocity (10.14):

$$v_{i,k}(t+1) = w \cdot v_{i,k}(t) + c_1 \cdot r_{1,k}(t) \cdot [y_{i,k}(t) - x_{i,k}(t)] + c_2 \cdot r_{2,k}(t) \cdot [\hat{y}_k(t) - x_{i,k}(t)] \tag{10.14}$$

where

1. r_1 and r_2 are two independent random sequences, $r_1 \sim (0,1)$ and $r_2 \sim (0,1)$, and contribute to the nature of stochastic algorithm.
2. c_1 and c_2 are constants, called acceleration coefficients, and influence the maximum step size for a particle in the same iteration, $c_1 > 0$ and $c_2 \leq 2$, where c_1 is the coefficient that regulates the maximum step toward y_i and c_2 toward $\hat{y}$.
3. w is the inertia weight and was introduced in (10.14) by [18] as one of the modifications in the original algorithm. In general the value of w is varied linearly at each iteration according to the following equation, being $iter_{max}$ the iterations maximum number, $iter$ denotes each iteration, w_{max} the maximum inertia weight, and w_{min} the minimum inertia weight.

$$w = w_{max} - \frac{w_{max} - w_{min}}{iter_{max}} \cdot iter \tag{10.15}$$

4. To the system not extrapolating the search space; the particle velocity is limited by (10.16). If the search space is defined by interval $[-x_{max}, x_{max}]$, the value v_{max} [19] is calculated by (10.17):

$$v_i = \begin{cases} v_{max}, & v_i > v_{max} \\ -v_{max}, & v_i < -v_{max} \end{cases} \tag{10.16}$$

$$v_{max} = kx_{max}; 0.1 < k \leq 1.0 \tag{10.17}$$

2. Update the position (10.18):

$$x_i(t+1) = x_i(t) + v_i(t+1) \tag{10.18}$$

Step 4: If the objective is not reached, return to *Step 1*.

10.7 Methodology for particle swarm optimization application to reactive power dispatch considering tangent-vector-based generation selection

In this section, the methodology for PSO algorithm is described to perform the reactive power dispatch to reduce the system electrical losses. For this purpose, the IEEE 118-bus test system is employed considering renewable generation, wind and solar photovoltaic, and variable demand profile during a day period.

The renewable generation is considered in PQ mode, which allowed its representation as negative load in the load flow problem. The generated powers are assessed for an entire day period. The wind sources contribution is obtained from a wind speed profile [20], while for solar photovoltaic generation, temperature and solar irradiation profiles [21] are considered.

For the allocation of these renewable sources, the system is considered in the base case and two candidate groups, respectively, considering voltage stability and loss sensitivity criteria, as in [22]. The first group consists of the critical buses of the system for allocation of wind sources, whereas the second is composed of the sensitive to losses for allocation of solar sources. The groups' identification is performed as depicted in Section 10.4.

After RES penetration, the loss reduction is performed by a two-steps process. First, the dispatchable generators identification is performed as described in Section 10.5. Second, these generators are provided to the PSO, and the optimization process is performed (Section 10.6).

This process is described in Fig. 10.3.

The *PV* and *QV* curves are plotted, and active and reactive load margins are determined for each interval before and after the optimization process. In addition, active power losses are also presented.

A comparison can be made by the means of dispatch benefit, allowing a superior performance once the increment for each generator is limited. This benefit corresponds to the ratio between the increases of the active load margin by the total dispatch increment.

10.8 Results and analysis

To verify the methodology effectiveness, five scenarios corresponding to distinct periods of the day are selected. These scenarios are presented later, where Pg_{wt} denotes the wind generation power, Pg_{pv} presents the solar generation power, and fc describes the demand profile.

- *Scenario 1*: corresponds to interval A1, dawn period, without renewables penetration, and $fc = 0.8922$ p.u;
- *Scenario 2*: corresponds to interval A11, morning period, with penetration of both renewables, $Pg_{wt} = 2.4$ p.u. and $Pg_{pv} = 0.0768$ p.u. and $fc = 1.0406$ p.u;
- *Scenario 3*: corresponds to interval A17, afternoon period, with penetration of both renewables, $Pg_{wt} = 2.4000$ p.u. and $Pg_{pv} = 1.1744$ p.u. and $fc = 0.8614$ p.u;

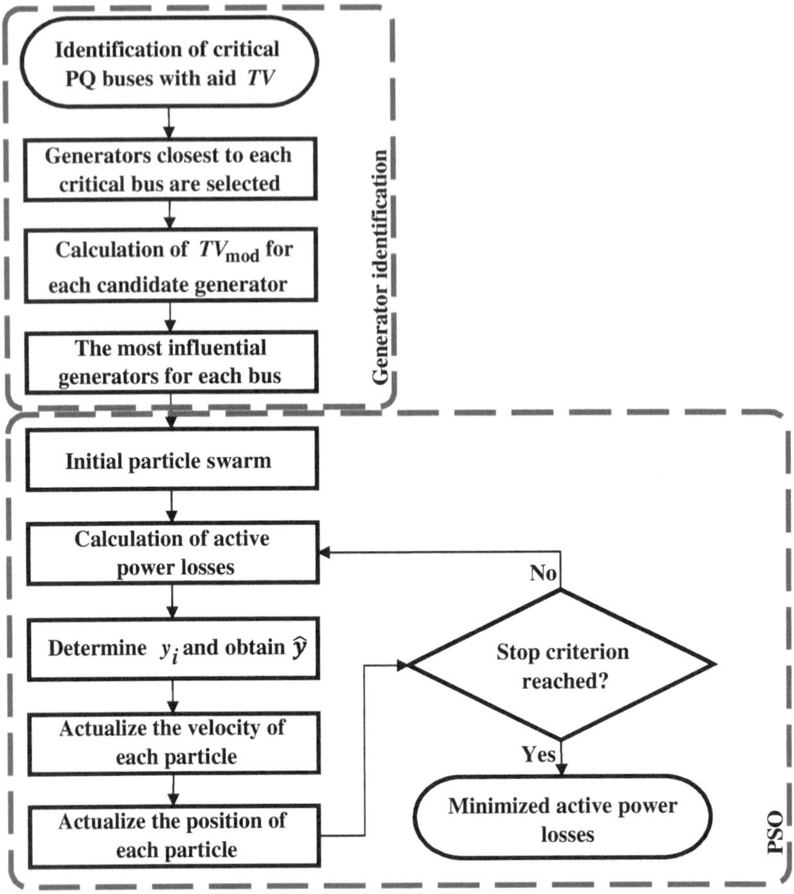

FIGURE 10.3 Flowchart of losses-reduction process.

- *Scenario 4*: corresponds to interval A19, afternoon period, with penetration only of solar generation, $Pg_{pv} = 0.2705$ p.u. and $fc = 1.2127$ p.u;
- *Scenario 5*: corresponds to interval A22, night time, with penetration only of wind generation, $Pg_{wt} = 2.4000$ p.u. and $fc = 1.3968$ p.u.

For each adopted scenario of the IEEE 118-bus system the eight critical buses of PQ type and appropriated generators for reactive power dispatch are identified. These results are shown in Table 10.1. As one may notice, all intervals present the closest generators to the critical bus as the effective ones for active losses minimization. This is expected as reactive power is a local problem.

The results before and after optimization of electrical losses and active load margin are presented in Table 10.2. To determine the reactive increment

TABLE 10.1 Buses of *PQ* and *PV* types.

Scenarios	Critical buses	Generation buses
A1	41, 39, 35, 43, 33, 37, 88, 53	40, 40, 36, 34, 15, 34, 85, 54
A11	41, 53, 109, 108, 58, 52, 39, 106	40, 54, 110, 105, 56, 54, 40, 107
A17	9, 5, 88, 117, 30, 3, 11, 2	10, 4, 85, 12, 8, 1, 12, 1
A19	117, 2, 3, 13, 14, 7, 16, 11	12, 1, 1, 15, 12, 6, 12, 12
A22	29, 14, 28, 2, 3, 16, 13, 20	31, 12, 27, 1, 1, 12, 15, 19

TABLE 10.2 Results of optimization process.

Optimization		A1 (p.u.)	A11 (p.u.)	A17 (p.u.)	A19 (p.u.)	A22 (p.u.)
Prior	System losses	1.40058	1.47235	1.37073	2.63705	3.55748
	Load margin	2.02946	1.86541	2.12143	1.41962	1.23357
Post	System losses	1.39076	1.46905	1.36297	2.63449	3.55069
	Load margin	2.03043	1.89265	2.20989	1.42162	1.23470
	Loss reduction	0.00982	0.00330	0.00776	0.00257	0.00678
	Margin increase	0.00097	0.02724	0.08846	0.00201	0.00113
	Total increment	2.00261	2.00842	1.37647	1.16579	1.35000
	Reactive power increment	0.36000	0.36000	0.12749	0.09000	0.09000
		0.36000	0.36000	0.09000	0.36000	0.09000
		0.20261	0.36000	0.36000	0.36000	0.09000
		0.36000	0.36000	0.09000	0.35579	0.36000
		0.36000	0.36000	0.36000	–	0.36000
		0.36000	0.20842	0.34897	–	0.36000
	Dispatch benefit	0.00049	0.01356	0.06427	0.00172	0.00084

in an actual power system scenario, upper and lower limits (9–36 MVar) are adopted for the redispatch of each generator.

It is possible to verify that the dispatch of reactive power for all intervals reached the objective function of loss reduction for the system. Analyzing the scenarios separately, A1 obtained greater loss reduction, whereas A17 presented the greater increase of load margin. Then, comparing the gains for each system, A17 is the scenario that presented the greater dispatch benefit.

The *PV* curves for the critical buses depicted in Table 10.1 concerning all intervals before and after optimization are illustrated in Figs. 10.4–10.14.

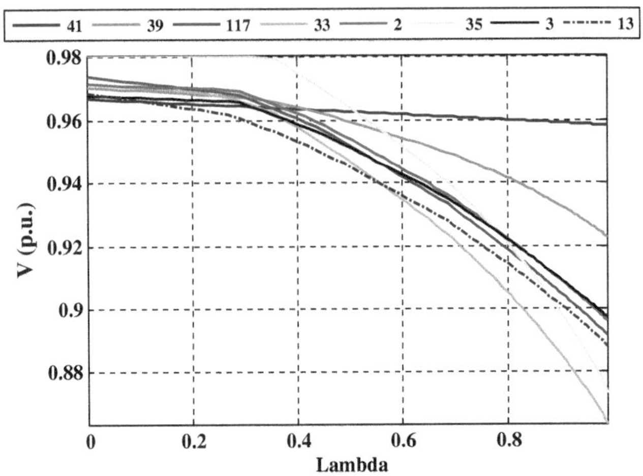

FIGURE 10.4 *PV* curves—base case.

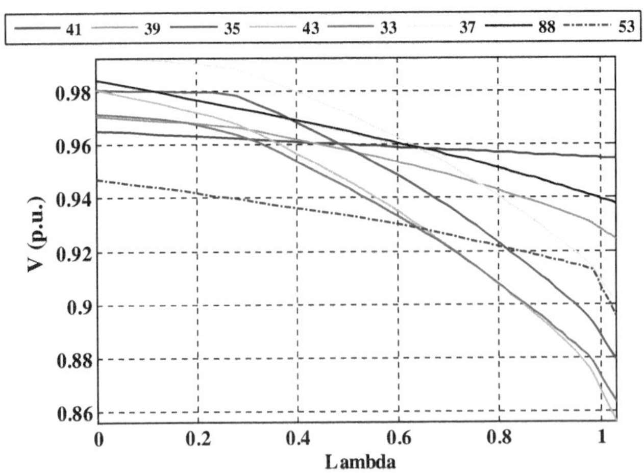

FIGURE 10.5 *PV* curves—A1, prior-dispatch.

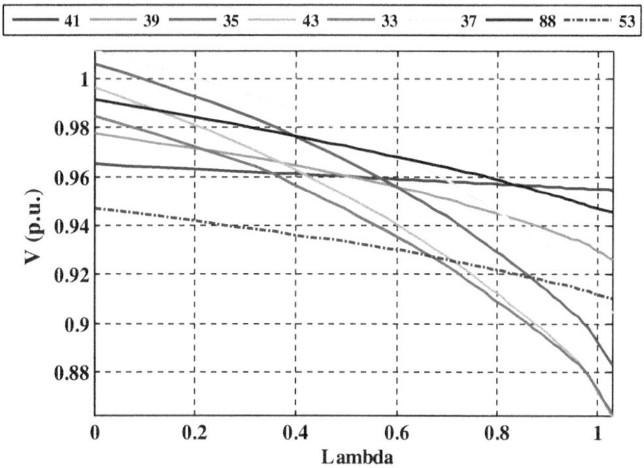

FIGURE 10.6 *PV* curves—A1, postdispatch.

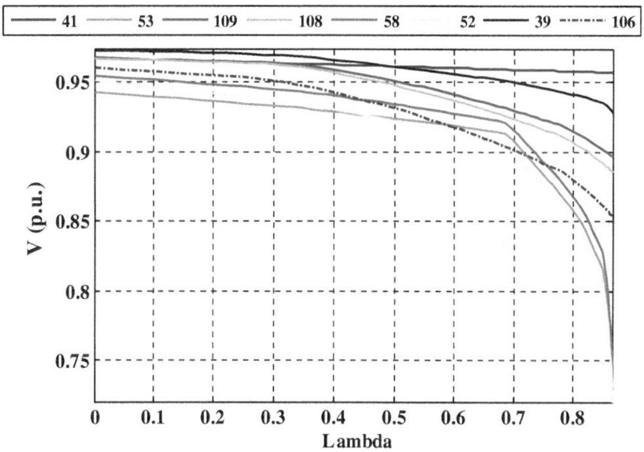

FIGURE 10.7 *PV* curves—A11, prior-dispatch.

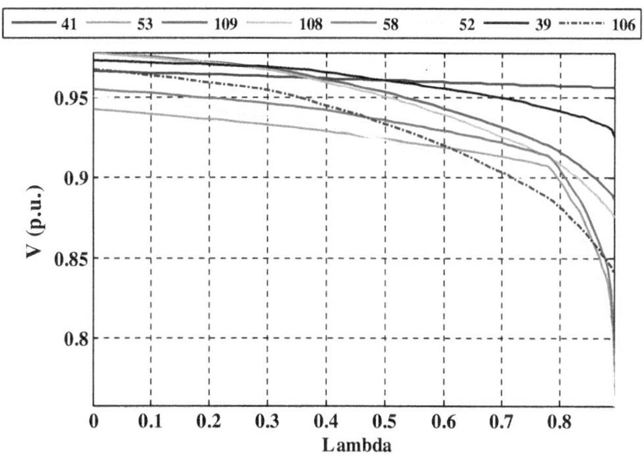

FIGURE 10.8 *PV* curves—A11, postdispatch.

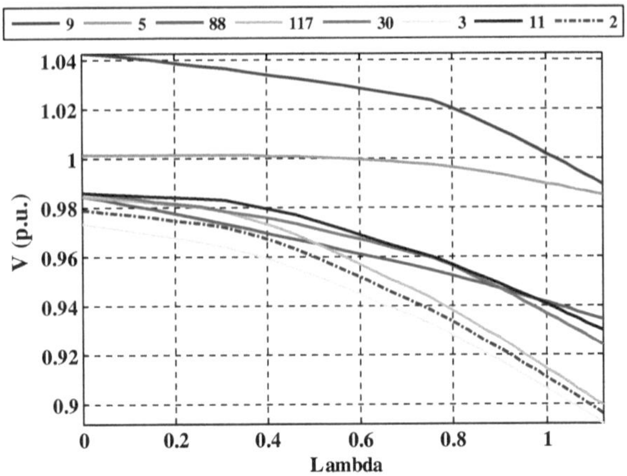

FIGURE 10.9 *PV* curves—A17, prior-dispatch.

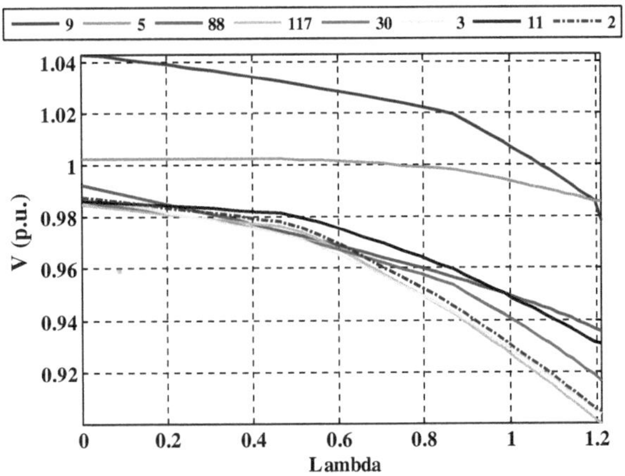

FIGURE 10.10 *PV* curves—A17, postdispatch.

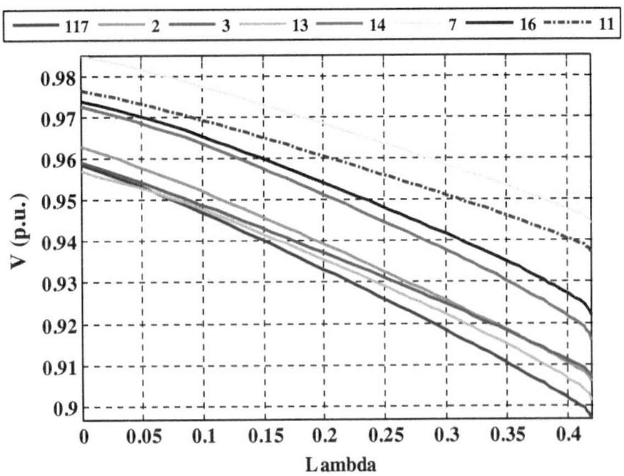

FIGURE 10.11 *PV* curves—A19, prior-dispatch.

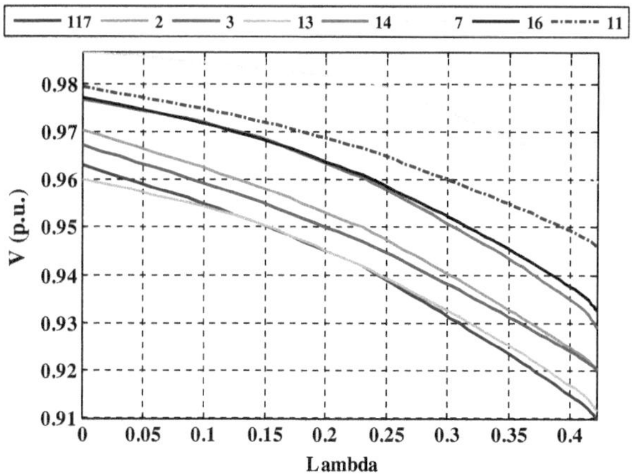

FIGURE 10.12 *PV* curves—A19, postdispatch.

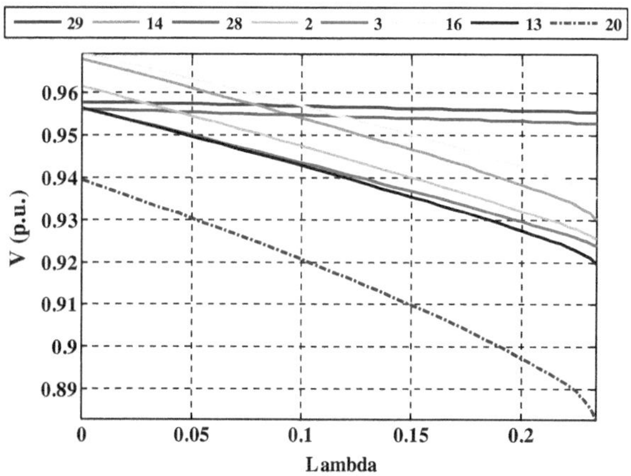

FIGURE 10.13 *PV* curves—A22, prior-dispatch.

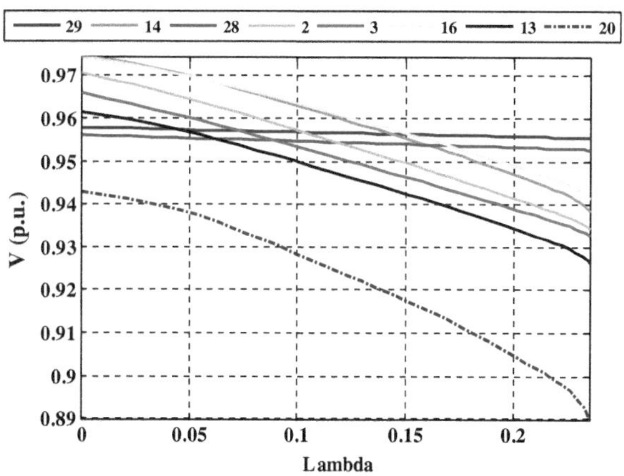

FIGURE 10.14 *PV* curves—A22, postdispatch.

These graphs allow one to observe the load margins and voltage levels behavior according to the respective increase in generation and load. To facilitate understanding, the legends present the buses in a decreasing sensitivity order.

The load margin values before redispatch for each scenario are presented in Table 10.3. Note that intervals with *fc* greater than 1 p.u., that is, A11, A19, and A22, presented a reduction of the load margin with respect to the base case, being A22 the scenario with greater reduction. Instead, A1 and A17 presented greater margins in comparison to the base case. These variations occur due to the different operative conditions faced by the system during the day.

It is important to highlight that the presence of renewables also increases the load margin, as it will reduce the amount of power consumed in the buses itself, and if surplus generation is available, they are exported to the transmission system.

Another analysis of interest consists of the reactive load margin. For this analysis, the critical buses are identified in Table 10.1 and the respective load margins, presented in Table 10.4. The reactive load margin values calculated and compared with the base case are shown in Table 10.5.

TABLE 10.3 Results of initial active load margin.

Intervals	Initial active load margin (p.u.)	Margin increase (p.u.)
Base case	1.98795	–
A1	2.02946	0.04151
A11	1.86541	−0.12254
A17	2.12143	0.13348
A19	1.41962	−0.56833
A22	1.23357	−0.75438

TABLE 10.4 Critical buses.

Intervals	A0	A1	A11	A17	A19	A22
Critical buses	41	41	41	9	117	29

A0 represents the base case.

TABLE 10.5 Results of initial reactive load margin.

	Initial reactive load margin (p.u.)					
Buses	*A0*	*A1*	*A11*	*A17*	*A19*	*A22*
41	− 4.35951	− 4.43132	− 4.52346	–	–	–
9	− 5.97804	–	–	− 6.00551	–	–
117	− 1.48549	–	–	–	− 1.40359	–
29	− 4.67670	–	–	–	–	− 4.41537
Increase (p.u.)		− 0.07181	− 0.16395	− 0.02747	0.08190	0.26133

TABLE 10.6 Reactive load margin-optimization.

Reactive load margin	A1 (p.u.)	A11 (p.u.)	A17 (p.u.)	A19 (p.u.)	A22 (p.u.)
Preoptimization	− 4.43132	− 4.52346	− 6.00551	− 1.40359	− 4.41537
Postoptimization	− 4.58260	− 4.64384	− 6.19161	− 1.43564	− 4.47146
Increase	− 0.15128	− 0.12038	− 0.18609	− 0.03206	− 0.05609

By analyzing the results obtained by the same critical buses for the base case, one can notice that the reactive load margin increases for the first three scenarios A1, A11, and A17, while for A19 and A22, it decreased.

The reactive load margin values are also determined and compared for these buses postoptimization, as shown in Table 10.6. Note that after compensation, all intervals had increased their reactive load margins. The larger increase occurred in A17, which has both renewables participating and fc less than 1 p.u. However, the intervals featuring fc greater than 1 p.u., except A11, had an increase smaller than 6%. Moreover, even after compensation, the values of the reactive load margin for A19 and A22 remained lower than the base case.

An important aspect observed is the displacement of the QV curve according to the variation of reactive load margin. Since the bus is of PQ type, the initial operating point is depicted by the unfilled marker located on the stable side (right) of the QV curve. This is visualized in Figs. 10.15–10.19.

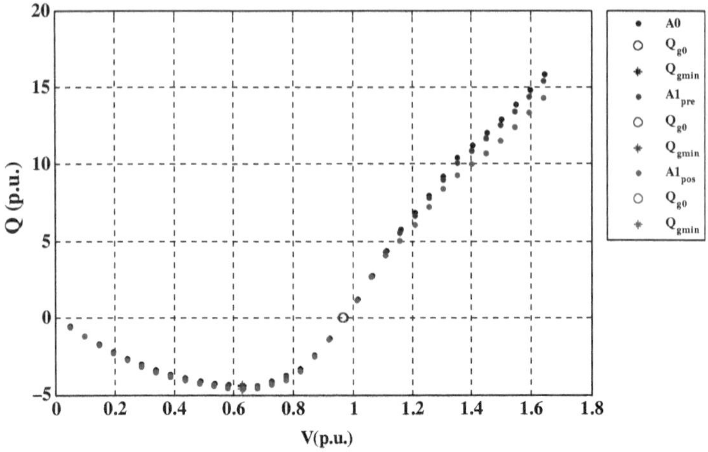

FIGURE 10.15 *QV* curves—A0, A1 prior and postdispatch.

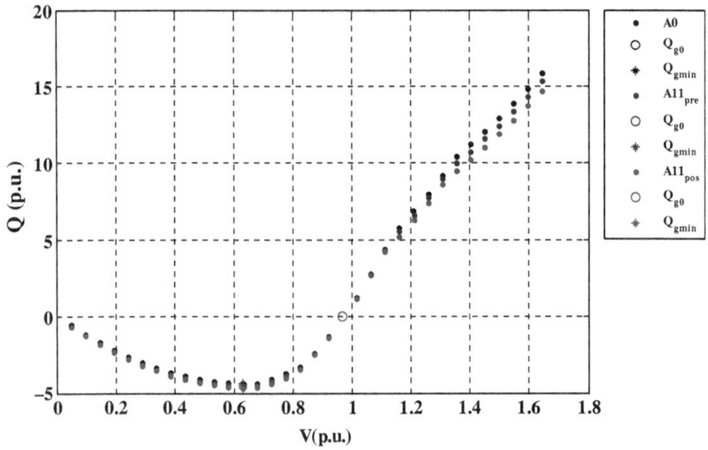

FIGURE 10.16 *QV* curves—A0, A11 prior and postdispatch.

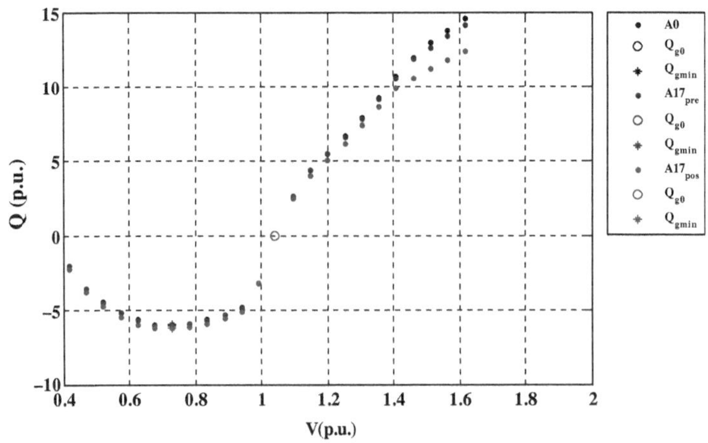

FIGURE 10.17 *QV* curves—A0, A17 prior and postdispatch.

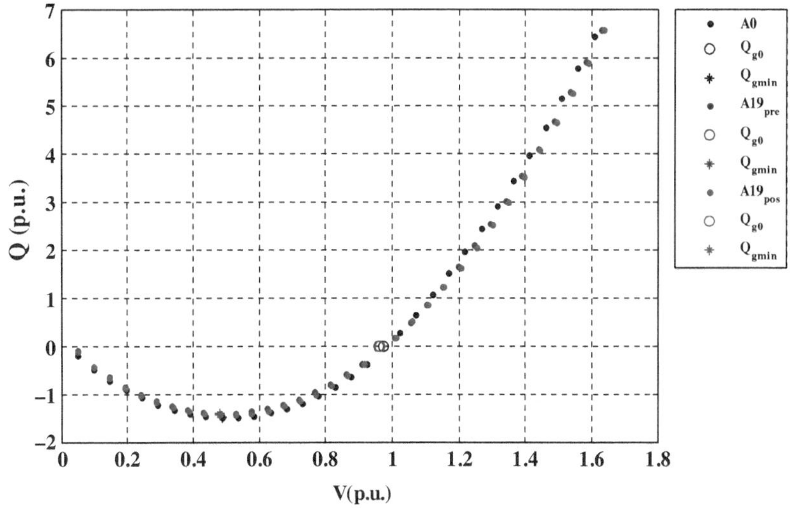

FIGURE 10.18 *QV* curves—A0, A19 prior and postdispatch.

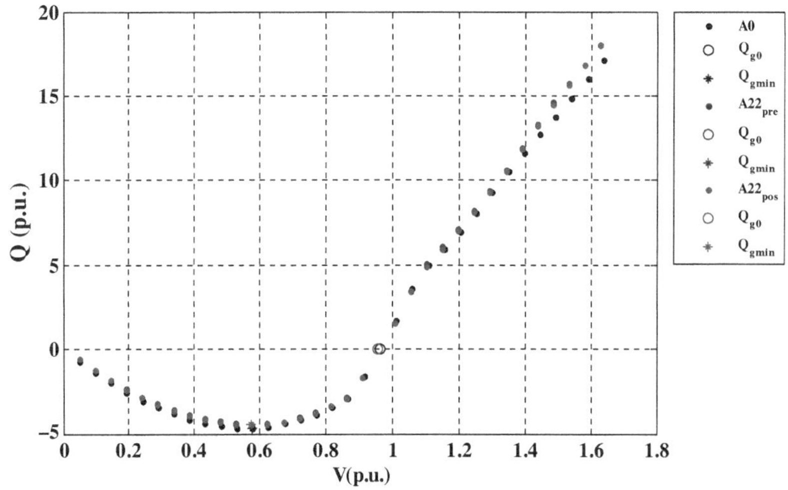

FIGURE 10.19 *QV* curves—A0, A22 prior and postdispatch.

10.9 Conclusion

This chapter presents several methods for decision-making and evaluation of reactive power dispatch in power systems planning and operation.

First, the placement of RES considering the system voltage stability and loss reduction as allocation criteria is developed. Following that, voltage collapse indexes based on *PV* and *QV* curves are presented to evaluate the system operating conditions and voltage safety margin. These indexes quantify how far the system is from voltage collapse, and how close is each bus of interest from the instability region.

Finally, a methodology for reactive power dispatch using PSO considering the tangent-vector for generating selection is proposed. The results show that an efficient dispatch of reactive power is achieved by the PSO method, which intelligently determines the adequate contribution of each generating to minimize system losses. At last, the results for the IEEE 118-bus test system indicated that for all scenarios, the active and reactive power margins increased after the proposed method application.

Acknowledgments

The authors thank to CNPq, CAPES, FAPEMIG, and INERGE for partially supporting this work. The author Maíra R. Monteiro is a scholarship holder of CNPq—Brazil. The author Yuri R. Rodrigues especially thanks CAPES Notice No. 18/2016 of the Full Doctoral Program Abroad/Process no. 88881.128399/2016-01.

References

[1] R.C. Leme, A.C.Z. Souza, J.C.S. Souza, K.L. Lo, Charging reactive power considering system security aspects, Int. J. Electr. Power Energy Syst. (2010) 203–4604.

[2] A.C.Z. Souza, Tangent vector applied to voltage collapse and loss sensitivity studies, Electr. Power Syst. Res. 47 (1) (1998) 65–70.

[3] K. Mahadevana, P.S. Kannanb, Comprehensive learning particle swarm optimization for reactive power dispatch, Appl. Soft Comput. 10 (2) (2010) 641–652.

[4] M.M. Rojas, A. Sumper, O.G. Bellmunt, A.A. Sundrià, Reactive power dispatch in wind farms using particle swarm optimization technique and feasible solutions search, Appl. Energy 88 (12) (2011) 4678–4686.

[5] B. Zhao, C.X. Guo, Y.J. Cao, A multiagent-based particle swarm optimization approach for optimal reactive power dispatch, IEEE Trans. Power Syst. 20 (2) (2005) 1070–1078.

[6] S. Ganguly, Multi-objective planning for reactive power compensation of radial distribution networks with unified power quality conditioner allocation using particle swarm optimization, IEEE Trans. Power Syst. 29 (4) (2014) 1801–1810.

[7] H.T. Yang, J.T. Liao, MF-APSO-based multiobjective optimization for PV system reactive power regulation, IEEE Trans. Sustain. Energy 6 (4) (2015) 1346–1355.

[8] V.N. Costa, M.R. Monteiro, A.C.Z. Souza, Particle swarm optimization applied to reactive power compensation, in: 17th International Conference on Harmonics and Quality of Power, Belo Horizonte, 2016.

[9] A.A. Esmin, G.L. Torres, A.C.Z. Souza, A hybrid particle swarm optimization applied to loss power minimization, IEEE Trans. Power Syst. 20 (2) (2005) 859–866.

[10] L.C.A. Ferreira, A.C.Z. Souza, S. Granville, J.W.M. Lima, Interior point method applied to voltage collapse problems and losses reduction, IEE Proc. Gener., Transm. Distrib. 149 (2) (2002) 165–170.

[11] P.F. Ribeiro, K. Schlansker, A Hybrid particle swarm and neural network approach for reactive power control, 2003. Available from: https://www.researchgate.net/publication/242452972_A_Hybrid_Particle_Swarm_and_Neural_Network_Approach_for_Reactive_Power_Control.

[12] A.C.Z. Souza, C.A. Cañizares, V.H. Quintana, New techniques to speed up voltage collapse computations using tangent vectors, IEEE Trans. Power Syst. 12 (3) (1997) 1380–1387.

[13] F.W. Mohn, A.C.Z. Souza, Tracing PV and QV curves with the help of a CRIC continuation method, IEEE Trans. Power Syst. 21 (3) (2006) 1115–1122.

[14] A.C.Z. Souza, Using PV and QV curves with the meaning of static contingency screening and planning, Electr. Power Syst. Res. 81 (7) (2011) 1491–1498.

[15] A.C.Z. de Souza, L.M. Honório, G.L. Torres, G. Lambert-Torres, Increasing the loadability of power systems through optimal-local-control actions, IEEE Trans. Power Syst. 19 (1) (2004) 188–194.

[16] J. Kennedy, R.C. Eberhart, Particle swarm optimization, in: Proceedings of IEEE International Conference on Neural Networks, Perth, Australia, vol. 4, 1995, pp. 1942–1948.

[17] R.C. Eberhart, J. Kennedy, A new optimizer using particle swarm theory. in: Proceedings of the Sixth International Symposium on Micro Machine and Human Science, Nagoya, Japan, 1995, pp. 39–43.

[18] Y. Shi, R.C. Eberhart, A modified particle swarm optimizer, in: IEEE International Conference of Evolutionary Computation, Anchorage, Alaska, May 1998.

[19] D. Corne, M. Dorigo, F. Glover (Eds.), New Ideas in Optimization, McGraw Hill, 1999.

[20] M.A. Abdullah, A.P. Agalgaonkar, K.M. Muttaqi, Probabilistic load flow incorporating correlation between time-varying electricity demand and renewable power generation, Renew. Energy 55 (2013) 532–543.

[21] W.D. Soto, S. Klein, W. Beckman, Improvement and validation of a model for photovoltaic array performance, Sol. Energy 80 (1) (2006) 78–88.

[22] M.R. Monteiro, A.C. Zambroni de Souza, B.I.L. Lopes, The influence of renewable generation in voltage collapse indexes, in: 6th International Conference on Clean Electrical Power, Santa Margherita Ligure, Italy, 2017.

Chapter 11

Decision-making-based optimal generation-side secondary-reserve scheduling and optimal LFC in deregulated interconnected power system

Hassan Haes Alhelou[1,2] and M.E.H. Golshan[1]

[1]*Department of Electrical and Computer Engineering, IUT, Isfahan, Iran,*

[2]*Department of Electrical Power Engineering, Tishreen University, Lattakia, Syria*

Nomenclatures and abbreviations

a_{ij}	participation factor between area i and j (p.u.)
B_i	frequency-bias parameter (p.u. MW/Hz)
c_1, c_2	generation cost vectors
c_{up}, c_{down}	reserve cost vectors
c_2	vector c_2 on the diagonal of matrix $[c_2]$
$d_{up}, \ d_{down}$	the distribution vectors
D	damping coefficient (p.u. Hz)
f	frequency deviation (Hz)
g	earth's gravitational field
H	inertia constant (p.u. s)
i, j	subscript referred to area i or j (1, 2, 3)
J	objective function
K_d	derivative coefficient
K_i	integral coefficient
K_R	the gain of the reheat system
K_p	proportional coefficient
K_{ps}	power system gain of area i (Hz/p.u. MW)
N_G, N_w	generating units and wind power plants
N_L, N_l, N_b	loads, lines, and buses
N	filter coefficient
P_w	the probability distribution of the wind power vector
P_m	the generation-load mismatch

Decision Making Applications in Modern Power Systems. DOI: https://doi.org/10.1016/B978-0-12-816445-7.00011-6

$\Delta P_{tie,ij}^{sch}$	the scheduled power flow through tie-lines
$\Delta P_{tie,ij}^{act}$	the actual power flow through tie-lines
$\Delta P_{Tie,ij}$	incremental change in tie-line power (p.u.)
ΔP_L	load power change
ΔP_m	mechanical power (p.u.)
R	governor speed regulation parameter (Hz/p.u. MW)
$R_{up,t}^i$, $R_{down,t}^i$	the probabilistically worst case up−down spinning reserves
R_t^i	the power correction term
ST	settling time (s)
t	time (s)
T_g	governor time constant (s)
T_{ij}	synchronizing coefficient of tie-line (p.u.)
T_{ps}	power system time constant (s)
T_R	reheat system time constant (s)
T_t	turbine time constant (s)
t^{sim}	simulation time (s)
$u_{c,i}$	the control signal
λ	the integral order
μ	the derivative order
Υ	a set of the indices corresponding to outages of all components
Υ_l	the set of branch outage index
Υ_L	the set of load outage index
Υ_G	the set of generators outage index
AGC	automatic generation control
AVR	automatic voltage regulator
FOPID	fractional-order PID
hGSA-PS	hybrid gravitational search and pattern search algorithm
ICA	imperialist competitive algorithm
LFC	load-frequency control
OPF	optimal power flow
PID	proportional−integral−derivative
RERs	renewable energy resources
SCADA	supervisory control and data acquisition
SC-OPF	security-constrained optimal power flow
TSO	transmission system operator
WAMS	wide-area measurement system
WDO	wind driven optimization

11.1 Introduction

A large modern power system under deregulation consists of several interconnected control areas, where each one is responsible for supplying its loads and keeping the scheduled power interchanges with its neighbor areas. These responsibilities gradually become more difficult when moving toward smart grid and deregulation concepts. Load-frequency control (LFC) is a technique adopted in the power system control center to guarantee the balance between generation and demand and consequently to maintain the frequency in its acceptable level in each area. Moreover, load-frequency controller should be able to maintain the exchanged active powers through different tie-lines at their scheduled values [1]. To this end the parameters of load-frequency controllers in each control area should be tuned optimally to achieve a suitable performance.

In the past, several control methods and techniques have been proposed for both generation-side reserve scheduling and LFC. In general the main aim of using secondary reserve–based LFC is to regulate frequency deviations caused by small disturbances such as the uncertainties of renewable sources and load fluctuation [2]. For LFC, proportional–integral and proportional–integral–derivative (PID) are widely adopted in industrial power systems due to their simplicity [3]. On the other hand, modern techniques are suggested in the literature for LFC, such as different fuzzy PID controller structures [4,5], two and three degree-of-freedom (DOF) integral-derivative controllers [6,7], and fractional-order PID (FOPID) controller [8,9]. It is worth mentioning that fractional calculus–based control technique, which is another type of controller that provides more DOFs, performs better in comparison with traditional PID controllers. In some researches [10,11], in order to eliminate the noise of the differentiation path in PID controller, PID with derivation filter controllers have been adopted.

Trial-and-error approach can be used to tune load-frequency controllers in power systems [1,2]. However, it is not an easy task to tune the controllers' parameters using trial-and-error approach. In addition, it might not lead to the optimal parameters. Hence, due to its importance in improving the control performance, a number of optimization methods have been used for the optimal tuning of load-frequency controllers in interconnected power systems. In general, methods such as evolutionary computing-based controllers' parameters tuning, model predictive control, and optimal control have been suggested for LFC in interconnected power systems [12–15]. From a survey on the literature, it can be seen that evolutionary algorithms have received a considerable attention from the researchers due to their good performance and simplicity. In this regard, a number of algorithms, such as particle swarm optimization [16], genetic algorithm [17], deferential algorithm [18], bacterial foraging optimization [18], firefly algorithm [19], imperialist competitive algorithm (ICA) [20], hybrid gravitational search and pattern search (hGSA-PS) algorithm [21], and many other algorithms [22–25], have been adopted for solving the problem of tuning load-frequency controllers' parameters. Recently, wind-driven optimization (WDO) algorithm is used for LFC in which it is verified that WDO is superior to other algorithms in improving the LFC performance [26]. Interested readers are referred to the recent literature survey [27].

The literature survey shows a knowledge gap regarding LFC design for future power systems considering the high penetration level of renewable energy resources (RERs) and their uncertainties. Moreover, it has been highlighted that the effect of electric vehicles in future interconnected power systems regarding the frequency behavior needs more investigations. Likewise, the traditional LFC schemes used for conventional power systems cannot achieve the requirements of the modern power systems under deregulation environments. Therefore more reliable and robust frequency control schemes are needed for future power systems.

Due to its importance in the reality, in this chapter, a probabilistic framework to design an $N-1$ secure day-ahead dispatch while determining the minimum cost reserves for power systems with wind power generation is introduced. A strategy where the reserves can be deployed as a corrective action is suggested. In order to construct a reserve decision scheme the steady-state behavior of the secondary frequency controller is considered. In this case, he deployed reserves are a piecewise linear function of the total generation-load mismatch. The chance constraints include not only probability of satisfying the transmission capacity constraints of the lines and generation limits but also the reserve capacity limits. A convex reformulation and a heuristic algorithm are proposed to achieve tractability. Likewise, in this chapter, a new fractional-order control scheme is suggested for future power systems with high penetration level of RER and electric vehicles. The DOF of the optimization problem of LFC is increased by using the fractional-order controllers, which leads to much better performance of LFC. In addition, the participation of electric vehicles (EVs) in providing secondary reserve for future smart grid is studied in this chapter along with a new participation method. Furthermore, the controllers' parameters are tuned via several evolutionary algorithms such as ICA and differential algorithm (DE). Moreover, several numerical analyzes are carried out to assess the performance of the proposed control scheme. Likewise, the effectiveness of EVs and RERs participation in LFC is examined. In addition, several objective functions are used to define the optimization problem, and their performances are compared. Finally, the robustness of the designed load-frequency controllers based on evolutionary algorithms is investigated by changing the parameters of power system under some conditions.

The rest of this chapter is organized as follows. Section 11.1 introduces this chapter. An overview of power system operation and decision-making is provided in Section 11.2. Decision-making application to reserve scheduling is introduced in Section 11.3. Section 11.4 introduces decision-making application to LFC. The power system under investigation with the simulation results is presented in Section 11.5. Section 11.7 concludes and proposes future research directions.

11.2 Power system operation and decision-making

Electrical power systems generally consist of generation, transmission, and distribution supplying the bulk of energy, which is critical both for domestic and industrial uses. Any disturbance to the power supply usually leads to detrimental effects both to the society and economical activities. It is key and important issue that the safe and reliable operation of power systems be guaranteed at all times. Therefore the primary function of a power system is to supply its customers with electrical energy as economically as possible with acceptable reliability and quality levels.

In general, reliability can be defined as the ability of the power system to provide the desired level of service continuously over an extended time horizon apart from only a few instances where this service is interrupted [28–30]. A power system is said to be reliable if the customer power requirements can be met on demand and to be secure if it can withstand unforeseen disturbances. Based on the definitions given to guarantee that the power system is reliable, it is a prerequisite that power system must be secured for the greatest part of its operation time. Power systems can be classified as dynamic time-varying systems; hence, to achieve a secure and reliable performance, certain conditions need to be met to guarantee that. Mainly frequency and voltage should be maintained within power system operation limits, equipment overloading should be avoided at all times, for example, if generators are overloaded, life span of the equipment can be significantly reduced.

In Ref. [30], five different states of power system operation are defined based on the ability of the system to withstand disturbances. These states of operation are as follows:

1. *Normal operation*

   Generally, the power system is said to be in a normal state if all operational limits are satisfied and is operating in a secure mode. Therefore during a disturbance, no system limits must be violated. In the first stage, when a disturbance occurs, which can be loss of generation or sudden load increase, the automatic voltage and frequency controllers are activated to keep the frequency and voltage within acceptable levels. The automatic control loops are also present in the other states of operation of the power system. In this state the economic operation of the power system is also of paramount importance. Therefore to minimize the operational costs, changes in generation of the power units are inevitable after a power disturbance, known as economic dispatch approach.
2. *Alert operation*

   A power system is said to be alert if the system that was previously operating with no limit violations fail to operate within its limits after a disturbance. In this condition, preventive control actions are employed to return the power system to the normal state since system security is at stake. Preventive control approaches try to balance the generation and load after a disturbance and the process may involve in increased system reserves, topological changes, etc.

3. *Emergency state*

   For a system operating in the alert state, if in case disturbance operational limits are violated, the system automatically enters the emergency state mode. In this case instead of preventive actions, corrective or emergency control actions are necessary to bring the power system back to the normal or alert state. Corrective control approaches may involve exciter control, fast generation reduction or increase, generation unit tripping, high voltage direct current (HVDC) modulation, system protection devices, load curtailment, etc.
4. *Extreme state*

   In this state the corrective action would have failed to operate satisfactorily, and as a result, a series of events occurs, and parts of the system may be disconnected. To prevent total system collapse or extensive blackouts, severe actions such as load shedding and controlled islanding should be implemented.
5. *Restorative state*

   When the electric system is rendered stable and the fault is cleared, the power system finally enters the restorative state. In this state, control actions are taken in steps to reconnect the lost parts of the system until the normal state in which the system was before a disturbance is achieved.

11.2.1 Real-time operation

Interconnected power systems are usually subdivided into different control areas, where each area may represent one country or part of a system for bigger countries. The transmission system operator (TSO) is the responsible entity for the security of a single control area. Nowadays, each area is monitored and controlled by the TSO through an IT infrastructure, commonly known as the supervisory control and data acquisition (SCADA). Recently, SCADA systems, however, are replaced by wide-area measurement system (WAMS) and control due to its superiorities.

The WAMS measures data using remote devices, which are installed at strategic points throughout the grid, and the information is gathered at one control center through communication channels. This data is processed by computer systems, and it gives the system's operating state in real time. Control commands, which are to be sent from the center back to the system, are determined based on the system state. The system is also equipped with local control devices, which helps to protect the equipment and to provide system-wide services after specific commands have been sent. Generally, voltage and frequency control and the security level assessment are the main tasks so as to keep the system in the normal state.

- *Voltage control*

  Voltage control of power systems is mainly provided by the generating units. Terminal voltage of generating units is controlled and maintained within specific limits by means of the automatic voltage regulator (AVR). This control loop is local, and the set point of the terminal voltage can be adjusted through the SCADA system or WAMS. Besides the AVR, other methods of voltage control include tap-changing transformers, static Var compensators, and synchronous condensers.
- *Frequency control*

  Frequency of an electric system depends on the active power balance. If a generation-load active power imbalance occurs, the frequency will either increase or decrease, thus moving away from its nominal operating value. In order to control the frequency the active power balance must be restored. Since some generating units, which can quickly increase or decrease their active power production, are therefore mainly used in frequency control.

  The frequency control is mainly divided in three control levels: primary; secondary, that is, automatic generation control (AGC) or LFC; and tertiary.

  Primary frequency control (PFC), here, a local proportional controller, is used to measure frequency deviations and adjust the active power production of the corresponding generating unit. The response of the PFC is usually quick and in the scale of seconds and fast generating units participated in this control action.

  In the AGC/LFC scheme, it is performed through the WAMS, and only preselected generating units participate in the control area. The aim is to make sure that frequency deviation is restored back to zero and also to maintain the power flow on the tie-lines that connect it with the other control areas at its prescheduled value. The response generally takes a few minutes; therefore it is possible for slower units to participate. The units that participate in this scheme must have predisturbance production set point where a sufficient margin from their capacity limits can be used once the frequency control is activated. Each generating unit can have a reserve capacity margin for primary and/or secondary frequency control.

  Tertiary frequency control, after the primary and secondary activation of the automatic frequency control loops, tertiary frequency control takes place. This is a manual process, and the main purpose is to release the deployed primary and secondary control reserves while performing an economic dispatch.
- *Security assessment*

  A commonly used security criterion measure is the $N-1$ security criterion, where the system is supposed to be secured if it can withstand a predefined set of credible single contingencies. The contingencies include outages that are likely to occur with higher probability, such as a single outage of a line, a transformer or a generator, and, in some cases, simultaneous component outages. As earlier indicated, the definition of security implies that both during and after a disturbance no operational limits violation should occur. Therefore security can be further classified as static security assessment and dynamic security assessment. The security of the system through the SCADA can either be continuously assessed or only when it is necessary. However, in practice, a static security assessment is performed where contingency screening is done. Here only the most severe contingencies are used for the dynamic security assessment, which involves dynamic simulations and evaluation of the transient state trajectory.

When the system is rendered insecure, preventive control actions, such as adjustment of the generation schedule to restore the system to a secure operation, are employed by the TSO. If a disturbance occurs before the security is restored, the state of the system may enter the emergency state, and additional actions should be taken.

11.2.2 Decision-making-based planning and economic operation

The actions that the TOS is performing when approaching a secure real-time operation can be divided into three main categories: day-ahead operation planning, short-term planning, and long-term planning [31].

Examples of long-term planning tasks are load forecast and identification of the new system conditions, investigation of system extensions, and control actions planning (preventive, corrective). Short-term planning has the following tasks: procurement of reserve power, approval of maintenance decisions, etc. Security assessment is not only limited to real-time operation but is also performed in the rest of the planning phases. In the latter case the evaluation of security is done over different possible scenarios of the system conditions. In day-ahead operation planning the responsibilities of the TSO are to schedule a day-ahead unit commitment, a generation dispatch, and make decisions about the reserve procurement in certain control areas, while minimizing the operational costs and satisfying the security requirements. The process of identifying the optimal generation dispatch while satisfying the network constraints is traditionally referred to as optimal power flow (OPF). The OPF problem has many variables that, in addition to generation dispatch, may also include set points for tap changers, phase shifters, generation terminal voltage, all of which can be adjusted to give a more optimal cost performance.

If static security constraints are included in OPF, it is called the security-constrained OPF (SC-OPF). In the SC-OPF the control variables that correspond to a disturbance-free scenario represent preventive control actions. If the optimal values of control variables are available, the system will result in a secure steady-state point, which depends only on the existing automatic control loops. Different control variables corresponding to each contingency represent corrective control actions that need to be taken for that contingency. Here, after a contingency, the system will result in a secure steady-state point of operation depending on both the existing automatic control loops and the new set points that result from specific corrective actions. Fast and slow corrective control actions must be clearly distinguished and are applied appropriately taking into account the time duration for which the components are allowed to be overloaded. For instance, power flow line limits can be separated into two levels. First, the steady-state limits should be met for a continuous operation. These limits may be violated for some minutes as long as they do not exceed the emergency limits, which, if violated, will result in catastrophic effects, such as line tripping. Devices with different time constants could be scheduled to offer corrective control dealing with different component limits. Generally, TSOs do not include security requirements in the day-ahead schedule optimization problem but only perform an a posteriori security analysis. In the event in which the security assessment shows that the system security will be compromised, control actions are employed until a secure dispatch is obtained. A priori-contingency power flow analysis can incorporate also new postcontingency set points for the devices that offer corrective control.

To satisfy the security requirements a sufficient reserve power must be available to balance the system after a contingency. Ideally, the reserves must be sufficient enough to supply power in the event that the largest generator in the system trips or must correspond to a percentage of the peak load. It can be seen that optimal day-ahead planning is of great importance to ensure that the system is secure. Due to the complexity of the problem, the different underlying market mechanisms and the level of system uncertainty, different alternative implementations have been already proposed; however, obtaining a satisfactory solution is still subject of ongoing research.

11.2.3 Operation and planning problems to be addressed

In planning for the day-ahead operation of power systems, a secure and economic schedule for the generating units and the reserves must be designed. This, however, comes at the expense of additional investment and operational costs, thus revealing the trade-off between a secure and an economic system operation. In an ideal setup, where the system is considered deterministic, there are ways to satisfy the minimum cost operating point while at the same time satisfying the desired security level. Power systems, however, are essentially stochastic since they are subjected to stochastic power flows, load uncertainty, unpredictable component outages, etc.

The operation of a power system under uncertainty has been a subject of key research. Regardless of the wide research, there is still no specific accepted approach or a unified basis to quantify the trade-off between a secure and an economic performance. As a consequence, decision-making in the presence of uncertainty has often resorted to ad hoc or rule-based methodologies, leading either to a design that is conservative in terms of cost for the desired security level or, if uncertainty is ignored in the design phase, to a solution where security might be at stake. Performance with ignored uncertainty might be acceptable if the level of uncertainty of the system is relatively low. But with the increase in renewable generation due to environmental concerns and the pursuit of sustainable energy sources system, uncertainty has generally increased. The main challenges with renewable energy sources (RESs), such as wind and photovoltaic power generation, are that they are nondispatchable, fluctuating, and uncertain. As a result, employing suboptimal measures to account for the stochastic nature of RES may result in an undesirable economic effect. On the other hand a deterministic design where uncertainties are ignored will lead to an unacceptable reliability level.

It is, therefore, necessary to develop a mechanism for optimal decision-making in the presence of uncertainty that takes into account the multiobjective nature of the problem, that is, the trade-off between security and economic operation. The increased share of RES results in an increased amount of required reserves, which may have an opposite effect both from an environmental and economic point of view. The latter raises the need of cheaper and environmental friendlier reserves providers. Demand-side resources have already been used to provide certain control services, but the full exploitation of their controllability has recently become an emerging research topic.

Demand response and storage resources could be utilized to offer ancillary services including reserve provision. Promising technologies, such as electric vehicles and thermostatically controlled loads, could contribute with a large amount of reserve capacity and hence allow for the integration of high shares of RERs.

However, these technologies include uncertainty, mainly introduced due to human behavior and weather conditions, rendering their successful exploitation challenging. Taking the uncertainty into account in the decision-making mechanism introduces additional operational costs compared with a deterministic solution. To alleviate this the controllability of certain network components other than the loads could also be exploited. These components could be utilized for preventive and corrective control actions. Some examples of controllable components are flexible alternative current transmission system (FACTS) devices, HVDC lines and transformers. These components do not provide reserve capacity, but their set point can be modulated in a post–disturbance situation, thus leading to lower operating costs.

This dissertation deals with the problem of developing a unified stochastic framework for optimal decision-making, taking the uncertainty due to

RES and the demand side into account, while exploiting the controllability of certain network components.

The main problems that need to be taken into account are as follows:

1. *Probabilistic security:* Probabilistic variants of deterministic SC-OPF problems need to be developed, providing enough flexibility to quantify the trade-off between security and economic system operation.
2. *Production and generation-side reserve scheduling:* Within a security-constrained probabilistic framework standard, day-ahead planning problems such as production and reserve scheduling need to be revisited.
3. *Exploiting demand response for reserve provision:* In an uncertain environment, demand-side resources should be taken into account a decision mechanism to provide ancillary services while reducing the cost that would occur is reserves were solely purchased from the generating units.
4. *Exploiting component controllability:* Corrective control actions offered by certain network components could result in a more economic operation of the network, especially in the cases where the level of uncertainty is increasing.
5. *Development of new algorithms and tools:* To address the problem of taking optimal decisions in the presence of uncertainty, new algorithms for stochastic scheduling with guaranteed performance need to be developed, and the (probabilistic) properties of the obtained solutions should be reinterpreted.

11.3 Decision-making application to reserve scheduling

Due to the ever increasing installed capacity of RES, for example, wind and photovoltaic, which are ever changing and are weather dependent, it is necessary to revisit certain operational concepts, such as $(N-1)$ security and reserve scheduling. In this framework the power required to balance the system is compensated by each generator with a fixed percentage, that is, fixed distribution vector; hence, the reserves of each generator are then determined by the worst-case value of the power mismatch. Here, the required reserves that the systems operator needs to purchase via the probabilistic approach can be determined but do not optimally distribute them to the generating units.

The aim of this section is to optimally allocate the reserve requirements to the generators.

In today's different electricity markets, the goal is to minimize the generation dispatch and the reserve costs, while satisfying the network constraints. The generation dispatch is determined by the energy market, while the network constraints are determined by transmission market, so that the network security is guaranteed, for example, the $N-1$ security criterion. Usually, a reserve capacity of units is predetermined and their optimality in dispatch is determined by the reserve market. These different markets can either be obtained sequentially in the unbundled market systems or in the same optimization problem in the integrated market systems [32]. In Ref. [33] the effectiveness and the advantages of both systems are assessed. In reality the sequential approach is usually applied. However, this approach gives a sub-optimal optimization solution to the overall objective, and feasibility issues can also arise. For example, if the reserve schedule is first determined while neglecting the $N-1$ criterion and all the reserves are allocated to the cheapest generator, therefore there is no feasible solution to an $N-1$ secure energy scheduling if this generator is tripped, since no other unit can provide the reserves that are required to compensate its production.

This is one of the worst-case scenarios, which shows that the reserves may not be adequate in unbundled market systems. However, in practice, heuristics are used to take care of such extreme issues. As a result, an integrated market mechanism allows us to ascertain the optimal solution to the overall problem. In this line a framework dealing with the cooptimization of energy and reserves, which takes into account network constraints and the $N-1$ security criterion is developed. In Refs. [32–37] the reserve optimization for a security-constrained market clearing context while maximizing the expected social welfare is presented. In Ref. [32] a multistage stochastic unit-commitment program, which models the uncertainty in generation by using reduction techniques to ensure tractability of the problem, is outlined. The limitation of these methods is that they do not guarantee reliability of the resulting solution.

This section presents a unified framework that simultaneously solves the problem of designing an $N-1$ secure day-ahead dispatch for the generating units, while determining the minimum cost reserves and the optimal way to deploy them. A probabilistic methodology which guarantees the satisfaction of the system constraints is used to account for wind power inconsistency. The security constraints emanating from the $N-1$ criterion are first integrated to a DC-OPF problem and formulate a stochastic optimization problem with chance constraints.

By modeling the steady-state behavior of the secondary frequency controller, LFC controllers, the reserves can be represented as a linear function of the total generation-load mismatch. Generation-load imbalance can be a result of difference between the actual wind and its forecast, or a generator load loss.

Different ways of reserve distribution, which are based on the type of mismatch offering an implementation of corrective security, are introduced in literature. The overall objective formulation includes both preventive and corrective control [33]. Preventive control actions are the generation dispatch and the reserve capacity determination, while the contingency-dependent reserves allocation in real-time operation is corrective control. The advantages of these strategies are their physical intuition and the decision variables, which do not grow with the number of uncertainty realizations as in Ref. [32–37]; thus the resulting solution is less conservative compared with [34].

11.3.1 Problem setup and reserve representation

In this work, we consider a power network consisting of N_G generating units, N_w wind power plants, N_L loads, N_l lines, and N_b buses. Taking into consideration that Υ is a set that includes the indices corresponding to outages of all components also including the index 0 that corresponds to the base case of no outage and denoted by $|\Upsilon|$ its cardinality. Υ_l, Υ_L, and Υ_G are the set of indices representing the branch, load, and generator outages, respectively.

The following assumptions are considered for the problem formulation:

1. A DC power flow approximation is considered.
2. High-accuracy load forecasts are assumed.
3. Line outages do not lead to multiple generator/load failures.
4. The on–off status of the generating units has been fixed a priori by solving a unit-commitment problem.

The first assumption is basic for these types of optimization problems while the second and third one are meant to simplify the presentation of the results and could still be captured by the proposed algorithm. If the last assumption is removed by incorporating the unit-commitment problem, the objective would give rise to a mixed-integer problem. This can be tackled using the probabilistically robust design that can deal with a specific class of nonconvex problems.

The results of generation-load mismatches in frequency deviations from the nominal and reserves are used to balance the mismatches. The process is achieved by the activation of the AGC, LFC, where its output is distributed to certain participating generators. The set point of each generator is changed by a certain percentage of the overall active power to be compensated. The existing setup of the AGC loop is shown in Fig. 11.1, demonstrating the role of the distribution vector.

This distribution vector results from the market that determines the secondary frequency control reserves, and it remains constant until the next market auction. However, this task is performed while neglecting the network constraints. Ideally, the distribution vector is the same for all possible outages but may differ between up-spinning and down-spinning reserves. Here different distribution vectors depending on the outage are also considered in addition to distinguishing between up-spinning and down-spinning reserves. An optimal reserve schedule, which takes into account the network security constraints, is determined over the distribution vectors. Using this approach, both the minimum cost reserves per generator and also a reserve

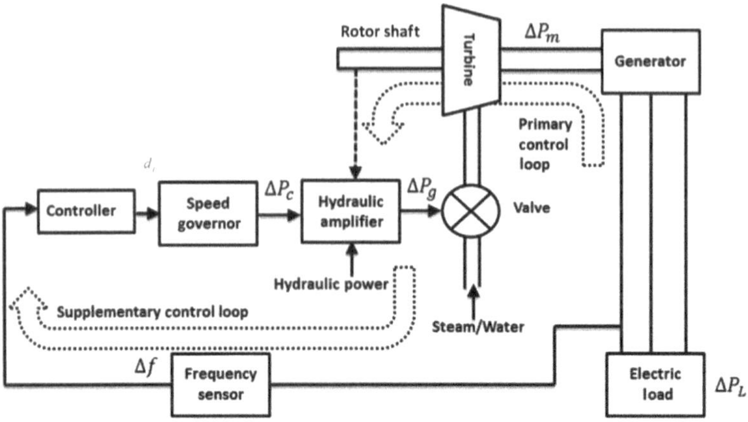

FIGURE 11.1 Schematic diagram of LFC/AGC system. *AGC*, Automatic generation control; *LFC*, load-frequency control.

strategy, which can be deployed in real-time operation, can be computed simultaneously.

This strategy makes use of the distribution vectors. Depending on the outage magnitude and the wind power deviation, the amount of power by which each generating unit should adjust its production can be determined. The proposed methodology is an alternative to other methods for reserve scheduling, which account implicitly for real-time response via their day-ahead decisions. A power correction term R^i is defined as a piecewise linear function of the total generation-load mismatch. This term shows the amount of the power that each generator should compensate for given an imbalance and is directly related to the reserves.

$$R^i(P_W) = d_{up}^i \max_+ \left(-P_m^i(P_W)\right) - d_{down}^i \max_+ \left(P_m^i(P_W)\right), \quad i \in \Upsilon \tag{11.1}$$

where $\max + (\cdot) = \max(\cdot; 0)$. Variable $P_m \in \mathfrak{R}$ denotes the generation-load mismatch, which for each outage is given by

$$P_m^i(P_W) = \sum_{k \in Z_w/K^i} \left(P_{w,k} - P_{w,k}^f\right) - c_L^i P_L + c_G^i P_G + c_w^i P_w, \quad \text{for all } i \in \Upsilon \tag{11.2}$$

d_{up}, $d_{down} \in \mathfrak{R}^{NG}$ are the distribution vectors. The sum of these elements must be equal to one and, if a generator is not contributing to the AGC, the corresponding element in the vector will be zero. To distinguish between up-spinning reserves and down-spinning reserves, the indices "up" and "down" are used. If P_m is negative, up-spinning reserves are provided, and the production of the generators is increased accordingly. In the opposite case the second term of (11.1) is active, and down-spinning reserves are provided. It should be noted however that the elements of d_{up}, $d_{down} \in \Re^{NG}$ may be non-positive. In the base case scenario where there are no outages, the power mismatch is negative $P_m < 0$, and some elements of d_{up} are also negative. In this case the network is congested, hence to relieve it, the generators corresponding to d_{up} negative should provide down-spinning reserves, while the rest of the units would provide up-spinning reserves.

11.4 Probabilistic security-constrained reserve scheduling

An optimization horizon $N_t = 24$ with hourly steps 1 is considered, the subscript t indicate the value of the quantities for a given time instance $t = 1, 2, 3, \ldots, T$. For the production cost a quadratic form is considered, and for the reserves a linear cost is also considered [38]. Let $c_1, c_2, c_{up}, c_{down} \in \Re^{NG}$ be generation and reserve cost vectors and $[c_2]$ denotes a diagonal matrix with vector c_2 on the diagonal

For each step t, the vector of decision variables is defined as follows:

$$x_t = \left[P_{G,t},\ \left[d^i_{up,t},\ d^i_{down,t},\ R^i_{up,t},\ R^i_{down,t}\right]_{i \in \Upsilon}\right] \in \Re^{NG+4NG(1+|\Upsilon|)} \tag{11.3}$$

where $R^i_{up,t}$, $R^i_{down,t}$ are the probabilistically worst-case up–down spinning reserves that the system operator needs to purchase for every $i \in \Upsilon$. Therefore the optimization problem is written as

$$\min_{\{x_t\}_{t=1}^{N_t}} \sum_{t=1}^{T} P^i_m(P_w) = \sum_{k \in Z_w/K^i} \left(c_1^T P_{G,t} + P_{G,t}^T [c_2] P_{G,t} + \sum_{i \in Z} \left(c_{up}^T R^i_{up,t} + c_{down}^T R^i_{down,t}\right)\right) \tag{11.4}$$

11.4.1 Deterministic constraints

These are constraints that correspond to a case where the wind power is equal to its forecast. Here the reserves are determined based on the generation-load mismatch that may occur due to an outage.

$$1^T\left(C_G P_{G,t} + C_W P^f_{w,t} - C_L P_{L,t}\right) = 0 \tag{11.5}$$

$$-\overline{P}^i_f \le A^i_t P^i_{inj,t}\left(P^f_{w,t}\right) \le \overline{P}^i_f \tag{11.6}$$

$$\underline{P}^i_G \le P^i_{G,t} + R^i_t\left(P^f_{w,t}\right) \le \overline{P}^i_G \tag{11.7}$$

$$-R^i_{down,t} \le R^i_t\left(P^f_{w,t}\right) \le R^i_{up,t} \tag{11.8}$$

$$R^i_{up,t},\ R^i_{down,t} \ge 0 \tag{11.9}$$

$$1^T d^i_{up,t} = 1, \quad 1^T d^i_{down,t} = 1 \tag{11.10}$$

11.4.2 Probabilistic constraints

These are constraints that deal with the uncertainty of the wind power forecast. The reserves are now characterized by both the generation-load mismatch that may occur due to the wind power forecast error and the outages. We thus have that for all $t = 1, 2, 3, \ldots, N_t$,

$$\begin{aligned} &\mathrm{P}(P_{w,t} \in \mathbb{R} \big| -\overline{P}^i_f \leq A^i P^i_{inj,t}(P_{w,t}) \leq \overline{P}^i_f \\ &\underline{P}^i_G \leq P^i_{G,t} + R^i_t(P_{w,t}) \leq \overline{P}^i_G \\ &-R^i_{down,t} \leq R^i_t(P_{w,t}) \leq R^i_{up,t}, \text{ for all } (i \in Z) \geq 1 - \varepsilon_t \end{aligned} \tag{11.11}$$

The probability is meant with respect to the probability distribution of the wind power vector $P_w \in \mathfrak{R}^w$. The last constraint in (11.11) is included to determine the reserves $R^i_{up,t}$, $R^i_{down,t}$ as the worst case, in a probabilistic sense, R^i_t is the power correction term. The reserves that the system operator will need to purchase are then determined as

$$R_{up,t} = \max_{i \in Z} R^i_{up,t} \tag{11.12}$$

$$R_{down,t} = \max_{i \in Z} R^i_{down,t} \tag{11.13}$$

which denote the worst-case values, given all the outages, of $R^i_{up,t}$, $R^i_{down,t}$, respectively. In (11.11) the same probability level is considered for each time step $t = 1, \ldots, N_t$; different probability levels per stage or a joint chance constraint for all stages can be captured by the proposed framework as well. In line with the formulation an additional AGC/LFC functionality is proposed. The system operator must monitor both the production of the tripped plant and the deviation of the wind power from its forecast. Thereafter, by using (11.1) as a look-up table the appropriate distribution vector, among those computed in the optimization problem, is selected as shown in Fig. 11.2.

The resulting problem given by (11.10) and (11.11) is a chance-constrained bilinear program whose stages are only coupled due to the temporal correlation of the wind power. Further coupling among the stages can be obtained if a unit-commitment problem was included or ramping constraints of the generating units and minimum up and down times were modeled [35]. The two major challenges faced when attempting to solve problem (11.3)–(11.11) are as follows: the first is due to the presence of bilinear terms that are a result of the products of $d^i_{up,t}$, $d^i_{down,t}$, and $P_{G,t}$ for $i \in \Upsilon_G$, the second is owing to the presence of the chance constraint.

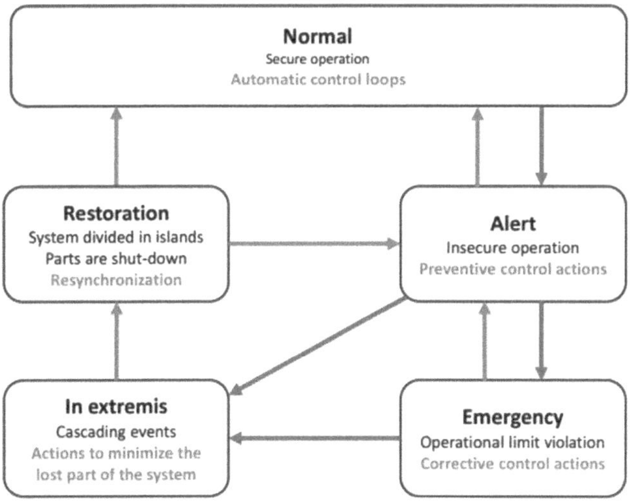

FIGURE 11.2 A schematic diagram summarizing the different states of operation [1].

11.5 Decision-making-based optimal automatic generation control in deregulated environment

11.5.1 An overview of the fractional calculus

Fractional calculus is a field of mathematics, which concerns about computing the integrations/differentiations with noninteger orders. By using fractional calculus methods a complexity of integrations/differentiations with noninteger orders can be solved. During the history, different definitions have been suggested to describe the problem of fractional calculus. The Grünwald–Letnikov definition, the Caputo definition, and the Riemann–Liouville definition are the well-established definitions for fractional calculus during the history [1–9,25]. In the field of engineering, Caputo definition is the mostly used for defining the problem of control based on fractional calculus [25]. The operator of integral/differential with order (α) and operation bounds (a, t) can be represented by $\left({}_aD_t^{\alpha}\right)$. According to Caputo definition, the fractional calculus operator is denoted by the sign of the order (α) as follows [25]:

$$
{}_aD_t^{\alpha} = \begin{cases} \dfrac{d^{\alpha}}{dt^{\alpha}} & \alpha \succ 0 \\ 1 & \alpha = 0 \\ \int_a^t (dt)^{-\alpha} & \alpha \prec 0 \end{cases} \tag{11.14}
$$

Taking in the consideration that m is the smallest integer that is larger than α, the fractional derivative is calculated based on Caputo definition as follows [25]:

$$ {}_aD_t^{\alpha}f_X(t) = \begin{cases} \dfrac{d^m}{dt^m} & \alpha = m \\ \dfrac{1}{\Gamma(m-\alpha)}\displaystyle\int_a^t \dfrac{D^m f_X(t)}{(t-\tau)^{\alpha-m+1}} & m-1 \prec \alpha \prec m \end{cases} \tag{11.15} $$

where

$$ \Gamma(m-\alpha) = \int_0^{\infty} t^{m-\alpha-1}\cdot \exp(-t)dt \tag{11.16} $$

To transform the computation from the time domain to the frequency domain, the Laplace transformation is used. In the fractional calculus the Laplace transform is given by the following equation [25]:

$$ \ell\{\,{}_aD_t^{\alpha}f_X(t)\} = s^{\alpha}F(s) - \sum_{k=0}^{m-1} s^{\alpha-k-1}f(0) \tag{11.17} $$

To implement and simulate the fractional-order calculus, the Laplace operator of the fractional order is approximated with integer-order transfer functions. The most known method to approximate fractional order to integer order is Oustaloup's method [8,9,25].

11.5.2 Load-frequency control and automatic generation control based on fractional calculus

Frequency control in power system consists of three subcontrol levels: (1) PFC that tries to stop the frequency decline before triggering the under frequency load shedding, (2) secondary frequency control known also as LFC, which aims to mitigate the frequency deviation through a suitable controller, and (3) tertiary frequency control, which aims to redispatch the generation units in order to achieve the most economic operation of the system [25].

In this section, we provide an overview of LFC in the deregulated power system modeling, which is given in Section 11.5.2.1, while a procedure of LFC controller design based on the fractional calculus is introduced in Section 11.5.2.2.

11.5.2.1 Load-frequency control under the deregulation environment

The power system face is changing by its transition from vertically integration utility (VIU) structure to deregulated one. In the first structure, that is,

VIU, all power system aspects such as generation, transmission, and distribution parts are under the control of one authority that is responsible for providing electrical energy with its ancillary services to consumers. On the other hand, in restructured environment/deregulation structure, the abovementioned parts are owned by different companies. In the deregulated systems, providing ancillary services are based on the negotiations between different generation companies (GENCOs) and distribution companies (DISCOs) based on electricity markets' negotiation [25].

In the restructured power systems, wherever a disturbance/change in the demand-side occurs, the new demand is satisfied by electric power provided from the GENCOS, which have contracted with the DISCOs of the load change. In this way, DISCOs should forecast the demand of their consumers and buy a sufficient electric energy for their consumers based on electricity market competitive. In this new environment, DISCOs have the choice to buy the energy from any GENCO. In addition, it can buy the energy from more than one GENCO at the same time to meet the requirements of its consumers [25].

Distributed power management (DPM) matrix is usually used to model the various contracts between DISCOs and GENCOs. In this matrix the columns models DISCOs and the rows correspond to GENCOs. Each entry of the matrix, which is called contract participation factor, *cpf*, stands for a specified contact between correspond DISCO and GENCO. It is clear that the sum of all entries of one column should be one, which means the provided electrical energy from all GENCOs for one DISCOs equal to the DISCO's demand in each time interval [25].

Let's consider a power system consists of n GENCOs and m DISCOs, then the DPM is modeled by the following equation [1]:

$$\text{DPM} = \begin{bmatrix} cpf_{11} & .. & cpf_{1n} \\ . & & . \\ & .. & \\ . & & . \\ cpf_{m1} & .. & cpf_{mn} \end{bmatrix} \tag{11.18}$$

It is worth mentioning that the change in the generated power from ith GENCO in a specific time can be calculated as follows:

$$\Delta P_{g,i} = \sum_{k=1}^{n_d} cpf_{ik} P_k \tag{11.19}$$

The abovementioned matrix can help with transmission system management. In this regard the scheduled power flow in specific transmission line or major tie-line between area i and area j can be calculated as follows:

$$\Delta P_{tie,ij}^{sch} = \sum_{k=n_{g,i}} \sum_{l=n_{d,j}} cpf_{lk} \Delta P_{Ll} - \sum_{k=n_{g,j}} \sum_{l=n_{d,i}} cpf_{lk} \Delta P_{Ll} \tag{11.20}$$

However, it is not easy to keep the power flow as the scheduled one. Therefore the actual power flow between two areas is measured as follows:

$$\Delta P^{act}_{tie,ij} = \frac{2\pi T_{ij}}{s}(\Delta f_i - \Delta f_j) \tag{11.21}$$

The difference between the actual and scheduled power flow between two areas determines the error in the transferred power as follows:

$$\Delta P^{err}_{tie,ij} = \Delta P^{act}_{tie,ij} - \Delta P^{sch}_{tie,ij} \tag{11.22}$$

In LFC studies, it is critical to determine the area control error of area i (ACE_i), which is very useful in generating LFC signal. The area control error can be determined as follows:

$$\text{ACE}_i = \beta_i \Delta f_i + \Delta P^{err}_{tie,i} \tag{11.23}$$

$$\Delta P^{err}_{tie,i} = \sum_{j=1\ \&\ j\neq i}^{n} \Delta P^{err}_{tie,ij} \tag{11.24}$$

11.5.2.2 Design of load-frequency controller based on the fractional calculus

AGC or LFC are in use in modern power system for removing or at least mitigating both frequency and tie-line power deviations. To this end, different types of PID controllers are utilized for controlling the frequency and tie-line power flow in such systems. Due to its superiority, the FOPID has been also adopted to regulate the frequency and tie-line power exchange deviations [25].

In this new controllers, that is, FOPID, apart from proportional (K_p), integral (K_i), and derivative (K_d) constants, they have additional integral order (λ) and the derivative order (μ); thus they have two further operators which add two more DOFs to the controller and make FOPID controller has better performance compared to the traditional PID controllers [25]. The LFC signal, $u_{c,i}$, used in each control area based on the FOPID is determined as follows:

$$u_{c,i} = \left(k_{p,i} + k_{D,i}s^{\mu} + \frac{K_{I,i}}{s^{\lambda}}\right)\text{ACE}_i \tag{11.25}$$

It should be noted that the further two variables, that is, λ and μ, provide much more accuracy and flexibility in designing the LFC controllers. Now, as a next step in the procedure, the controller variables should be optimally tuned. In the next discussion, we will show how these important variables can be tuned using evolutionary computing methods [25].

11.5.3 Optimal tuning of the controller parameter

11.5.3.1 The proposed objective function

In this chapter an objective function considering the settling time (ST) and damping of the frequency oscillations of both frequency and tie-lines power flow is used as follows:

$$J = \omega_1 \cdot \sum_{i=1}^{NA} \int_0^{t_{sim}} \left(\alpha_i \Delta f_i(t) + \sum_{\substack{j=1 \\ j^1 i}}^{NA} \alpha_{ij} \Delta P_{tie_{i-j}}(t) \right) \cdot t \cdot dt + \omega_2 \cdot \frac{1}{\min\left(\left\{(1-\zeta_i),\ i=1\ldots n\right\}\right)} + \omega_3 \cdot \left(\sum_{i=1}^{NA} \left(\mathrm{ST}(\Delta f_i(t)) + \sum_{\substack{j=1 \\ j^1 i}}^{NA} \mathrm{ST}(\Delta P_{tie_{i-j}}(t)) \right) \right) \tag{11.26}$$

where ST is the time at which the final value of the signal settles to less than a specific value. The weight (ω) of each term of objective function shows the importance of each term in the objective function.

Based on the adopted FOPID controller, the optimization problem can be described as follows:

$$\begin{aligned} &\min\{J\} \\ &s.t. \\ &K_p^{\min} \le K_p \le K_p^{\max} \\ &K_I^{\min} \le K_I \le K_I^{\max} \\ &K_D^{\min} \le K_D \le K_D^{\max} \\ &\lambda^{\min} \le \lambda \le \lambda^{\max} \\ &\mu^{\min} \le \mu \le \mu^{\max} \end{aligned} \tag{11.27}$$

11.5.3.2 Imperialist competitive algorithm–based fractional-order proportional–integral–derivative controller's optimization

ICA is a sociopolitical metaheuristic, inspired by the history of colonization and competition among imperialists, to capture more colonies. The set of countries, which are the solutions in ICA, is partitioned to form several empires. Each empire consists of a single Imperialist and several other weaker countries, called colonies [20]. Two competition mechanisms are used in the algorithm, which are the intraempire competition and the interempire

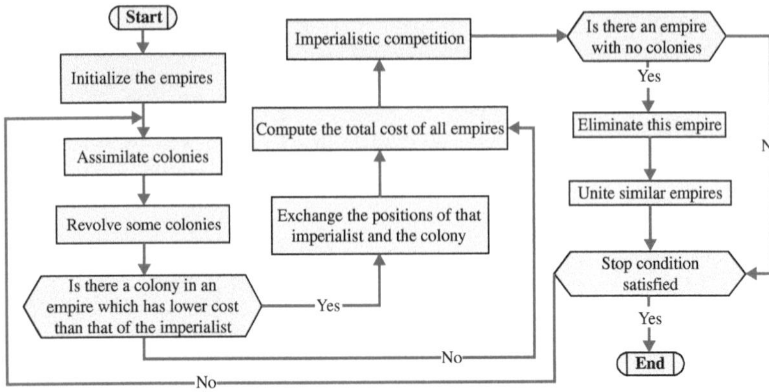

FIGURE 11.3 Imperialist competitive algorithm–based LFC's parameters tuning. *LFC*, Load-frequency control.

competition, which are the competition among the members of an empire and the competition among empires respectively. The power of each colony or imperialist is determined from the cost of the optimization algorithm as shown in Fig. 11.3. Countries with the least cost function become the imperialists, and they form empires by taking control of countries with higher cost functions which become colonies in their empires. The process involves assimilation where the imperialists gets stronger and gain full control for the colonies or revolution where some colonies become stronger than the imperialists. Among the imperialist competition to gain full control of other colonies exists [20]. The process stops when powerless empires have been completely eliminated.

The ICA algorithm can be used in the implementation of the optimal under frequency load shedding (UFLS) scheme. With an objective defined as in (11.26) where the goal is minimization of the overall objective function, the lessor the number of imperialist the better the solution as shown in Fig. 11.3. This algorithm enjoys several advantages like it works well with nonlinear systems, and the solution obtained is a global one.

11.6 Case study

11.6.1 The studied deregulated power system

In this chapter the performance of the adopted LFC controllers is shown on a large-scale power system. Therefore IEEE 39-bus system is considered for investigating the superiority of FOPID controllers. The single-line diagram of IEEE 39-bus system is depicted in Fig. 11.4. This system is widely used in dynamic studies and usually divided into three subareas. Moreover, it is assumed that there is a flexible demand such as electric vehicles which is

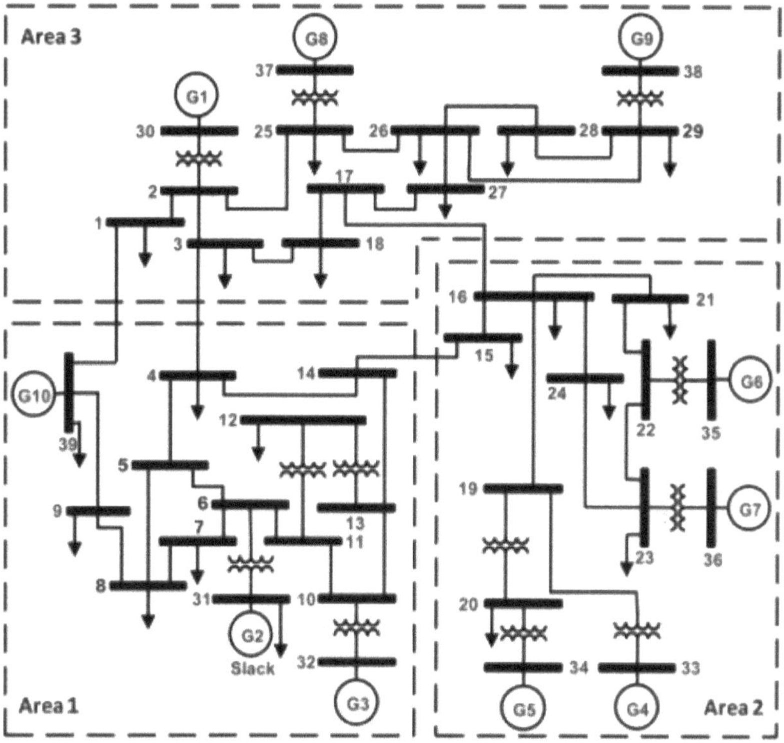

FIGURE 11.4 The power system under investigation.

considered to be 5% of the total demand. The used data in the simulation can be found in Refs. [24–26].

11.6.2 Simulation results and discussions

For evaluating the adopted fractional-order LFC technique, the optimal values of controllers' parameters used to control different areas are first determined using objective function (11.26) [26]. As demonstrated earlier, ICA is utilized to obtain the optimal values of the controllers' parameters by solving LFC optimization problem. Table 11.1 presents optimal values of the controllers obtained by using ICA to solve (11.27).

In order to confirm robustness of the adopted fractional-order control method, the performance of the controllers is evaluated and compared with other control methods. In the comparison stage, load disturbance magnitudes and types such as signal disturbance in one area, signal disturbance in all areas, and multidisturbance in the different control areas are taken into account [26]. In addition, several simulations are applied to power system

TABLE 11.1 The controllers' parameters value based on imperialist competitive algorithm (ICA) and others.

Cont. no.	Parameters	ICA	hGSA-PS	DE
Controller 1	K_p	1.2734	1.4735	−2
	K_i	−0.47637	−1.53653	−2
	K_d	−0.354746	−0.1836	1.01673
	λ	0.326	0.5064	0.7043
	μ	0.63	0.3838	0.65
Controller 2	K_p	−1.87535	1.3434	1.1189
	K_i	−0.85221	−0.888	−1.998
	K_d	2	−2	1.675
	λ	0.5	0.5	0.6767
	μ	0.133	0.35	0.6767
Controller 3	K_p	−2	−0.786	−2
	K_i	−1.84575	−1.6	−2
	K_d	−1.46660	−0.9898	−2
	λ	0.60	0.5	0.333
	μ	0.72	0.3535	0.333

DE, Differential algorithm; *hGSA-PS*, hybrid gravitational search and pattern search; *ICA*, imperialist competitive algorithm.

under investigation to prove the robustness of the adopted method against the varying of power system loading and parameters.

As to assess the adopted control strategy in case of occurrence of disturbances in all control areas, the performance of FOPID controllers in the case of a 0.01 p.u. step increase in the demand of all areas is investigated [26]. The performance of ICA in tuning the parameters of load-frequency controllers is compared to hGSA-PS and DE algorithms. Fig. 11.5 shows that the maximum frequency deviation of all areas in the case of using ICA algorithm is highly decreased compared to hGSA-PS and DE algorithms. Fig. 11.6 shows that the maximum deviation of the tie-lines power is highly decreased in compare to the method proposed in hGSA-PS [25,26].

In order to evaluate the contribution of EVs in supporting the frequency control in power systems, it is assumed that EVs can provide some secondary reserve. It is assumed that the participation of EVs in LFC is 15%.

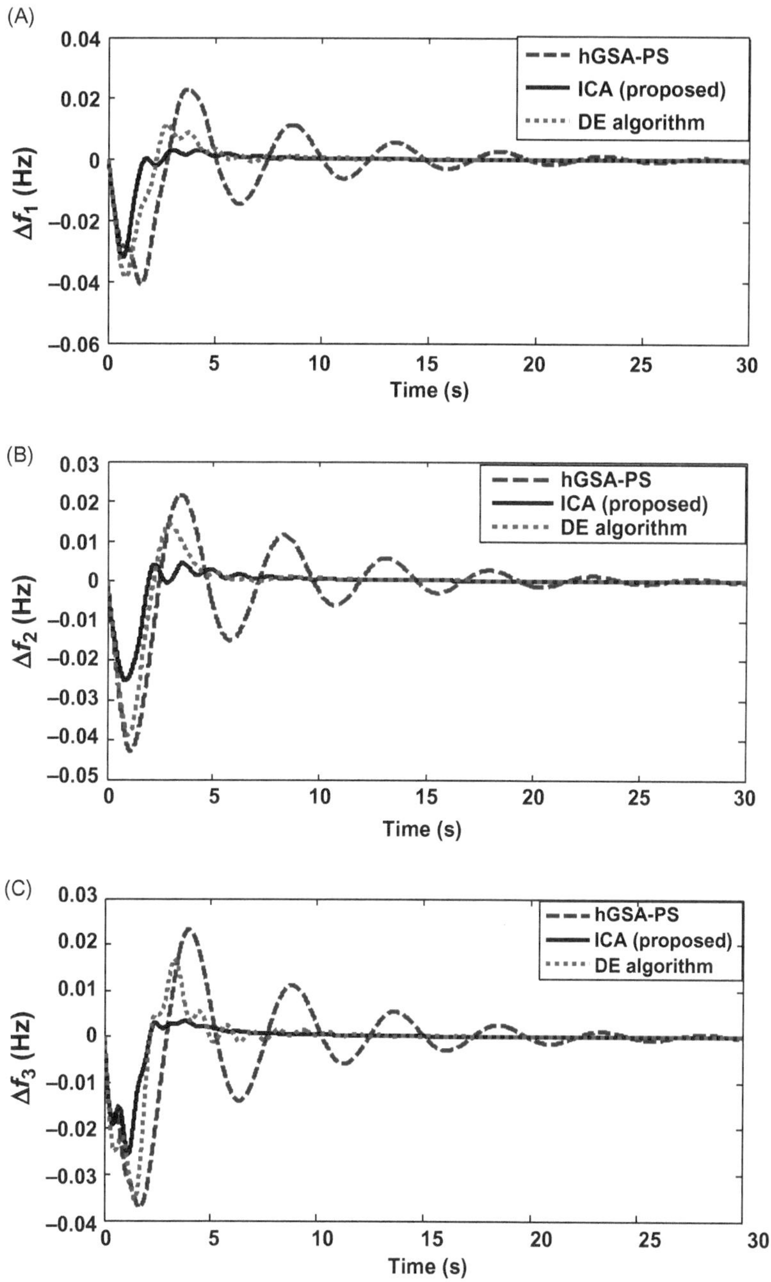

FIGURE 11.5 The frequency deviation in the different areas: (A) area 1, (B) area 2, and (C) area 3.

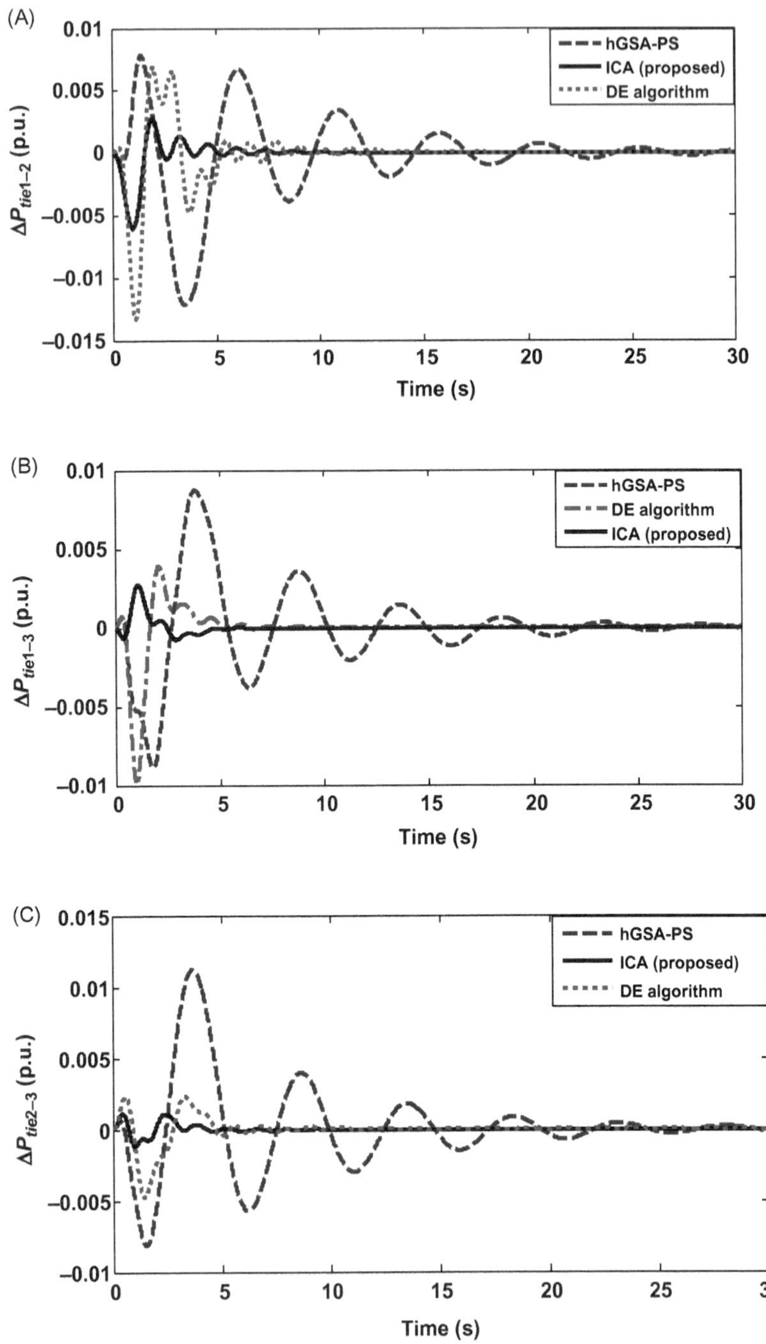

FIGURE 11.6 The tie-line power deviation in the different areas: (A) tie-lines 1–2, (B) tie-lines 1–3, and (C) tie-lines 2–3.

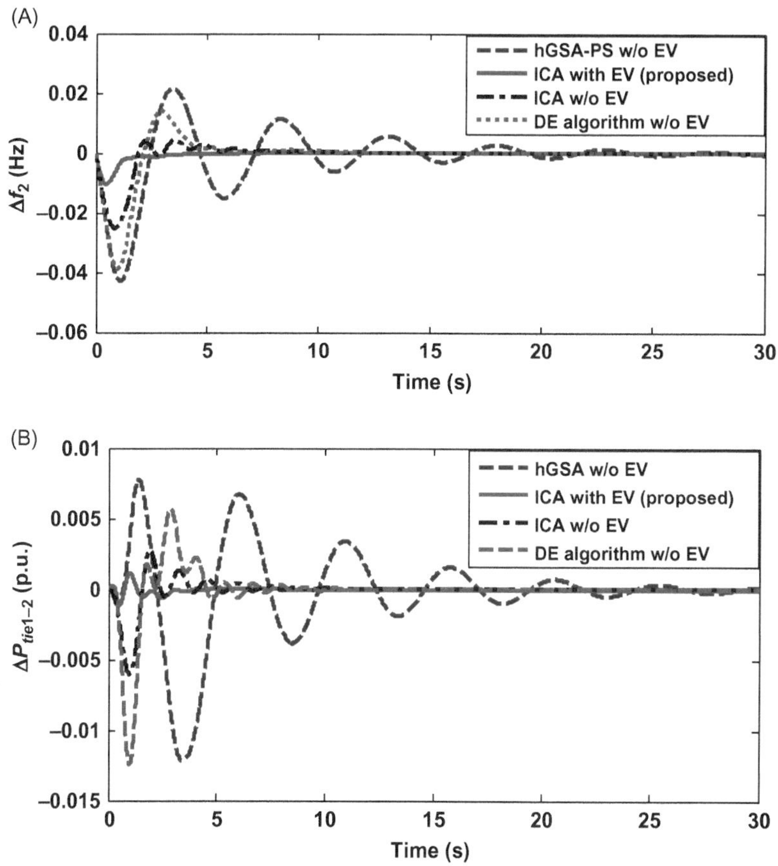

FIGURE 11.7 The frequency and tie-line power deviations: (A) frequency and (B) tie-line power.

To show the advantages of EV participation in LFC, the studied system is undertaken a simulation when a 0.01 p.u. step increase in the demand of all areas is suddenly happened considering the participation of EVs. Fig. 11.7A shows that the maximum frequency deviation in the different areas in case of using EVs is highly decreased compared to case study without EVs. Fig. 11.7B shows that the maximum deviation of the tie-lines power is highly decreased in compare to the conventional LFC without EVs participation [26].

To verify the robustness of the adopted method in this chapter, the response of the utilized controllers for this power system, to a 1% p.u. step increase in the total demand, in the case of change in T_{ij}, and both of T_t and T_g, is investigated.

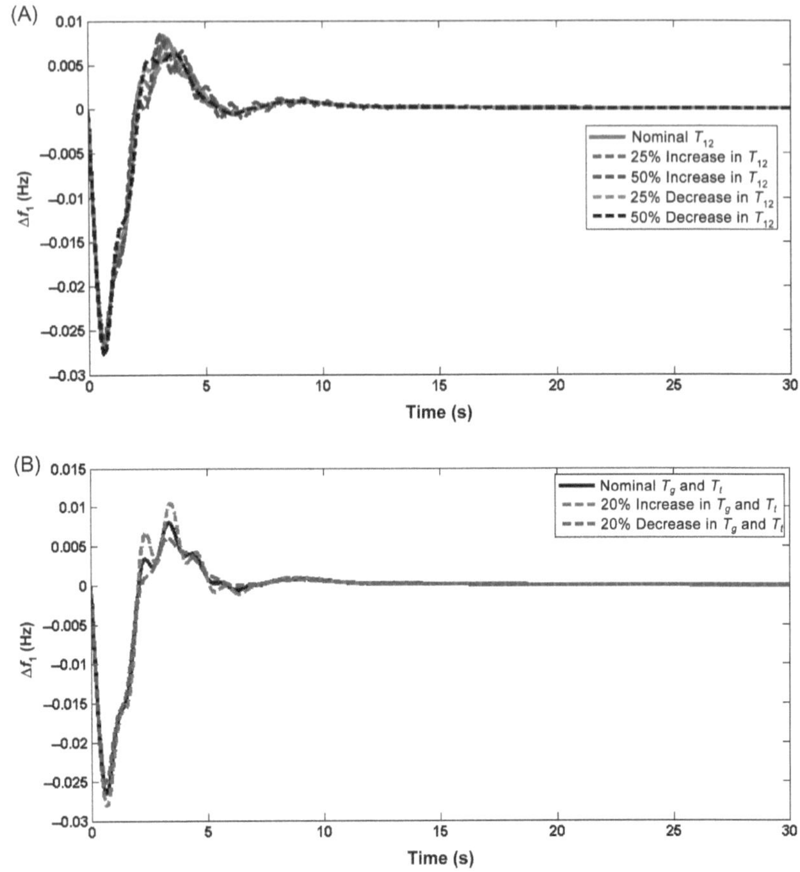

FIGURE 11.8 The frequency deviation in the different areas: (A) due to tie-line coefficient variations and (B) due to governor–turbine parameter uncertainties.

The robustness of the used control method in the case of changes, −0.2 and 0.2 per units, in the studied power system parameters, that is, both T_t and T_g, is verified by Fig. 11.8. Also in the case of changes, −0.2, 0, and +0.2 per units, in the time constant of the governor and turbines of area 1, the adopted control method shows better performance.

The maximum frequency deviation and the ST of the frequency deviation have not been affected in case of the changes, −0.5, −0.25, 0, +0.25, and +0.5, in tie-line synchronizing coefficient as shown in Fig. 11.8. These results prove the superiority and robustness of the adopted control strategy where its controllers' parameters are tuned using ICA algorithm.

11.7 Conclusion

In this chapter the generation-side reserve scheduling and LFC issues in modern power systems were studied. An overview of power system stability and security was first introduced. Based on the power system security constraints, the optimization problem of secondary-reserve scheduling and optimal tuning of frequency controllers parameters in modern power system was also presented. Furthermore, some scenarios have been applied to show the superiority of the presented methods. Moreover, the effectiveness of the fractional calculus–based control scheme was investigated to show the importance of decision-making methods in these topics. The main findings and recommendations in this chapter are as follows:

- A comprehensive framework including both generation-side and LFC is required for future power systems.
- The secondary reserve should be scheduled for modern power systems considering the fluctuation of both generation-side and demand-side participations.
- The demand side, in comparison with generation side, has more flexibility in providing ancillary services.
- The robustness of load-frequency controllers can be guaranteed by the optimal tuning of them using evolutionary techniques.

As a research direction for future works in these important topics, it is recommended to propose a suitable LFC framework considering the high fluctuations form both demand side and generation side such as renewable power variations and load fluctuations. Also, it is very important to consider the inertia reduction due to the increase of renewable energy share in power systems in reserve scheduling and frequency control of future power systems. Furthermore, it is suggested that to study the effects of emerging technologies such as distributed generating units and their scheduling in the performance and availability of required reserve in power systems. Moreover, new issues such as cyberattacks should be addressed for future smart power systems.

References

[1] H. Bevrani, Robust Power System Frequency Control, Vol. 85, Springer, New York, 2009.

[2] H.A.S.H. Shayeghi, H.A. Shayanfar, A. Jalili, Load frequency control strategies: a state-of-the-art survey for the researcher, Energy Convers. Manage. 50 (2) (2009) 344–353.

[3] C.K. Shiva, V. Mukherjee, Comparative performance assessment of a novel quasi-oppositional harmony search algorithm and internal model control method for automatic generation control of power systems, IET Gener. Transm. Distrib. 9 (11) (2015) 1137–1150.

[4] B.K. Sahu, et al., A novel hybrid LUS–TLBO optimized fuzzy-PID controller for load frequency control of multi-source power system, Int. J. Electr. Power Energy Syst. 74 (2016) 58–69.

[5] R.K. Sahu, S. Panda, P.C. Pradhan, Design and analysis of hybrid firefly algorithm-pattern search based fuzzy PID controller for LFC of multi area power systems, Int. J. Electr. Power Energy Syst. 69 (2015) 200–212.

[6] S. Debbarma, L.C. Saikia, N. Sinha, Automatic generation control using two degree of freedom fractional order PID controller, Int. J. Electr. Power Energy Syst. 58 (2014) 120–129.

[7] A. Rahman, L.C. Saikia, N. Sinha, Load frequency control of a hydro-thermal system under deregulated environment using biogeography-based optimised three-degree-of-freedom integral-derivative controller, IET Gener. Transm. Distrib. 9 (15) (2015) 2284–2293.

[8] S.A. Taher, M.H. Fini, S.F. Aliabadi, Fractional order PID controller design for LFC in electric power systems using imperialist competitive algorithm, Ain Shams Eng. J. 5 (1) (2014) 121–135.

[9] S. Sondhi, Y.V. Hote, Fractional order PID controller for load frequency control, Energy Convers. Manage. 85 (2014) 343–353.

[10] M.H. Fini, G. Yousefi, H.H. Alhelou, A Comparative study on the performance of many-objective and single-objective optimisation algorithms of tuning load frequency controllers in multi-area power systems, IET Gener. Transm. Distrib. (2016).

[11] R.K. Sahu, S. Panda, S. Padhan, A novel hybrid gravitational search and pattern search algorithm for load frequency control of nonlinear power system, Appl. Soft Comput. 29 (2015) 310–327.

[12] T.H. Mohamed, et al., Decentralized model predictive based load frequency control in an interconnected power system, Energy Convers. Manage. 52 (2) (2011) 1208–1214.

[13] F. Liu, et al., Optimal load-frequency control in restructured power systems, IEE Proc.—Gener. Transm. Distrib. 150 (1) (2003) 87–95.

[14] B. Mohanty, S. Panda, P.K. Hota, Controller parameters tuning of differential evolution algorithm and its application to load frequency control of multi-source power system, Int. J. Electr. Power Energy Syst. 54 (2014) 77–85.

[15] H. Alhelou, M.E. Haes, Hamedani Golshan, J. Askari-Marnani, Robust sensor fault detection and isolation scheme for interconnected smart power systems in presence of RER and EVs using unknown input observer, Int. J. Electr. Power Energy Syst. 99 (2018) 682–694.

C.S. Chang, W. Fu, F. Wen, Load frequency control using genetic-algorithm based fuzzy gain scheduling of PI controllers, Electr. Mach. Power Syst. 26 (1) (1998) 39–52.

[16] B. Mohanty, S. Panda, P.K. Hota, Differential evolution algorithm based automatic generation control for interconnected power systems with non-linearity, Alexandria Eng. J. 53 (3) (2014) 537–552.

[17] P. Bhatt, R. Ranjit, S.P. Ghoshal, Optimized multi area AGC simulation in restructured power systems, Int. J. Electr. Power Energy Syst. 32 (2010) 311–332.

[18] E.S. Ali, S.M. Abd-Elazim, Bacteria foraging optimization algorithm based load frequency controller for interconnected power system, Int. J. Electr. Power Energy Syst. 33 (3) (2011) 633–638.

[19] K. Naidu, et al., Application of firefly algorithm with online wavelet filter in automatic generation control of an interconnected reheat thermal power system, Int. J. Electr. Power Energy Syst. 63 (2014) 401–413.

[20] H. Shabani, B. Vahidi, M. Ebrahimpour, A robust PID controller based on imperialist competitive algorithm for load-frequency control of power systems, ISA Trans. 52 (1) (2013) 88–95.

[21] R.K. Sahu, S. Panda, S. Padhan, A novel hybrid gravitational search and pattern search algorithm for load frequency control of nonlinear power system, Appl. Soft Comput. 29 (2015) 310–327.
[22] H.S.H. Alhelou, M.E.H. Golshan, M.H. Fini, Multi agent electric vehicle control based primary frequency support for future smart micro-grid, in: Smart Grid Conference (SGC), 2015, IEEE, 2015.
[23] H.H. Alhelou, M.E.H. Golshan, Hierarchical plug-in EV control based on primary frequency response in interconnected smart grid, in: 2016 24th Iranian Conference on Electrical Engineering (ICEE), IEEE, 2016.
[24] A. Ghafouri, J. Milimonfared, G.B. Gharehpetian, Fuzzy-adaptive frequency control of power system including microgrids, wind farms, and conventional power plants, IEEE Syst. J. 12 (2017) 2772–2781.
[25] H.H. Alhelou, M.E. Hamedani-Golshan, E. Heydarian-Forushani, A.S. Al-Sumaiti, P. Siano, Decentralized fractional order control scheme for LFC of deregulated nonlinear power systems in presence of EVs and RER, in: 2018 International Conference on Smart Energy Systems and Technologies (SEST), IEEE, September 2018, pp. 1–6.
[26] H. Haes Alhelou, M.E. Hamedani Golshan, M. Hajiakbari Fini, Wind driven optimization algorithm application to load frequency control in interconnected power systems considering GRC and GDB nonlinearities, Electr. Power Compon. Syst. 46 (2018) 11–12.
[27] H. Alhelou, M.E. Hamedani-Golshan, R. Zamani, E. Heydarian-Forushani, P. Siano, Challenges and opportunities of load frequency control in conventional, modern and future smart power systems: a comprehensive review, Energies 11 (10) (2018) 2497.
[28] R. Billinton, Power System Reliability Evaluation, Gordon and Breach, New York, 1970.
[29] P. Kundur, J. Paserba, V. Ajjarapu, G. Andersson, A. Bose, C. Canizares, et al., Definition and classification of power system stability, IEEE Trans. Power Syst. 19 (3) (2004) 1387–1401.
[30] K. Carlsen, L.H. Fink, Operating under stress and strain, IEEE Spectr. 15 (1978) 48–53.
[31] M. Zima, M. Bockarjova, Lecture Notes Operation, Monitoring and Control Technology of Power Systems, ETH Zurich, 2007.
[32] F.D. Galiana, F. Bouffard, J.M. Arroyo, J.F. Restrepo, Scheduling and pricing of coupled energy and primary, secondary, and tertiary reserves, Proc. IEEE 93 (11) (2005) 1970–1983.
[33] F. Bouffard, F.D. Galiana, A.J. Conejo, Market-clearing with stochastic security – Part I: formulation, IEEE Trans. Power Syst. 20 (4) (2005) 1818–1826.
[34] F. Bouffard, F. Galiana, Stochastic security for operations planning with significant wind power generation, IEEE Trans. Power Syst. 23 (2) (2008) 306–316.
[35] J.M. Morales, A. Conejo, J. Pérez-Ruiz, Economic valuation of reserves in power systems with high penetration of wind power, IEEE Trans. Power Syst. 24 (2) (2009) 900–910.
[36] B. Stott, O. Alsac, Optimal power flow – a brief anatomy, in: Proceeding of XII Symposium of Specialists in Electric Operational and Expansion Planning, 2012.
[37] K. Margellos, T. Haring, P. Hohayem, M. Schubiger, J. Lygeros, G. Andersson, A robust reserve scheduling technique for power systems with high wind penetration, in: International Conference on Probabilistic Methods Applied to Power Systems, 2012.
[38] H. Bevrani, P.R. Daneshmand, Fuzzy logic-based load-frequency control concerning high penetration of wind turbines, IEEE Syst. J. 6 (1) (2012) 173–180.

Chapter 12

Heuristic methods for the evaluation of environmental impacts in the power plants

Jandecy Cabral Leite[1], Jorge de Almeida Brito Júnior[1], Manoel Henrique Reis Nascimento[1], Carlos Alberto Oliveira de Freitas[1], Milton Fonseca Júnior[2], David Barbosa de Alencar[1], Nadime Mustafa Moraes[3] and Tirso Lorenzo Reyes Carvajal[1]

[1]*Research Department, Institute of Technology and Education Galileo of Amazon - ITEGAM, Manaus, AM, Brazil,* [2]*Mauá Generation Department, Generation Eletrobras Amazonas GT Manaus, Amazonas, Brazil,* [3]*Research Department, University of the State of Amazonas (UEA), Manaus, Amazonas, Brazil*

12.1 Introduction

Developing countries seek self-sustainability in the face of current global energy scenario. The reasons for this inclination arise due to the scarcity of fossil fuels in the coming decades, added to the high thermal energy production costs and rising energy consumption, either by developmental reasons or by the misuse of power available. The action strategies to the current national scene need to be updated both with respect to the difficulties observed in the economic and environmental spheres, such as investment in the research of new instruments, methods, and criteria to ensure the effective contribution of the electricity sector in the process of seeking a self-sustainable development [1]. For [2] the electric power generation and shipping problems, the absolute minimum cost is not the only criterion to be fulfilled. Apart from that, environmental considerations have become a major concern for power generation. The limited economic dispatch (ED) problem can be environmentally classified as a multiobjective optimization and a nonlinear programming problem. According [3] they argue that since the beginning of the 1970s the dispatch of thermal generation has been proposed as an effective means of dealing with the problem of air pollution. The latest restrictive legislation has led to the adoption of pollution-limiting techniques and/or the use of cleaner fuels. However, an order with restricted emissions

Decision Making Applications in Modern Power Systems. DOI: https://doi.org/10.1016/B978-0-12-816445-7.00012-8

is even more necessary when the weather conditions are adverse to the diffusion of effluents. The authors present a dynamic dispatch procedure, which is able to hold the integral nature of the restrictions of issue. So, the environmental economic dispatch (EED) in thermal plants is a very important task to ensure the power demand, which is to make a distribution to all the mill engines, ensuring that the cost is minimal.

In this chapter a model and a mathematical method for EED tools using evolutionary algorithms (EAs) [nondominated sorting genetic algorithm (GA) II (NSGA-II)] to reduce both the cost of energy production in thermal power plants (TPPs) and the environmental impact are applied. The identification of different ways of evaluating the emissions produced by power plants suggests mathematical models and computational tools to be used for the assessment of the economic (cost generation and fuel consumption) and environmental (emissions) variables, considering the pollution generated as well as the permissibility of each pollutant in the atmosphere to allow the construction of different simulation scenarios. It also formulates the optimization of bi-objective EED problem, using a computational tool (NSGA-II-EA) analysis for the selection of the configuration of independent and dependent variables of the mathematical model, considering the demanded power and the environmental impacts.

12.2 Materials and methods

12.2.1 Heuristic optimization techniques

The use of heuristic methods increases to quickly get tools to give solutions to actual problems. It is important to note that these methods do not guarantee the best optimization solution found, although the purpose is to find the solution next to the optimal solution in a reasonable time. Fig. 12.1 shows the classification of global optimization methods [4,5].

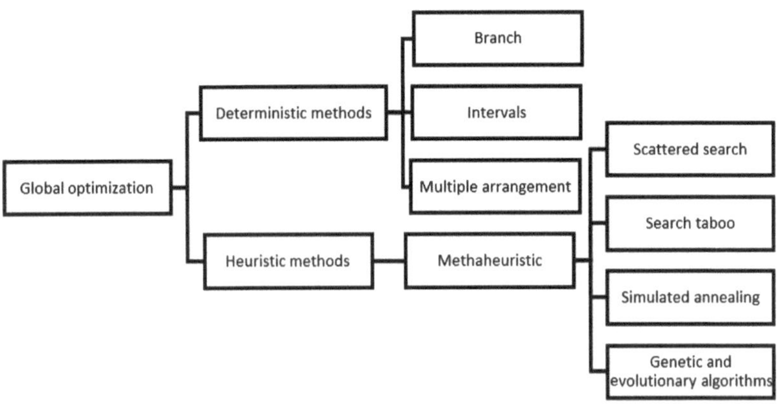

FIGURE 12.1 Global optimization methods.

The heuristic optimization techniques can be of exhaustive and nonexhaustive types. The comprehensive or exhaustive techniques, such as algorithmic schemes, Backtracking and Branch & Bound, have the advantage of finding the optimal solution always, using the worst case—the entire solution space is huge. It is difficult to narrow the search by the use of heuristic techniques and, therefore, may result in inefficient algorithms for medium-to-large problems.

The nonexhaustive techniques are known by the name of metaheuristics, which can be algorithmic schemes based on different ideas in many outlets, occasions, and the workings of nature, which is a common approach of problem-solving by successive improvements of a solution or set of solutions, with an exploration of broader solution space and with some random factor [6,7].

In this work the metaheuristic techniques, specifically GAs, will be used. It is taken into consideration that the types of optimization problems have a very complex resolution space; therefore exploring it completely may not be feasible for certain applications. In this type of technique, what is done is to work with a solution or a set of solutions for new responses that are closer to the optimal in order to avoid the great places and, iteratively, to achieve a high-quality convergence. In this way, it is possible to guarantee the quality of the solution, as this will comply with the criteria found.

12.2.2 Genetic algorithms

GAs are adaptive heuristic search algorithms that are based on evolutionary ideas of natural selection and genetics. As such, they represent an intelligent exploitation of a random search used to solve optimization problems. Although randomized, GAs are not random, instead, exploit historical information to direct the search for the best performance region within the search space. The basic techniques of GAs are designed to simulate processes in natural system necessary for evolution, especially those that follow the principles established by Charles Darwin first—"survival of the fittest," where, in nature, in competition among individuals for scarce resources, the more capable individuals dominate over the weak.

It is better than conventional techniques of artificial intelligence (AI) that is more robust. Unlike older systems, AI, they don't break easily even if the inputs change slightly, or in the presence of reasonable noise. In addition, when searching a large state space, multimodal state space, or n-dimensional surface, a GA can provide significant benefits on the types of most typical search engine optimization techniques (linear programming, heuristic depth-first search width, and praxis) [9].

GAs mimic the survival-of-the-fittest individuals from every successive generation of a problem to solve. Each generation consists of a population of strings of characters that are similar to chromosome that we see in DNA.

Each individual is a point in a search space and a possible solution. The individuals of the population are made to go through a process of evolution.

GAs are based on analogy with the genetic structure and behavior of chromosomes within a population of individuals using the following bases [8]:

- The individuals in a population compete for resources and mates.
- The most successful individuals in each "competition" will produce more offspring than those who have a poor performance.
- The genes of individuals "good" spread throughout the population so that two good parents sometimes produce offspring that are better than either parent.
- Thus each succeeding generation becomes more suitable to their environment.
- The simplest form of GA has the following three types of operators [10]:
- Selecting and playing: This operator drains chromosomes among the population to make the play. The more capable is the chromosome, the more often will be selected to reproduce.
- Crossing: This is an operator who has to choose a place of function and change the sequences before and after that position between two chromosomes, to create a new offspring (e.g., 10,010,011 and 11,111,010 chains can cross after the third place to produce offspring 10011010 and 11110011), and mimics the biological recombination between the haploid organisms.
- Mutation: This operator produces random variations in a chromosome (e.g., the chain can exchange 00011100 its second position for the current 01,011,100). The mutation can take place in each position of a bit in a string, with a probability typically very small (e.g., 0.001). As can be seen, the GAs are different from traditional methods of search and optimization in four key areas.
- They seek a population of points, not a single point. Maintaining a population of well-adapted sampling points, the probability of falling into a false peak is reduced.
- Employing the objective function and it doesn't need derivatives or other information complementary, because sometimes they are very hard to be achieved. Thus they gain in efficiency and generality.
- They use stochastic transition rules, not deterministic. The GAs use random operators to guide the search to the best spots; it may seem strange, but the nature is full of precedents in this regard.

12.2.3 Nondominated sorting genetic algorithm II

For the development the multiobjective algorithms require mathematical methods optimization on a population of solutions because the NSGA-II was chosen as proposed, due to its diversity and reliability characteristics. However, an overview should be maintained to enable the use of other procedures, such as ant colonies, simulated annealing, and the particle swarm.

The NSGA-II, the first version based on GAs, is classified as an elitist type, since it incorporates a preservation mechanism of the dominant solutions through several generations of a GA.

The process starts from a set of size N solutions (couple) obtained randomly or methodically. Later generations are determined using modified mechanisms of selection, crossover, and mutation defined by classic GA.

12.2.3.1 Selection process, crossover, and mutation

On the current population (pair), randomly selected N pairs of solutions are selected. Each pair competes in a tournament in winning alternative that belongs to the category of best quality. If the dominance of alternatives belongs to the same front, then winning it introduces a greater degree of diversity to all that are under construction. The winners of each tournament are allowed only for seed; the crossover and mutation are handled in the same way as shown by the classic GA.

Thus what is expected is that the genetic information of the dominant alternative be present in the following generations and attract the rest of the population to their respective neighborhoods.

12.2.3.2 Stacking operator

The multiobjective algorithms seek to find a big number of solutions that belong to the Pareto front. Therefore it is necessary that the population be kept as much diverse as possible. The stacking operator quantifies the space around an alternative that is not occupied by any other solution. This is due to calculating the perimeter of the cuboid formed by neighboring solutions that have the same category of the alternative dominance i, which is described by the following equation:

$$d_i = \sum_{m=1}^{M} \left| \frac{f_m^{(I_{i+1}^m)} - f_m^{(I_{i-1}^m)}}{f_m^{\max} - f_m^{\min}} \right| \tag{12.1}$$

where I^m is a vector indicating the nearby alternative solution alternative i, $f_m^{\max}$ and $f_m^{\min}$ are the maximum and minimum values in the function of the solution space object m, respectively, and M is the number of optimized objective functions.

12.2.3.3 Selection by tournament second stacking operator

This procedure replaces the selection used in traditional GA. They consist of comparing two solutions; each one of them has two attributes:

- A range of nondomination r_i, according to the Pareto front.
- A local stacking distance, d_i.

The selection returns to winning solution i based on two fundamental criteria:

- If j has better hierarchy, $r_i < r_j$.
- If j has the same hierarchy, but i has a better stacking distance, $d_i > d_j$.

12.2.3.4 Determination of final set descending

Before finalizing a generation of algorithm, a process of preselection and preservation of elite solutions is performed, which involves getting the set of solution parents and offspring obtained through the selection of operators, crossover, and mutation.

Thus the present population increases to double the initial population of individuals. It is necessary to classify the full set of fronts in their respective dominance and preserve individuals who belong to the best quality fronts, as is shown in Fig. 12.2.

If it is not possible to enter all the alternatives of a particular forward, then those individuals are disposed with a smaller distance to the crowd.

12.2.3.5 Pseudocode for the nondominated sorting genetic algorithm II

The steps used in NSGA-II are as follows:

1. Generate a population of size N.
2. Identify the dominance of fronts and evaluate stacking distances on every front.
3. Using selection, crossover, and mutation generates a downward population, the same size as P.
4. Parents and children together in a set of $2N$ rank the dominance fronts.
5. Determine the final set down by selecting the fronts of the best features or hierarchy. If exceeded the threshold population of N, eliminate solutions with the shortest distance across stacking the last selected.
6. With the fulfillment of convergence criterion, the process ends, if not, return to step 3.

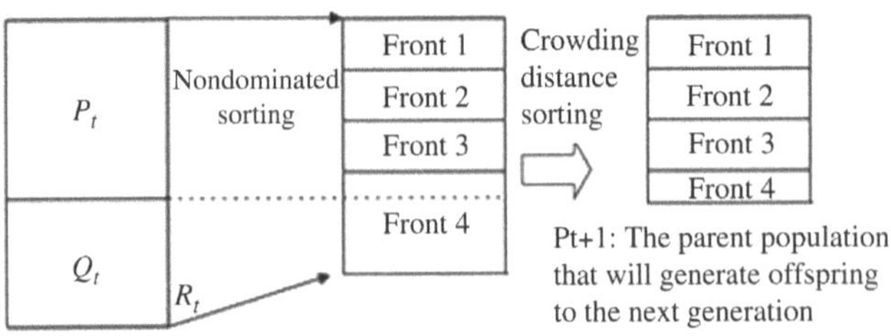

FIGURE 12.2 Determination of new population. In the figure, P_t is the current population, Q_t is the offspring population, and R_t is the population after recombination.

In this chapter the EED problem will be used as NSGA-II with two objective functions—one is the cost of fuel and the other, emission index.

12.2.4 The emission ratio as a parameter to assess the environmental contamination

The production of energy by fossil fuels, industrial processes, and means of transport has a great influence on the environment, due to the deforestation and emissions (CO_2, NO_x, SO_x, C_xH_y, particulates, etc.); it is considered the main anthropogenic sources of pollution.

The Kyoto Protocol resulted from the meeting of 160 nations in 1997 in Japan to reduce emissions of gases that cause the greenhouse effect (CO_2, CH_4, etc.) and encourage the development of new technologies and the implementation of clean sources power. Since then, the right to trade emissions (primarily CO_2 resulting from the burning of fossil fuels, whose use in developed countries is intensive) is gaining strength as a political strategy.

The air pollutants originate mainly from incomplete combustion of fossil fuels. Those are classified into two types: primary and secondary. Primary pollutants are those emitted directly from sources to the atmosphere, highlighting particulate matter (smoke, dust, and mist), carbon monoxide (CO), carbon dioxide (CO_2), nitrogen oxides (NO and NO_2), sulfur compounds (H_2S and SO_2), hydrocarbons, and chlorofluorocarbons [11–13].

With the introduction of emissions and ecological taxation market for the electricity sector, the development of decision-making methods for emissions trading or emission restrictions is becoming increasingly important, and many studies may decide to program generators for operation [14–18].

Although there are many studies on CO_2 restrictions, they focus primarily on the problem of deciding the output level of each generator during the ED.

However, to obtain an optimal solution, it is important to consider not only the dispatch level of each generating unit, but also the schedule (on/off), since the power minimum output, restrictions, and start-up influence the final solution of the cost/emission. So it is essential to consider the restriction problem of each unit in decision-making methods. In addition, most generating unit studies, including CO_2 restrictions, are focusing on programming the solution that maximizes earnings per unit [24,25] but not in optimal solutions Pareto in reducing CO_2 [14,26,27].

According [28], CO_2 emission allowances are usually given for a period of 1 year, while the time frame for programming is scheduled of 24 hours for several days, and restrictions have an effect only when the value of CO_2 emissions became high. According [29] they believe that maximum profit is

TABLE 12.1 Data to determine the emission index of gas engines.

Pollutant	Primary standard CONAMA (g/m^3)	Specific weight permissible	Value of influence
Total particulate matter	240	0.00592885	0.99407115
Carbon monoxide	40,000	0.98814229	0.01185771
Nitrogen dioxide	320	0.00790514	0.99209486
Hydrocarbons	160	0.00395257	0.99604743
Total	40,480	1	3

important but the trade-off of cost reduction and CO_2 should not be taken into consideration.

12.2.5 Emission index of gas engines

To evaluate the environmental pollution caused by gas engines, the emission rate is established by the author of this work, considering the value of the first data table. To develop the mathematical expression, the emission index limits were considered and the air quality was determined by CONAMA and the weighted value of each pollutant in the air quality [30] expressed in Table 12.1.

The influence of the amount of CO_2 to 1 is considered. The equation for calculating the emission index from gas engines is expressed by (12.2).

$$I_{emg} = CO_2 + 0.99407115MP + 0.01185771CO + 0.99209486NO_2 + 0.99604743C_XH_y \tag{12.2}$$

As gas engines also emit nitrogen monoxide, it was decided to include them in the expression with the same amount of influence as NO_2, in the following formula:

$$I_{emg} = CO_2 + 0.99407115MP + 0.01185771CO + 0.99209486(NO_2 + NO) + 0.99604743C_XH_y \tag{12.3}$$

To calculate the emission index or rate of emissions, all emission values must be in the same system of units, which is necessary to perform conversions of the same according to the companies that make the control of these

TABLE 12.2 Conversion factors.

Parameter	Initial units	Multiply by	End units
PM	mg/m^3	1	mg/m^3
Nitrogen dioxide (NO_2)	mg/m^3	1	mg/m^3
Nitrogen monoxide (NO)	mg/m^3	1	mg/m^3
Carbon dioxide (CO_2)	%	18,000	mg/m^3
Carbon monoxide (CO)	ppm	1.25	mg/m^3
Hydrocarbons (C_xH_y)	%	17,960	mg/m^3

PM, Particulate matter.

TABLE 12.3 Molecular weights.

Sustenance	Molar weight (g/mol)
W	12
O_2	32
O	16
CO_2	44
CO	28
N	14
N_2	28
H	1
Methane (CH_4)	16
Hexane (C_6H_{14})	86

emissions. Table 12.2 shows the emission values as the thermal plant and the conversion factors.

To perform conversions, molecular weights of the components were considered, according to the following procedures and amounts (see Table 12.3):

$$mg/m^3 = \frac{\text{ppm} \times \text{PM}}{24.45} \tag{12.4}$$

Thus the expression to calculate the emission rate of gas engines is given by

TABLE 12.4 Typical emissions from gas engines.

Typical emissions of a gas engine (UGGN 12)	Original *U*	3% mg/m^3
Particulate material (mg/m^3)	76.57	76.57
Nitrogen dioxide (mg/m^3)	315.07	315.07
% Oxygen (mg/m^3)	12.3	80,490.7975
Carbon dioxide (CO_2)% to mg/m^3	4.8	86,400
Carbon monoxide CO (ppm mg/m^3)	286	327.525562
Nitrogen monoxide (mg/m^3)	105	105
Hydrocarbons (C_xH_y) (ppm mg/m^3)	861.64	1688.8144
Nitrogen oxides (NO_x as NO_2) (ppm mg/m^3)	213	400.44

$$I_{emg} = 18,000\ \mathrm{CO_2} + 0.99407115\ \mathrm{MP} + 0.01185771 \times 1.25\ \mathrm{CO} + 0.99209486(\mathrm{NO_2} + \mathrm{NO}) + 0.99604743 \times 17,960\mathrm{C_xH_y}\ \text{in mg}/m^3 \tag{12.5}$$

In expression (12.5), CO_2 and C_xH_y are expressed in % in ppm CO and the other data in mg/m^3.

Table 12.4 shows the typical emissions of a plant gas engine in Manaus.

12.2.6 Index engine emissions of heavy fuel oil

In this case the developed procedure was the same as for gas engines but taking into account the emissions of such engines (see Table 12.5).

Table 12.6 shows the conversion values:

$$I_{\mathrm{emHFO}} = 18,000 \times \mathrm{CO_2} + 0.9941\mathrm{MP} + 1.25 \times 0.0265\mathrm{CO} + 0.992(\mathrm{NO_2} + \mathrm{NO}) + 0.991\mathrm{SO_2} + 0.9961 \times 17,960\ \mathrm{C_xH_y}\ (g/m^3) \tag{12.6}$$

Table 12.7 shows the typical emissions of a motor heavy fuel oil (HFO), the plant in Manaus.

12.2.7 Contamination caused by plant

The thermal plant studied lies in Manaus and has a generating capacity of 173 MW. It contains 23 Jenbacher gas engines of 3.5 MW and 5 engines of HFO of 18.5 MW; but for the use of optimization engines, 10 were used.

TABLE 12.5 Data to determine the emission index of the heavy fuel oil engines.

Pollutant	Primary standard CONAMA (g/m^3)	Specific weight permissible	Value of influence
Total particulate matter	240	0.0059	0.9941
Carbon monoxide	40,000	0.9735	0.0265
Nitrogen dioxide	320	0.0078	0.9922
Sulfur dioxide	365	0.0089	0.9911
Hydrocarbons	160	0.0039	0.9961
Total	*41,085*	*1*	*4*

TABLE 12.6 Conversion factors in the case of engines heavy fuel oil.

Parameter	Initial units	Multiply by	End units
PM	mg/m^3	1	mg/m^3
Nitrogen dioxide (NO_2)	mg/m^3	1	mg/m^3
Nitrogen monoxide (NO)	mg/m^3	1	mg/m^3
Carbon dioxide (CO_2)	%	18,000	mg/m^3
Carbon monoxide (CO)	ppm	1.25	mg/m^3
Sulfur dioxide	mg/m^3	1	mg/m^3
Hydrocarbons (C_xH_y)	%	17,960	mg/m^3

PM, Particulate matter.

To analyze the contamination caused by the plant, we studied the data of exhaust emissions from the years 2011 and 2012. In addition, to compare the contamination of gas engines with the contamination of HFO engines, data from HFO engines at the same level of oxygen were converted to the data from gas engines.

To convert data to different % oxygen, the following expression was used:

$$C_c = C_{\mathrm{GAS}} \times \frac{(21 - O_{\mathrm{REF}})}{(21 - O_{\mathrm{MED}})} \tag{12.7}$$

where C_c is the corrected concentration expressed % to specified oxygen, C_{GAS} is the concentration of gas corrected (values obtained with checks),

TABLE 12.7 Typical emissions of an engine heavy fuel oil (HFO).

Typical emissions from a motor HFO (MAN 1)	U original	7% mg/m^3	3% mg/m^3
Particulate material (mg/m^3)	156.65	156.65	201.407143
Sulfur dioxide (mg/m^3)	287.42	287.42	369.54
% Oxygen (mg/m^3)	13.7	179,304.703	230,534.619
Nitrogen dioxide (mg/m^3)	315.07	315.07	405.09
Carbon dioxide (CO_2)% to mg/m^3	5.5	98,977.5051	127,256.792
Carbon monoxide CO (ppm mg/m^3)	66.66	76.3386503	98.1496933
Nitrogen monoxide (mg/m^3)	1167	1167	1500.42857
Nitrogen oxides (NO_x as NO_2) (mg/m^3)	1843	1843	2369.57143
Total hydrocarbons (C_xH_y) from the % (mg/m^3)	0.03	588	756

TABLE 12.8 Emissions in the different systems of units.

Gas	mg/m^3 ppm
CO	1.25
At the	1339
NO_2	2054
NO_x (as NO_2)	2054
SO_2	2857

O_{REF} is the oxygen reference, that is, it is noted as the measurements, O_{MED} is the average oxygen during measurements.

It was necessary to establish the same system of units for some values as presented in Table 12.8.

For the engine emissions to index HFO, the plant was carried out similarly to the gas engine, that is, measurements were carried out in the chimney values of engines of different pollutants and statistically processed results. With the average values, emission index for each motor HFO plant was calculated, shown in Fig. 12.3.

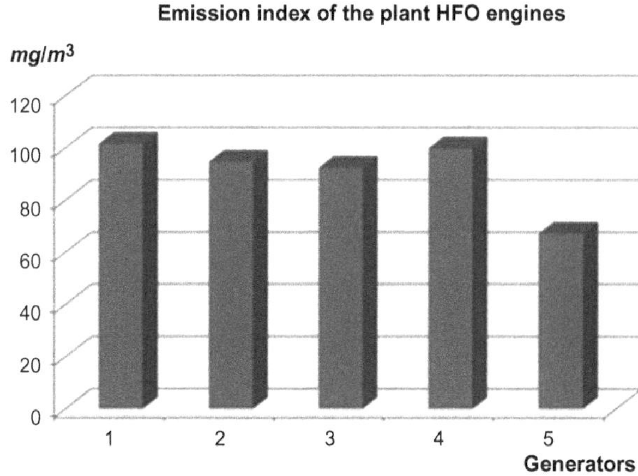

FIGURE 12.3 The emission index of the plant HFO engines. *HFO*, Heavy fuel oil.

The figure can be seen that the emission rate of the HFO power plant engines has values very close to each other, which are almost similar, that is, are not so different as in the case of gas engines. In this case, the contamination of these motors is lower than that of the gas engines; the explanation for this contradiction is that the technical state of the gas engine is lower than the roadworthiness of motor HFO.

12.2.8 Specific emission index

To better evaluate, emission indexes were divided by the power generated by the engines, thus obtained specific emission index. Table 12.9 shows a comparison between the specific index emissions from gas engines and engines HFO, and Fig. 12.4 shows a comparison according to the emission power supplied to the gas engines and fuel oil engines.

The graph in Fig. 12.4 shows the specific emission index for each type of pollutant.

In the graph of Fig. 12.4, it can be seen that in this case, the motors HFO contaminate the environment more than the gas engine, especially the emissions of carbon dioxide; these results are in agreement with those established in the literature but the rest of emissions should behave similarly. Gas engines emit more NO_2 than the engines HFO, but this fact has to do with two things, the first is the LENOX device that has these engines that regulate them for maximum efficiency, and this is achieved when the NO_2 emissions are highest. The other aspect that influences the technical state of gas engines is mentioned above.

12.2.9 Permissible values of emission Index

In the literature referred to for EED, only permissible values or restrictions for the emission of TPPs appear, which reinforces the need of emission using

TABLE 12.9 Comparison of emissions between gas engines and heavy fuel oil (HFO) engines in relation to power supplied.

Specific emission index g/m^3, kW	HFO	Gas
Particulate matter	10.6564	22.32
Sulfur dioxide	19.5523	0
Oxygen	121.975	234.66
Nitrogen dioxide	21.4333	91.85
Carbon dioxide (CO_2)	673.16	251.89
Carbon monoxide (CO)	6.41	11.8
Nitrogen monoxide (NO)	79.3877	30.61
Nitrogen oxides (NO_x as NO_2)	125.374	116.74
Total hydrocarbons (C_xH_y)	40	49.36

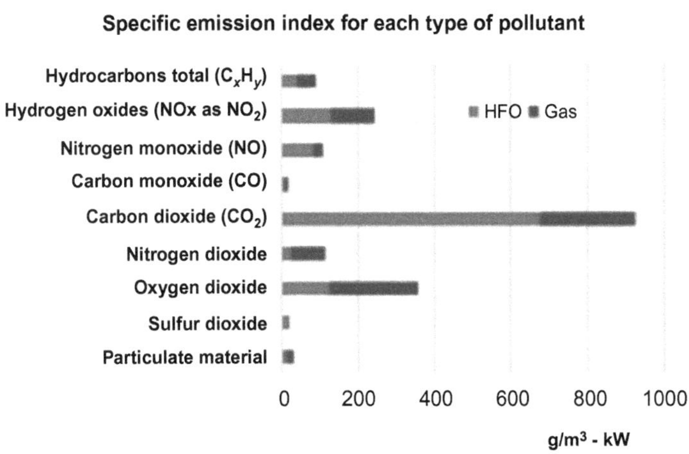

FIGURE 12.4 Specific emission index for each type of pollutant.

a parameter, which constricts the operation of a TPP. In most cases the authors estimate the amount of emissions and sometimes convert these values in cash [35] Developed by emission factors for the case of the use of coal, but not limited to these emissions [36]. Developed a complex mathematical procedure to determine the allowable CO_2 emissions, but this procedure being very complex includes only CO_2 emissions. According [37] they made an inventory of NO_x and CO through several years of observation and concluded that they should be restricted.

The emission index developed in this work presents a great advantage because it brings restrictions, which was developed from the damage that the emissions cause. However, if desired greater precision, you can set a maximum rate of emissions that would be the sum of all allowable values for [38] adding permissible values of CO_2 emissions, which are not provided by this standard.

According to these considerations, the maximum pollutant emission index (MPEI) would be

$$MPEI = 58.48\ \text{mg}/m^3$$

It can be seen from the analyses carried out in this chapter that the various engine emission indexes ever exceed this value, which shows something that is already known and that the thermal plants greatly harm the environment.

12.2.10 Obtaining primary data

For the emission index from gas engines, plant measurements were performed in the chimney values of engines of different pollutants and statistically processed results. With the average values, the emission index for each gas engine of the plant was calculated.

To get the raw data in expectation to calculate emission index, it develops the following:

- They were placed at all plant engines operating at different power levels with respect to maximum power (20%, 30%, 40%, 50%, 60%, 70%, 80%, 90%, and 100%).
- For each of the power levels, emissions of various pollutants were measured, both in volume and type of pollutants in mg/m^3.
- For each of the power levels of each motor, emission rate according to Eqs. (12.5) or (12.6) was calculated, depending on whether it is a gas engine or HFO engine.
- With the emission index of each engine, power curve versus emission index was obtained. In the graphs of this chapter, indexes emissions at full power were placed.
- With the curve of the emission index of each engine and using a regression software, equation index emissions from all power plant engines were obtained. We also calculated the coefficients *d*, *e*, and *f* to be used in the function emission index.

12.2.11 Price of carbon emissions

Carbon credits are a financial instrument envisaged in the Kyoto Protocol to try to mitigate the threats that cause the greenhouse effect. Each credit is equivalent to 1 t of carbon dioxide that was allowed to emit into the atmosphere. They may be generated through the mechanisms established in the Kyoto Protocol. As the mechanism exists, different types of credits are provided [39,40].

In other words, these credits are used to make it easy to calculate the amount of gases that are released into the air and offset their emissions. It is part of an international plan, probably the largest, that has been created in human history, to reduce global warming and effects. It is even the total amount of emissions that can be released by a company or business. If there is an excess amount of gases that are emitted, there is a monetary value assigned to that excess and can be traded, especially for projects that offset pollution, that is, to renew dioxide that has been emitted into the atmosphere, such as reforestation projects (usually in poor or developing countries).

It is well known that some companies think if they can bribe, they would be illegally allowed to pollute. In addition, there are credits that are bought and sold in international markets. So, this may be the object of speculation and does not have to be used to care for the environment.

By convention, 1 t of carbon dioxide (CO_2) represents 1 carbon credit. This credit can be negotiated in international market. Reducing the emission of other gases, also generators of greenhouse effect, it can also be converted to carbon credits, using the concept of carbon equivalent (carbon dioxide equivalent) [41].

A ton of CO_2 equivalent corresponds to a carbon credit. The CO_2 equivalent is the result of multiplying the tons of greenhouse effect emitted by its global warming potential. The global warming potential of CO_2 was set to 1. The global warming potential of methane is 21 times greater than CO_2 potential, so the CO_2 equivalent of methane is equal to 21. Therefore a reduced ton of methane corresponds to 21 carbon credits [18,42,43].

Global warming potential of greenhouse gases is as follows [44, 45]:

- CO_2—carbon dioxide = 1
- CH_4—methane = 21
- N_2O—nitrous oxide = 310
- HFCs—hydrofluorocarbons = 140–11,700
- PFCs—perfluorocarbons = 6500–9200
- SF_6—sulfur hexafluoride = 23,900

Since 2008 the price of carbon credits traded to sell to the developed countries in America (CO_2 Certificate of Emission Reduction) fell 98%, from 23 euros per ton to only 35 cents of euro per ton. The value of securities is traded in the domestic market in Europe—European Union Allowance. In turn, it fell from 30 to 4 euros [40]. Basically, considering the concept of supply and demand, there is currently an excess of credit carbon, which is a problem. In fact, the latest figures estimate that the market is saturated in about 1700 million tons of carbon credits.

Second [45], the cost per ton of emissions of CO_2 has varied between 9 and 24 euros. These authors as mentioned earlier also make an equivalence between the tons of other pollutants and tons of CO_2. Considering the above criteria, you can calculate the cost using the emission index by the following approximation:

$$\text{Cost}_{\text{emissions}} = 24 \times I_{em} \text{ in euros} \tag{12.8}$$

12.3 A mathematical model for the optimization of EED considering the emission index

Optimizing the EED is one of the most important tasks in power plants with internal combustion engines. The ED energy with a single goal cost of fuel only considers the one objective, that is, the question of generation. It has given way to multiobjective orders because of the environmental issues that arise from emissions from thermal plants. The purpose of this chapter is to analyze a new solution optimizing the EED by the technique of NSGA-II but using the new concept of emission index instead of using emissions as a cost or as much of greenhouse gases.

The EED of the problem is to minimize the total cost of generation and emission levels while at the same time to satisfy the demand of generation plants.

Thermal power generation is one of the sources of significant carbon dioxide (CO_2), sulfur dioxide (SO_2), and nitrogen oxides (NO_x) that create air pollution [13]. The classic problem generating ED is to provide the required amount of power at the lowest cost, to meet the demand and operational restrictions.

This is a very complex problem to be solved for its high dimensionality, a nonlinear objective function, and many restrictions. Various techniques, such as Integer Programming [46], Dynamic Programming [47], Newton's method for [48], and the functions of Lagrange by [49], have been used to solve the problem EED generation.

To solve the EED problem, other optimization methods, such as the method of simulated annealing (simulated annealing goal attainment) pointed to by [50], particle swarm used by [51], the Game Theory used by [52], and the approach using the technique for order preference by a similarity of the ideal solution (TOPSIS) [53].

Various methods have also been developed on the basis of mathematical approaches to offer a quicker solution to the ED [54]. EAs have also been applied to the ED of the problem in question [55]. The research has also been developed to minimize the costs, including emission restrictions to solve the ED of generation and selection of generators [56]. Recently, it has been successfully employed by a combination of gravitational search algorithms modified by NSGA-II [57] and NSGA-III [58] that are convenient to solve the generation of EED optimization problems.

Therefore all previous models for EED only take into account the emissions and the amount of emissions. This work presents the degree of influence of each type of emissions on the environment. In the coming sessions, we develop a novel mathematical model that accurately classifies emissions according to their impact on the environment and this will be one of the functions to be optimized within the template.

12.3.1 Mathematical model for environmental economic dispatch

In the mathematical formulation of the multiobjective problem of EED, two important goals in a thermal system of power generation have to be considered, which are economic and environmental impacts [52,59,60].

12.3.1.1 Minimizing costs

The fuel cost of a thermal unit is considered as an essential criterion for economic viability. The fuel cost curve is assumed to be approximated by a quadratic function of the output power of the generator P_i [52,59,61,62]. The function to be used to minimize the cost is

$$F_1(P_i) = \sum_{i=1}^{n} \left(a_i + b_i P_i + c_i P_i^2\right) \ \$/\mathrm{h} \tag{12.9}$$

where a_i, b_i, c_i, and P_i are the fuel cost coefficients of the ith generating unit, and n is the number of generators and the active power of each generator.

However, despite the great financial benefit of classical dispatch strategy described by Eq. (12.9), whose fuel cost versus power generated curve is shown in Fig. 12.5, it tends to produce high amount of SO_2 and NO_x.

The fuel cost function of each thermal generating unit considering the valve-point effect is expressed as the sum of a quadratic function and a sine function [64,65]. The total cost of fuel in terms of active power can be expressed as

$$F_1 = \sum_{m=1}^{M} \sum_{s=1}^{N_S} t_m \left[a_s + b_s P_{sm} + c_s P_{sm}^2 + \left|d_s \sin\left\{e_s\left(P_s^{\min} - P_{sm}\right)\right\}\right|\right] \tag{12.10}$$

12.3.1.2 Minimizing the environmental impact

The generators with fossil fuels are the main source of emissions of nitrogen, oxides, and other pollutants. Currently, there are strong constraints of environmental protection agencies to reduce the emission of nitrogen oxides (NO_x) being important from the point of view of environmental conservation.

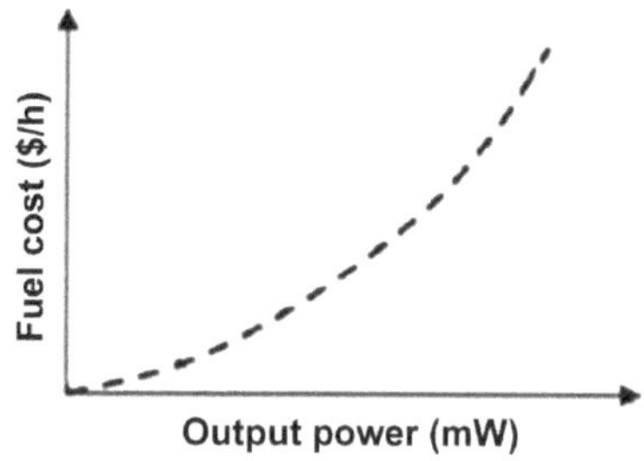

FIGURE 12.5 Cost of fuel versus output power.

There are various alternatives to consider and minimize the environmental impact of power plants, which are as follows:

- A dispatch alternative strategy that must meet environmental requirement is to minimize the cost of operation under environmental restrictions.
- Control of emissions may be included in conventional ED, adding the environmental cost to generation costs [2,66]. Emissions are modeled as a cost to the environment, which are later added to the cost of generation. The objective function is expressed as follows:

$$\text{minimize } C = w_0 \cdot F + w_1 \cdot E_S + w_2 \cdot E_N \tag{12.11}$$

where E_S and E_N is the emission function of SO_2 and NO_x, respectively. w_0, w_1, and$w_2$are the cost of weight in relation to the fuel (F) and the emissions of SO_2 and NO_x, respectively. F is the function of the cost of fuel, which is another variation to consider emissions into a single objective function, where particular weightage is given to NO_x and SO_2 emissions.

The functions of the function in emission cost curves of the active power generated included in function (12.11) can be expressed as follows:

$$E_s = \sum_{i=1}^{n} (d_i + e_i P_i + f_i P_i^2) \tag{12.12}$$

$$E_N = \sum_{i=1}^{n} (g_i + h_i P_i + k_i P_i^2) \tag{12.13}$$

where d_i , e_i ,f_i , g_i , h_i, and k_i are the estimated parameters based on the results of the emission tests generating unit, and P_i is the power of each generator.

In this model, when the emission weights are 0, the objective function becomes a classic problem of ED. In this case the goal is to minimize costs and total system output. For SO_2 emission, the weights w_0 and w_2 are equal to 0 and w_1 is equal to 1. For SO_2, the goal is to minimize the emission. For NO_x emission, the weights w_0 and w_1 are 0 and w_2 is equal to 1, where the problem lies in the minimization of NO_x emissions. On the contrary, when the weights are not 0, minimizing both the cost of fuel and emissions at the same time becomes the problem.

For [67], the amount of emission of each generator is given as a function of its output, which is the sum of a quadratic function and an exponential function. The total emission system can be expressed as

$$F_2 = \sum_{m=1}^{M}\sum_{s=1}^{N_S} t_m\left[\alpha_s + \beta_s P_{sm} + \gamma_s P_{sm}^2 + \eta_s \exp(\delta_s P_{sm})\right] \tag{12.14}$$

where $\alpha_s, \beta_s, \gamma_s, \eta_s,$ andδ_s are the coefficients of the emission characteristics of each generator, and P_{sm} is the power of each generator.

According [68], the multiobjective problem of dispatch emissions and combined economic can be converted into an optimization problem of a single goal by introducing a factor h penalty price as follows:

$$\text{Minimize } F = F_C + h_i \times \text{EC} \tag{12.15}$$

where F_C is the fuel cost function and EC is the total amount of emissions.

Expression (12.15) is subject to the equations and power flow restrictions. The price of the penalty factor h combines the issue with the cost of fuel and F is the total operating cost in $/h. The price penalty factor is the ratio of the maximum cost of fuel and the emission maximum of the corresponding generator h_i [68]:

$$h_i = \frac{F_C\left(P_{g_i}^{max}\right)}{EC\left(P_{g_i}^{max}\right)} \tag{12.16}$$

where F_C is the fuel cost function, EC is the total amount of emissions and g_i is the power in generator ith.

The emissions that are considered most important in the power generation industry due to their effects on the environment are sulfur dioxide (SO_2) and nitrogen oxides (NO_x) [13,69]. These emissions can be modeled by associating functions with emission power production for each unit. One approach to represent the emissions of SO_2 and NO_x is to use a combination of polynomial terms [68,70]:

$$\text{EC}\left(P_g\right) = \sum\left(\alpha_i P_{g_i}^2 + \beta_i P_{g_i} + \gamma_i\right) + \varepsilon_i \exp\left(\lambda_i P_{g_i}\right) \tag{12.17}$$

where $\alpha_i, \beta_i, \gamma_i, \varepsilon_i,$ andλ_i are the emission characteristics of the coefficients of the total power generated, P_g, which is the power of each generator.

Second [71], the total emission $F_2(P_i)$ of air pollutants such as sulfur dioxide, SO_2, and nitrogen oxides, NO_x, caused by the combustion of fuel in thermal units may be expressed as

$$F_2(P_i) = \sum_{i=1}^{n}\left(d_i + e_i P_i + f_i P_i^3\right)\ \text{m}^3/\text{h} \tag{12.18}$$

where d_i, e_i, and f_i are the coefficients of emission characteristics for each generating unit.

12.3.1.3 Load dispatch restrictions considering emissions

In this section, a number of restrictions are considered:

- An equal restriction of active power balance generated
  The following equation is the power balance constraint [72,73]:

$$\sum_{i=1}^{n} P_i - P^D - P^L = 0 \tag{12.19}$$

where P_i is the output power of each i generator, P^D is the load demand, and P^L are transmission losses.

In other words, the total power generation has to meet the total demand, P^D, and the loss of active power transmission lines, P^L:

$$\sum_{i=1}^{n} P_i = P^D + P^L \tag{12.20}$$

The calculation of power losses involves the solution of the load flow problem, which has equal restrictions on active and reactive power in each bar as follows [74]:

$$P^L = \sum_{i=1}^{n} B_i P_i^2 \tag{12.21}$$

To model the transmission loss, each function generator loss through the derivatives of formula Kron coefficients for loss is set to output.

$$P^L = \sum_{i=1}^{N} \sum_{j=1}^{N} P_{Gi} B_{ij} P_{Gj} + \sum_{i=1}^{M} B_{0i} P_{Gi} + B_{00} \tag{12.22}$$

where B_{ij}, B_{0i}, andB_{00} are the power loss coefficient in the transmission line. A reasonable accuracy can be obtained when the actual operating conditions are close to the base case where the coefficients B were obtained [75].

- An inequality constraint in terms of generation capacity
  For stable operation, the active power generated by each generator is limited by the upper and lower limits. These restrictions in the generation limits are expressed by

$$P_{\min,i} \leq P_i \leq P_{\max,i} \tag{12.23}$$

where P_i is the output power of the generator, $i, P_{\min,i}$ is the minimum power of the generator, i, and $P_{\max,i}$ is the maximum generator power.

- An inequality constraint in terms of fuel supply

At each time interval the amount of fuel supplied to each generator F_{im} must be within its lower limit, $F_i^{\min}$, and its upper limit, $F_i^{\max}$, [59] such that

$$F_i^{\min} \le F_{im} \le F_i^{\max}, \quad i \in N, \ m \in M \tag{12.24}$$

where F_{im} is the fuel supplied to the engine in the interval m, $F_i^{\min}$ is the minimum quantity of fuel supplied to the machine, $F_i^{\max}$ is the maximum quantity of fuel supplied to the machine.

- An inequality constraint in terms of fuel storage limits

Each unit of fuel storage volume in each interval, V_{im}, should be within its lower limit, $V_{\min}$, and the upper limit, $V_{\max}$, [59] so that

$$V_{\min} \le V_{im} \le V_{\max} \tag{12.25}$$

$$V_{im} = V_{(m-1)} + F_{im} - t_m\left[\eta_i + \delta_i P_i + \mu_i P_i^2\right] \quad i \in N, \ m \in M \tag{12.26}$$

where $\eta_i, \delta_i, \text{and} \mu_i$ are the fuel consumption coefficients for each generating unit.

Although a strong review in the literature is made on the restrictions of emissions comparing them with a ceiling which cannot be achieved, there were no mathematical expressions of equal or unequal restriction emissions.

12.3.1.4 Objective functions

The objective function used to minimize the cost of fuel was expressed in Eq. (12.9). It is important to note that to apply this equation; the coefficients $a_i, b_i, \text{and} c_i$ of each engine were first calculated by putting all the engines of power plants operating at different power values, which result in the power curve versus the cost of each engine. Subsequently, regression equation methods and their respective coefficients were obtained.

The function used to minimize the emission index is given by the following equation:

$$I_{em}(P_i) = \sum_{i=1}^{n} \left(d_i + e_i P_i + f_i P_i^3\right) \ \text{mg/m}^3 \tag{12.27}$$

where $d_i, e_i, \text{and} f_i$ are the coefficients of the characteristics of the emission index for each unit.

12.3.2 Order environmental economic load: case studies

12.3.2.1 Problem formulation

Two plants to the case studies were chosen to examine the feasibility of the proposed solution; we used a set of 10 thermal generating units of TPP in

TABLE 12.10 Characteristic data of the case study of the plant generators.

Generator	c_i (\$/MW)	b_i (\$/W)	a_i (\$)	P_{min} (MW)	P_{max} (MW)
PG1	0.15247	38.53973	756.79886	0.76	3.36
PG2	0.10587	46.15916	451.32513	0.76	3.36
PG3	0.02803	40.3965	1049.9977	0.76	3.36
PG4	0.03546	38.30553	1243.5311	0.76	3.36
PG5	0.02111	36.32782	1658.5596	0.76	3.36
PG6	0.01799	38.27041	1356.6592	0.76	3.36
PG7	0.02682	45.27041	1260.6592	0.76	3.36
PG8	0.02700	46.27041	1266.6592	0.76	3.36
PG9	0.02754	47.27041	1287.6592	0.76	3.36
PG10	0.02799	48.27041	1290.6592	0.76	3.36

$$B_m = \begin{bmatrix} 0.000049 & 0.000014 & 0.000015 & 0.000015 & 0.000016 & 0.000017 & 0.000017 & 0.000018 & 0.000019 & 0.000020 \\ 0.000014 & 0.000045 & 0.000016 & 0.000016 & 0.000017 & 0.000015 & 0.000015 & 0.000016 & 0.000018 & 0.000018 \\ 0.000015 & 0.000016 & 0.000039 & 0.000010 & 0.000012 & 0.000012 & 0.000014 & 0.000014 & 0.000016 & 0.000016 \\ 0.000015 & 0.000016 & 0.000010 & 0.000040 & 0.000014 & 0.000010 & 0.000011 & 0.000012 & 0.000014 & 0.000015 \\ 0.000016 & 0.000017 & 0.000012 & 0.000014 & 0.000035 & 0.000011 & 0.000013 & 0.000013 & 0.000015 & 0.000016 \\ 0.000017 & 0.000015 & 0.000012 & 0.000010 & 0.000011 & 0.000036 & 0.000012 & 0.000012 & 0.000014 & 0.000015 \\ 0.000017 & 0.000015 & 0.000014 & 0.000011 & 0.000013 & 0.000012 & 0.000038 & 0.000016 & 0.000016 & 0.000018 \\ 0.000018 & 0.000016 & 0.000014 & 0.000012 & 0.000013 & 0.000012 & 0.000016 & 0.000040 & 0.000015 & 0.000016 \\ 0.000019 & 0.000018 & 0.000016 & 0.000014 & 0.000015 & 0.000014 & 0.000016 & 0.000015 & 0.000042 & 0.000019 \\ 0.000020 & 0.000018 & 0.000016 & 0.000015 & 0.000016 & 0.000015 & 0.000018 & 0.000016 & 0.000019 & 0.000044 \end{bmatrix}$$

FIGURE 12.6 Symmetric matrix with the transmission loss coefficients.

the city of Manaus and the test system with 10 generating units [76]. The characteristics of the generators are shown in Table 12.10. For the determination of the coefficients, a_i, b_i, and c_i, a trial operation test was conducted by running generators for different powers and measuring the fuel consumed. Then the power curves versus fuel costs were plotted and a regression method was acquired. Demand for energy used was 20 MW in the case of 10 generators.

The transmission loss coefficients (B_m) are given by a square matrix of dimension $n \times n$, where n is the number of engines. The loss matrix B_m, for a plant with 10 units (all figures should be multiplied by e^{-2}), as shown in Fig. 12.6 having the symmetric matrix defined by the following: a square matrix, $S = [a_{ij}]$, is symmetric if and only if $S^T = S$. If $S = [a_{ji}]$ is a

TABLE 12.11 Emission coefficients for the 10 generators of the plant.

Generator	f_i [(Mg/m^3 h)/(MW)]	e_i [(Mg/m^3 h)/(MW)]	d_i (Mg/m^3 h)
PG1	0.00419	1.32767	73.85932
PG2	0.00419	0.32767	13.85932
PG3	0.00683	−0.54551	40.2669
PG4	0.00683	−0.54551	40.2669
PG5	0.00461	−0.51116	42.89553
PG6	0.00461	−0.51116	42.8955
PG7	0.00461	−0.51116	42.8955
PG8	0.00461	−0.51116	42.8955
PG9	0.00061	−0.51116	10.8955
PG10	0.00461	−0.51116	42.8955

All values are multiplied by e^{-2}.

symmetric matrix, the elements arranged symmetrically with respect to the main diagonal are equal, $a_{ij} = a_{ji}$. In this case the product of a square matrix S by its transpose S^T is also a symmetric matrix.

Table 12.11 shows the emission coefficient for 10 generators of the plant.

To develop the whole optimization process, NSGA-II was used, known as GA elitist ordination, and not dominated, which has the following characteristics [77,78]:

The multiobjective optimization problem [56,79], considered in this chapter is defined as

$$\text{Minimize}[F_1(P),\ F_2(P)] \tag{12.28}$$

where $F_1(P)$and$F_2(P)$ are the objective functions to be minimized over admissible decision set, that is, the vector P.

In this case the function $F_1(P)$ of Eq. (12.10) and the function $F_2(P)$ of Eq. (12.18) are used.

There are two stages to solve multiobjective problems: determining the set of nondominated solutions and selecting the best feasible solution. The execution procedure is explained in the following steps [79]:

Step-1: Power demand being supplied by the plant ($P_d = 20$ MW).

Step-2: The selection of the minimum number of more efficient generators that satisfy the active power demand.

Step-3: Set the parameters of the algorithm:

Population size;

Number of generations.

Step-4: Initialize the population, P_t.
Step-5: Create a young population or descendants Q_t of the current population P_t
Step-6: Combine the two populations Q_t and P_t to form R_t where $R_t = P_t \cup Q_t$.
Step-7: Find the nondominated Pareto fronts $F_i \text{and} R_t$.
Step-8: Start the new population $P_{t+1} = 0$ and the count for inclusion $i = 1$.
Step-9: While $P_{t+1} + F_i \leq N_{\text{pop}}$ do: $P_{t+1} \leftarrow P_{t+1} \cup F_i$, where $i \leftarrow i + 1$.
Step-10: Order the last front F_i using the distance agglomeration in descending order and choose the first elements $\left(N_{\text{pop}} - P_{t+1}\right) \text{of} F_i$.
Step-11: Use the selection of operators, crossover, and mutation to create the young population or the descendants of the new population Q_{t+1}.

12.3.3 Analysis and discussion of results

The solution report presents the input parameters to run the program, such as the energy demand, the minimum and maximum power of the engines and the results of the total cost of fuel, total power loss, and optimal power for each machine in the plant to meet the load demand.

Table 12.12 shows the results of the case study of the plant located in the city of Manaus (first case study). These results were obtained after the execution of the program for a power demand of 20 MW.

As can be seen from Table 12.12, there is a certain difference between the levels of emission of generators, and the power demand is distributed among all generators with lower values assigned to the generators 2 and 10. It can also be seen that the power is not always the maximum power that is related to the maximum emission.

Table 12.13 shows the results for the case study of the IEEE test system [75]. These results were obtained after the execution of the program for a 1036 MW power demand, which is the power between the maximum and minimum power of this system. This system has 10 units.

As is shown in Table 12.13, there are some differences between the emission index of the engine test system.

This is mainly due to the power difference between the engines of this system. This also leads to different emission index of generators.

Fig. 12.7 shows the trade-off between emission index and the fuel cost of the first case study after the application of NSGA-II, generated by MATLAB.

Fig. 12.8 shows the trade-off between emission index and the fuel cost of the test system IEEE 118-bars after applying NSGA-II, generated by MATLAB.

TABLE 12.12 Nondominated sorting genetic algorithm II (NSGA-II) final programming of Manaus test system.

Solutions to environmental economic dispatch using NSGA-II			
Power demand		20 MW	
Minimum power		0.76 MW	
Maximum power		3.36 MW	
Power losses		0.135 MW	
Fuel cost		6684.72 \$/h	
Power values and each generator emission index			
Power (Pm_i)	MW	Emission index (Em_i)	g/m^3
Pm_1	1.72	Em_1	41.24
Pm_2	3.01	Em_2	42.96
Pm_3	2.35	Em_3	41.64
Pm_4	0.92	Em_4	43.37
Pm_5	0.76	Em_5	43.29
Pm_6	0.76	Em_6	43.29
Pm_7	0.76	Em_7	43.29
Pm_8	3.34	Em_8	44.78
Pm_9	3.02	Em_9	44.57
Pm_{10}	3.36	Em_{10}	43.12
Total	20.00	Total	431.55

Pm_i is the signed power of every ith generator and Em_i is the emission level of each ith generator.

Fig. 12.9 shows a comparison of the active power generator for each plant in the first case study where it can be seen that the generators (12.6)–(12.8) produce less power.

Fig. 12.10 shows the comparison graph of active power output of each generator to the test system.

Fig. 12.11 shows a graph comparing the cost of each generator of the first case study, in which it is noted that the highest cost is incurred by generators (12.9) and (12.10).

Fig. 12.12 shows the graph comparing the cost of each generation to the test system.

TABLE 12.13 Nondominated sorting genetic algorithm II (NSGA-II) Final Programming of IEEE test system.

Solutions to environmental economic dispatch using NSGA-II	
Power demand	1036 MW
Minimum power	10 MW
Maximum power	470 MW
Power losses	0.0377 MW
Fuel cost	55,485.25 $/h

Power values and each generator emission index		
Power	MW	Emission index (g/m^3)
Un1	0	0
Un2	0	0
Un3	293.62	4,291,234.13
Un4	299.99	4,376,989.15
Un5	157.25	3,333,188.65
Un6	156.38	3,330,974.55
Un7	75.78	3,320,000.42
Un8	57.66	3,308,745.27

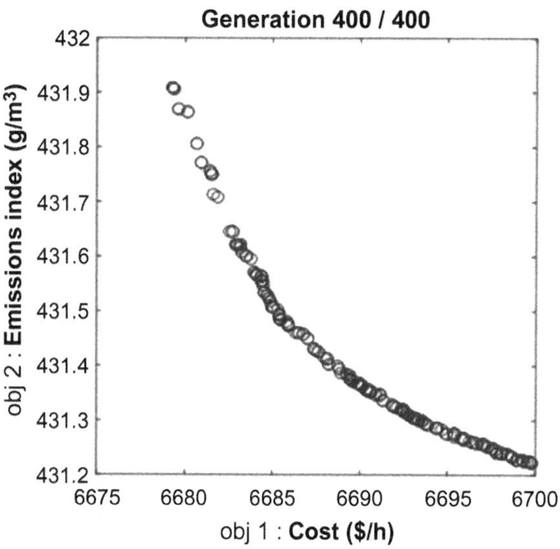

FIGURE 12.7 Trade-off between emission level and the cost of fuel after the application of the NSGA-II, 10 generator system. *NSGA-II*, Nondominated sorting genetic algorithm II.

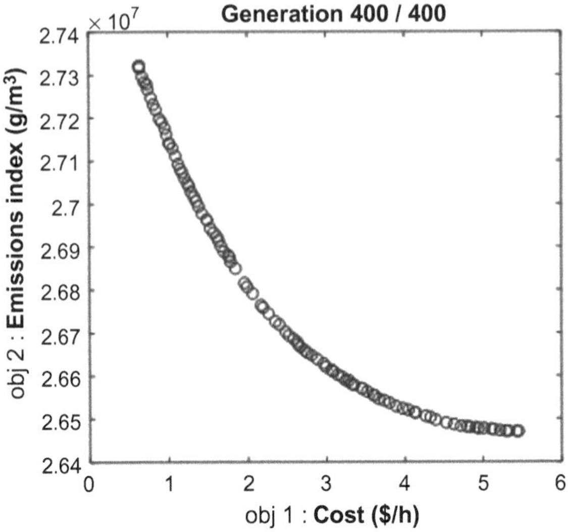

FIGURE 12.8 Trade-off between emission level and the cost of fuel after applying the NSGA-II for the test system 118-bars IEEE. *NSGA-II*, Nondominated sorting genetic algorithm II.

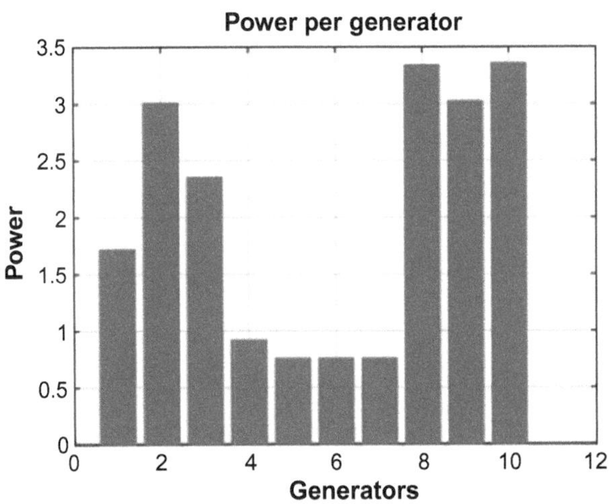

FIGURE 12.9 Power of each generator in the first case study.

Fig. 12.13 shows the comparison graph of the emission index of each generator of the first case study, and it was observed that generators (12.9) and (12.10) are generating the highest emission rates.

Fig. 12.14 shows a graph comparing the emission index of each generator to the test system where it can be seen the difference between the emission index of each generator according to their respective power.

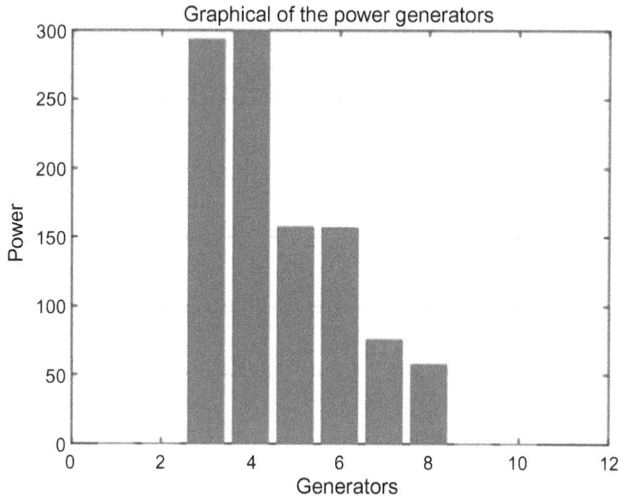

FIGURE 12.10 Output powers of the generator to the test system.

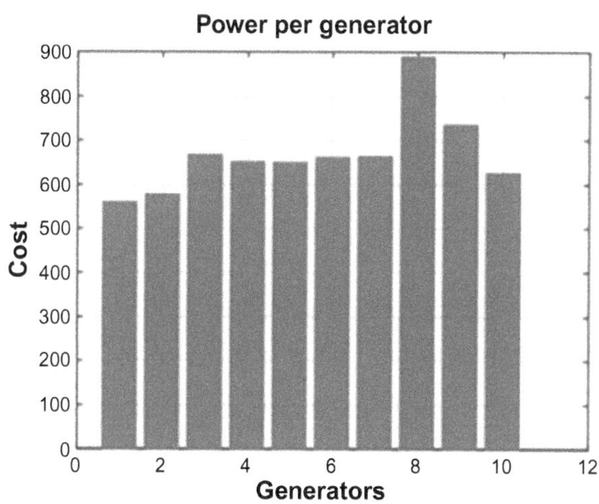

FIGURE 12.11 Generation costs of each generator of the first case study.

The developed procedure was applied satisfactorily to both cases in a TPP Manaus and testing system. These two cases were used to validate this approach.

In the case of the TPP, emission indexes are not so different as in the case of the 118-bus test system. It is widely known that this system of power generators is very different. In both the cases the power allocated to each generator corresponds to the values that guarantee the minimum cost of TPP and at the same time the minimum emission level is guaranteed.

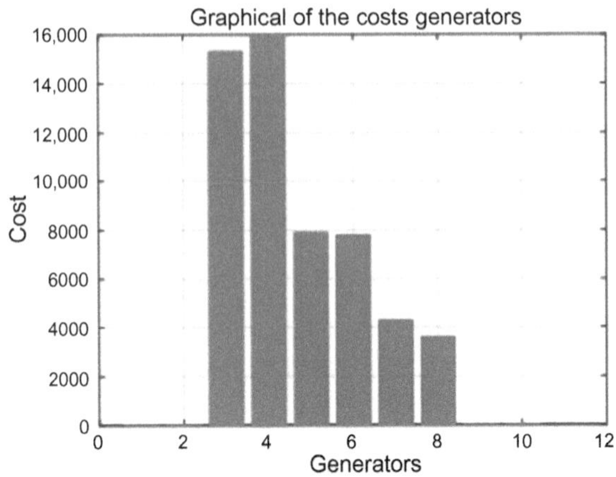

FIGURE 12.12 Cost of each generation to the test system.

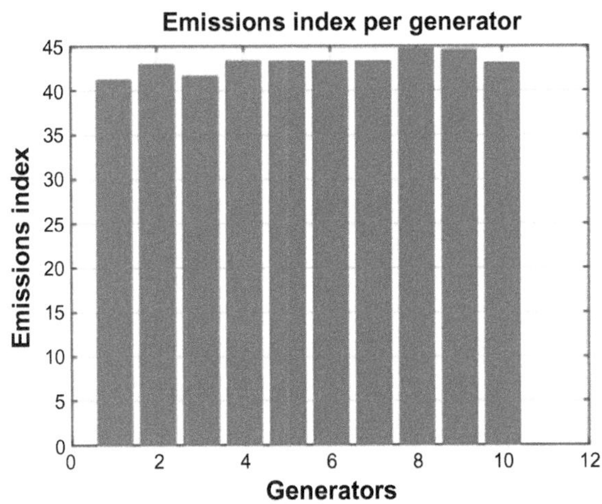

FIGURE 12.13 Generators of emission index of the first case study.

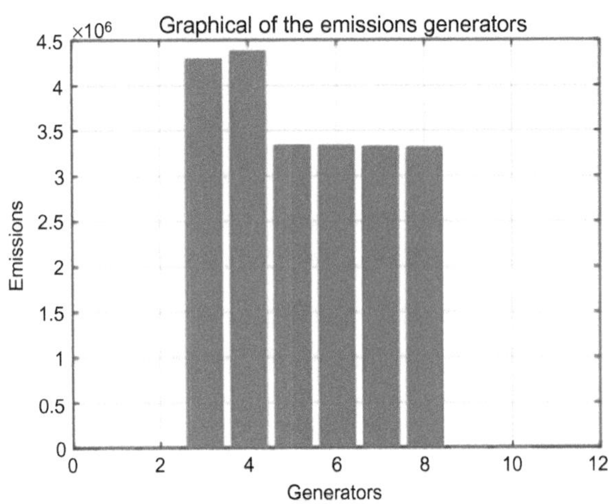

FIGURE 12.14 Issue of generators indexes for the test system.

12.4 Conclusions

The EED optimization problem in electric distribution systems is formulated as a multiobjective problem that considers the economic benefits in the operation of electric networks and the reduction of environmental pollution by inserting the emission index calculation on the system in relation to the minimization of the function emission index. In addition, the formulation presented considers the relevant restrictions imposed by Brazilian standards in relation to electrical and environmental specifications.

From the results obtained in this study, the model and mathematical method for the EED using EA NSGA-II tools reduce the cost of energy production from TPPs and environmental impact. The use of NSGA-II allows the computational tool to establish the solution to this formulation. It has determined the Pareto optimal solutions to the problem and allows the professional to determine the most effective solutions.

According to the analysis of the old EED in relation to the methodology used in this work forward, a new mathematical approach to assess emissions from generators and at the same time reducing the cost of fuel is a new possibility of identifying the different ways of evaluating the emissions produced by power plants, in relation to mathematical models with the implementation of computational tools to evaluate the economic and environmental variables, considering the permissibility of each pollutant in the atmosphere.

The mathematical procedure developed has been applied to the case study of a power-generating plant in the city of Manaus, Amazonas, and also to the test system. The relevant results of this study based on examples and practical analyses show the advantages and validate all developed procedures.

It was seen from the case study, the value of the emission index varies for different plant engines. Their values vary from 54 to 102 g/m^3 for maximum power.

The gas engine emission rate in the mill case study is quite different from all the engines of the plant. Furthermore, gas engines emit particulate matter, something that, according to the literature, is not permissible. In the indicated situation, due to technical conditions of gas engines, a huge number of burning oil particles are mixed with gas and so exist in the exhaust gas.

In the case of HFO engines, the emission index difference among various engines is not as significant as in the case of gas engines. It can be seen that, in general, the HFO engines have a specific emission index lower than that of gas engines, that is, engines emit less pollutants HFO in relation to the power they deliver. For the 10-engine test system, the results showed a discrepancy among emission levels in relation to the characteristics of the respective generators as the power supplied.

An extensive literature review of the EED was presented, among which numerous techniques solve the problem in reducing emissions due to power generation. However, all the techniques for this purpose were searched for a solution that requires less investment to run, which is ED with minimal emissions.

A new methodology was developed to evaluate the environmental pollution caused by a TPP. This method, unlike the other ones in the literature, does not assign a cost value to emissions but use a general index that considers not only the cost but also the impacts on the environment caused by the production. To make comparisons among different engines and fuels, the concept of specific emission index was developed, which is simply the emission index divided by the power generated by each engine.

The results of the current study, considering the emission index and using the NSGA-II optimization procedure, were significant and can be applied to any TPP that makes the use of the new approach possible to give support to professionals in the field to reduce the cost and emission involved in generation.

References

[1] A.I.S. Kumar, K. Dhanushkodi, J.J. Kumar, C.K.C. Paul, Particle swarm optimization solution to emission and economic dispatch problem, in: Paper Presented at the IEEE Conference Tencon, 2003.

[2] A. Vourc'h, M. Jimenez, Enhancing environmentally sustainable growth in Finland, in: OECD Economics Department Working Papers, No. 229, 2000, OECD Economics Department Working Papers, Finland, 2000.

[3] L. Albright, Albright's Chemical Engineering Handbook, CRC Press, 2008.

[4] P.R. Arévalo, Optimización del diseño y rediseño de procesos químicos complejos bajo incertidumbre mediante cooperación de técnicas de programación matemática y metaheurísticas, Universidad Politécnica de Madrid, 2005.

[5] N. Aversano, T. Temperini, El Calentamiento Global: Bonos de Carbono, una alternativa. Modelización y Simulación de Sistemas Económicos, 2006. Available from: <https://www.ingenieriaquimica.org/articulos/bonos_de_carbono>.

[6] A.K. Awopone, A.F. Zobaa, W. Banuenumah, Techno-economic and environmental analysis of power generation expansion plan of Ghana, Energy Policy 104 (2017) 13–22. Available from: https://doi.org/10.1016/j.enpol.2017.01.034.

[7] J.F. Bard, Short-term scheduling of thermal-electric generators using Lagrangian relaxation, Oper. Res. 36 (5) (1988) 756–766.

[8] M. Basu, A simulated annealing-based goal-attainment method for economic emission load dispatch of fixed head hydrothermal power systems, Int. J. Electr. Power Energy Syst. 27 (2) (2005) 147–153.

[9] M. Basu, Dynamic economic emission dispatch using nondominated sorting genetic algorithm-II, Int. J. Electr. Power Energy Syst. 30 (2) (2008) 140–149.

[10] M. Basu, Economic environmental dispatch using multi-objective differential evolution, Appl. Soft Comput. 11 (2) (2011) 2845–2853. Available from: https://dx.doi.org/10.1016/j.asoc.2010.11.014.

[11] M. Basu, Fuel constrained economic emission dispatch using nondominated sorting genetic algorithm-II, Energy 78 (0) (2014) 649–664. Available from: https://dx.doi.org/10.1016/j.energy.2014.10.052.

[12] C. Palanichamy, N.S. Babu, Day-night weather-based economic power dispatch, IEEE Trans. Power Syst 17 (2) (2002) 469–475.

[13] J.C.O. Carretero, D.G. Cánovas, F.D.Q. Pereira, Técnicas Heurísticas para Problemas de Diseño en Telecomunicaciones, 2008.

[14] J.P.S. Catalao, S.P.S. Mariano, V.M.F. Mendes, L.A.F.M. Ferreira, Profit-based unit commitment with emission limitations: a multiobjective approach, Power Tech. (2007). 1417–1422.

[15] S. Chaturvedi, Right to Pollute? An Understanding on the Implications of International Carbon Trading Market, 2014.

[16] F. Chen, J. Zhou, C. Wang, C. Li, P. Lu, A modified gravitational search algorithm based on a non-dominated sorting genetic approach for hydro-thermal-wind economic emission dispatching, Energy 121 (2017) 276–291. Available from: https://doi.org/10.1016/j.energy.2017.01.010.

[17] Conama, Padroes de Qualidade do ar. Portal Saude., 2017. Available from: <https://portal.saude.gov.br/portal/arquivos/pdf/conama_03_90_padroes_de_qualidade_do_ar.pdf>, 1990 (retrieved 01.01.17).

[18] R. Conama, 357, de 17 de Março de 2006. Conselho Nacional do Meio Ambiente-CONAMA, 357, 2006.

[19] J. Curtis, M.Á. Lynch, L. Zubiate, Carbon dioxide (CO_2) emissions from electricity: the influence of the North Atlantic Oscillation, Appl. Energy 161 (2016) 487–496. Available from: https://doi.org/10.1016/j.apenergy.2015.09.056.

[20] N. Daryani, K. Zare, Multiobjective power and emission dispatch using modified group search optimization method, Ain Shams Eng. J. (2015). Available from: https://doi.org/10.1016/j.asej.2016.03.001.

[21] E. Delarue, K. Van den Bergh, Carbon mitigation in the electric power sector under cap-and-trade and renewables policies, Energy Policy 92 (2016) 34–44. Available from: https://doi.org/10.1016/j.enpol.2016.01.028.

[22] D. Ashish, D. Arunesh, P. Surya, A.K. Bhardwaj, A traditional approach to solve economic load dispatch problem of thermal generating unit using MATLAB programming, Int. J. Eng. Res. Technol. (IJERT) 2 (9) (2013) 3147–3152.

[23] J. Dhillon, S.K. Jain, Multi-objective generation and emission dispatch using NSGA-II, IACSIT Int. J. Eng. Technol. 3 (5) (2011) 460.

[24] J. English, T. Niet, B. Lyseng, K. Palmer-Wilson, V. Keller, I. Moazzen, et al., Impact of electrical intertie capacity on carbon policy effectiveness, Energy Policy 101 (2017) 571–581. Available from: https://doi.org/10.1016/j.enpol.2016.10.026.

[25] B.A.D. Fernández, K. Dowsland, Diseño de heurísticas y fundamentos del recocido simulado, Inteligencia Artif.: Rev. Iberoam. Inteligencia Artif. 7 (19) (2003) 93–102.

[26] C.A. Floudas, Deterministic Global Optimization: Theory, Methods and Applications, vol. 37, Springer Science & Business Media, 2013.

[27] L.L. Garver, Power generation scheduling by integer programming—development of theory, IEEE Trans. Power Apparatus Syst. PAS 82 (3) (1963) 730–735.

[28] A. Ghasemi, M. Gheydi, M.J. Golkar, M. Eslami, Modeling of Wind/Environment/Economic Dispatch in power system and solving via an online learning meta-heuristic method, Appl. Soft Comput. 43 (2016) 454–468. Available from: https://doi.org/10.1016/j.asoc.2016.02.046.

[29] G. Grande-Acosta, J. Islas-Samperio, Towards a low-carbon electric power system in Mexico, Energy Sustain. Dev. 37 (2017) 99–109.

[30] G. Granelli, M. Montagna, G. Pasini, P. Marannino, Emission constrained dynamic dispatch, Electr. Power Syst. Res. 24 (1) (1992) 55–64.

[31] D.W. Green, Perry's Chemical Engineering Handbook, McGrawHill Professional Publ, 2007.

[32] H.Y. Yamin, Q. Ei-Dwairi, S.M. Shahidehpour, A new approach for GenCos profit based unit commitment in day-ahead competitive electricity markets considering reserve uncertainty, Electr. Power Energy Syst. 29 (2007) 609–616.

[33] B. Hassler, B.C. Mcdonald, A. Borbon, K. Civerolo, C. Granier, G.J. Frost, et al., The application of long-term observations of NO_x and CO to constrain a global emissions inventory, in: Paper Presented at the IGAC 2016 Science Conference (International Global Atmospheric Chemistry), 2016.

[34] J.S. Holladay, J. LaRiviere, The impact of cheap natural gas on marginal emissions from electricity generation and implications for energy policy, J. Environ. Econ. Manage. 85 (2017) 205–227. Available from: https://doi.org/10.1016/j.jeem.2017.06.004.

[35] B. Hong, E. Slatick, Carbondioxide emission factors for coal, Q. Coal Rep. (1994) 7.

[36] I. Kockar, A. Conejo, R. Mcdonald, Influence of the emissions trading scheme on generation scheduling, Electr. Power Energy Syst. 3 (1) (2009) 465–473.

[37] L. Jebaraj, C. Venkatesan, I. Soubache, C.C.A. Rajan, Application of differential evolution algorithm in static and dynamic economic or emission dispatch problem: a review, Renew. Sustain. Energy Rev. 77 (2017) 1206–1220. Available from: https://doi.org/10.1016/j.rser.2017.03.097.

[38] X. Jiang, J. Zhou, H. Wang, Y. Zhang, Dynamic environmental economic dispatch using multiobjective differential evolution algorithm with expanded double selection and adaptive random restart, Int. J. Electr. Power Energy Syst. 49 (2013) 399–407.

[39] A.M. Jubril, O.A. Olaniyan, O.A. Komolafe, P.O. Ogunbona, Economic-emission dispatch problem: a semi-definite programming approach, Appl. Energy 134 (2014) 446–455.

[40] M.P. Kasmaei, Despacho ótimo de potências ativa e reativa de sistema elétricos multi-áreas considerando restrições físicas, econômicas e ambientais = : environmentally constrained active-reactive optimal power flow-a compromising strategy for economic-emission dispatch and a multi-area paradigm, 2015.

[41] A.G. Kök, K. Shang, S. Yücel, Impact of electricity pricing policies on renewable energy investments and carbon emissions, Manage. Sci. 64 (2016) 1–493.

[42] E.T. Lau, Q. Yang, G.A. Taylor, A.B. Forbes, P.S. Wright, V.N. Livina, Optimisation of costs and carbon savings in relation to the economic dispatch problem as associated with power system operation, Electr. Power Syst. Res. 140 (2016) 173–183. Available from: https://doi.org/10.1016/j.epsr.2016.06.025.

[43] Y. Li, M. Li, Q. Wu, Energy saving dispatch with complex constraints: prohibited zones, valve point effect and carbon tax, Int. J. Electr. Power Energy Syst. 63 (2014) 657–666.

[44] X. Liu, B. Lin, Y. Zhang, Sulfur dioxide emission reduction of power plants in China: current policies and implications, J. Cleaner Prod. 113 (2016) 133–143. Available from: https://doi.org/10.1016/j.jclepro.2015.12.046.

[45] G. Lobos, O. Vallejos, C. Caroca, C. Marchant, El Mercado de los Bonos de Carbono ("bonos verdes"): Una Revisión, RIAT, 1 (1) (2005).

[46] X. Ma, Y. Wang, C. Wang, Low-carbon development of China's thermal power industry based on an international comparison: review, analysis and forecast, Renew. Sustain. Energy Rev. 80 (2017) 942–970. Available from: https://doi.org/10.1016/j.rser.2017.05.102.

[47] N.M. Moraes, Modelo matemático para otimização multiobjetivo do despacho econômico ambiental de usinas térmicas usando o NSGA-II. UNIVERSIDADE FEDERAL DO PARÁ INSTITUTO DE TECNOLOGIA PROGRAMA DE PÓS-GRADUAÇÃO EM ENGENHARIA ELÉTRICA, 2017.

[48] E. Muela, J. Secue, Environmental economic dispatch with fuzzy and possibilistic entities, Rev. Fac. Ingeniería 59 (2012) 227–236.

[49] M. Muslu, Economic dispatch with environmental considerations: tradeoff curves and emission reduction rates, Electr. Power Syst. Res. 71 (2) (2004) 153–158.

[50] N. Mustafa-Moraes, U. Holanda-Bezerra, J.L. Moya-Rodríguez, J. Cabral-Leite, The emission index as a parameter for assessing the environmental pollution from thermal power plants. Case study, Dyna 83 (199) (2016) 218–224.

[51] R. Muthuswamy, M. Krishnan, K. Subramanian, B. Subramanian, Environmental and economic power dispatch of thermal generators using modified NSGA-II algorithm, Int. Trans. Electr. Energy Syst. 25 (8) (2015) 17.

[52] E. Naderi, A. Azizivahed, H. Narimani, M. Fathi, M.R. Narimani, A comprehensive study of practical economic dispatch problems by a new hybrid evolutionary algorithm, Appl. Soft Comput. (2017). Available from: https://doi.org/10.1016/j.asoc.2017.06.041.

[53] N.I. Nwulu, X. Xia, Multi-objective dynamic economic emission dispatch of electric power generation integrated with game theory based demand response programs, Energy Convers. Manage. 89 (0) (2015) 963–974. Available from: https://dx.doi.org/10.1016/j.enconman.2014.11.001.

[54] I. Pavic, T. Capuder, I. Kuzle, Low carbon technologies as providers of operational flexibility in future power systems, Appl. Energy 168 (2016) 724–738. Available from: https://doi.org/10.1016/j.apenergy.2016.01.123.

[55] B. Purkayastha, N. Sinha, Optimal combined economic and emission load dispatch using modified NSGA-II with adaptive crowding distance, Int. J. Inform. Technol. Knowl. Manage. 2 (2) (2010) 553–559.

[56] B.Y. Qu, J.J. Liang, Y.S. Zhu, Z.Y. Wang, P.N. Suganthan, Economic emission dispatch problems with stochastic wind power using summation based multi-objective evolutionary algorithm, Inform. Sci. 351 (2016) 48–66. Available from: https://doi.org/10.1016/j.ins.2016.01.081.

[57] B.Y. Qu, Y.S. Zhu, Y.C. Jiao, M.Y. Wu, P.N. Suganthan, J.J. Liang, A survey on multi-objective evolutionary algorithms for the solution of the environmental/economic dispatch problems, Swarm Evol. Comput. (2017). Available from: https://doi.org/10.1016/j.swevo.2017.06.002.

[58] I.J. Raglend, S. Veeravalli, K. Sailaja, B. Sudheera, D.P. Kothari, Comparison of AI techniques to solve combined economic emission dispatch problem with line flow constraints, Electr. Power Energy Syst. 32 (2010) 592–598.

[59] N.G. Rahul Dogra, H. Saroa, Economic load dispatch problem and Matlab programming of different methods, in: International Conference of Advance Research and Innovation (ICARI-2014), 2014.

[60] A. Rajan, T. Malakar, Optimum economic and emission dispatch using exchange market algorithm, Int. J. Electr. Power Energy Syst. 82 (2016) 545–560. Available from: https://doi.org/10.1016/j.ijepes.2016.04.022.

[61] S. Rebennack, B. Flach, M.V. Pereira, P.M. Pardalos, Stochastic hydro-thermal scheduling under CO_2 emissions constraints, IEEE Trans. Power Syst. 27 (1) (2012) 58–68.

[62] P.T. Rodríguez-Piñero, Introducción a los algoritmos genéticos y sus aplicaciones: Universidad Rey Juan Carlos, Servicio de Publicaciones, 2003.

[63] S. Sahoo, K. Mahesh Dash, R.C. Prusty, A.K. Barisal, Comparative analysis of optimal load dispatch through evolutionary algorithms, Ain Shams Eng. J. 6 (1) (2015) 107–120. Available from: https://dx.doi.org/10.1016/j.asej.2014.09.002.
[64] K. Sastry, D.E. Goldberg, G. Kendall, Genetic Algorithms Search Methodologies, Springer, 2014, pp. 93–117.
[65] S. Sayah, A. Hamouda, A. Bekrar, Efficient hybrid optimization approach for emission constrained economic dispatch with nonsmooth cost curves, Int. J. Electr. Power Energy Syst. 56 (2014) 127–139.
[66] D.C. Secui, A new modified artificial bee colony algorithm for the economic dispatch problem, Energy Convers. Manage. 89 (2015) 43–62.
[67] P.K. Sharma, C.P. Verma, Carbon Credit Accounting. Dear Academicians & Research Scholars, Journal of Management Value & Ethics consistently stepping up towards their target groups day by day. Our Journal published many research papers from across the world, some of them from USA, Thailand, Indonesia, Saudi Arabia and other counties. Some of the most renowned academic personalities from USA, Uzbekistan, 2013.
[68] R. Sivaraj, T. Ravichandran, A review of selection methods in genetic algorithm, Int. J. Eng. Sci. Technol. 1 (3) (2011) 3792–3797.
[69] W.L. Snyder, H.D. Powel, J.C. Rayburn, Dynamic programming approach to unit commitment, IEEE Trans. Power Syst. 2 (2) (1987) 339–350.
[70] J. Talaq, F. El-Hawary, M. El-Hawary, A summary of environmental/economic dispatch algorithms, IEEE Trans. Power Syst. 9 (3) (1994) 1508–1516.
[71] V.L. Toache, J.R. Amado, G.T. Bertollini, S.G. Sánchez, Bonos de carbono: financiarización del medioambiente en México Carbon credits: Mexico's environment financialization, Estud. Soc. Rev. Alimentación Contemp Desarrollo Reg. 25 (47) (2016) 189–214.
[72] D.G. Victor, J.C. House, S. Joy, A Madisonian approach to climate policy, Science 309 (5742) (2005) 1820–1821.
[73] K. Wang, Y.-M. Wei, Z. Huang, Potential gains from carbon emissions trading in China: a DEA based estimation on abatement cost savings, Omega 63 (2016) 48–59.
[74] L. Wang, C. Singh, Environmental/economic power dispatch using a fuzzified multi-objective particle swarm optimization algorithm, Electr. Power Syst. Res. 77 (12) (2007) 1654–1664. Available from: https://dx.doi.org/10.1016/j.epsr.2006.11.012.
[75] T. Yalcinoz, H. Altun, Environmentally constrained economic dispatch via a genetic algorithm with arithmetic crossover, in: Paper Presented at the African Conference in Africa, IEEE AFRICON, Sixth, 2002.
[76] X. Yuan, H. Tian, Y. Yuan, Y. Huang, R.M. Ikram, An extended NSGA-III for solution multi-objective hydro-thermal-wind scheduling considering wind power cost, Energy Convers. Manage. 96 (2015) 568–578.
[77] R. Zhang, J. Zhou, L. Mo, S. Ouyang, X. Liao, Economic environmental dispatch using an enhanced multi-objective cultural algorithm, Electr. Power Syst. Res. 99 (2013) 18–29.
[78] Y. Zhang, D.-W. Gong, Z. Ding, A bare-bones multi-objective particle swarm optimization algorithm for environmental/economic dispatch, Inform. Sci. 192 (2012) 213–227.
[79] Z. Zhu, J. Wang, M.H. Baloch, Dynamic economic emission dispatch using modified NSGA-II, Int. Trans. Electr. Energy Syst. (2016). Available from: https://doi.org/10.1002/etep.2228.

Chapter 13

Maintenance management with application of computational intelligence generating a decision support system for the load dispatch in power plants

Milton Fonseca Júnior[1], Jandecy Cabral Leite[2], Tirso Lorenzo Reyes Carvajal[2], Manoel Henrique Reis Nascimento[2], Jorge de Almeida Brito Júnior[2] and Carlos Alberto Oliveira Freitas[2]

[1]*Mauá Generation Department, Generation Eletrobras Amazonas GT Manaus, Amazonas, Brazil,* [2]*Research Department, Institute of Technology and Education Galileo of Amazon - ITEGAM, Manaus, AM, Brazil*

13.1 Introduction

Most of the Brazilian thermoelectric parks remain completely closed for months whenever the hydrological situation is favorable. As in recent time, average hydroelectric generation has been 90% of its generation capacity for the system [1], idleness has prevailed in the thermal park as the plants can only be activated when the hydroelectric reservoirs go below 50% of their maximum volume. The contrasting fact with respect to those of other countries is striking. In most of the countries, combined-cycle coal or gas-fired power plants typically do not experience long-term idleness; instead, they operate at the base level of the system, being dispatched almost continuously. On the other hand, thermals that are used in other electrical systems for the generation of tip, with daily activation or at least in good part of the working days, such as open or thermal cycle gas engines with motors, in Brazil can remain idle for long, because they are not necessary in normal or favorable hydrologic situations.

It is necessary to ensure the supply of electricity to consumers within standards of continuity and reliability. Although the lack of investments in the industry causes the loss of product quality, the excessive investment makes the product very costly, which discourages its consumption [2,3].

Decision Making Applications in Modern Power Systems. DOI: https://doi.org/10.1016/B978-0-12-816445-7.00013-X

One of the most important aspects to sustain the quality and reliability of the electric power supply is to be able to perform an optimal load dispatch [2,4]. The great majority of the works presented in the literature develop the load dispatch of the thermal plants, considering that all the engines of the plant have a favorable technical state, but this is not always the case, so in this chapter, a method is presented for the predispatch of load that takes into account the technical state of the plant's motors through diagnosis and making use of fuzzy logic.

The development of a computational tool to support the decision of cargo dispatch according to the operating conditions of the engines and generators of thermal plants is proposed, which are classified in relation to the probabilities of failure by a fuzzy system developed in this chapter, from indicators obtained from lubricant oil analysis, vibration analysis, and thermography of power generation equipment. The study is based on the principle of operation and operational conditions of the equipment to be dispatched for the generation in a thermal plant, besides its particularities as specific consumption and the quality of pollutants sent by each equipment.

13.2 Maintenance systems and their application in thermoelectric plants

The ability of a generation source to meet an energy demand can be influenced by unexpected units of power generating units. The tests were even more advanced to repair preventive maintenance measures but were not revised in the 1990s with maintenance and maintenance work on engines and generators.

In recent times, condition-based maintenance (CBM) has been introduced in industrial systems to preventively maintain the right equipment at the right time relative to its current "operating condition." The good state of operation of a generator can be represented mainly by conventional indicators, such as oil temperature, harmonic data, and vibration. Monitoring the motors/generators and their diagnosis is also important so that the dispatch of cargo does not have unexpected interruptions.

Most energy generation unit–scheduling packages are considered preventive maintenance schedules for units over an operational planning period of 1 or 2 years in order to defray the operating total, while meeting the requirements—system power and maintenance restrictions. This process consists of verifying the generating units that must be stopped from production. The generating units should be regularly examined for safety. It is important for a failure in a power generating unit that can be used in the machines. Therefore the fixation and the key point are used in the proposed methodology. The issue is addressed as an optimization problem. The model is developed by determining the objective function, which is a net power reserve of the unit [5].

They point out that CBM is a strategy that collects and evaluates information in real time and recommends maintenance decisions based on the current condition of the system. Since the last decade research on CBM has been growing rapidly due to the considerable development of computer-enabled monitoring technologies. Research studies have proven that CBM, if properly planned, can be effective in improving the reliability of equipment at reduced costs [6].

It presents an algorithm for deciding preferences in maintenance activities for power supply sections of distribution systems. A component measure of importance, known as a diagnostic importance factor (DIF), was used for this purpose. A methodology was developed to calculate a weighted cumulative DIF for each feed section, which represents a quantitatively relative significance for the prioritization of maintenance activities. The developed methodology includes the distributed generation effect and the loads. It was implemented in two distribution systems, so that, at the end, sorted lists of feed sections for maintenance activities are obtained [7].

CBM is an increasingly applicable policy in the competitive market that acts as a means of improving the reliability and efficiency of equipment. Not only does maintenance have a close relationship with security, but its costs also make the issue even more attractive to researchers [8].

Proper maintenance can increase the company's productivity and increase its value in the market. The main study provided a robust model that can strategically evaluate important available technologies and may exclude outdated and/or inappropriate technologies. There are many researches in this field in which the number of models has been proposed, such as the maintenance management system, maintenance performance measurement, and maintenance performance indicators, but the details of the effectiveness of the predictive maintenance indicator, specifically based on maintenance and conditions with maintenance and management requirements using the analytical hierarchy process, are hardly available in the literature [9].

Basically, the process consists of monitoring parameters that characterize the state of operation of the equipment. The methods employed involve techniques and procedures for measuring, monitoring, and analyzing these parameters. It can be related as oil analysis, ferrography, thermography, and vibration analysis.

Motor operation data in conjunction with vibration, oil, and temperature analysis data are collected periodically at the plant and are used in an integrated way to feed a fuzzy rule–based system, which returns the predispatch scheduling of the plant for the period of interest, taking into account the state of operation of the machines, Fig. 13.1.

Thermal power plants consist of a set of mechanical and electrical systems that require constant monitoring of energy production. The data obtained through the monitoring actions are necessary in the operation, maintenance, and evaluation of the performance of the plants. For this purpose, it is often called distributed control systems (DCS).

The obsolescence of this equipment (DCS) increases the risks of unavailability of the generating units, mainly in thermoelectric plants, with a high degree of mechanical wear, due to the high temperatures and the chemical agents used for the production of electric energy [10].

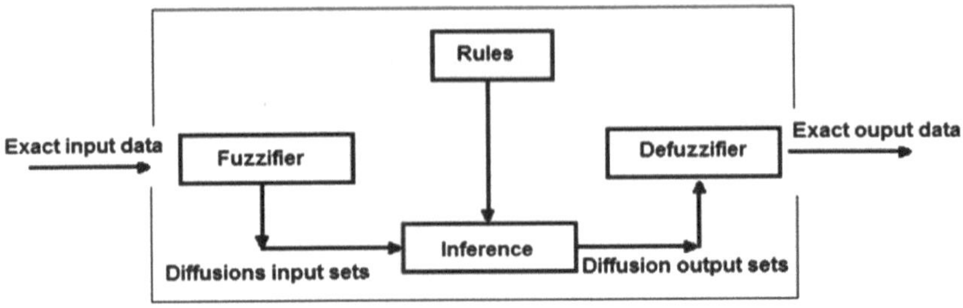

FIGURE 13.1 System based on fuzzy logic.

MTBF (mean time between failures) or mean period between failures is a value assigned to a particular device or a device to describe its reliability. This assigned value indicates when a device failure may occur. The higher this index, the greater the reliability of the equipment, and, consequently, the maintenance will be evaluated in efficiency issues.

The average mean time for repair (MTTR) is a measure on the basis of repairable item maintenance. It represents that the average time required to repair a component failure or mathematically expressed equipment is the corrective maintenance.

Oil analysis: The initial purpose of oil analysis for a lubricated assembly or a hydraulic system is to economize by optimizing the intervals between the exchanges. As the analyses carried out resulted in indicators that report on the wear of the lubricated components, the second objective of this process became the defect control for predictive maintenance [11].

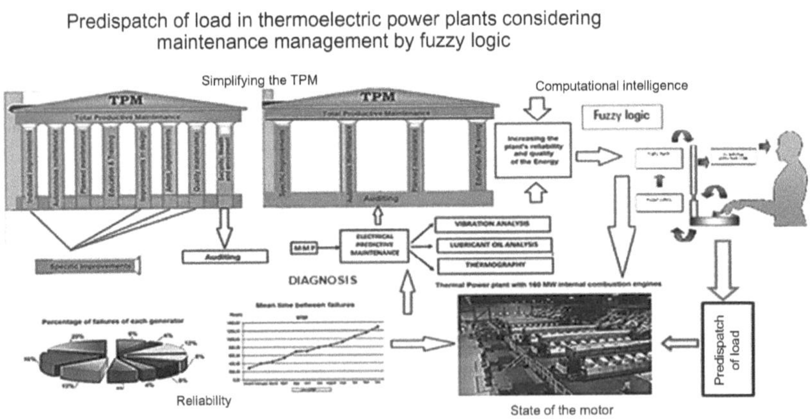

Methodology.

- In the upper left, you can see the simplification of the eight pillars of the total productive maintenance (TPM), for four pillars.
- The left-center part shows the diagnostic activities that allow one to know the technical state of the motors, to know whether or not they can be used in the predispatch of load.

- In the lower left, we show the reliability analyses, which together with the diagnosis allow us to know when it is possible for each motor to fail to consider it in the predispatch of load.
- The right part shows the application of the fuzzy logic, to perform the predispatch of load, according to the fuzzy rules that meet the technical state of the engines.

13.3 Fragments used for implantation end methodology TPM program

This study presents a new solution proposal, which includes the predispatch of load focused on the operational conditions of the machines using computational intelligence, specifically fuzzy logic. Such an implementation is characterized by incorporating some innovations, such as good maintenance management through TPM program for decision-making, considering performance indicators of the generating units with respect to vibration, lubricating oil, temperature, being it possible to say if the generating unit will operate and maintain reliability or will get into maintenance due to poorly diagnosed performance.

1. *Pilar of specific improvements* (*Recommended group*: Coordinators of ME, MA (Plant Managers), MP and SMA):

   Purpose: To maximize the overall efficiency of the equipment and the operation through the analysis and elimination of operational losses (Table 13.1).
2. *Automatic maintenance pillar* (*Recommended group*: Managers, Supervisors and Operators of each Plant):

   Objective: To enable operators to keep their workplaces clean, organized, inspecting their equipment, following operating procedures, lubricating, identifying abnormalities, labeling and attempting to eliminate hard-to-reach places and sources of dirt (Table 13.2).
3. *Planned maintenance pillar* (*Recommended group*: PM Coordinator, Service Managers and Supervisors and each Plant):

   Objective: To create a maintenance management corporate model for all engines and auxiliary equipment of the plants and external clients to optimize interventions and reduce maintenance costs, ensuring the performance of auxiliary engines and equipment (Table 13.3).
4. *Pillar of education and training* (*Recommended group*: This pillar is corporate and only depends on HR):

   Objective: To support the other pillars, analyzing the qualification of participants and the need for training. Responsible for communication, TPM disclosure, event planning, compliance with the Basic Program Guidelines to facilitate documentation, reduction of dissemination costs and support materials (Table 13.4).

TABLE 13.1 Pillar of specific improvements.

Pillar of specific improvements		Evaluation/progress/ criteria
Evaluation/progress/ criteria	**Background/objective**	**Expected condition**
1. Elaborate complete and detailed flow of the operation, identifying the various auxiliary engines and equipment, their respective priorities and main risks. Note the current conditions so that you can compare after the improvements implemented	To increase the knowledge of the whole operation and to standardize the knowledge of the participants of the working group, using the tools of quality	Working group formed, operational flow completed in a clear and didactic way, equipment, priorities and main risks identified and being known by all participants
2. Identify the generation capacities in megawatt (MW) of each engine/ plant—standard and real—and the current losses of the operation, quantifying through the Pareto chart	Identify the distortions between the actual and expected (standard or standard) of each engine/ plant. Identify fuel and lubricant/engine/plant consumptions, knowing the performance of each one to be able to act on improvements	Motors, auxiliary equipment, and operations identified with their nominal and actual capacities Criterion to analyze and identify the main losses of the operation, stratify and classify graphically in A, B, and C (Pareto)
3. Investigate losses in detail according to the priority grades I, II, III of the chart, presenting alternatives for reducing or eliminating current losses found for later comparison	Allow to identify the fundamental causes of each selected loss, the actual operating conditions of each motor/auxiliary equipment (clearances, paint, leaks, instrumentation, working environment conditions, qualification of operators, necessary and available tools, etc.)	Use of the Analysis and Problem Solving Method (APSM) tools to analyze and solve identified losses. PDCA, fishbone, 5W2H Methodology being used to investigate and eliminate losses
4. Prepare detailed action plan for the chosen losses and develop a schedule of activities, following the APSM methodology	Organize the various activities necessary to eliminate identified losses, in order of priority (from highest to lowest) and investment (from lowest to highest)	Plan of Action prepared by the working group with actions, responsible, deadlines and progress of the activities chosen in the item above, through the APSM tools Put the action plan into practice and compare the results before and after

(Continued)

TABLE 13.1 (Continued)

Pillar of specific improvements		Evaluation/progress/ criteria
Evaluation/progress/ criteria	**Background/objective**	**Expected condition**
5. Standardize operational procedures, ensuring that engines and auxiliary equipment are operated within the required conditions of pressure, temperature, speed, rpm, etc.	After achieving the expected results, standardize the procedures that should be followed by all operators	Interim operational standard completed and being used by the operators in each engine and auxiliary equipment
6. Analyze the existing operational reports and make the necessary modifications to improve the quality of the annotated information, including maintenance stops by motor or auxiliary equipment, lack of spare parts, labor problems, transportation, etc.	Improving the quality of information to assist in the investigation of losses and their eliminations	Performance of the operation/motor and auxiliary equipment being evaluated by comparing the indicators and objectives defined for each engine/plant. Information of the operational reports being provided with quality and accompanied by the managers, supervisors, and operators. No data distortion

Members of the audit pillar checklist should meet monthly to discuss the MTBF goals and the monthly MTTR and other activities corresponding to the Maintenance Management Program (Table 13.5).

13.4 Predictive maintenance using computational (fuzzy logic) decision support tool in preload dispatch

An application of fuzzy logic is justified by the ability to anticipate the possibilities of making the predispatch time of the load on the operational tasks of the equipment.

This study deals with the application of fuzzy logic to load dispatch, but with a particularity that is to perform said predispatch of load taking into account the technical state of the engines, evaluated by different variables related to maintenance. In the first part the development of the fuzzy rules and of the whole procedure of inference is exposed, and in the second part, all the tests to evaluate the maintenance and the technical state of the motors.

TABLE 13.2 Automatic maintenance pillar.

Automatic maintenance pillar		Valuation/progress/ criteria
Evaluation/progress/ criteria	**Background/objective**	**Expected condition**
1. Determine the procedure and how to identify abnormalities through labels. Determine labeling procedures, label types, and colors 2. Train participants to identify abnormalities of motors, auxiliary equipment, and work area through labeling	Eliminate abnormalities of motors, auxiliary equipment, installations, workplace, accumulated dirt, eliminate unused materials in the operation, visually identify the abnormal conditions that need to be repaired, maintain the ideal working conditions that meet industrial safety	At initial cleaning, operators and personnel involved must be trained to identify abnormalities in motors, auxiliary equipment, facilities, and workplaces through *stickers* Areas, engines, auxiliary equipment, and facilities must be clean and maintained in this condition, no longer tolerating any signs of clutter and dirty locations. Use the 5 S's Locations that are not meeting this requirement should at least be flagged and their future repair be included in a timely, responsible action plan
3. Perform initial cleaning on all motors, auxiliary equipment and operating areas, determining ideal working conditions (no leakage, good flooring, motors, auxiliary equipment, and facilities, painted and corrosion-free, with necessary signaling, conditions security, etc.)	In this initial cleaning, the conditions of motors, auxiliary equipment, and installations such as loose bolts, lack of fixings and protections, damaged parts and temporary repairs, lack of signaling, etc., identified each with a label and providing the necessary repairs must be observed. The label must only be removed after approval of the service performed	
4. Prepare planning/ schedule to carry out the necessary activities of removal of the labels placed in places that presented abnormalities	Monitor the activities performed and measure the results after the improvements implemented	After label placement, a control should be created indicating the type of problem, the number of labels placed and removed, and the areas involved in the abnormalities, such as maintenance, operation, safety, and environment. Identification, simple and objective control

(*Continued*)

TABLE 13.2 (Continued)

Automatic maintenance pillar		Valuation/progress/ criteria
Evaluation/progress/ criteria	**Background/objective**	**Expected condition**
5. Establish the basic conditions of engines, auxiliary equipment and facilities, workplaces, floors, walls, lighting, painting, signaling, temperatures, etc.	Ensure operation within the ideal standards required	Ideal conditions for motors, auxiliary equipment, signed installations and work areas, with industrial safety colors, nameplates, lighting, and cleaning

TABLE 13.3 Planned maintenance pillar.

Planned maintenance pillar		Evaluation/progress
Evaluation/progress/criteria	**Background/ objective**	**Expected condition**
1. Elaborate and approve methodology to prioritize engines, auxiliary equipment and facilities in A, B, and C and disclose to all Operation and maintenance management program (OMMP) coordinators. Determine form of identification and approve with the steering committee	Standardize how to prioritize engines, auxiliary equipment and facilities as the company needs, with a focus on *business*	Complete prioritization worksheet containing pertinent questions from the areas involved in the operation (operation, maintenance, engineering, safety, and environment)
2. Determine how and when the meeting involving operation, engineering, maintenance, safety, and environment will be made to define all the engines, auxiliary equipment, and facilities of each plant in A, B, or C	Identify the company's *business* priorities to facilitate the deployment of a maintenance management model	Meeting to evaluate and classify in A, B, and C all engines, auxiliary equipment, and facilities of the company, marked or performed with the areas involved

(*Continued*)

TABLE 13.3 (Continued)

Planned maintenance pillar		Evaluation/progress
Evaluation/progress/criteria	**Background/ objective**	**Expected condition**
3. After completion, visually list and identify priorities A, B, or C to facilitate supervision	Facilitate service and decision in the most appropriate action to be taken, according to priority	All motors, auxiliary equipment, and facilities, classified in A, B, and C with the visual identification labels, according to the model approved and adopted by the company
4. Identify the current state of each engine, auxiliary equipment, and installation, inspect/review and make necessary repairs to maintain in perfect operational conditions	Rescue the ideal operating conditions of engines, auxiliary equipment, and facilities, improving availability, reliability, and maintenance	Inspection/revision planning in engines, equipment, and facilities A to redeem desired conditions Action plan defined with: activities, materials, deadlines, time provided in each repair activity and maintenance team
5. Elaborate the most indicated maintenance procedures for each engine, auxiliary equipment, and facilities, as recommended in the master plan Consider those in the technical manuals, MaMa2i and create those that do not exist and are necessary	Define a maintenance philosophy to be used in equipment A, B, and C according to priority	Equipment A, B, and C classified and with the type and recommended maintenance plan completed
6. Start the required activities for each engine, auxiliary equipment, and installation A. Follow the MaMa2i plan and add nonexisting services to the system	Organize maintenance and update the data of each engine, auxiliary equipment, and installation A, creating history, technical inspection standards and maintenance procedures Follow template created by engineers for reference	Planning and schedule of activities for engines, auxiliary equipment, and facilities A completed and started

TABLE 13.4 Pillar of maintenance education and training.

Pillar of maintenance education and training		**Evaluation/progress**
Evaluation/progress/criteria	**Background/ objective**	**Expected condition**
1. List all employees who have already been trained and those who require basic Operation and maintenance management program (OMMP) training to participate in the work groups	Level the knowledge of all participants before starting to develop the activities in the working groups	All employees participating in the OMMP program identified to receive the basic training provided by the Pillar Coordinators
2. Elaborate and make available in the network a basic training to minister to all the employees and in the integration of new ones	Standardize the material and information passed to employees	Teaching material for the basic training, completed, approved by the steering committee and made available to the Pillar Coordinators
3. Determine the dates of the training of each pillar and the person in charge of ministering	Organize a schedule of activities to monitor and audit the development of the TPM	Planning/schedule of training to be performed, indicating employees, dates and Instructors TPM training for integration of new employees, completed to be incorporated by HR
4. After the training, disseminate the number of participants to serve as an evaluation indicator of the pillar in the TPM program	To measure the degree of PMS development and to present the technical state (TE) pillar indicators	Constant and updated dissemination of the number of employees trained and hours of training performed
5. Make competency map of all participants in the working groups, to identify the qualification, knowledge and needs	Identify the need for training, planning and implementation in order to allow the activities of the other OMMP pillars to proceed	Worksheet of skills and qualification of the maintainers and operators completed, indicating the basic and specific training required

(Continued)

TABLE 13.4 (Continued)

Pillar of maintenance education and training		Evaluation/progress
Evaluation/progress/criteria	**Background/ objective**	**Expected condition**
6. Elaborate internal/external training plan, one-point training, APSM, and lectures to adapt the knowledge need of each work group participant	Level the knowledge of the working group participants so they can take on other activities without any problem, according to the steps of the ME, autonomous maintenance (AM), planned maintenance (PM), and audit maintenance system (AMS) pillars	Planning to carry out the training identified in the previous item, including lectures, one-point training, APSM, etc.

TABLE 13.5 Checklist pillar auditoria.

Activities name	Responsibility	Check list
Check cleaning of areas	Plant manager	Check leaks, state of conservation, paint, and signage
Check the binder where the "cleaning pattern" of the area is located	Supervisors	Check if the cleaning pattern plug is placed in an easy to read location
Check if the tags are in control	Plant manager	Check in the label control software if there is any movement of placement of new labels
Check action plan to remove labels	Plant manager and supervisors	Check in the label control software if there is an action plan for removing the labels
Check signage of plants	Plant manager and supervisors	Check standardization of signaling
Check Maintenance Management Program (MMP) frame of the plant	Engineering	Check if the information is up to date

This tool served as a basis for the resolution of the real problem of preshipment of cargo to satisfy the rationalized methods of just in time of the thermal plant on the operational conditions of the equipment.

A system based on fuzzy logic, as shown in Fig. 13.1, can have its action schematized by the following constituent elements: fuzzifier; rules, or knowledge base; inference, or logical decision-making, and defuzzifier [12].

In the first part the development of the fuzzy rules and of the whole procedure of inference is exposed and in the second part, all the tests to evaluate the maintenance and the technical state of the motors. This tool served as the basis for the resolution of the real problem of preshipment of cargo to satisfy the rationalized methods of just in time of the thermal plant on the operational conditions of the equipment [13,14].

The interfuzz aims to model the mode of reasoning, trying to imitate the ability to make decisions in an environment of uncertainty and imprecision. In this way, fuzzy logic is an intelligent technology, which provides a mechanism to manipulate imprecise information—concepts of small, high, good, very hot, cold—and that allows one to infer an approximate answer to a question based on an inexact, incomplete, or not fully reliable knowledge.

Development of a computational tool to support the cargo dispatch according to the location of motors and generators for thermal energy analyzes the main thermoelectric generation variables for the entire predictive maintenance process.

All variables are inserted considering the intervals determined in the rules of inference as shown below.

The computational interface was useful for the search of some preselected characteristics to enable its implementation. Tables 13.6–13.12 show such characteristics and the respective purposes.

In this context, the following groups of information and data are abstracted: the input values, called crisp, the linguistic variables, and the fuzzy variables. The fuzzy logic is justified in the solution of this case study in

TABLE 13.6 Manufacturer vibration levels.

Class	1–[N] Normal	2–[P] Permissible	3–[A] Alert	4–[C] Critical (mm/s)
(Class I)	(0.18–0.71)	(0.71–1.80)	(1.80–4.50)	(Above 4.50)
(Class II)	(0.18–1.10)	(1.10–2.80)	(2.80–7.10)	(Above 7.10)
(Class III)	(0.18–1.80)	(1.80–4.50)	(4.50–11.2)	(Above 11.2)
(Class IV)	(0.18–2.80)	(2.80–7.10)	(7.10–18.0)	(Above 18.0)
	A	B	C	D

TABLE 13.7 Vibration severity rating relevance function.

Zone	Qualification	Operation of machines
Zone *A*	[N] Normal *0.18–2.80 mm/s*	Commissioned machines should generally operate in this area
Zone *B*	[P] Permissible *2.80–7.10 mm/s*	It is acceptable for unrestricted operation for long periods
Zone *C*	[A] Alert *7.10–18.0 mm/s*	Unsatisfactory for continuous operations for long periods
Zone *D*	[C] Critical *above 18.0 mm/s*	It is sufficient to cause damage to the machine at any time

TABLE 13.8 Lubricating oil.

Class	1—[N] Normal	2—[A] Alert	3—[C] Critical
(Water% volume)	(% ≤ 0.2)	(0.3)	(Above 03)
(Micron iron content)	(% ≤ 49)	(50)	(Above 51)
(Micron copper content)	(% ≤ 1)	(20)	(Above 21)
	A	B	C

function of the input variables with better representation in fuzzy sets. The variables due to the dimension of the universe of study were divided into 04 (three), 03 (two) inputs, and 01 (one) output, all independent of each other.

- *The input variable "vibration analysis"*

  For the determination of each variable, it was convenient to divide them into strips to approximate the actual situation to be checked. The calculation of these ranges on a scale according to Tables 13.6–13.12 is shown next.

  As the first level of variation "vibration level," let us consider that better variable levels were subdivided into four variables, normal, permissive, alert, and critical, each corresponding to the classification of vibration, velocity, and displacement levels measured in the equipment.

- *The input variable lubricating oil*

  The "level of analysis of the lubricating oil" can be presented, for example, with the water content in the oil, solid, and nonlubricated particle content (iron and obre), the energy sources of the dispatch of load for generation of energy. The levels of analysis of the command type were subdivided into three variables, correspondence and information quality [9].

TABLE 13.9 Function of pertinence of the severity according to the oil.

Zone	Qualification	Operation of machines
A	[N] Normal	Commissioned machines should generally operate in this area
Water% volume	$\% \leq 0.2$	
Micron iron content	$\% \leq 49$	
Micron copper content	$\% \leq 19$	
B	[A] Alert	Unsatisfactory for continuous operations for long periods
Water% volume	0.3	
Micron iron content	50	
Micron copper content	20	
C	[C] Critical	It is sufficient to cause damage to the machine at any time
Water% volume	Above 0.3	
Micron iron content	Above 51	
Micron copper content	Above 21	

TABLE 13.10 thermography to determine hot spots.

(Zone)	(Thermography)	
(A)	*([N] Normal less or equal 94.0 F)*	
B	*[P] Permissible*	
(B/C)	(94.0 F)	(164.2 F)
C	*[A] Alert*	
(C/D)	(164.2 F)	(199.3 F)
(D)	*[C] Critical above 199.3 F*	

TABLE 13.11 Function of pertinence of the classification of thermography.

Zone	Qualification	Operation of machines
Zone *A*	[N] Normal ($T \leq 34.5°F$)	Commissioned machines should generally operate in this area
Zone *B*	[P] Permissíble (34.5°F < $T \leq$ 73.5°F)	It is acceptable for unrestricted operation for long periods
Zone *C*	[A] Alert (73.5°F < $T \leq$ 93°F)	Unsatisfactory for continuous operations for long periods
Zone *D*	[C] Crítical ($T >$ 93°F)	It is sufficient to cause damage to the machine at any time

TABLE 13.12 Variable "engine technical status" (ETS).

ETS for operating conditions		Operation of machines
Normal	76 at 100%	Commissioned machines should generally operate in this area
Permissible	51 at 75%	It is acceptable for unrestricted operation for long periods.
Alert	26 at 50%	Unsatisfactory for continuous operations for long periods
Critical	0 at 25%	It is sufficient to cause damage to the machine at any time

- *The input variable thermography analysis level*

  Thermographic analysis is a third input variable, which can be used as a load dispatch tool for power generation. The levels of analysis thermographic were just subdivided into four variables, each corresponding to the dynamic memory, the use of images thermal plants in the alert in the reason. Materials and methods: Infrared radiation is a base of studies on thermal images, which has a function of capturing this radiation, interpreting, and generating a quantitative image of the body temperature studied [15].
- *Output variable "Technical condition of the motor"*

The "estimated technical state of the engine" is the output variable of the system, in relation to vibration (oil, water, iron, and copper). Table 13.12

describes the operating state of the generating units. The variable under study, as well as the variable "Level," was transferred to a percentage scale of 100, where "EXCELLENT" corresponds to the range of maximum values and the variable "BAD" corresponds to the range of minimum values up to zero. This value gives a greater range of possibilities, making the case study more precise. Thus we can deduce that the equation that defines the estimation is. These variables are used for decision-making involving the predispatch of the unit generators, as shown in Table 13.6.

13.5 Fuzzy simulation

The fuzzy simulation containing the system variables was performed using the MATLAB version 8.0 tool, and the fuzzy model applied in this simulation was Mamdani. This model is characterized by adopting the semantic rules used for the processing of inferences and is commonly referred to as maximum–minimum inference. Such an inference model applies well to this type of problem since it uses union and intersection operations between sets. The implementation is done by the Mamdani model applied to this case study. All variables are entered considering the intervals determined in the rules of inference. Fig. 13.2 shows the variables "vibration," "water," "thermography," "iron," "copper."

All variables are entered considering the intervals determined in the rules of inference. Figs. 13.3–13.6 show the variables "vibration," "water," "thermography," "iron," and "copper."

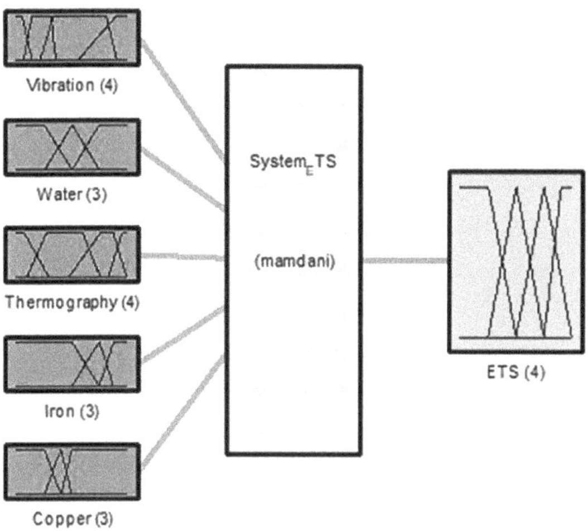

FIGURE 13.2 Mamdani's model.

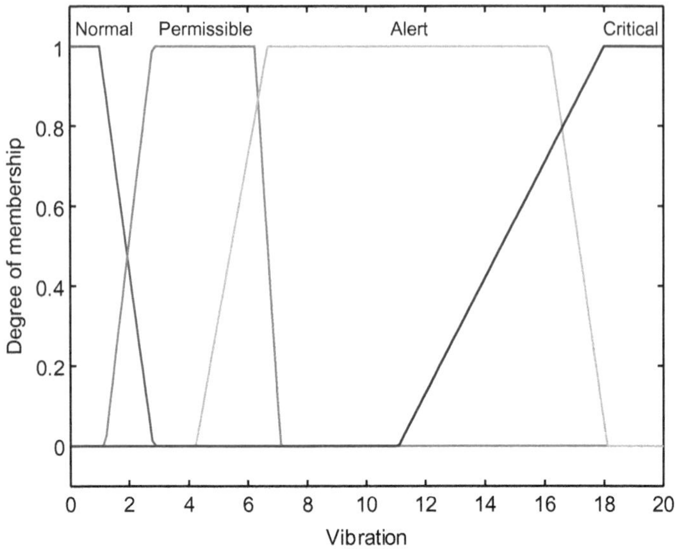

FIGURE 13.3 Vibration level.

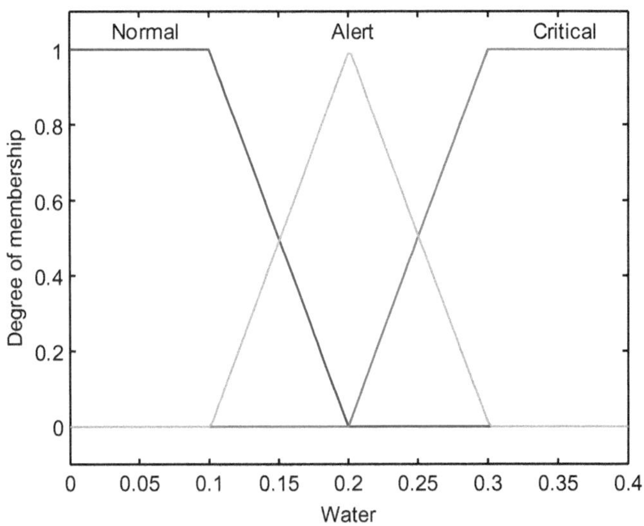

FIGURE 13.4 Water.

The first input variable is vibration. According to Tables 13.6 and 13.7, the variable "vibration" is represented in Fig. 13.3.

The second input variable is water produced by the generating units. According to Tables 13.8 and 13.9, the variable water is shown in Fig. 13.4.

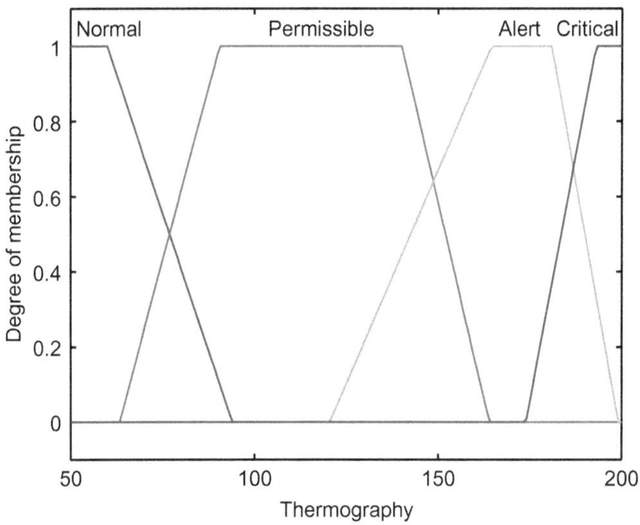

FIGURE 13.5 Thermography.

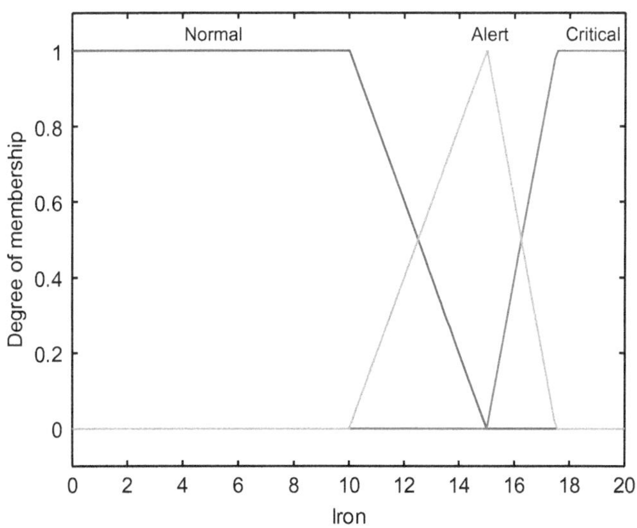

FIGURE 13.6 Iron.

The third input variable is the thermography, produced by the generating units. According to , the variable thermography is shown in Fig. 13.5.

The fourth input variable is iron produced by the generating units. According to Tables 13.8 and 13.9, the variable "iron" is shown in Fig. 13.6.

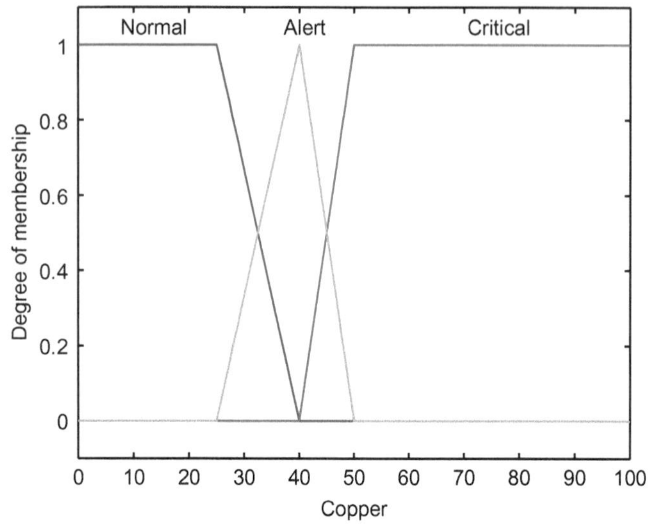

FIGURE 13.7 Copper.

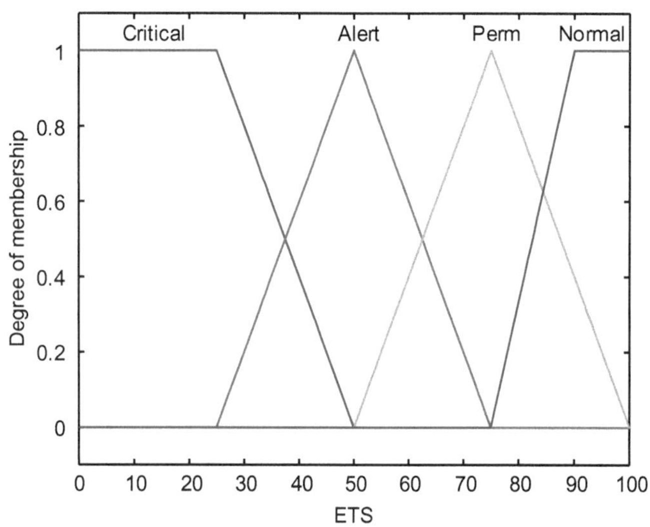

FIGURE 13.8 Output variable: technical status.

The fifth input variable is copper, produced by the generating units. According to Tables 13.8 and 13.9, the variable "copper" is shown in Fig. 13.7.

The Motor Technical State is a product of the relationship between the input variable and output variable, which compose the pertinence functions expressed in the curves of Fig. 13.8.

After editing the pertinence functions of all variables, the implemented rules are arranged in Table 13.7, as shown in Fig. 13.9 for the visualization of the linguistic variables, thus forming antecedents and subsequent ones based on the fuzzy inference rules.

To better understand the screen expressed, Fig. 13.10 shows all the possibilities that the simulation can produce. The movement of the red lines determines the other rule to be evaluated.

Figs. 13.11–13.14 show the results of the inference rules from the 3D surface of the graph. In blank are present all the forms of execution that can exist within the simulation.

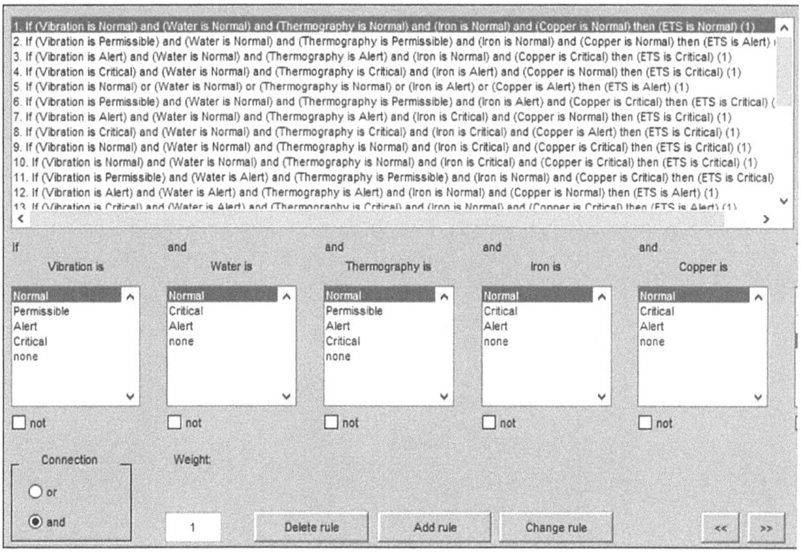

FIGURE 13.9 Implemented inference rules.

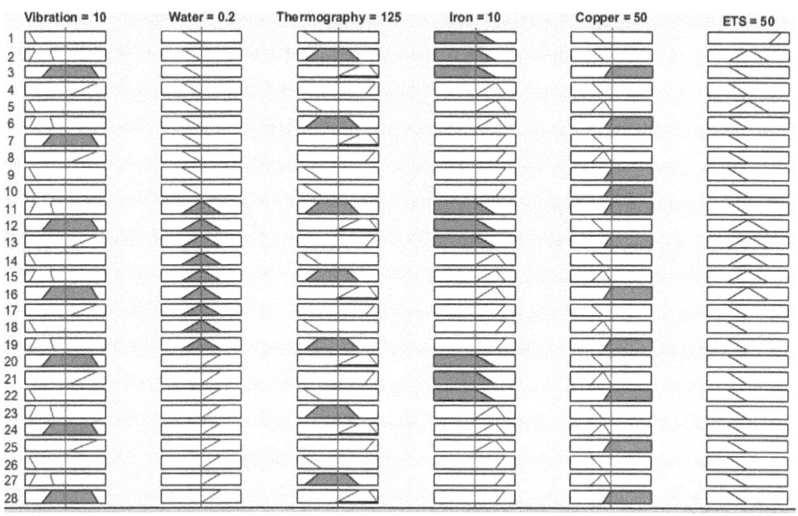

FIGURE 13.10 The input and output variables.

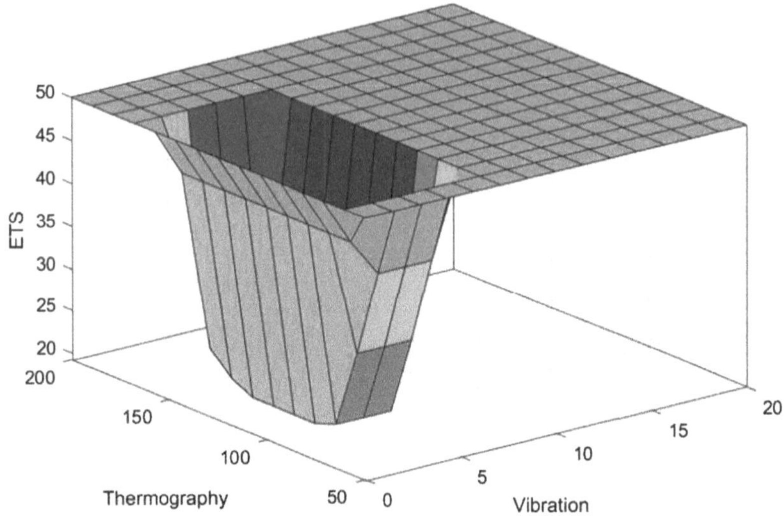

FIGURE 13.11 Thermography × vibration.

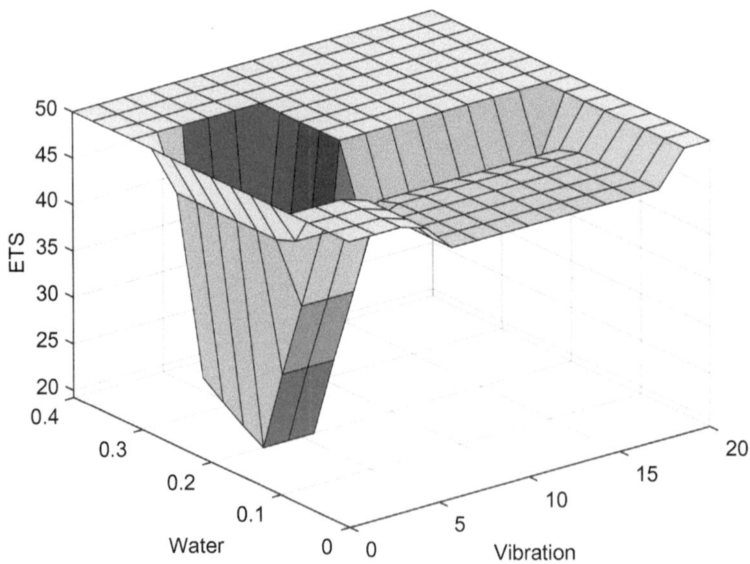

FIGURE 13.12 Water versus vibration.

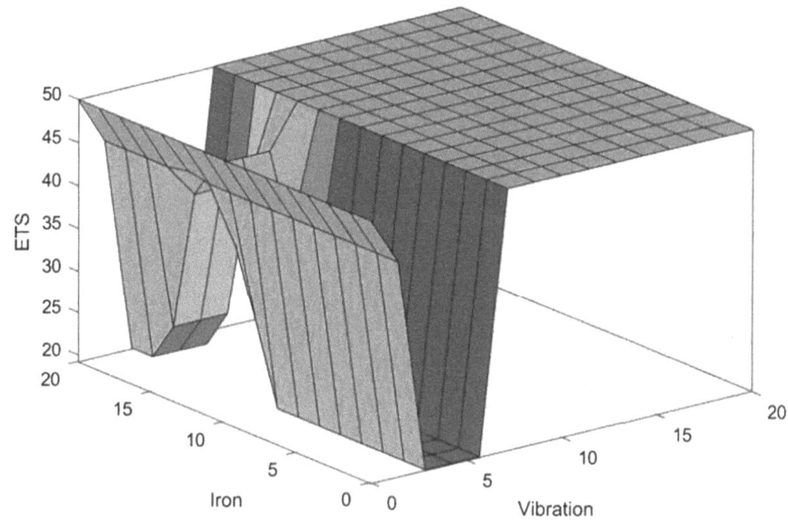

FIGURE 13.13 Iron × vibration.

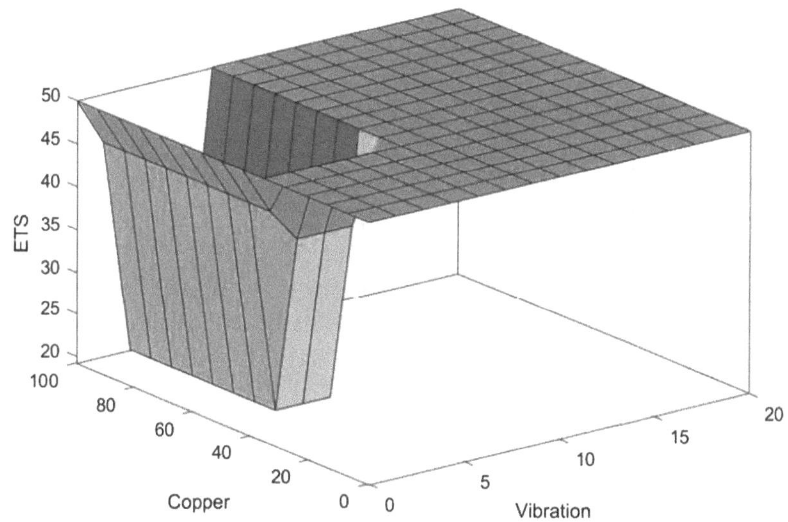

FIGURE 13.14 Copper versus vibration.

13.6 Case study (fuzzy logic with predictive maintenance)

1. *Vibration analysis*

Equipment status control is performed based on a calculated global value for the vibration signal measured at critical points on the machine surface. Since this value is due to a response signal from the structure to the dynamic excitation of the equipment operation, it represents a measure of the amplitude level of its vibration signal. In the case of the application for predictive maintenance, the international technical standards, among them the ISO, define two criteria for adoption of a global value (Fig. 13.15).

2. *Analysis of water content in lubricating oil*

The determination of the presence and content of water in the case of study was carried out through the distillation by drag. The sample is subjected to heating for distillation under controlled conditions, thus verifying the water content in the lubricating oil.

The graph shows the results of the analysis of water content, done periodically as predictive, showing normal levels, since the tolerable is 0.3% (Fig. 13.16).

a. *Analysis of metal content in lubricating oil*

The graph made by direct reading (iron and copper) ferrography, which was carried out based on the extraction of the magnetizable

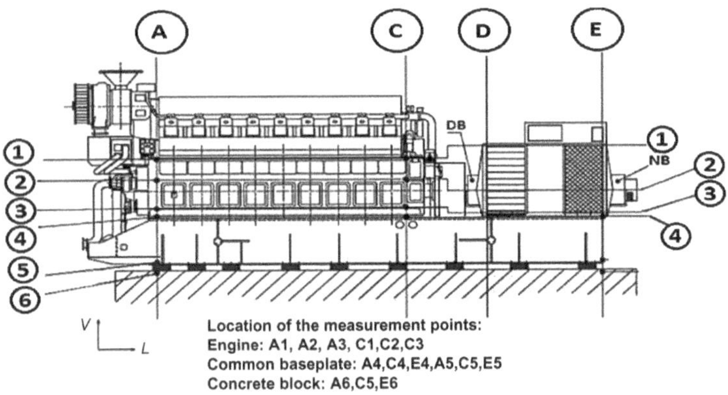

FIGURE 13.15 Measurement points in the vibration analysis (2018).

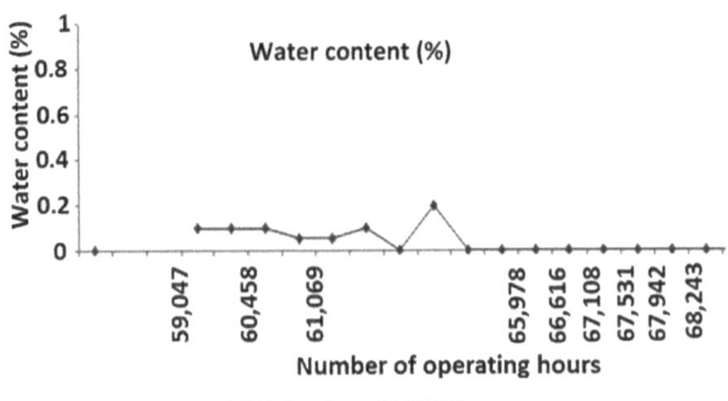

FIGURE 13.16 Water content and lubricating oil (2018).

contaminant particles, contained in the lubricant, through the action of a magnetic field (Figs. 13.17 and 13.18).

3. *Thermography*

   In addition to the use of the supervision system provided by the 9 FLUKE software, a thermovision is used, as shown in Fig. 13.19, for measurement in low or high voltage electrical systems, temperature

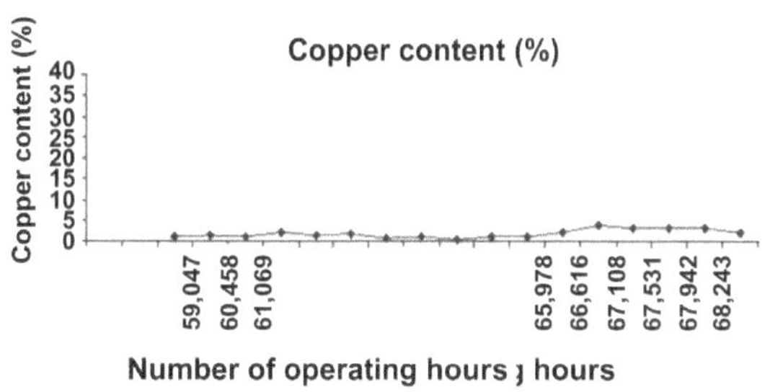

FIGURE 13.17 Copper content in the lubricating oil (2018).

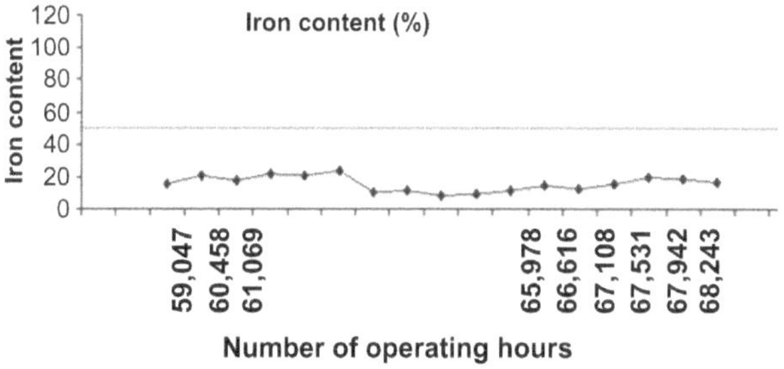

FIGURE 13.18 Iron content in the lubricating oil (2018).

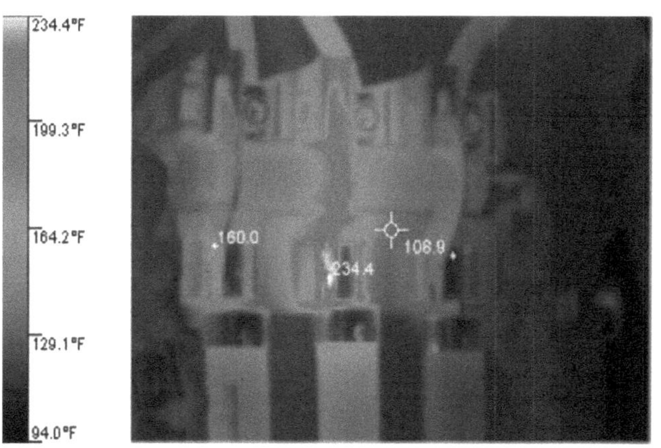

FIGURE 13.19 Copper content in the lubricating oil (2016).

variations caused by excess electric current in the furnace motor/generator 01 be with the hot spot and it will not be able to predispatch cargo.

a. *Fuzzy logic goes into the operating conditions of the equipment*

In Fig. 13.20, we can identify the anomalies likely to occur in the electric motor of the generator 1 for all effects of temperature caused by excessive electric current. The heating screen of Fig. 13.21 indicates that the engine/generator 1 cannot be related for preloading under operating plants. The other motors and generators are in the normal comfort area (A) and can be classified for normal operation of the diffuse rule, such as the motors and tuners 2, 3, 4, 5, 6, 7, 8, 9, and 10. 5 and 25 mark as normal operations without interruption and execution of restriction only for the motor/generator 1. Activate the excluded points in your electronic connection.

In Fig. 13.20 the parameters for the location of the equipment, according to the engine/generator 1, are not allowed to operate because they are not in good operating condition. The other engines and generators are located in the area A (N) Normal, 2, 3, 4, 5, 6, 7, 8, 9, and 10 and are able to position themselves according to the needs of the organization. Fig. 13.21 informs, which engines are conditional ideal for preshipment of cargo under operating conditions.

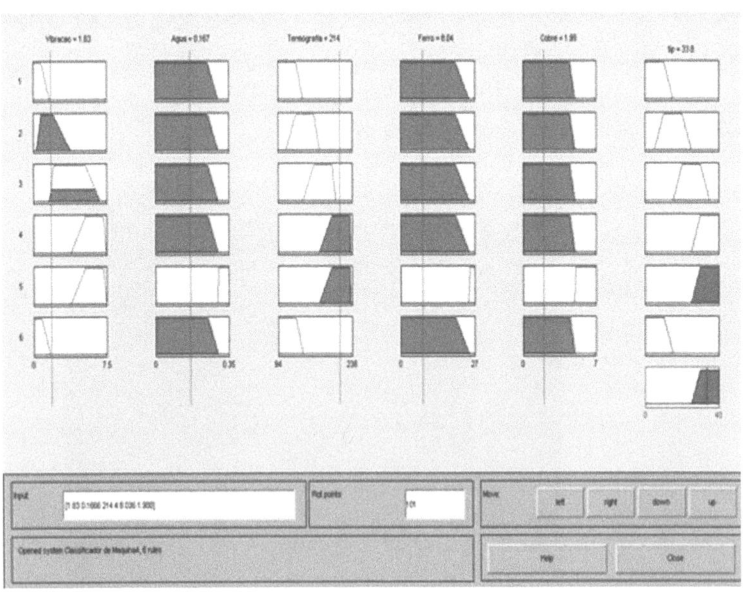

FIGURE 13.20 Copper content in the lubricating oil (2016).

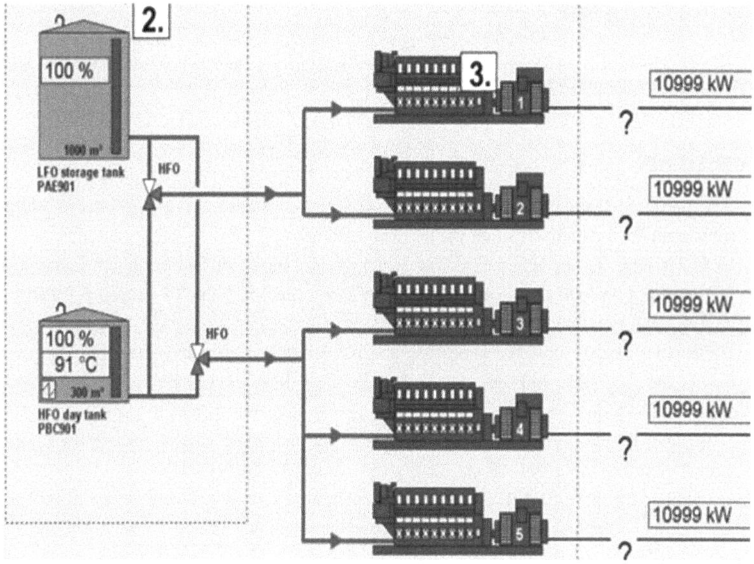

FIGURE 13.21 Technical state of the engine for load dispatch.

13.6.1 Results achieved

The objective of this work was to analyze the maintenance management system and its optimization through nebulous logic for the development of an intelligent system of support and decision-making for an ideal load dispatch demand.

The interface of the developed computational tool achieved the simplicity desired by the users themselves, as well as the ease of learning in their operation. According to the facts presented in this paper, it was possible to show that, currently, a predictive maintenance program and a total maintenance program are indispensable for large companies. This is to provide reliability to processes and equipment, detecting problems still in the initial phase. Programs of this type provide good maintenance planning for the maintenance industry. Thus, the company grows with regard to meeting deadlines, resulting in an increase in customer satisfaction.

In the present study, the gains from the two plans mentioned earlier could be assessed based on the information from the case study, we verified the reduction of corrective maintenance and we verified the results with the increase of the MTBF and the decrease of the MTTR. The observed case can be implemented in any power generation machine that uses the fuel oil and, consequently, the use of oil stock, independent of the tank capacity and storage tank scales, which have only the standards of this system. The case study presented here can be implemented in any thermoelectric plant, independent of the loads to be dispatched, since the variables of this system are common to all.

Acknowledgment

The authors gratefully acknowledge the support of this research by FAPEAM, UFPA, ELETROBRAS, and ITEGAM.

References

[1] S.G. Silva, Sistema interligado nacional: Análise das penalizações impostas às transmissoras com foco na aplicação da parcela variável. 2016.

[2] M.A. Haiping, et al., Multi-objective biogeography-based optimization for dynamic economic emission load dispatch considering plug-in electric vehicles charging, Energy 135 (2017) 101–111. ISSN 0360-5442.

[3] J. Pless, H. Fell, Bribes, bureaucracies, and blackouts: towards understanding how corruption at the firm level impacts electricity reliability, Res. Energy Econ. 47 (2017) 36–55. ISSN 0928-7655.

[4] N. Kumar, A non convex cost function based optimal load dispatch using tlbo algorithm, J. Eng. Sci. Technol. Rev. 10 (1) (2017) 155–159. ISSN 1791-2377.

[5] N. Alam, M.F. Karim, S.A. Islam, A.N. Ahsan, A 0/1 mixed integer linear programming approach to establish an effective preventive maintenance policy for power plant, Int. J. Ind. Syst. Eng. 25 (4) (2017) 478–498.

[6] S. Alaswad, Y. Xiang, A review on condition-based maintenance optimization models for stochastically deteriorating system, Reliab. Eng. Syst. Saf. 157 (2017) 54–63.

[7] R. Arya, Ranking of feeder sections of distribution systems for maintenance prioritization accounting distributed generations and loads using diagnostic importance factor (DIF), Int. J. Electr. Power Energy Syst. 74 (2016) 70–77.

[8] A. Azadeh, S.M. Asadzadeh, N. Salehi, M. Firoozi, Condition-based maintenance effectiveness for series–parallel power generation system—a combined Markovian simulation model, Reliab. Eng. Syst. Saf. 142 (2015) 357–368.

[9] R. Baidya, S.K. Ghosh, Model for a predictive maintenance system effectiveness using the analytical hierarchy process as analytical tool, IFAC-PapersOnLine 48 (3) (2015) 1463–1468.

[10] M. Fonseca, et al., Pre-dispatch of Load in Thermoelectric Power Plants Considering Maintenance Management Using Fuzzy Logic, IEEE Access, 2018.

[11] M. Fonseca Junior, et al., Maintenance tools applied to electric generators to improve energy efficiency and power quality of thermoelectric power plants, Energies 10 (8) (2017) 1091.

[12] S. Larguech, et al., Fuzzy sliding mode control for turbocharged diesel engine, J. Dyn. Syst. Meas. Contr. 138 (1) (2016) 011009. ISSN 0022-0434.

[13] B.P. Gonçalves, et al., Avaliação de impactos harmônicos na rede elétrica através dos indicadores THD e fator de potência utilizando lógica Fuzzy, Rev. Bras Energ. 19 (1) (2013) 9–27.

[14] E.L. Nogueira, M.H.R. Nascimento, Inventory control applying sales demand prevision based on fuzzy inference system, J. Eng. Technol. Ind. Appl. 3 (2017) 31–36. <https://doi.org/10.5935/2447-0228.20170060>.

[15] D. Lopez-Perez, J. Antonino-Daviu, Application of infrared thermography to failure detection in industrial induction motors: case stories, IEEE Trans. Ind. Appl. 53 (3) (2017) 1901–1908. ISSN 0093-9994.

Chapter 14

Integration of fixed-speed wind energy conversion systems into unbalanced and harmonic distorted power grids

Alp Karadeniz[1], Murat E. Balci[1] and Shady H.E. Abdel Aleem[2]
[1]*Department of Electrical and Electronics Engineering, Balikesir University, Balikesir, Turkey,* [2]*Mathematical, Physical and Engineering Sciences Department, 15th of May Higher Institute of Engineering, Cairo, Egypt*

14.1 Introduction

Nowadays, the share of electricity generated by renewables worldwide is growing in response to technical, economic, and environmental developments as well as political and social initiatives, driven by the declining cost of technology, fossil fuel resource depletion, subsidies offered by governments, and the increased awareness toward global warming rise and greenhouse gas emissions [1]. Low-carbon distributed generation (DG) units, such as wind and solar energy conversion systems, which are the key enablers in the transition of the energy market to renewable energy, are mainly employed in power distribution networks (DNs) since they offer different benefits, such as voltage profile improvement, reliability and power quality enhancement, power loss reduction, and energy efficiency increase [2].

In the near past the DG units, mainly wind and solar energy, were weaker to affect the traditional power grids, which mostly rely on fossil fuels to generate electricity. Today, the increasing rate of DG penetration into distribution systems has transformed the electricity sector into a smart decentralized one that is more driven by a mix of technologies and decentralized operators. Thus unfortunately, many challenges have also raised along with the DG systems. Specifically, excessive penetration levels (PLs) or inappropriate DG rating may cause adverse effects in the DNs, such as increased power losses, thermal overloading of transformers and feeders, over and under voltages, protection failure, and increased fault levels as well as system security and reliability

Decision Making Applications in Modern Power Systems. DOI: https://doi.org/10.1016/B978-0-12-816445-7.00014-1

issues [2,3]. As a result, in the literature [4,5], great efforts were presented to solve the problem of optimal sizing and proper placement of DG units in DNs, such as analytical, numerical, and heuristic-based methods. In addition, for the determination of the optimal DG PL, bus rms voltage limits and current limits related to the allowable thermal loading capabilities of the lines/cables and transformers were traditionally considered constraints in the literature [6].

Today, extensive employment of nonlinear loads and large-scale grid-connected DG units has led to considerably distorted voltages and currents in the distribution systems [7–9]. Accordingly, in the recent studies [2,4,10–13], the harmonic distortion limits placed in the international standards have been considered additional constraints for optimal planning of the DG units with power electronic–based interfaces. These studies clearly show that harmonic distortion considerably limits the hosting capacity of the system for those kinds of DG units.

In addition, the distribution systems always have unbalanced voltages and currents due to the unequal distribution of single-phase loads over the three phases, asymmetry of lines, and/or unbalanced power system faults [14]. Apart from that, single-phase and nondispatchable DG units of residential customers can be another important reason for the unbalance problem since they are often randomly distributed in the system [15–17]. Thus the voltage unbalance should be considered a constraint for the determination of the system's DG hosting capacity [18–20].

Induction machines are widely used as motors in consumer-connected devices and as generators in the small-to-medium-sized fixed-speed wind energy conversion systems (FSWECSs) due to their advantages such as small size, low cost, and high reliability [21]. Under unbalanced and/or distorted grid voltages the induction machines have poor power factor, torque pulsations, and extra losses resulting in overheating problem [21–24]. The torque pulsations and overheating problems lead to their fast aging. Accordingly, in the literature [14,25], to avoid the loss of life related to extra losses under these voltage conditions, it is suggested that their permissible loading ratio, the so-called derating factor (*DF*), should be limited by equalizing the highest phase current to the rated value. However, few studies paid attention to the maximum loading ratios or permissible PL (*PPL*) of induction generators (IGs) [26,27].

To mitigate unbalance and/or harmonic distortion in the systems with DG units, there are several custom power devices, such as dynamic voltage restorers, static synchronous compensators, active power filters, and unified power quality conditioners [28–30]. On the other hand, when compared to these compensation devices, passive harmonic filters [2,12,13] and Steinmetz compensator [31–33] are simple and cost-effective solutions for harmonic distortion and unbalance mitigation, respectively.

In this paper, first, an algorithm is presented to determine *PPL* of a FSWECS with squirrel cage IG (SCIG) operating in the unbalanced and non-sinusoidal distribution system. The simulated system is modeled through

Matlab/Simulink, where the well-known $d-q$ model of the SCIG [14] is utilized. Second, a parallel combination of a Steinmetz compensator (SC) and a single-tuned harmonic filter (STF) is proposed for minimization of the voltage unbalance factor (VUF) and average total voltage harmonic distortion ($THDV_{Mean}$) and maximization of PPL and displacement power factor (DPF). The VUF and individual voltage harmonic distortion and $THDV$ limits, stated in the standards and the desired DPF and rms bus voltage ranges, are regarded as constraints of the optimal compensator design problem. And then, for the nonsinusoidal and unbalanced test system, one of the widely preferred and reliable metaheuristic optimization methods, particle swarm optimization (PSO) algorithm [34–36], was chosen to employ the proposed optimal compensator. Optimal STF, which has the problem formulation of the proposed compensator except the voltage unbalance–related objectives and constraints, and optimal SC, which has the problem formulation of the proposed compensator except the voltage harmonic distortion–related objectives and constraints, were employed using the same optimization algorithm. Finally, the results of all three compensators were comparatively evaluated regarding their performances on the PPL improvement, the unbalance and harmonic distortion mitigation of the point of common coupling (PCC) voltage, and DPF correction.

14.2 Problem statement and description

Steinmetz circuit is a compensator that consists of passive elements as inductors and capacitors to mitigate load unbalance and improve power factor in three-phase systems with unbalanced linear loading and symmetric sinusoidal supply voltage conditions [37]. For those systems, it has closed form of design expressions to fully compensate power factor and unbalanced part of the load current. However, these expressions are not valid for unbalanced or asymmetrical voltage conditions [31–38]. In addition, an SC is not able to mitigate harmonic distortion but may amplify it [39,40]. Therefore, in some studies, it was designed using optimization algorithms to improve loads' power factor and solve voltage unbalance problem [33]. In view of that, for mitigation of both unbalance factors and harmonic distortion levels, SCs should be supplemented by harmonic filters. But, to the best of the authors' knowledge, no study on the optimal design of both shunt-connected compensators exists in the literature.

Apart from that, a lot of studies have investigated the impacts of various wind energy systems on power quality performance of DNs, and it was found that the power quality performance of systems mainly depends on the type of the wind system. As a result, in this study, the determination of PPL of IGs in FSWECS is investigated in an unbalanced and nonsinusoidal system, shown in Fig. 14.1. The system comprises distorted and unbalanced utility side, which are represented by hth harmonic Thevenin three-phase voltage

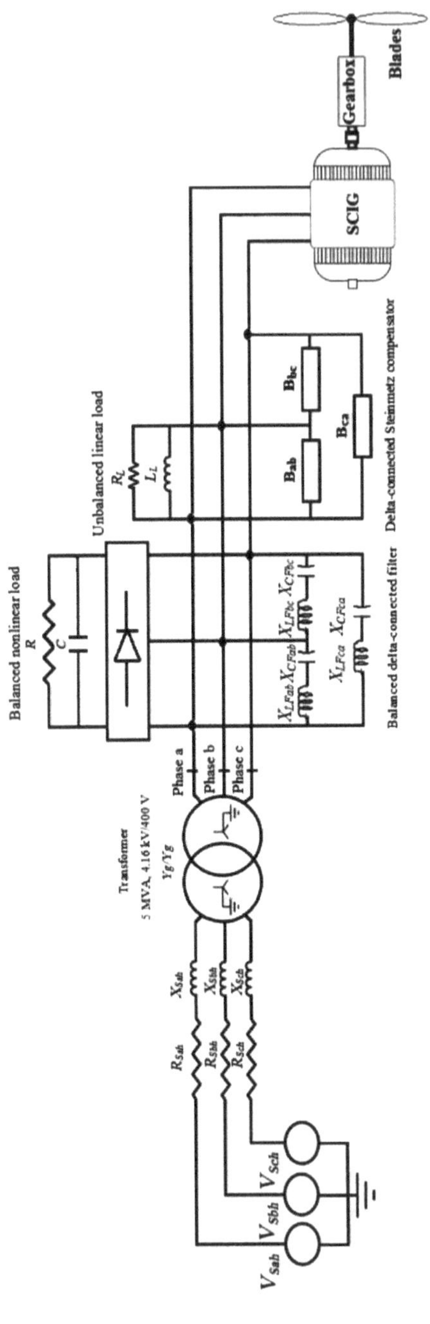

FIGURE 14.1 The system under study.

sources ($V_{Sah}, V_{Sbh}, \text{and} V_{Sch}$), the hth harmonic Thevenin impedance of the utility side, a power transformer, an unbalanced and nonlinear load group, which consists of a single-phase linear load and a six-pulse uncontrolled rectifier, a group of identical FSWECSs, a delta-connected single-tuned harmonic filter (STF) with nonidentical branches, and a delta-connected Steinmetz compensator (SC). It should be mentioned that the Thevenin impedance of the utility side is represented as its short-circuit impedance.

In the following section, for the considered unbalanced and nonsinusoidal system, dynamic modeling of FSWECS is presented; further, the proposed algorithm is to find the *PPL* of the FSWECSs under the same system conditions as introduced and discussed.

14.2.1 Modeling of the fixed-speed wind energy conversion systems

The dynamic modeling of the FSWECS is provided for analyzing its permissible PL under unbalanced and nonsinusoidal conditions. As shown in Fig. 14.1, it consists of blades, gearbox, and SCIG. The mechanical power transmitted from the gearbox to the shaft of the SCIG is written in terms of the air density (ρ), the area swept by the rotor (A), power coefficient (C_p), and wind speed (u) [21]:

$$P_M = \frac{1}{2}\rho A u^3 C_p(\lambda) \tag{14.1}$$

The power coefficient (C_p) depends on the tip–speed ratio (λ), which is determined as follows:

$$\lambda = \frac{\omega_m R}{u} \tag{14.2}$$

where ω_m and R are the rotor angular velocity and rotor radius, respectively.

In the analysis, SCIG is modeled using the well-known $d-q$ equivalent circuits [14,41] shown in Fig. 14.2. Here, it should be noted that in the literature, there are several studies [42,43], which show that the results of $d-q$ model and experiments are in close agreement for the performance analysis of the induction machine under unbalanced supply voltages. It can also be mentioned that the same model was considered for the transient analysis of the induction motors and doubly fed IGs under distorted supply voltages in many studies [44–46].

For the model, Eqs. (14.3)–(14.6), the instantaneous stator voltages (V_{qs} and V_{ds}) and instantaneous rotor voltages referred to the stator side (V'_{qr} and V'_{dr}) can be expressed in terms of the magnetic fluxes (φ_{qs}, φ_{ds}, φ'_{qr}, and φ'_{dr}), stator currents (i_{qs} and i_{ds}), rotor currents referred to the stator side (i'_{qr} and i'_{dr}), stator resistance (R_S), rotor resistance referred to the stator side (R'_r), the reference frame angular velocity (ω), and the electrical angular velocity (ω_r), thus

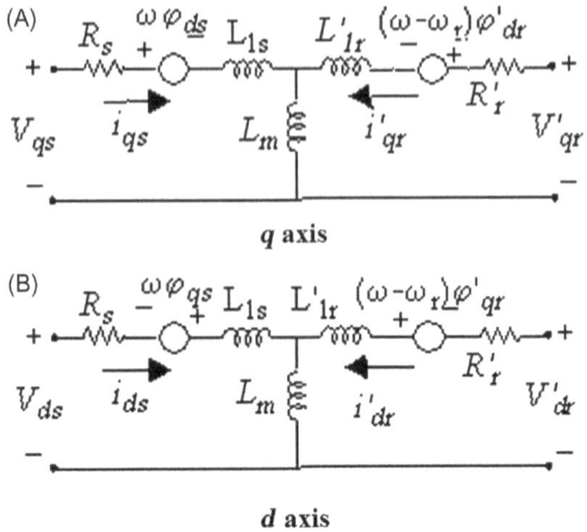

FIGURE 14.2 Equivalent circuits of the induction machine, (A) q-axis equivalent circuit and (B) d-axis equivalent circuit.

$$V_{qs} = R_s i_{qs} + \frac{d}{dt}\varphi_{qs} + \omega\varphi_{ds} \tag{14.3}$$

$$V_{ds} = R_s i_{ds} + \frac{d}{dt}\varphi_{ds} - \omega\varphi_{qs} \tag{14.4}$$

$$V'_{qr} = R'_r i'_{qr} + \frac{d}{dt}\varphi'_{dr} - (\omega - \omega_r)\varphi'_{dr} \tag{14.5}$$

$$V'_{dr} = R'_r i'_{dr} + \frac{d}{dt}\varphi'_{dr} + (\omega - \omega_r)\varphi'_{qr} \tag{14.6}$$

In addition, the magnetic fluxes of the stator and rotor on the $d-q$ axes are written as follows:

$$\varphi_{qs} = L_s i_{qs} + L_m i_{qr} \tag{14.7}$$

$$\varphi_{ds} = L_s i_{ds} + L_m i'_{dr} \tag{14.8}$$

$$\varphi'_{qr} = L'_r i'_{qr} + L_m i_{qs} \tag{14.9}$$

$$\varphi'_{dr} = L'_r i'_{dr} + L_m i_{ds} \tag{14.10}$$

where L_s, L'_r, and L_m stand for the total stator inductance, total rotor inductance referred to the stator side and magnetization inductance, respectively. The two inductance values (L_s, L'_r) can be found by summation of the corresponding leakage inductance (L_{ls} and L'_{1r}) and L_m, thus

$$L_s = L_{1s} + L_m \tag{14.11}$$

$$L'_r = L'_{1r} + L_m \tag{14.12}$$

The electromagnetic torque (T_e) and its relation with the mechanical torque (T_m) of the shaft can be expressed as follows:

$$T_e = 1.5p\left(\varphi_{ds}i_{qs} - \varphi_{qs}i_{ds}\right) \tag{14.13}$$

$$T_e - T_m = 2H\frac{d}{dt}\omega_m \tag{14.14}$$

where p, H, and ω_m are pole pairs of the machine, inertia constant, and rotor angular velocity, respectively.

The relationships among the line voltages and the $d-q$ voltage components are given in (14.15) and (14.16), thus

$$\begin{bmatrix} V_{qs} \\ V_{ds} \end{bmatrix} = \frac{1}{3}\begin{bmatrix} 2cos\,\theta & cos\,\theta + \sqrt{3}sin\,\theta \\ 2sin\,\theta & sin\,\theta - \sqrt{3}cos\,\theta \end{bmatrix}\begin{bmatrix} V_{abs} \\ V_{bcs} \end{bmatrix} \tag{14.15}$$

$$\begin{bmatrix} V'_{qr} \\ V'_{dr} \end{bmatrix} = \frac{1}{3}\begin{bmatrix} 2cos\,\beta & cos\,\beta + \sqrt{3}sin\,\beta \\ 2sin\,\beta & sin\,\beta - \sqrt{3}cos\,\beta \end{bmatrix}\begin{bmatrix} V'_{abr} \\ V'_{bcr} \end{bmatrix} \tag{14.16}$$

where θ is the angular position of the reference frame, and β is the difference between θ and the angular position of the rotor (θ_r).

Hence, by using the $d-q$ current components, the phase currents of the induction machine can be calculated as follows:

$$\begin{bmatrix} I_{as} \\ I_{bs} \end{bmatrix} = \begin{bmatrix} cos\,\theta & sin\,\theta \\ \dfrac{-cos\,\theta + \sqrt{3}sin\,\theta}{2} & \dfrac{-\sqrt{3}cos\,\theta - sin\,\theta}{2} \end{bmatrix}\begin{bmatrix} I_{qs} \\ I_{ds} \end{bmatrix} \tag{14.17}$$

$$\begin{bmatrix} I'_{ar} \\ I'_{br} \end{bmatrix} = \begin{bmatrix} cos\,\beta & sin\,\beta \\ \dfrac{-cos\,\beta + \sqrt{3}sin\,\beta}{2} & \dfrac{-\sqrt{3}cos\,\beta - sin\,\beta}{2} \end{bmatrix}\begin{bmatrix} I'_{qr} \\ I'_{dr} \end{bmatrix} \tag{14.18}$$

$$I_{cs} = -I_{as} - I_{bs} \tag{14.19}$$

$$I'_{cr} = -I'_{ar} - I'_{br} \tag{14.20}$$

14.2.2 Determination of the permissible penetration level

Determination of the maximum loading ratio or *PPL* for induction machines is based on one of the following procedures [14]:

1. Use of a standardized DF curve (or expression) in terms of voltage unbalance level based on the National Electrical Manufacturers Association. This is more common for induction motors.

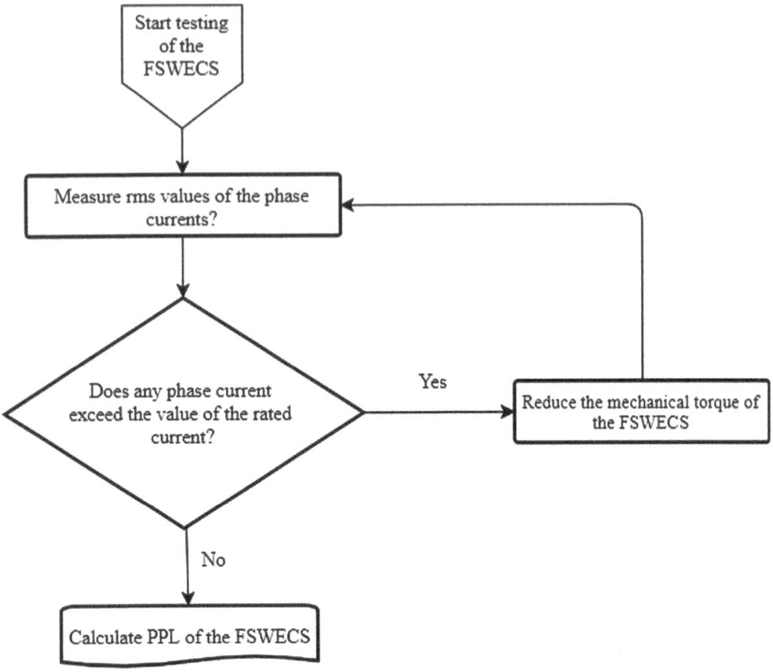

FIGURE 14.3 *PPL* determination algorithm. *PPL*, Permissible penetration level.

2. Measuring the loss under the existed unbalanced and distorted supply voltage, hence equalizing the measured one to the rated loss by reduction of the loading ratio.
3. Measuring the highest rms stator phase current under the existed unbalanced and distorted supply voltage, hence equalizing the measured one to the rated current by reduction of the loading ratio.

In this work, to find the highest permissible active power (P_{Gmax}) of the IGs under voltage unbalance and harmonic distortion conditions, whose mechanical input power is adjusted by pitch control, the highest stator phase current (I_{Gmax}) is intentionally reduced to the rated current (I_{GR}). Hence, *PPL* is found as follows:

$$PPL(\%) = \left(\frac{P_{Gmax}}{P_{GR}}\right) \times 100 \tag{14.21}$$

where P_{GR} is the rated active power of the IGs. The flowchart of the *PPL* determination algorithm is shown in Fig. 14.3.

14.2.3 Modeling of the Steinmetz compensator

The three-phase equivalent circuit of the SC is shown in Fig. 14.4. In this figure the SC's admittances are denoted as

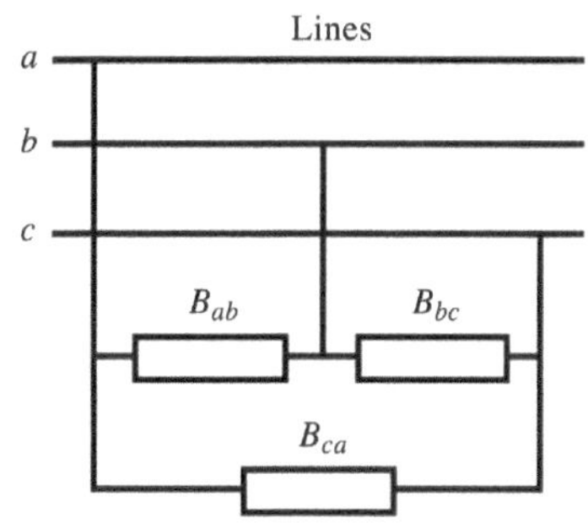

FIGURE 14.4 Three-phase equivalent circuit of the Steinmetz compensator at the fundamental frequency.

$$Y_{ab}^{comp} = jB_{ab},\, Y_{bc}^{comp} = jB_{bc} \text{and} Y_{ca}^{comp} = jB_{ca} \tag{14.22}$$

The negative and positive values of a susceptance indicate whether its value is inductive or capacitive, respectively.

As mentioned earlier, the studied system has an unbalanced background voltage and the load side, which consists of three-phase balanced loads, single-phase loads unequally distributed over three-phase, and a group of FSWECSs with SCIG.

In this study the SC is employed to reduce unbalance level of the grid voltage as well as improve *DPF*. *VUF* is the ratio between magnitudes of the fundamental frequency negative (V_1^-) and positive-sequence voltages (V_1^+), thus

$$VUF(\%) = \frac{V_1^-}{V_1^+} \times 100 \tag{14.23}$$

DPF is the ratio of total fundamental harmonic active power to total fundamental harmonic apparent power:

$$DPF = \frac{P_1}{S_1} = \frac{\sum_{m=a,b,c} P_{m1}}{\sqrt{\left(\sum_{m=a,b,c} P_{m1}\right)^2 + \left(\sum_{m=a,b,c} Q_{m1}\right)^2}} \tag{14.24}$$

where fundamental active and fundamental reactive powers for $m = a$, b, c phases are

$$P_{m1} = V_{m1} I_{m1} \cos(\theta_{m1}) \text{ and } Q_{m1} = V_{m1} I_{m1} \sin(\theta_{m1}) \tag{14.25}$$

14.2.4 Modeling of the single-tuned harmonic filter

The three-phase equivalent circuit of the STF is shown in Fig. 14.5.

For the harmonic tuning orders (h_{tab}, h_{tbc}, and h_{tca}) of each branch of the STF, the relations between the fundamental frequency reactances of the capacitor and inductor in each branch can be written as

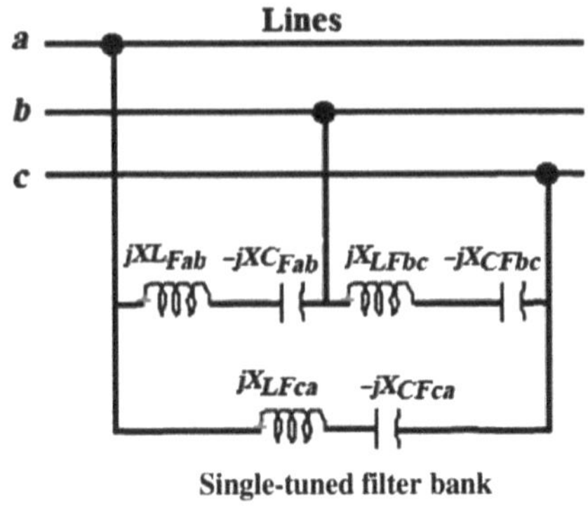

FIGURE 14.5 Three-phase equivalent circuit of a delta-connected STF at the fundamental frequency. *STF*, Single-tuned harmonic filter.

$$X_{LFab} = \frac{X_{CFab}}{h_{tab}^2}, X_{LFbc} = \frac{X_{CFbc}}{h_{tbc}^2} \text{and} X_{LFca} = \frac{X_{CFca}}{h_{tca}^2} \tag{14.26}$$

The three capacitive reactances ($X_{CFab}, X_{CFbc}, \text{and} X_{CFca}$) are able to adjust *DPF* to its desired interval. In addition, to mitigate the total harmonic distortion values of the phase voltages ($THDV_a$, $THDV_b$, and $THDV_c$) and their mean value ($THDV_{Mean}$) expressed as follows:

$$THDV_m = 100\frac{\sqrt{\sum_{h\geq 2} V_{mh}^2}}{V_{m1}} \text{ and } THDV_{Mean} = \frac{\sum_{m=a,b,c} THDV_m}{3} \tag{14.27}$$

the values of the tuning harmonic orders ($h_{tab}, h_{tbc}, \text{and} h_{tca}$) should be decided. Thus regarding Eq. (14.26), three inductive reactance values of the STF branches ($X_{LFab}, X_{LFbc}, \text{and} X_{LFca}$) are found. In practice the STFs have a better harmonic mitigation capability than other types of passive filters. However, they may result in parallel resonance in the system. Therefore during their design stage, it should be checked whether resonance risks are safely avoided or not [47].

14.3 Problem formulation and solution algorithm

In the system, SC may amplify the harmonic distortion level, and STF may increase the unbalance level. Due to this, the design of the proposed compensator consisting of SC and STF is an optimal design problem. Regarding the harmonic distortion and unbalance levels of the grid voltages, desired fundamental frequency reactive power compensation level, and the *PPL* of the system, for the problem formulation of the proposed optimal compensator, the objective and constraints are presented in this section.

14.3.1 Objective function

For the optimal sizing of the proposed compensator, the objective function (*OF*) is considered as minimization of *VUF*, and $THDV_{Mean}$ of the phase voltages and maximization of *PPL* and *DPF* using the weight factors (k_1, k_2, k_3, and k_4). *OF* can be written as follows:

$$Min\, f(STF_P, SC_P, P_{Gmax}) = (k_1 \times VUF) + (k_2 \times THDV_{Mean}) + \left(k_3 \times \frac{1}{PPL}\right) + \left(k_4 \times \frac{1}{DPF}\right) \tag{14.28}$$

Because the sensitivity of *PPL* to voltage imbalance and harmonic distortion is case dependent; therefore the first two subobjectives were included in the *OF* to ensure maximization of the *PPL* at any level.

14.3.2 Nonequality constraints

For the problem formulation of the proposed optimal compensator sizing, nonequality constraints are detailed as follows:

- Rms voltage limits

  The rms values of the positive-sequence component of the fundamental bus voltage (V_1^+) should be within a desired range as ± 10% of the per unit (p.u.) nominal voltage value. Thus the rms voltage constraint is given as

$$0.9 < V_1^+ < 1.1 \text{ p.u.} \tag{14.29}$$

- Harmonic distortion limits

  The individual and total harmonic distortions of the phase voltages measured at the PCC are taken into considerations as per the IEEE Standard 519 [48]. Accordingly, the harmonic distortion limits can be expressed regarding the voltage rms level as

$$IHDV_m(\%) = \frac{V_{mh}}{V_{m1}} \times 100 \leq IHDV_{max} \tag{14.30}$$

$$THDV_m(\%) \leq THDV_{max} \tag{14.31}$$

  where V_{mh} and V_{m1} denote the *h*th harmonic and fundamental harmonic voltages for $m = a$, b, and c phases, respectively. $IHDV_m$ is the individual harmonic distortion of voltages for the a, b, and c phases. $IHDV_{max}$ and $THDV_{max}$ are the maximum permissible individual and total harmonic voltage distortion reported in the IEEE Standard 519, respectively.

- VUF limit

  According to the standard IEC 61000-2-4: 2002 [49], VUF in a DN should not exceed 2%. As a result, the VUF limit is expressed as

$$VUF(\%) \leq 2 \tag{14.32}$$

- *DPF* limits

  Its value measured at the PCC should have a lagging value between 95% and 100% according to the IEEE Standard 519. Thus a *DPF* constraint is formulated as follows:

$$95(\text{lagging}) \leq DPF(\%) \leq 100 \tag{14.33}$$

- The thermal limit (current limit) for the IG of FSWECS

  Maximum one of the rms phase currents (I_{Ga}, I_{Gb}, I_{Gc}) of the FSWECSs should not exceed the rated current (I_{GR}):

$$Max(I_{Ga}, I_{Gb}, I_{Gc}) \leq I_{GR} \tag{14.34}$$

14.3.3 Particle swarm optimization algorithm

PSO is a stochastic optimization algorithm that was developed by Dr. Eberhart and Dr. Kennedy in 1995 [34–36]. PSO algorithm is based on bird swarms that communicated with each other, for instance, to find food. In this algorithm a bird is expressed as a candidate of solution, and all birds in the group together are called "swarm."

The best solution is found regarding fitness function output. Velocity and position values of each agent are randomly initialized, and then they are updated using random variables and mathematical equations in terms of global best (*gbest*), position of the best fitness value in the whole process, and the personal best value of each agent (*pbest*). The velocity values are accelerated over the *gbest* and *pbest* values by using the following equations:

$$v_{id}{}^{k+1} = w \times v_{id}{}^{k} + c_1 \times U \times (pbest_{id}{}^{k} - x_{id}{}^{k}) + c_2 \times U \times \left(gbest_{d}{}^{k} - x_{id}{}^{k}\right) \tag{14.35}$$

$$x_{id}{}^{k+1} = x_{id}{}^{k} + v_{id}{}^{k+1} \tag{14.36}$$

where d denotes a column vector in m-dimension ($d = 1, 2, \ldots, m$), i denotes a row vector in the n-dimension ($i = 1, 2, \ldots, n$), k is the iteration number, $v_{id}{}^{k}$ is the velocity value of ith agent at the kth iteration, $x_{id}{}^{k}$ is the current position of ith agent at the kth iteration; $pbest_{id}{}^{k}$ is the personal best value of the ith agent at the kth iteration, $gbest_{d}{}^{k}$ is the global best value of the ith agent at the kth iteration, U is the random values between 0 and 1, w is the inertia weight, c_1 and c_2 are the weighting factors, they are selected as 2 in this study. w decreases in the interval (0.9–0.4) as given in the following equation:

$$w = w_{max} - \frac{w_{max} - w_{min}}{iter_{max}} \times iter \tag{14.37}$$

where w_{max} and w_{min} are bounds of w, $iter_{max}$ is the maximum iteration number, and $iter$ is the current iteration number.

14.4 Simulation results and discussion

For the simulations the test system, given in Fig. 14.1, has three-phase unbalanced and nonsinusoidal utility voltages with the fundamental harmonic phasor components, unbalance level, and total harmonic distortion measured as $V_{Sa} = 2.31\angle 0^\circ$ kV, $V_{Sb} = 2.31\angle -120^\circ$ kV, $V_{Sc} = 2.42\angle 120^\circ$ kV, $VUF_S = 1.65\%$, and $THDV_S = 3.8\%$, respectively.

The harmonic components of the three-phase utility voltages are presented in Table 14.1.

In addition, utility side of the system has a star–star connected distribution transformer, of which ratings are 5 MVA and 4.16/0.4 kV, and the balanced supply line, of which fundamental frequency impedance parameters are $R_S = 0.01\ \Omega$ and $X_S = 0.1\ \Omega$. The transformer is modeled by using its well-known T-equivalent circuit, and its ratings and circuit parameters are shown in Fig. 14.6.

The fundamental frequency impedance parameters of the single-phase linear load are $R_L = 0.11\ \Omega$ and $X_L = 0.67\ \Omega$, and its P_1 and DPF values as 480 kW and 98.64%, respectively. The six-pulse uncontrolled rectifier has circuit parameters as $R = 0.4\ \Omega$ and $C = 1000\ \mu$F, and its P_1 and DPF values as 650 kW and 99.70%, respectively. The SCIGs, which are used in the five FSWECSs, have power, voltage, current and rotor speed ratings as 110 kW, 400 V, 182 A and 1487 rpm. The ratings and circuit parameters of the SCIG's model are shown in Fig. 14.7.

As mentioned earlier, the consumer side has two compensators as a STF and a SC, which will be optimally designed to improve the *PPL*, *VUF*, $THDV_{Mean}$, and *DPF* values. Fig. 14.8 shows the system under study in Matlab/Simulink platform.

For the *PLs* of the FSWECs into the system from 0% to 60%, the variation of the total rms phase currents (I_{Ga}, I_{Gb}, I_{Gc}) injected by them, total harmonic distortion values of the phase voltages ($THDV_a$, $THDV_b$, $THDV_c$) at the PCC, and the *VUF* and *DPF* values at the PCC are plotted in Figs. 14.9–14.11, respectively. In Fig. 14.9 the dashed vertical line points

TABLE 14.1 Harmonic components of the three-phase utility voltage.

h	V_{Sah} (V)	V_{Sbh} (V)	V_{Sch} (V)
5	$70\angle 0^\circ$	$70\angle -5\cdot 120^\circ$	$73\angle 5\cdot 120^\circ$
7	$46\angle 0^\circ$	$46\angle -7\cdot 120^\circ$	$49\angle 7\cdot 120^\circ$
11	$23\angle 0^\circ$	$23\angle -11\cdot 120^\circ$	$24\angle 11\cdot 120^\circ$
13	$23\angle 0^\circ$	$23\angle -13\cdot 120^\circ$	$24\angle 13\cdot 120^\circ$

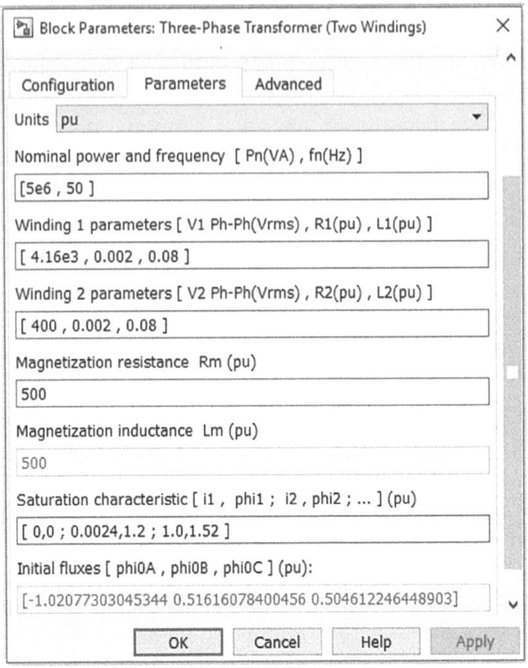

FIGURE 14.6 Ratings and circuit parameters of the transformer model.

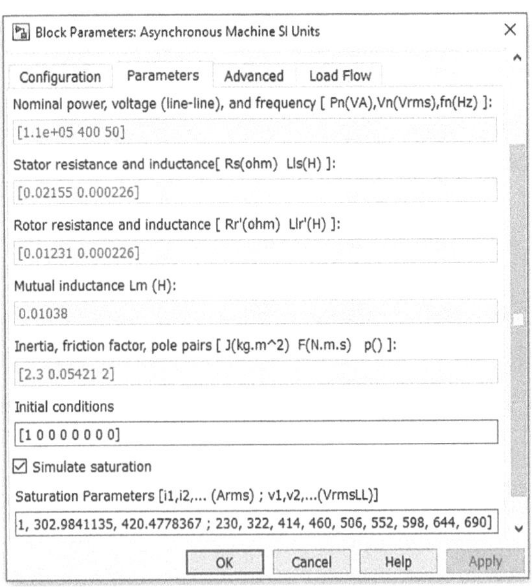

FIGURE 14.7 Ratings and circuit parameters of the SCIG used in the FSWECS. *FSWECS*, Fixed-speed wind energy conversion system; *SCIG*, squirrel cage induction generator.

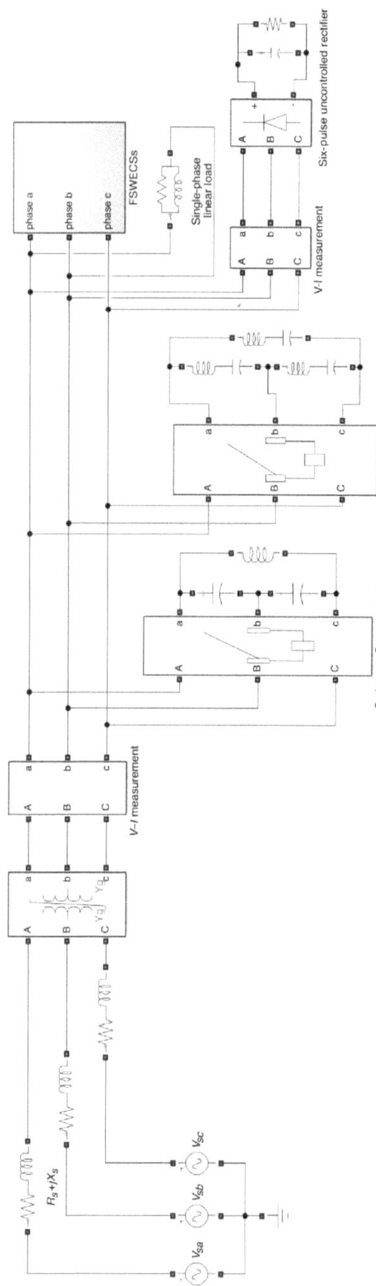

FIGURE 14.8 The system under study in Matlab/Simulink platform.

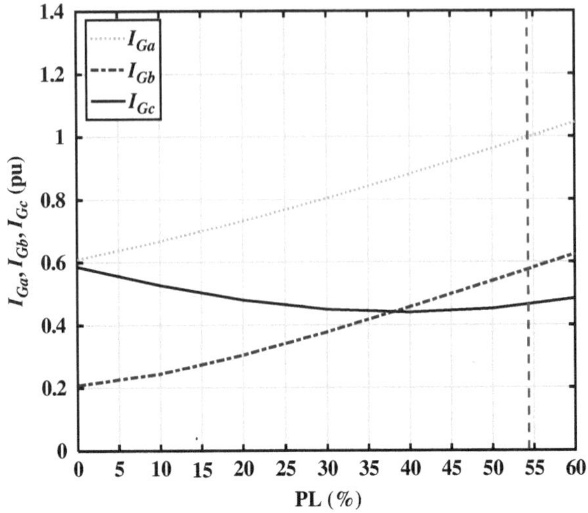

FIGURE 14.9 Variation of the total rms phase currents injected by the FSWECSs with their penetration level into the system. *FSWECSs*, Fixed-speed wind energy conversion systems.

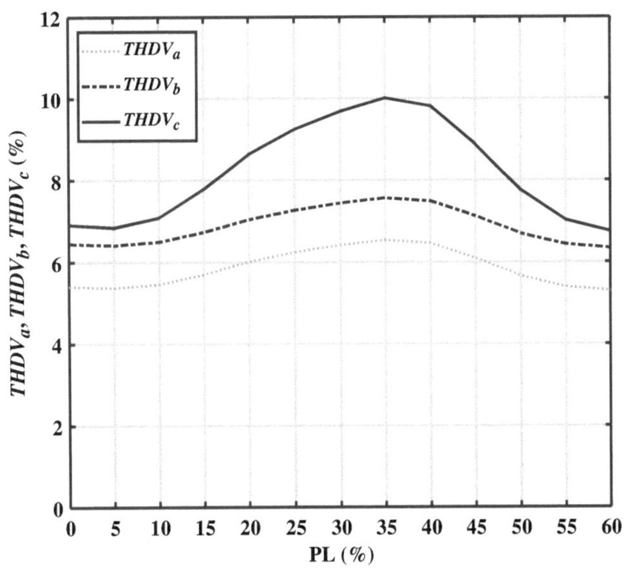

FIGURE 14.10 Variation of the total harmonic distortion values of the phase voltages at the PCC with the penetration level of the FSWECSs into the system. *FSWECSs*, Fixed-speed wind energy conversion systems; *PCC*, point of common coupling.

out that the *PPL* of the FSWECSs is 54.50% where one of their phase currents reach the rated currents ($I_{Gc} = 1$ p.u.) under the considered test system conditions. It is seen from Fig. 14.10 that $THDV_a$, $THDV_b$, and $THDV_c$ are considerably affected by the *PL*, and they have values under the limit of

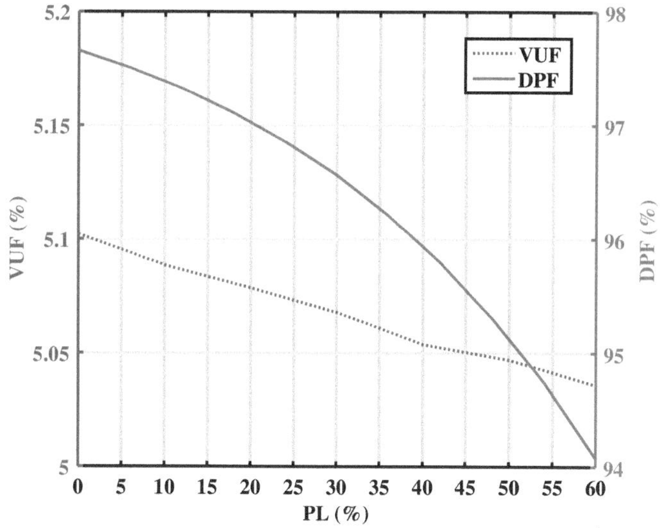

FIGURE 14.11 Variation of the voltage unbalance factor and displacement power factor values at the PCC with the penetration level of the FSWECSs into the system. *FSWECSs*, Fixed-speed wind energy conversion systems; *PCC*, point of common coupling.

IEEE Standard 519 as 8% for the *PPL* value of the FSWECSs. On the other hand, for the *PPL* case, *VUF* and *DPF* values are measured as 5.04% and 94.60%, and these values are not compatible with their limits defined in the international standards. In addition to that, it can be mentioned that *DPF* is considerably reduced by the increment of the FSWECSs' *PL* value.

14.4.1 Performance evaluation of the proposed compensator

Regarding a need for improvement of *PPL* and mitigation of *VUF* and *THDV* values of the PCC voltages, the proposed optimal compensator (SC + STF) design is obtained for the system. The parameters of the compensator and the values of the power quality indices and *PPL* after the compensator connection to the system are given in Table 14.2. It should be mentioned that the selected weights to the *k*th factors that led to the global optimal values are $k_1 = 0.2941$, $k_2 = 0.05882$, $k_3 = 0.5883$, and $k_4 = 0.05882$.

It can be seen from Table 14.2 that in the proposed compensator, the STF part has three unbalanced branches, of which impedance parameters are determined as $X_{LFab} = 0.160\ \Omega$, $X_{CFab} = 3.991\ \Omega$, $X_{LFbc} = 0.172\ \Omega$, $X_{CFbc} = 4.323\ \Omega$, $X_{LFca} = 0.102\ \Omega$, and $X_{CFca} = 4.180\ \Omega$, and the SC part has susceptance values as $B_{Sab} = 2.513\ \Omega^{-1}$, $B_{\mathrm{Sbc}} = 4.806\ \Omega^{-1}$, and $B_{\mathrm{Sca}} = -5.305\ \Omega^{-1}$. It achieves $THDV_{Mean}$ and *VUF* values as 2.67% and 0.75%, respectively. It can also be mentioned that $THDV_a$, $THDV_b$, and $THDV_c$ are well below the *THDV* limit of the IEEE Standard 519 as 8%,

TABLE 14.2 Results of the proposed optimal Steinmetz compensator and single-tuned harmonic filter, optimal Steinmetz compensator and optimal single-tuned harmonic filter designs according to particle swarm optimization algorithm.

Parameters	Optimal SC + STF	Optimal SC	Optimal STF
X_{LFab} (Ω)	0.160	–	0.200
X_{CFab} (Ω)	3.991	–	5.052
X_{LFbc} (Ω)	0.172	–	0.062
X_{CFbc} (Ω)	4.323	–	5.450
X_{LFca} (Ω)	0.102	–	0.079
X_{CFca} (Ω)	4.180	–	3.978
B_{Sab} (Ω^{-1})	2.513	2.605	–
B_{Sbc} (Ω^{-1})	4.806	4.335	–
B_{Sca} (Ω^{-1})	− 5.305	− 5.221	–
$THDV_a$ (%)	3.78	9.62	3.57
$THDV_b$ (%)	1.82	9.51	5.40
$THDV_c$ (%)	2.43	13.72	6.50
$THDV_{Mean}$ (%)	2.67	10.95	5.15
VUF (%)	0.75	0.69	5.09
V^+ (p.u.)	1.02	1.01	1.00
DPF (%)	99.99	99.49	98.11
PPL (%)	97.42	96.23	54.92

DPF, Displacement power factor; *PPL*, permissible penetration level; *THDV*, total voltage harmonic distortion; *VUF*, voltage unbalance factor.

and V_1^+ is kept between 0.9 and 1.1 p.u. Apart from that, it improves *PPL* from 54.50% to 97.42% and *DPF* from 94.50% to 99.99%, respectively.

In order to demonstrate the necessity of the cooperative employment of the SC and STF, results of optimal SC and optimal STF designs are given in Table 14.2. The optimal SC design is attained by considering the proposed compensator's problem formulation that except objectives and constraints related to voltage harmonic distortion. Its susceptances are found as $B_{Sab} = 2.605\ \Omega^{-1}$, $B_{Sbc} = 4.335\Omega^{-1}$, and $B_{Sca} = -5.221\ \Omega^{-1}$. It attains better *VUF*, *PPL*, and *DPF* values measured as 0.69%, 96.23%, and 99.49%, respectively, with respect to the uncompensated system. In addition, it keeps V_1^+ in the acceptable limit between 0.9 and 1.1 p.u. However, $THDV_a$,

$THDV_b$, and $THDV_c$ deteriorated to 9.62%, 9.51%, and 13.72%, respectively, when the optimal SC is inserted into the system.

On the other hand, optimal STF design is provided regarding the proposed compensator's problem formulation, except objectives and constraints related to voltage unbalance. It has impedance parameters as $X_{LFab} = 0.200\ \Omega$, $X_{CFab} = 5.052\ \Omega$, $X_{LFbc} = 0.062\ \Omega$, $X_{CFbc} = 5.450\ \Omega$, $X_{LFca} = 0.079\ \Omega$, and $X_{CFca} = 3.978\ \Omega$. Using the optimal STF design, $THDV_{Mean}$ is mitigated to 5.15%, and *DPF* is increased to 98.11%. In addition, it achieves $THDV_a$, $THDV_b$, and $THDV_c$ values as 3.57%, 5.40%, and 6.50%, respectively. However, voltage unbalance slightly worsens in the system compensated with the optimal STF design ($VUF = 5.09\%$) when compared to the uncompensated system ($VUF = 5.04\%$). Lastly, it attains *PPL* value measured as 54.92%.

14.4.2 Sensitivity analysis of the proposed optimal compensator design under variation of utility and load-side conditions

The proposed optimal compensator design is determined for the rated loading level and utility voltage with $VUF_S = 1.65\%$ and $THDV_S = 3.8\%$ conditions in the test system. However, the loading condition and utility voltage may be changed in the practical systems. Accordingly, to analyze performance of the proposed compensator, of which parameters are presented in Table 14.2, under varying utility voltage and load-side conditions, it is tested for three cases of the test system as follows:

- *Case 1*: 50% loading level and the utility voltage with $VUF_S = 1.65\%$ and $THDV_S = 3.8\%$
- *Case 2*: 100% loading level and the rated sinusoidal balanced utility voltages
- *Case 3*: 50% loading level and the rated sinusoidal balanced utility voltages

For the abovementioned cases the power quality indices and *PLL* of the FSWECSs, which are measured before and after the connection of the proposed compensator in the system, are given in Tables 14.3 and 14.4. It is seen from the same tables that, for all cases, the proposed compensator provides considerable improvement of $THDV_{Mean}$, *DPF*, and *PPL*. In addition to that, it achieves to keep $THDV_a$, $THDV_b$, $THDV_c$, and V_1^+ within their desired ranges.

Lastly, for *Cases 1* and *2*, *VUF* is also mitigated by the proposed compensator. However, for *Case 3*, *VUF* is almost the same with and without the compensator. Besides, it should be noted that values of the capacitors should be readjusted in light loading conditions (*Cases 1* and *3*) to avoid over compensation. This can be performed by capacitors switching.

TABLE 14.3 The power quality indices and *PPL* measured in the three cases of the system before compensation.

Parameters	Case 1	Case 2	Case 3
$THDV_a$ (%)	3.91	3.77	4.24
$THDV_b$ (%)	4.31	4.90	4.65
$THDV_c$ (%)	5.80	6.26	7.52
$THDV_{Mean}$ (%)	4.67	4.97	5.47
VUF (%)	2.89	4.50	2.26
V^+ (p.u.)	1.01	1.03	1.04
DPF (%)	86.91	96.26	85.91
PPL (%)	75.40	61.90	85.00

DPF, Displacement power factor; *PPL*, permissible penetration level; *THDV*, total voltage harmonic distortion; *VUF*, voltage unbalance factor.

TABLE 14.4 The power quality indices and *PPL* measured in the three cases of the system after compensation using the proposed compensator.

Parameters	Case 1	Case 2	Case 3
$THDV_a$ (%)	2.30	2.82	1.66
$THDV_b$ (%)	1.00	1.01	0.60
$THDV_c$ (%)	1.59	2.19	1.28
$THDV_{Mean}$ (%)	1.63	2.00	1.18
VUF (%)	2.08	0.65	2.34
V^+ (p.u.)	1.03	1.05	1.06
DPF (%)	95.93	99.99	95.81
PPL (%)	84.70	99.50	89.00

DPF, Displacement power factor; *PPL*, permissible penetration level; *THDV*, total voltage harmonic distortion; *VUF*, voltage unbalance factor.

14.5 Conclusion

This paper at first presented an algorithm to find the *PPL* of the FSWECS under unbalanced nonsinusoidal voltage and current conditions. Second, a compensator design, which consists of SC and STF, was suggested for maximization of *PPL* and *DPF*, and minimization of the $THDV_{Mean}$ and *VUF*. The *VUF* and harmonic distortion limits, stated in the standards and the desired *DPF* and rms bus voltage ranges, were regarded as constraints of the optimal design problem. The PSO algorithm was employed to find the optimal parameters of the proposed compensator.

The simulation results were presented for the system without and with three different compensators as the proposed optimal compensator (SC + STF), optimal SC, and optimal STF designs. It should be noted that optimal SC design was attained by considering the proposed compensator's problem formulation except the objectives and constraints related to voltage harmonic distortion. In addition, optimal STF design was provided regarding the proposed compensator's problem formulation except the objectives and constraints related to voltage unbalance.

It can clearly be mentioned from the analysis that the proposed one achieved the highest *PPL*, and it provided better collective mitigation of $THDV_{Mean}$, *DPF*, and *VUF* when compared to the other two compensators.

Finally, performance of the proposed optimal compensator was tested under varying utility voltage and load-side conditions. The results showed that the proposed compensator provides $THDV_{Mean}$, *PPL*, and *DPF* enhancement under low loading levels and ideal sinusoidal utility voltages even if it was optimally designed for the rated loading level and considerably distorted and unbalanced utility voltage conditions. In addition to that, it avoided *VUF* increment for low loading level or ideal utility voltage conditions.

References

[1] S.H.E. Abdel Aleem, A.F. Zobaa, H.M. Abdel Mageed, Assessment of energy credits for the enhancement of the Egyptian Green Pyramid Rating System, Energy Policy 87 (2015) 407–416.

[2] S. Sakar, M.E. Balci, S.H.E. Abdel Aleem, A.F. Zobaa, Integration of large-scale PV plants in non-sinusoidal environments: considerations on hosting capacity and harmonic distortion limits, Renew. Sustain. Energy Rev. 82 (2018) 176–186.

[3] F.H. Guan, D.M. Zhao, X. Zhang, B.T. Shan, Z. Liu, Research on distributed generation technologies and its impacts on power system, in: SUPERGEN'09, Nanjing, China, April 2009, pp. 1–6.

[4] S.M. Ismael, S.H.E. Abdel Aleem, A.Y. Abdelaziz, A.F. Zobaa, State-of-the-art of hosting capacity in modern power systems with distributed generation, Renew. Energy 130 (2019) 1002–1020.

[5] P.S. Georgilakis, N.D. Hatziargyriou, A review of power distribution planning in the modern power systems era: models, methods and future research, Electr. Power Syst. Res. 121 (2015) 89–100.

[6] Conti S., Raiti S., Tina G., Vagliasindi U. Distributed Generation in LV distribution networks: voltage and thermal constraints, in: 2003 IEEE Bologna PowerTech – Conf., Bologna, Italy, June 2003, pp. 413–418.

[7] F.D. Kanellos, N.D. Hatziargyriou, The effect of variable-speed wind turbines on the operation of weak distribution networks, IEEE Trans. Energy Convers. 17 (2002) 543–548.

[8] A. Chidurala, T.K. Saha, N. Mithulananthan, Harmonic impact of high penetration photovoltaic system on unbalanced distribution networks – learning from an urban photovoltaic network, IET Renew. Power Gen. 10 (2016) 485–494.

[9] R. Senra, W.C. Boaventura, E.M.A.M. Mendes, Assessment of the harmonic currents generated by single-phase nonlinear loads, Electr. Power Syst. Res. 147 (2017) 272–279.

[10] I.N. Santos, V. Ćuk, P.M. Almeida, M.H.J. Bollen, P.F. Ribeiro, Considerations on hosting capacity for harmonic distortions on transmission and distribution systems, Electr. Power Syst. Res. 119 (2015) 199–206.

[11] C.A.N. Pereira, J.A.P. Lopes, M.A.C.C. Matos, Assessment of the distributed generation hosting capacity incorporating harmonic distortion limits, in: 2018 Int. Conf. on Smart Energy Systems and Technologies (SEST), Sevilla, 2018, pp. 1–6.

[12] S. Sakar, M.E. Balci, S.H.E. Abdel Aleem, A.F. Zobaa, Increasing PV hosting capacity in distorted distribution systems using passive harmonic filtering, Electr. Power Syst. Res. 148 (2017) 74–86.

[13] M.M. Elkholy, M.A. El-Hameed, A.A. El-Fergany, Harmonic analysis of hybrid renewable microgrids comprising optimal design of passive filters and uncertainties, Electr. Power Syst. Res. 163 (2018) 491–501.

[14] M.S. Kurt, M.E. Balci, S.H.E. Abdel Aleem, Algorithm for estimating derating of induction motors supplied with under/over unbalanced voltages using response surface methodology, IET J. Eng. 6 (2017) 627–633.

[15] F. Shahnia, R. Majumder, A. Ghosh, G. Ledwich, F. Zare, Voltage imbalance analysis in residential low voltage distribution networks with rooftop PVs, Electr. Power Syst. Res. 81 (2011) 1805–1814.

[16] A. Rodriguez-Calvo, R. Cossent, P. Frías, Integration of PV and EVs in unbalanced residential LV networks and implications for the smart grid and advanced metering infrastructure deployment, Int. J. Elec. Power 91 (2017) 121–134.

[17] F. Shahnia, A. Ghosh, G. Ledwich, F. Zare, Voltage unbalance improvement in low voltage residential feeders with rooftop PVs using custom power devices, Int. J. Elec. Power 55 (2014) 362–377.

[18] E. De Jaeger, A. Dubois, B. Martin, Hosting capacity of LV distribution grids for small distributed generation units, referring to voltage level and unbalance, in: 22nd Int. Conf. on Electricity Distribution (CIRED), Stockholm, 2013, pp. 1–4.

[19] D. Schwanz, S.K. Rönnberg, M. Bollen, Hosting capacity for photovoltaic inverters considering voltage unbalance, in: 2017 IEEE Manchester Power Tech, Manchester, 2017, pp. 1–6.

[20] O. Lennerhag, G. Pinares, M.H.J. Bollen, G. Foskolos, T. Gafurov, Performance indicators for quantifying the ability of the grid to host renewable electricity production, CIRED - Open Access Proc. J. (2017) 792–795.

[21] A. Koksoy, O. Ozturk, M.E. Balci, M.H. Hocaoglu, A new wind turbine generating system model for balanced and unbalanced distribution systems load flow analysis, Appl. Sci. 8 (2018) 502–520.

[22] P. Donolo, G. Bossio, C. De Angelo, G. Garcia, M. Donolo, Voltage unbalance and harmonic distortion effects on induction motor power, torque and vibrations, Electr. Power Syst. Res. 140 (2016) 866–873.

[23] E.B. Agamloh, S. Peele, J. Grappe, An experimental evaluation of the effect of voltage distortion on the performance of induction motors, in: 2012 Annual IEEE Pulp and Paper Industry Technical Conf. (PPIC), Portland, OR, 2012, pp. 1–7.

[24] J.P.G. De Abreu, A.E. Emanuel, Induction motor thermal aging caused by voltage distortion and imbalance: loss of useful life and its estimated cost, in: 2001 IEEE Industrial and Commercial Power Systems Technical Conf., New Orleans, LA, USA, 2001, pp. 105–114.

[25] C.Y. Lee, W.J. Lee, Effects of nonsinusoidal voltage on the operation performance of a three-phase induction motor, IEEE Trans. Energy Convers. 14 (1999) 193–201.

[26] A.H. Ghorashi, S.S. Murthy, B.P. Singh, B. Singh, Analysis of wind driven grid connected induction generators under unbalanced grid conditions, IEEE Trans. Energy Convers. 9 (1994) 217–223.

[27] R.C. Bansal, T.S. Bhatti, D.P. Kothari, Bibliography on the application of induction generators in nonconventional energy systems, IEEE Trans. Energy Convers. 18 (2003) 433–439.

[28] S.K. Khadem, M. Basu, Power quality in grid connected renewable energy systems: role of custom power devices, in: ICREPQ'10, Granada, Spain, 2010, pp. 1–6.

[29] A.D. Falehi, M. Rafiee, Maximum efficiency of wind energy using novel Dynamic Voltage Restorer for DFIG based Wind Turbine, Energy Rep. 4 (2018) 308–322.

[30] B. Singh, S. Singh, GA-based optimization for integration of DGs, STATCOM and PHEVs in distribution systems, Energy Rep. 5 (2019) 84–103.

[31] W. Qingzhu, W. Mingli, C. Jianye, Z. Guiping, Model for optimal balancing single-phase traction load based on Steinmetz's method, in: 2010 IEEE Energy Conversion Congress and Exposition, Atlanta, GA, 2010, pp. 1565–1569.

[32] C. Arendse, G. Atkinson-Hope, Design of a Steinmetz symmetrizer and application in unbalanced network, in: UPEC2010, Cardiff, Wales, 2010, pp. 1–6.

[33] O. Jordi, L. Sainz, M. Chindris, Steinmetz system design under unbalanced conditions, Eur. Trans. Electr. Power 12 (2002) 283–290.

[34] A. Karadeniz, M.E. Balci, Comparative evaluation of common passive filter types regarding maximization of transformer's loading capability under non-sinusoidal conditions, Electr. Power Syst. Res. 158 (2018) 324–334.

[35] H.H. Zeineldin, A.F. Zobaa, Particle swarm optimization of passive filters for industrial plants in distribution networks, Electr. Power Compon. Syst. 39 (2011) 1795–1808.

[36] M.R. Al Rashidi, M.E. El-Hawary, A survey of particle swarm optimization applications in electric power systems, IEEE Trans. Evol. Comput. 13 (2009) 913–918.

[37] T.J.E. Miller, Reactive Power Control in Electric Systems, John Wiley & Sons, New York, USA, 1982.

[38] I.A. Sirotin, Fryze's compensator and Fortescue transformation, Przegląd Elektrotechniczny 87 (2011) 101–106.

[39] G. Chicco, M. Chindris, A. Cziker, P. Postolache, C. Toader, Analysis of the Steinmetz compensation circuit with distorted waveforms through symmetrical component-based indicators, in: IEEE Bucharest Power Tech Conf., 2009, pp. 1–6.

[40] L. Sainz, L. Monjo, S. Riera, J. Pedra, Study of the Steinmetz circuit influence on AC traction system resonance, IEEE Trans. Power Deliv. 27 (2012) 2295–2303.

[41] S. Shah, A. Rashid, M.K.L. Bhatti, Direct quadrate (d–q) modeling of 3-phase induction motor using Matlab/Simulink, Can. J. Elect. Comput. E 3 (2012) 237–243.

[42] A. Jalilian, R. Roshanfekr, Analysis of three-phase induction motor performance under different voltage unbalance conditions using simulation and experimental results, Electr. Power Compon. Syst. 37 (2009) 300–319.

[43] I. Temiz, C. Akuner, H. Calik, Analysis of balanced three-phase induction motor performance under unbalanced supply using simulation and experimental results, Electron. Electr. Eng. – Kaunas: Technologija, 3, 2011, pp. 41–45.

[44] A. Raina, A. Khosla, Effect of harmonics on performance characteristics of three phase induction motor, Int. J. Sci. Tech. Adv. 2 (2016) 221–226.

[45] C. Liu, F. Blaabjerg, W. Chen, D. Xu, Stator current harmonic control with resonant controller for doubly fed induction generator, IEEE Trans. Power Electron. 27 (2012) 3207–3220.

[46] M.I. Martinez, A. Susperregui, G. Tapia, L. Xu, Sliding-mode control of a wind turbine-driven double-fed induction generator under non-ideal grid voltages, IET Renew. Power Gen. 7 (2013) 370–379.

[47] S.H.E.A. Aleem, M.T. Elmathana, A.F. Zobaa, Different design approaches of shunt passive harmonic filters based on IEEE Std. 519-1992 and IEEE Std. 18-2002, Recent Patents Electr. Electron. Eng. 6 (2013) 68–75.

[48] IEEE 519, IEEE Recommended Practices and Requirements for Harmonic Control in Electrical Power Systems, 2014.

[49] IEC 61000-2-4, Electromagnetic Compatibility (EMC) – Part 2-4: Environment – Compatibility Levels in Industrial Plants for Low-Frequency Conducted Disturbances, 2002.

Chapter 15

Impact of demand-side management system in autonomous DC microgrid

Rajeev Kumar Chauhan[1] and Kalpana Chauhan[2]

[1]*Department of Electronic and Instrumentation Engineering, Galgotias College of Engineering and Technology, Greater Noida, India,* [2]*Department of Electrical and Electronics Engineering, Galgotias College of Engineering and Technology, Greater Noida, India*

15.1 Introduction

According to the Central Electricity Authority report, India is running short of around 11% peak demand and more than 24% transmission and distribution losses, which is the cause of poor grid availability in many remote locations in India. This is the motivation for the residential building consumers to install their own captive power plant to improve the electricity reliability in the building. The rapid decrement in the photovoltaic (PV) system cost and government of India subsidized the rooftop PV plant, motivates the residential consumers to mount the PV plant on their rooftop. The electronic loads (DC appliances) have become the popular choice of the residential buildings due to high efficiency of the DC appliances [1]. The PV system is a naturally DC power source. Therefore the DC−AC) and AC−DC converters are needed to integrate the direct current (DC) appliances and PV system to the traditional alternating current (AC) power distribution system [2]. This increases the conversion losses and overall capital costs as well as decreases the efficiency and reliability of power distribution system [3]. In this chapter a capital cost−based comparative analysis of the AC and DC power distribution structures is conducted: (1) traditional AC system: AC distribution system with AC compatible appliances, (2) hybrid AC system: AC distribution system with DC appliances; (3) conventional DC distribution system: DC distribution system with DC compatible appliances (high efficient); and (4) DC distribution system with demand-side management (DSM) scheme: DC distribution system with DC compatible appliances and load is managed as per the DSM scheme.

Decision Making Applications in Modern Power Systems. DOI: https://doi.org/10.1016/B978-0-12-816445-7.00015-3

In this chapter a DSM scheme has been proposed for the autonomous DC microgrid for the building. The battery bank (BB) responds to the changes in a power imbalance between PV power and demand within an autonomous DC microgrid. The control objective of the DSM scheme is to use the PV energy more efficiently. The DSM scheme shifts the deferrable load from nonsunny to sunny hours and decreases the building demand during nonsunny hours. In this way, it decreases the charging/discharging cycles (i.e., use) of the batteries. This is reducing the power losses in the battery and improves the system efficiency. The DSM scheme reduces the size of the PV plant, energy storage (ES), and capital cost of the system.

The DSM is the form of programs that are implemented by the utility companies to control the energy consumption at the customer side [4]. The main tasks of the DSM are to change the use of load quantitatively by implementing and planning for the utility or monitoring the consumer activities of load utilization. Load management and consumer side local generation are other tasks of the DSMs [5]. The DSM approaches are employed to use the available energy efficiently without installing new generation, transmission infrastructure [6]. The decentralized DSM mechanism to defragment of home loads is based on grid prices and utility companies offer the incentives to use the devices optimally which is in switch to reduce the heating [7]. The heuristic-based evolutionary algorithm is used for day-ahead load shifting to reduce the peak demand and reshape demand curve [8]. The DSM promotes distributed generation in order to avoid long-distance transport. The DSM facilitates the consumption of locally generated energy immediately whenever it is available for local loads [9]. The main benefit of a demand-side management system is that its less expensive nature to intelligently influence a load and the ability to build a new power plant or install some electrical storage device. [10].

The most obvious features of DSM scheme are load scheduling of the deferrable load from nonsunny to sunny hours to direct utilization of PV power and approaching the desired state of charge (SOC) by controlling the "cycle-based load" and other controllable load that makes it more advance. The prototype hardware has been set up for the home including PV plant and BB. The microcontroller is connected to the current and voltage sensors to achieve the PV, battery power, and home demand. In addition, it is also connected to the electronic switches (i.e., relays) to automatically switch "ON"/"OFF" the appliance and loads of the building. The controller schedules the deferrable load (e.g., washing machine, pump), changes operation patterns of the cycle-based appliances (i.e., refrigerator), and switches "ON"/"OFF" the controllable load (i.e., LED bulb) as per PV generation. The DSM reduces the charging/discharging cycles of the batteries to improve the system efficiency by reducing the power losses in the batteries. It also decreases the peak demand that is the main cause of the large size of PV plant and an ES.

So the small size of PV plant and ES is required with the DSM scheme. In other words, it reduces the capital cost of the system.

15.2 Analysis of AC microgrid and DC microgrid

The converter stage, energy demand, and size of power sources (battery and PV plant) are considered in this study for comparative analysis of the AC microgrid and DC microgrid.

15.2.1 Converter stages

The DC loads are increasing in the residential as well as commercial building due to the high efficiency of the DC load as compared to the AC compatible load. It forced to interconnect the AC–DC converters in the case of the traditional distribution system (AC system) to supply the DC loads [11,12]. These conversion stages at the load end are one of the causes to decrease the system efficiency and increase the system cost. The details of the conversion stages with AC system in an autonomous DC microgrid for residential building can be found in Table 15.1.

TABLE 15.1 Source and appliances internal converters and their efficiency and cost.

Source	Traditional AC system					
	Power	**Type of converters**	**Efficiency (%) [13]**	**Number**	**Price (Rupees)**	**Total cost (Rupees)**
PV plant	DC	DC–AC	89	1	7000	7000
Battery	DC	AC–DC–AC	83	1	7000	7000
Loads						
LED bulbs	DC	AC–DC	84	5	50	250
Tube lights	DC	AC–DC	84	5	50	250
Fans	DC	AC–DC	85	3	150	450
Pump	DC	AC–DC	85	1	450	450
Washing machine	DC	AC–DC	85	1	450	450
Refrigerator	DC	AC–DC	85	1	450	450
Total converters cost (Rupees)						16,300

AC, Alternating current; *DC*, Direct current; *LED*, Light Emitting Diode

15.2.2 Energy demand

The energy demand (E_d) of the DC microgrid consists of the energy consumption in the appliances, total energy losses ($E_{\text{loss},t}$).

$$E_d = \sum_{t=1,j=1}^{T,n_a} P_{aj}(t)\tau_h + E_{\text{loss},t} \tag{15.1}$$

where P_{aj} is the power consumption in jth appliance, n_a is the number of appliances, T is the scheduling horizon of the optimization, and τ_h is the optimum time step 1 minute.

The total energy loss consists of the energy losses in the converters ($E_{\text{loss},c}$), line, and energy loss ($E_{\text{loss},b}$) in the battery during charging/discharging due to chemical reactions in it. As the system is prototyped and the cable length is very small, the line losses are considered negligible. The total energy losses ($E_{\text{loss,ac}}$) in the AC system can be expressed as

$$E_{\text{loss},t} = \sum_{t=1,j=1}^{T,n_{ca}} P_{\text{loss},cj}(t)\tau_h + P_{\text{loss},b}(t)\tau_h \tag{15.2}$$

where $n_c = (n_a + n_s)$ is the number of converters in the autonomous microgrid with AC distribution system, and n_s is the number of power sources.

While total energy demand with the DC system includes the energy consumption in the appliances and energy losses in the battery, the energy consumption can be expressed as

$$E_d = \sum_{t=1,j=1}^{T,n_a} P_{aj}(t)\tau_h + E_{\text{loss},b} \tag{15.3}$$

where $E_{\text{loss},b}$ is the energy loss in the batteries.

15.2.3 Estimation of photovoltaic and battery size

The PV plant is the only power generator in this autonomous DC microgrid. Moreover, the ES (i.e., BB) is used to store the surplus PV plant energy as well as to supply the surplus energy demand of the building. Therefore estimation of their size is required to get the uninterrupted power supply. The power rating of the PV plant and BBs can be calculated as follows:

PV size: The PV Watt-hours needed per day are the combination of the total power consumption in the appliances and power losses in the system. The size of the PV plant depends on the Watt-hours needed from the PV plant and a number of average sun hours per day in the whole year [14] and can be expressed as

$$\text{PV module size} = \frac{\text{PV Watt-hours needed per day}}{\text{Average sun hours per day in year}} \tag{15.4}$$

Battery capacity: The battery size should be large enough to store the sufficient energy to supply the load in the nonsunny hours and cloudy days.

$$\text{Battery capacity } (A\ h) = \frac{k_a}{\eta_b \times \text{DOD}_b \times V_{nl}} \times n_{au} \tag{15.5}$$

where k_a is the Watt-hours needed for the appliances per day, η_b is the battery efficiency (0.85), DOD_b is the depth of discharge (0.5), V_{nl} is the nominal voltage of the battery, n_{au} is the number of autonomy (2 days), and average winter temperature is 80°F.

The size of the BB, PV plant, and the system cost for an autonomous DC microgrid with the consideration of the different distribution systems can be found in Table 15.2. In the case of AC distribution system including AC appliances (ACDSA), the AC technology–based appliances are used, which are less efficient as compared to DC appliances. The DC–AC and AC–DC–AC converters are used to interconnect the PV plant and BB with the AC bus, which is the cause of poor efficiency of the system. Therefore building energy demand with ACDSA is 4.2 kW h, which is the highest one as compared to the other four cases and 1.5 times higher than the direct current distribution system including direct current appliances (DCDSA).

TABLE 15.2 Description of PV plant and battery bank and their cost.

Source/Load	ACDSA	ACDSD	DCDSD	DCDSD–DSM
Building energy demand (kW h)	4.2	3.1	2.7	2.7
Battery AH needed per day	910	672	585	434
Number of battery (150 A h)	7	5	4	3
PV plant (kW)	1.590	1.175	1.023	0.7
Number of PV module (100 W)	16	12	11	7
PV cost (Rupees)	160,000	120,000	110,000	70,000
Battery cost (Rupees)	94,500	67,500	54,000	40,500
System cost (Rupees)	254,500	187,500	164,000	110,500

ACDSA, AC distribution system including AC appliances; *ACDSD*, AC distribution system including DC appliances; *DCDSD*, DC distribution system including DC appliances; *DCDSD–DSM*, DC distribution system including DC appliances with demand-side management; *PV*, photovoltaic.

Therefore it required large size of PV plant and BB to provide the uninterrupted power supply, which is the cause of highest capital cost.

In the case of DC distribution system including DC appliances (DCDSD) the high-efficient appliances are used as well as the conversion staged at the source as well as load end has been removed [15,16]. Therefore the energy demand for DCDSD is smaller as compared to ACDSA and AC distribution system including DC appliances. The small size of PV plant and the BB is sufficient with DCDSD. In the case of DCDSD–DSM the DSM scheme shifts the operating time of the deferrable appliances that help to reduce the charging and discharging of the BB and save the energy loss in the BB. In this way, there is a further improvement in the system efficiency. Therefore a small amount of AH and PV Watt-hours per day is needed, which is the cause of less numbness (i.e., size of ES) of the battery and PV modules to supply the uninterrupted power supply to the building. The capital cost of the DCDSD–DSM system is the smallest in comparison to other three cases because it depends on the size of PV plant and BB.

15.3 State of charge of battery bank

In general the SOC of the battery is the ratio of its present capacity ($\mu(t)$) to the nominal capacity (μ_n) of the battery [17]. The SOC can be defined as

$$\psi(t) = \frac{\mu(t)}{\mu_n} \tag{15.6}$$

The Coulomb counting method measures the charging/discharging current of a battery and integrates it over time in order to estimate SOC [16,17]. The SOC can be calculated as

$$\psi(t) = \psi(t-1) + \frac{i(t)}{\mu_n}\Delta t \tag{15.7}$$

where $\psi(t)$ and $\psi(t-1)$ are the SOC of the battery at time instant t and $t-1$, $i(t)$ is the battery current at time instant t, and Δt is the time interval (considered 1 minute).

15.4 Autonomous DC microgrid

The conceptual diagram, hardware of the experimental setup, and control and monitoring unit of the DC microgrid for the building are discussed in this section.

15.4.1 Conceptual diagram of DC microgrid

The conceptual layout of the prototype autonomous DC microgrid of the residential building is shown in Fig. 15.1. Two solar charge controllers (SCCs)

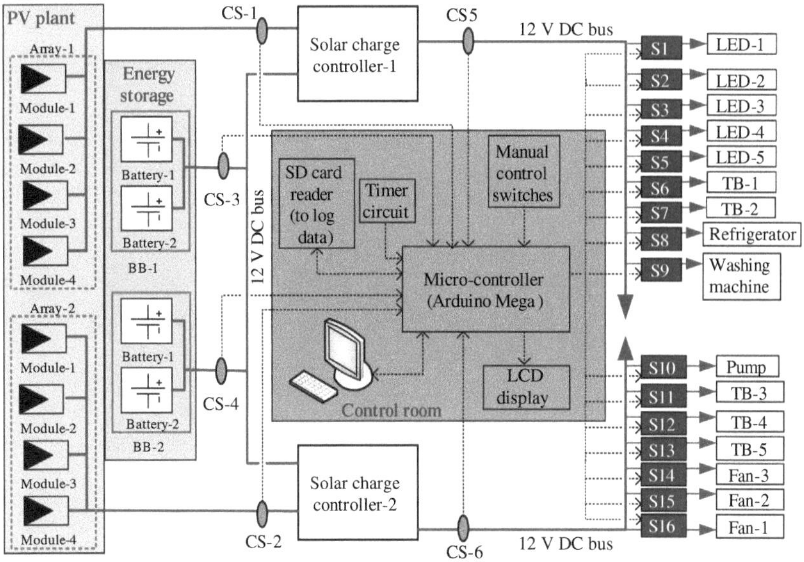

FIGURE 15.1 Conceptual diagram of autonomous DC microgrid.

are connected in parallel to supply the load. In addition, both the SCCs connect to the ES that consists of two BBs, BB-1 and BB-2, in parallel to store the surplus energy or balance the surplus demand. These types of connections for the batteries increase the reliability of the DC microgrid. For example, the PV plant of the SCC-1 is an outage, then the SCC-1 connects their load to the ES and SCC-2 via 12 V DC bus between both the SCCs to feed the surplus energy to the ES. The voltage and SOC of the batteries are monitored in real time, which gives the energy stored in the batteries. It helps to find out the battery back-up period to supply the future demand. The current sensors, CS-1 and CS-2, are mounted in the service mains of array-1 and array-2 of the PV plant to monitor the injected currents of array-1 and array-2 to the controllers, while the current sensors, CS-3 and CS-4, mounted on the battery terminals of the SCC-1 and SCC-2 to monitor the current sharing of BB-1 and BB-2 with SCC-1 and SCC-2 during surplus load and generation, respectively. The loads on both the controllers are supplied separately and monitored by the current sensors, CS-5 and CS-6, respectively. All the sensors are connected to the microcontroller (Arduino Mega) and retrieve the values at every 1 minute. The electronic switches, S_1-S_{16}, are used to electronically switch "ON"/"OFF" the building load.

15.4.2 Hardware setup of DC microgrid

The hardware setup of the prototype autonomous DC microgrid consists of building load, ES, including PV plant as shown in Fig. 15.2A and B. The

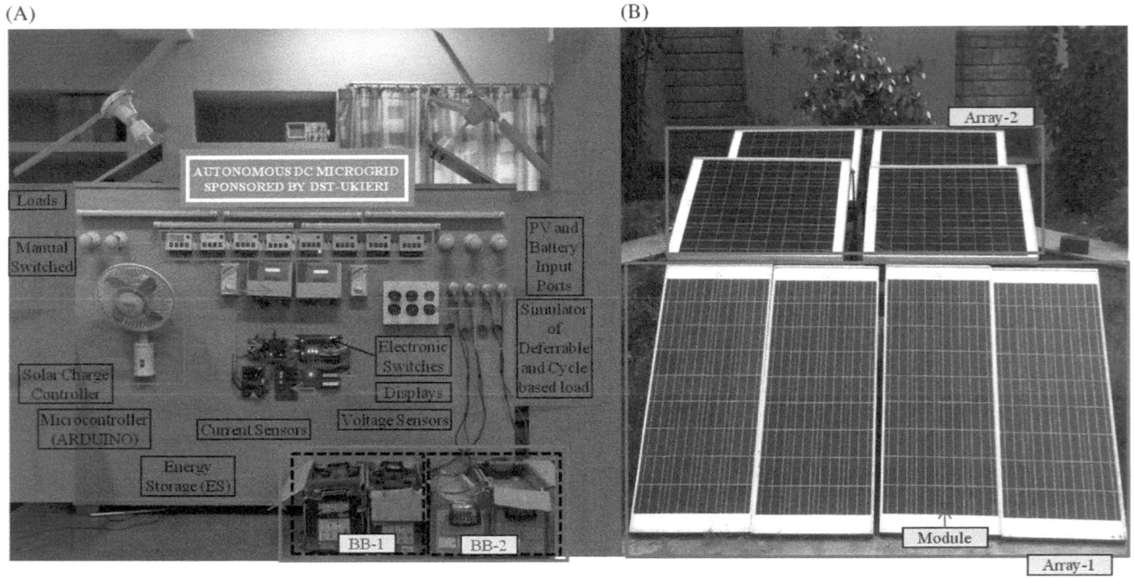

FIGURE 15.2 Experiential setup of DC microgrid (A) controller, load, and battery bank, (B) photovoltaic plant.

building load includes DC appliances, such as the brushless DC fans, DC tube lights, and LED bulbs. Moreover, there are five DC bulbs used to make the simulator of deferrable and cycle-based load. The simulators of the washing machine and the water pump are in the parallel combination of two DC bulbs of 40 W, while one DC bulb of 60 W is used to design the simulator of the refrigerator. The specifications of the microgrid components, such as appliances, PV plant, ES, and SCCs, can be found in Table 15.3. The considered location of the appliances can be found in Table 15.4. The manual switches are used to manually switch "ON"/"OFF" the appliance. Two speed regulators are used to control the speed of the ceiling fans. The building load is supplied by the PV plant and the BB. The ES consists of two BBs, BB-1 and BB-2, while the BB is the parallel combination of two batteries. The PV plant consists of two arrays, and each array is the parallel combination of four modules of 100 W each.

15.4.3 Control and monitoring unit of DC microgrid

The layout of the control and monitoring unit of the DC microgrid is shown in Fig. 15.3. It consists of five boards. The board-1 consists of a microcontroller (Arduino mega), real-time clock, SD card reader, and power supply,

TABLE 15.3 Parameters of microgrid components.

Components	Specifications	Quantity	Total power
PV	I_{sc} = 5.95 A, V_{oc} = 22.95 V, Power = 100 W	8	800 W
Battery	12 V, 150 A h, C20	4	600 A h
Solar charge controller	12 V, 40 A	2	
Connecting load			
Fan	12 V, 30 W	3	90 W
LED bulb	12 V, 5 W	5	25 W
Tube light	12 V, 8 W	5	40 W
Refrigerator	12 V, 60 W	1	60 W
Washing machine	12 V, 80 W	1	80 W
Water pump	12 V, 80 W	1	80 W

PV, Photovoltaic.

TABLE 15.4 Location of appliances in the building [18].

Location	Controllable load				Deferrable load	
	Fan	Tube light	LED bulb	Refrigerator	Washing machine	Water pump
Kitchen	–	√	√	√	–	–
Bed room-1	√	√	√	–	–	–
Bed room-2	√	√	√	–	–	–
Living room	√	√	√	–	–	–
Bathroom	–	–	√	–	√	√
Balcony	–	√	–	–	–	–

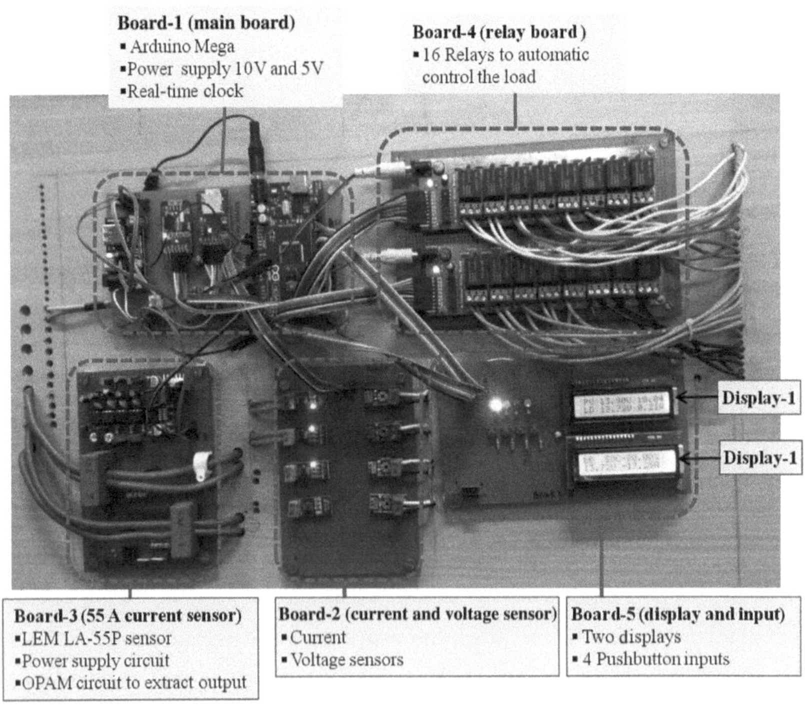

FIGURE 15.3 Control and monitoring unit for DC microgrid.

while the green push button is used to switch "ON/OFF" the supply of the main board. The microcontroller executes the control algorithm and creates the digital control signal in the form of "0" and "1" to actuate the relays to switch "ON"/"OFF" their respective appliances. The real-time clock gives an accurate measure of the current time in the circuit as the Arduino mega board itself does not have a time clock. The values extracted from this circuit are used to put a time stamp on the recorded data on the SD card. The current and voltage sensors are mounted on the board-2 and board-3 to measure the current for the load, PV plant, and ES and their bus voltage. The data logging algorithm automatically created a new file with a date stamp and stored the current and voltage for each minute in a single row of the text file delimited by a space. The 16 relays are mounted on the board-4, which automatically switch "ON/OFF" the appliances as per the control action generated by the Arduino Mega based on the control algorithm. The push buttons and LCD displays are mounted on the board-5. One of the push buttons is used to start and stop the data logging process, while another is used to convert the relay operations from automatic to manual mode. The display-1 is used to display the measured current and voltage of PV plant and load, while the measured SOC, current, and voltage of the battery (i.e., ES) are displayed on the display-2.

15.5 Demand-side management algorithm

The flowchart of the control algorithm for the DSM scheme is shown in Fig. 15.4. The PV plant feed their generated power to the DC microgrid irrespective of the building demand at different time instants. The smart sensors are mounted at service mains of the load, PV plant, and BB to monitoring the power of PV plant, load, and ES in real time. The control objective of the DSM scheme is to maximize the direct use of the PV power in the building and maintain the desired SOC of the ES to supply the future load of the building. Therefore the DSM scheme schedules the deferrable loads (i.e., washing machine and pump) during the regular sunny hours (i.e., when the PV generation is higher than the critical load of the building) without affecting the comfort of the consumer. In this way the DSM scheme maximizes the direct use of power generation of PV plant. This approach reduces the charging/discharging cycles of the battery (i.e., power losses in the BB) and improves the efficiency of the DC microgrid. The controllable loads such as DC tube lights are switched "OFF," and the refrigerator operates with the "control cycle" mode to reduce the building demand and increase the charging or decrease the discharging of the BB to achieve the desired SOC as soon as possible.

The desired SOC shows the minimum requirement of stored energy in the battery to supply the future load during the time interval in which the PV plant does not generate any power (i.e., nighttime) or produces a small

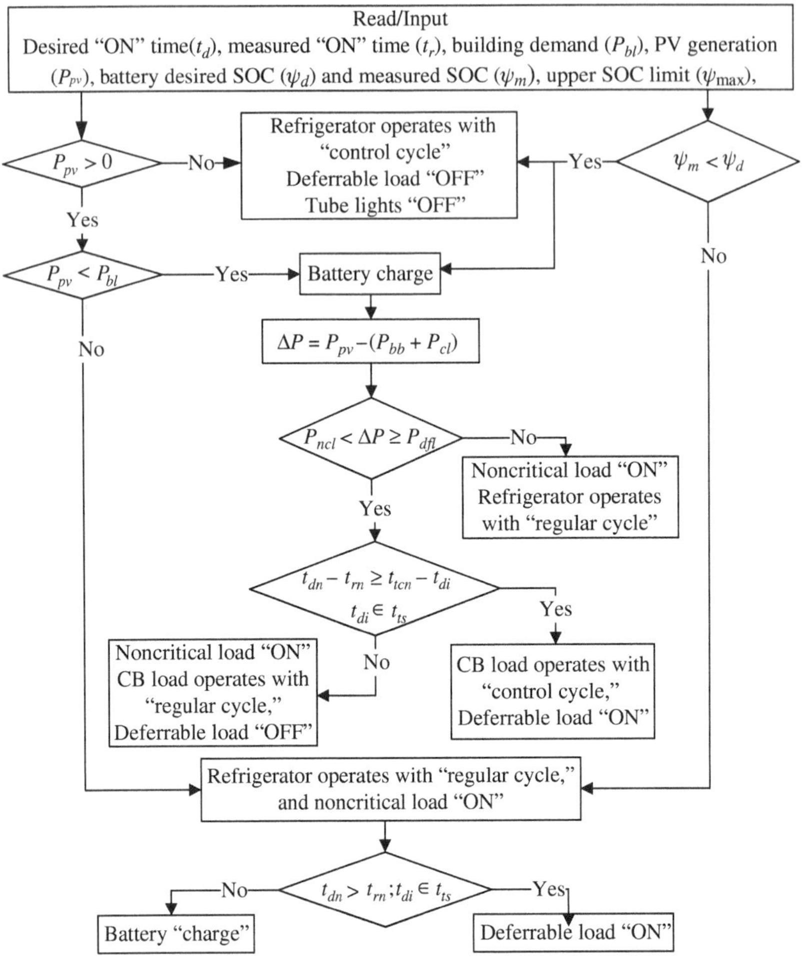

FIGURE 15.4 Flowchart of control algorithm for demand-side management.

amount of power as compared to the building demand. In this case the ES is responsible to balance the building demand. Therefore the SOC level of ES decreases at every time instant. As the battery SOC deficits from the desired level, the critical loads (i.e., LED bulbs and fans) remain switched "ON," while the refrigerator operates with "control cycle" mode. Besides that, other loads get switched "OFF." In this way the battery discharging rate decreases, which helps to achieve its SOC at desired level. Moreover, when the PV starts generation and the battery SOC is still less than its desired level, then the battery starts charging while the surplus PV power (i.e., difference of actual PV power and battery charging rate) feeds to the DC bus to supply the building load.

In case the surplus PV power is less than the building demand while the difference of desired time (t_d) to complete its task and appliance "ON" time (t_r) is greater than or equal to the difference of task completion time instant (t_{tc}) and decision time instant (t_{di}) then the deferrable load is switched "ON," while the refrigerator operates with "control cycle" and other noncritical loads get switched "OFF." The priority of the deferrable load is based on the desired "ON" time of the loads to complete the task. For example, at 10:00 hours time instant the PV plants produce approximately 90 W of the surplus power, which activates the controller to switch "ON" the deferrable load up to 90 W, while the status of the deferrable loads is as follows: the washing machine (70 W) should run for 90 minutes up to 13:00 hours to wash the cloth as per the consumer desires (instructions), and pump (80 W) must run for 30 minutes to 10:50 hours. The differences of the decision time instant and task termination time instants are 180 and 50 minutes for washing machine and pump, respectively. It verifies that the pump has more priority than the washing machine; therefore pump gets switched "ON," and washing machine remains switched "OFF."

When the PV power generation is greater than the building demand (i.e., demand of critical and noncritical loads, including deferrable loads) and the decision time instant belongs to the task slot of the deferrable load while its task is not completed, the deferrable load remains switched "ON," and the surplus PV power feeds to the battery. On the other hand, if the decision instant does not belong to the task slot of the deferrable load, then it remains switched "OFF."

15.6 Results and discussions

The actual hardware setup of an autonomous DC microgrid has been built to validate the feasibility of the DSM scheme for the building. The PV plant, ES, and DC appliances are connected to the unipolar DC bus–based distribution system. The two SCCs are connected in parallel to connect the array-1 and array-2 of the PV plant with the load and ES. Each SCC has three terminals to connect with the PV plant, ES, and load. The BB terminals of both the SCCs are connected to each other via 12 V DC bus. This type of connection between BB terminals of SCCs connects the batteries in parallel and provides the common point of connection to both the SCCs to connect with the ES. The beauty of these connections is that the ES can store the surplus power of both the SCCs and supply their surplus demand. The DC bus power can be expressed as

$$\sum_{j=1}^{n_a} P_{aj}(t) = \sum_{j=1}^{n_{pv}} P_{pvj}(t) \pm \sum_{j=1}^{n_b} P_{bbj}(t) \tag{15.8}$$

where P_{bbj} is the power of jth battery, n_b is the number of batteries, P_{pvj} is the power of jth PV array, n_{pv} is the number of PV arrays, P_{aj} is the power consumption in jth appliance, and n_a is the number of appliances.

15.6.1 Performance results of demand-side management scheme with sufficient photovoltaic power

In this case the PV energy generation is smaller than the building energy demand for particular day; therefore the proposed DSM scheme is load shedding noncritical load and operates the cycle-based load in a "control cycle mode" to achieve desired SOC of the battery storage unit. The building demand curve with the conventional scheme (manual load control), DSM scheme (automatic load control), and the PV power generation for a typical day is shown in Fig. 15.5. The peak demand of the building is 297 and 285 W with a conventional and DSM scheme, respectively.

The building peak demand with the conventional scheme is 297 W during 07:21–07:30 hours time interval (nonsunny period), and it is supplied by the battery, while the DSM scheme reduces the building demand up to 215 W in this time interval because the deferrable load water pump has been switched "OFF." In the sunny hours the building peak demand with DSM scheme is 285 W during 17:47–17:50 hours, while the building peak demand with conventional scheme is 200 W during 18:00–18:10 hours.

When the PV generation is not available or lower than the building demand, the surplus demand is supplied by the BB; therefore DSM scheme reduces the demand as compared to the conventional scheme. It verifies that the DSM scheme shifts the building demand from nonsunny to sunny hours and reduces the BB charging/discharging power (rate) as compared to the conventional scheme as shown in Fig. 15.6. This is the one cause of battery

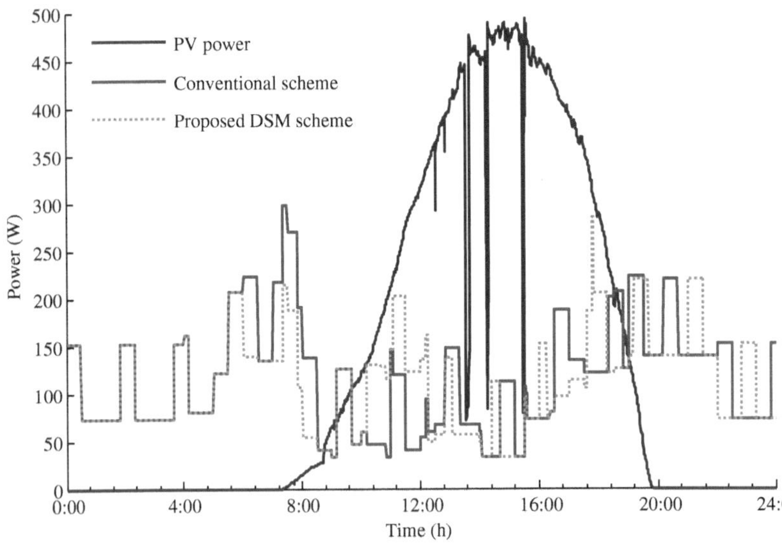

FIGURE 15.5 PV power and building demand curve with conventional and DSM scheme. *DSM*, Demand-side management; *PV*, photovoltaic.

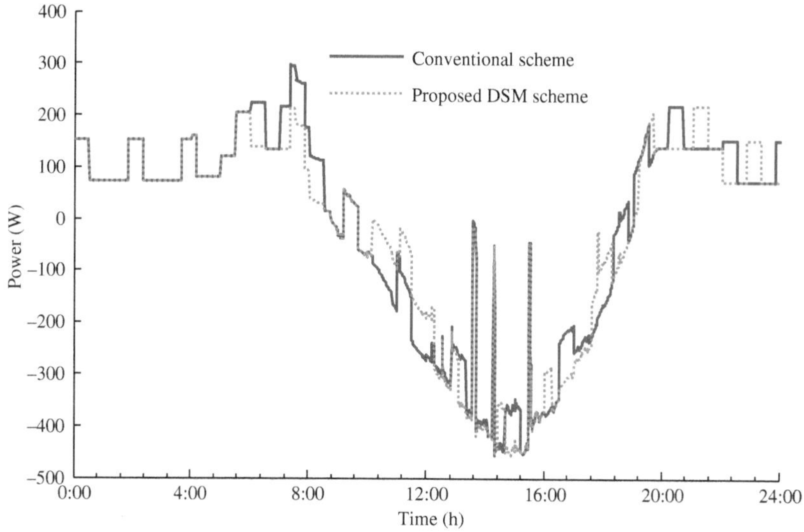

FIGURE 15.6 Battery power curve with conventional and DSM scheme. *DSM*, Demand-side management.

heating, poor health, and short life. It verifies that the DSM scheme keeps the battery cool, healthy, and increases their life as compared to the conventional scheme.

Mode I: Battery as source (regular operation): The SOC of the battery gives the information about the energy stored in the battery to supply future load. The desired SOC of the battery represents the minimum energy required at different time instants to supply the load. At the starting of the day at a time instant 00:00, the desired SOC level is 49.5% to supply the future load of the building while both conventional and DSM schemes have 50% SOC level as shown in Fig. 15.7. This is the reason the DSM scheme keeps the load pattern as per the conventional scheme. The battery supplies the same amount of load with conventional and DSM scheme during 00:00–05:59 hours time interval as shown in Fig. 15.6.

Mode II: PV plant and battery as source (deferrable load scheduling): During the early morning the PV generation is lower than the building demand and most of the load is supplied by the battery. For example, at time instant 06:00 hours, the pump is switched "ON" with the conventional scheme, which is the cause to achieve lower SOC than the desired SOC with the conventional scheme during 06:00–06:30 hours time interval as shown in Fig. 15.8C by the grey line, while the pump task completion time instant is 10:50 hours and operating time to fill the tank is 30 minutes. Therefore the pump has to run for 30 minutes to complete

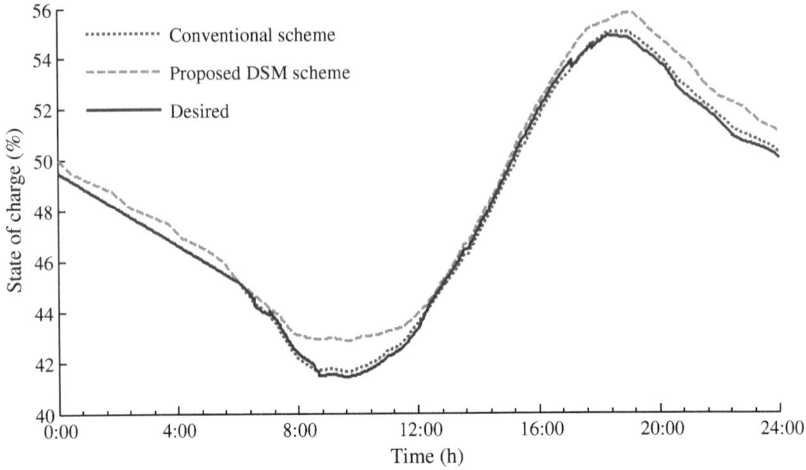

FIGURE 15.7 State of charge of the battery bank.

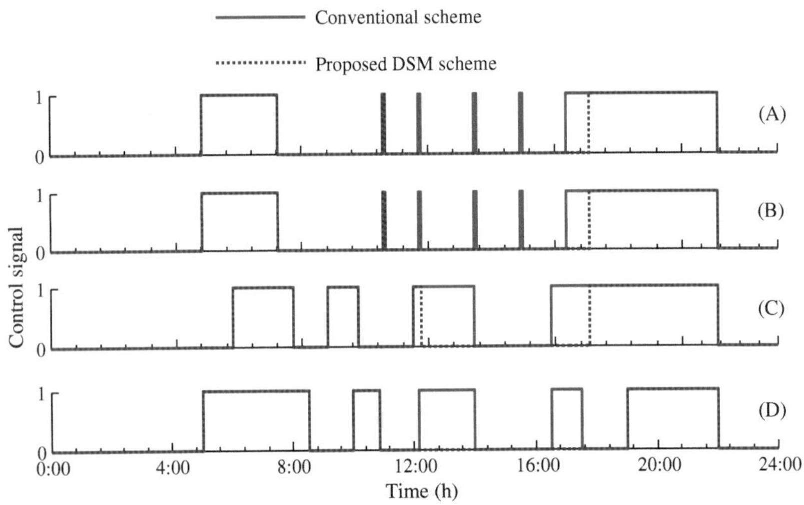

FIGURE 15.8 Control signal for controllable loads (LED bulb) mounted in building at (A) bedroom-1, (B) bedroom-2, (C) living room, and (D) kitchen.

their task before the 10:50 hours. The DSM scheme keeps critical and another controllable load (as per Fig. 15.7) connected as per conventional scheme and waits to switch "ON" the pump for the time interval in which the PV generation is higher than the combination of critical load and pump load. As the PV generation is higher than the critical load of the building at time instant 10:10 hours, the water pump is switched "ON" and remains "ON" to 10:40 hours (for 30 minutes), and the surplus PV power is stored in the battery as shown in Fig. 15.5.

Similarly, at time instant 07:00 hours the washing machine is switched "ON" with the conventional scheme and remains "ON" during 07:00–08:30 hours time interval as shown in Fig. 15.8B by a red line, which is the cause to achieve lower SOC than the desired SOC with a conventional scheme during 07:00–08:30 hours time interval as shown in Fig. 15.7, while the washing machine task completion time instant is 12:30 hours with 90 minutes operating time. The DSM scheme keeps the washing machine at the second priority and shifts the washing machine operating time from 07:00–08:30 hours to the future. As the PV generation at 10:41 hours becomes higher than the building demand (critical load and deferrable load washing machine), the washing machine is switched "ON." At time instant 10:59 hours the PV generation becomes lower than the building demand (critical load and deferrable load like washing machine) and remains lower up to 11:04 hours, so that the washing machine remains switched "OFF" during 10:59–11:04 hours time interval while the battery is charged by the surplus power to achieve the SOC higher than the desired. As the washing machine operation time is 90 minutes, and it runs only for 18 minutes, it has to be switched "ON" again to complete its task. As the PV generation becomes higher than the building demand at time instant 11:05 hours, the washing machine is switched "ON" and remains switched "ON" to 12:18 hours to complete the task as shown in Fig. 15.9B by a blue line.

Mode III: *PV with low SOC of BB* (*control of noncritical and cycle-based load*): During 12:16–17:40 hours time interval the PV generation is greater than the building demand but the battery SOC is lower than the desired with the conventional scheme. The DSM scheme reduces the building demand by switching "OFF" the controllable loads (LED bulbs mounted in bedroom-1, bedroom-2, living room, and kitchen as shown in

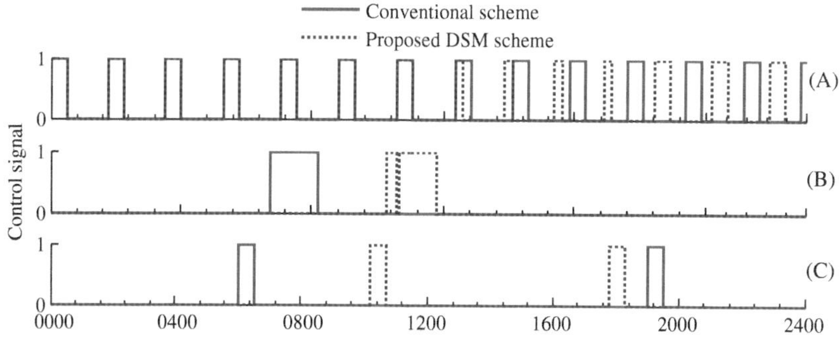

FIGURE 15.9 Control signal for cycle-based loads and deferrable loads (A) refrigerator, (B) washing machine, and (C) water pump.

Fig. 15.8A–D by dotted lines), and cycle-based load (refrigerator) operates with the control cycle mode (with a small switch "ON" time, i.e., 15 minutes "ON" time at the place of 30 minutes "ON" time) as shown in Fig. 15.9A by a dotted line.
Mode IV: *PV and SOC higher (prescheduling of deferrable load)*: During 17:41–18:10 hours time interval the PV generation is higher than the building demand (demand with a conventional scheme and future deferrable load, i.e., pump), and the battery SOC is higher than the desired. In addition, the task starts and completion time limit (evening time) for the pump is 17:30–19:30 hours, and the operating time is 30 minutes. In the DSM scheme the microcontroller switches "ON" the pump at the time instant 17:41 and keeps it "ON" during 17:41–18:10 hours as shown in Fig. 15.9C. The surplus PV generation is stored in the battery.

15.6.2 Performance results of demand-side management scheme with insufficient photovoltaic power

In this case the PV energy generation is sufficient enough to fulfill the building energy demand for a particular day. Therefore the proposed DSM scheme is load shedding noncritical load and operates the cycle-based load in a "control cycle mode" to achieve a desired SOC of the battery storage unit. The building power demand curve with the conventional scheme, DSM scheme, and the PV power generation for a typical day is shown in Fig. 15.10. The peak demand of the building is 297 and 220 W with conventional and DSM schemes, respectively. The battery charging and discharging power curve with conventional and proposed DSM scheme for this scenario is shown in Fig. 15.11. The peak discharging power of the battery is 327 and 243 W with conventional scheme and DSM scheme, respectively, during nonsunny hours. It shows that the small size of the battery is sufficient with the DSM scheme to supply the peak load of the building during nonsunny hours.

The state of the charge of the ES with the conventional scheme remained higher than the desired level, excluding the 06:00–12:20 hours as shown in Fig. 15.12. The DSM scheme achieves desired SOC of the ES unit by shifting the operating hours of the deferrable load to achieve the desired SOC. The operating duration of the washing machine has been shifted from 07:00–08:30 to 10:35–12:10 hours time interval as shown in Fig. 15.13B. The morning operating time of the water pump has been postponed from 08:00–08:30 to 12:10–12:40 hours time interval and evening operating time of the water pump is proponed from 19:05–19:35 to 17:45 to 18:15 hours and as shown in Fig. 15.13C. The refrigerator operates in the "regular control mode" same as conventional scheme as shown in Fig. 15.13A. The SOC of the ES unit is achieved by controlling the deferrable load of the building;

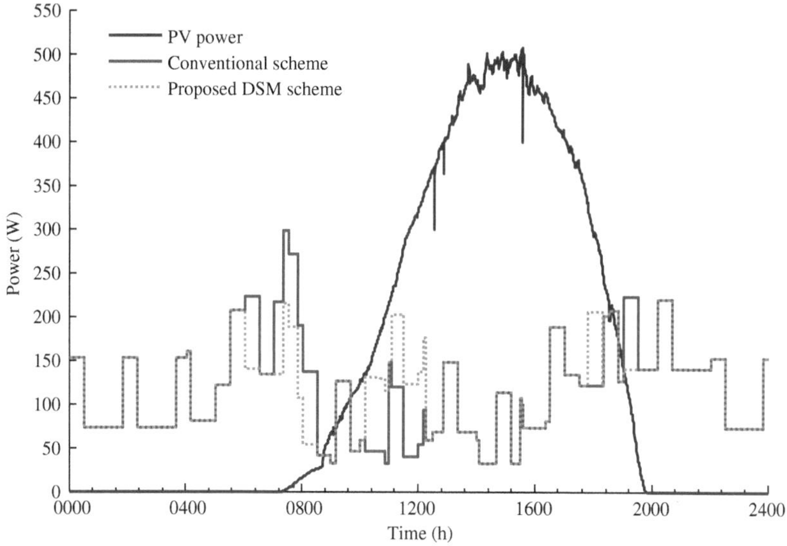

FIGURE 15.10 PV power and building demand curve with conventional and DSM scheme. *DSM*, Demand-side management; *PV*, photovoltaic.

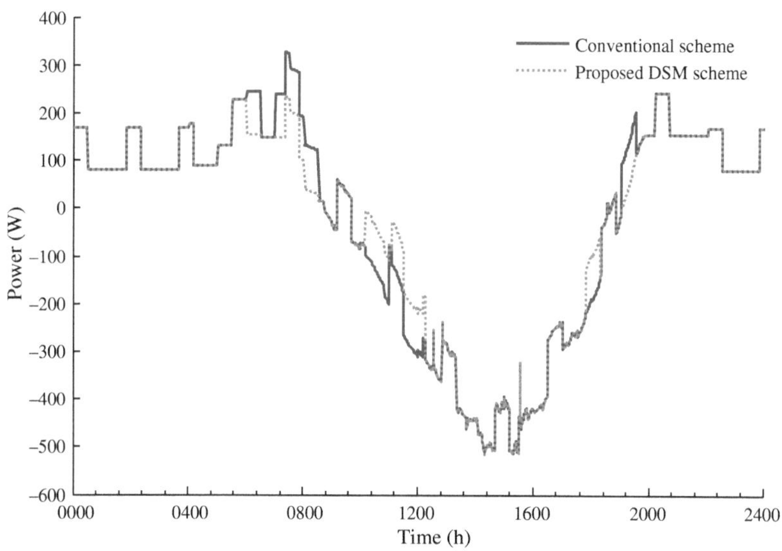

FIGURE 15.11 Battery power curve with conventional and DSM scheme. *DSM*, Demand-side management.

therefore the other controllable load, such as LED bulbs, mounted at the various locations in the building is switched "ON/OFF" same as per the conventional scheme (Fig. 15.14).

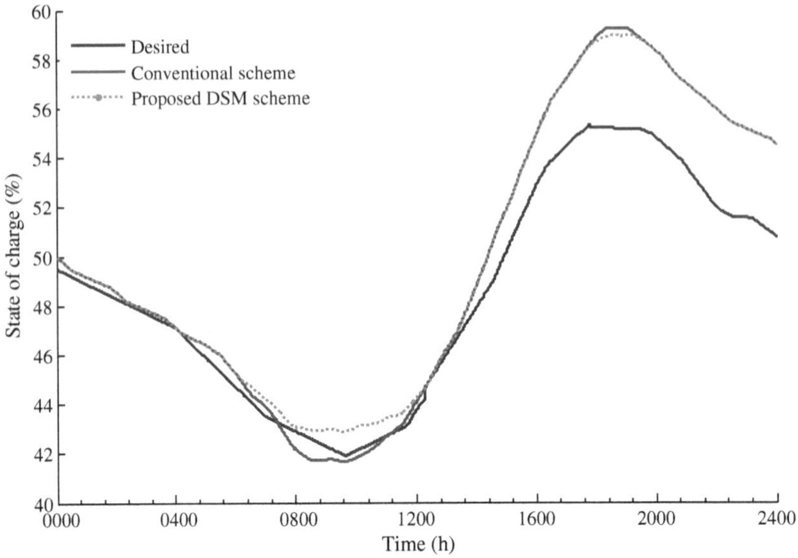

FIGURE 15.12 State of charge of the battery bank.

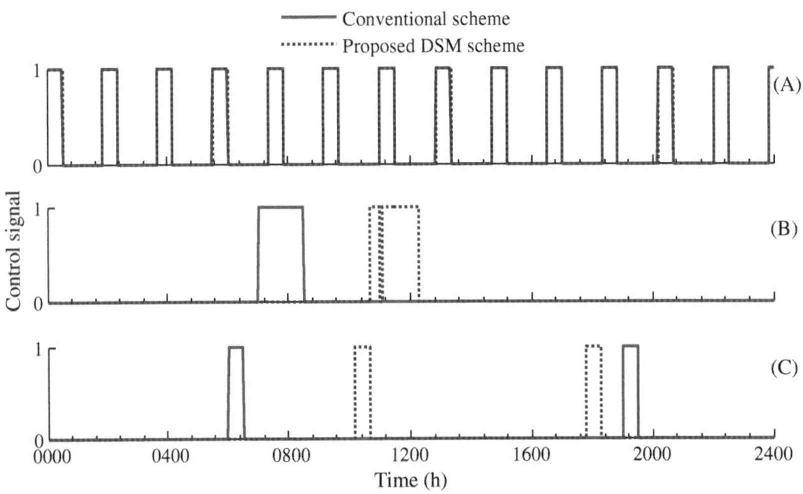

FIGURE 15.13 Control signal for cycle-based loads and deferrable loads (A) refrigerator, (B) washing machine, and (C) water pump.

15.7 Conclusion

The chapter presented a DSM scheme for the PV plant and battery-based residential building. The DSM scheme schedules the operation time of the deferrable loads during the regular sunny hours (i.e., when PV power

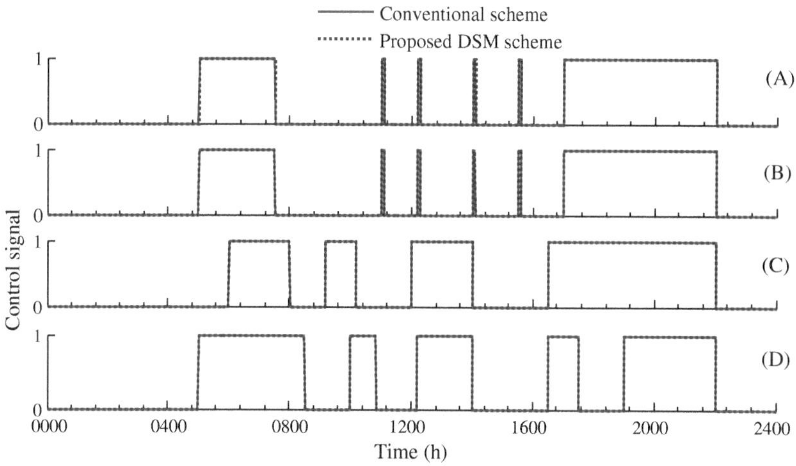

FIGURE 15.14 Control signal for controllable loads (LED bulb) mounted in building at (A) bedroom-1, (B) bedroom-2, (C) living room, and (D) kitchen.

generation is higher than the critical load of the building) to directly use the PV power without affecting the consumer comfort. Moreover, DSM scheme maintains the desired SOC of the ES to supply the future load of the building. This reduces the charging/discharging cycles of the battery as well as reducing the power loss in the battery during their charging/discharging time interval and improves the system efficiency. The Arduino creates the control signal based on DSM scheme and sends to the relays to automatically switch "ON/OFF" the building appliances. The performance of the DSM scheme has been tested in terms of efficient utilization of the PV energy by scheduling the deferrable load from nonsunny hour to the regular sunny hours. It also monitors the battery SOC and keeps it at the desired or higher level. When the SOC droops from the desired level, the controllable appliances (LED bulbs) switched "OFF" while the cycle-based load (refrigerator) had shifted from regular cycle mode to the control cycle mode. The peak demand of the building has also been reduced significantly during sunny and nonsunny hours. The energy exchange band of the battery and the capital cost of the system have been reduced in the DSM scheme significantly as compared to the conventional scheme.

References

[1] R.K. Chauhan, K. Chauhan, Performance analysis of power distribution systems with weakly grid connected rural homes in India, Energy Build. 172 (2018) 307–316.

[2] R.K. Chauhan, B.S. Rajpurohit, DC distribution system for energy efficient buildings, in: Proc. IEEE, 18th National Power System Conference, IIT Guwahati, India, 2014, pp. 1–6.

[3] R.K. Chauhan, B.S. Rajpurohit, N.M. Pindoriya, DC power distribution system for rural applications in: Proc. Eighth National Conference on Indian Energy Sector, AMA, Ahmedabad, 2012, pp. 108–112.

[4] A. Jossen, J. Garche, D.U. Sauer, Operation conditions of batteries in PV applications, Solar Energy 76 (6) (2004) 759–769.

[5] X. Hu, S. Li, H. Peng, F. Sun, Robustness analysis of state of charge estimation methods for two types of li-ion batteries, J. Power Sour. 217 (1) (2012) 209–219.

[6] Reducing electricity consumption in houses, Ontario Home Builders Association, Energy Conservation Committee Report and Recommendations, May 2006.

[7] W. Clark Gellings, The concept of demand side management for electric utilities, Proc. IEEE 73 (10) (1985).

[8] P. Samadi, H.M. Rad, R. Schober, V.W.S. Wong, Advanced demand side management for the future smart grid using mechanism design, IEEE Trans. Smart Grid 3 (3) (2012) 1170–1180.

[9] S.D. Ramchurn, P. Vytelingum, A. Rogers, N. Jennings, Agent based control for decentralised demand side management in the smart grid, in: Proc. 10th International Conference on Autonomous Agents and Multi-Agent Systems, vol. 1, May 2011, pp. 5–12.

[10] T. Logenthiran, D. Srinivasan, T.Z. Shun, Demand side management in smart grid using heuristic optimization, IEEE Trans. Smart Grid 3 (3) (2012) 1244–1252.

[11] A. Jhunjhunwala, Solar-DC: India towards energy independence, Curr. Sci. 111 (4) (2016) 599–600.

[12] S.B. Jeyaprabha, A.I. Selvakumar, Optimal sizing of photovoltaic/battery/diesel based hybrid system and optimal tilting of solar array using the artificial intelligence for remote houses in India, Energy Build. 96 (1) (2015) 20–52.

[13] K. Chauhan, R.K. Chauhan, Optimization of grid energy using demand and source side management for DC microgrid, J. Renew. Sustain. Energy 9 (3) (2017) 035101–035115.

[14] R.K. Chauhan, B.S. Rajpurohit, S.N. Singh, F.M. Gonzalez-Longatt, Real time energy management system for smart buildings to minimize the electricity bill, Int. J. Emerg. Electr. Power Syst. 18 (3) (2017) 1–15.

[15] W.Y. Chang, The state of charge estimating methods for battery: a review, ISRN Appl. Math. 2013 (2013) 1–7.

[16] C. Phurailatpam, R.K. Chauhan, B.S. Rajpurohit, F.M. Gonzalez-Longatt, S.N. Singh, Demand side management system for island DC microgrid for future buildings, in: Proc. IEEE International Conference on Sustainable Green Buildings and Communities, IIT Madras, India, 2016, pp. 1–6.

[17] K.S. Ng, C.S. Moo, Y.P. Chen, Y.C. Hsieh, Enhanced Coulomb counting method for estimating state of charge and state of health of lithium ion batteries, Appl. Energy 86 (9) (2009) 1506–1511.

[18] R.K. Chauhan, C. Phurailatpam, B.S. Rajpurohit, F.M. Gonzalez-Longatt, S.N. Singh, Demand side management system for autonomous DC microgrid, Technol. Econ. Smart Grids Sustain. Energy 2 (4) (2017) 1–11.

Further reading

R.K. Chauhan, K. Chauhan, Building automation system for grid-connected home to optimize energy consumption and electricity bill, J. Build. Eng. 21 (2019) 409–420.

Chapter 16

Multistage and decentralized operations of electric vehicles within the California demand response markets

Bin Wang, Rongxin Yin, Doug Black and Cy Chan
Lawrence Berkeley National Laboratory, Berkeley, CA, United States

16.1 Introduction

There has been an increasing deployment of electric and plug-in hybrid electric vehicles (EVs) in the market, which has great potential to significantly reduce the consumption of carbon fuels, air pollution, and carbon emissions. Regarding the integration of EVs into the electric power system, an early study [1] proposes a conceptual framework composed of the grid technical operation and the electricity markets environment. Case study results indicate that the adoption of advanced centralized EV charging control strategies allows the integration of a larger number of EVs in the grid, without significant amount of grid upgrades. In addition, since EV battery is capable of providing fast compensation to the grid, the adoption of local controls allows a better operation performance in islanded operation mode and in cases with more penetration of renewable energy source (e.g., wind and photovoltaic). From the high-level perspective of distributed loads in the power system, the studies [2,3] discuss conceptual frameworks for achieving controllability of electric loads to provide power system control services as well as grid services [i.e., demand response (DR), capacity, ancillary service]. It points out that uncoordinated load control of EVs can create a new load peak after the original peak due to the similar switching on time in the early morning. Economic analysis of optimal integration of EVs in microgrid has also been investigated in recent work [4–10], which consider all kinds of distributed energy resource (DER), such as distributed generators, EVs, grid, and load in the case studies.

Decision Making Applications in Modern Power Systems. DOI: https://doi.org/10.1016/B978-0-12-816445-7.00016-5

Regarding EV load control, a number of recent studies aim to understand the adaptation needs of the existing operational control mechanisms to realize smart charging and often propose novel planning and control approaches. These approaches can be grouped into direct and indirect control approaches [11]. In direct control approaches the control actions are realized without the vehicle owner in the control loop. Often, load aggregations are created to increase the size of the resource, so it can offer economic benefits to the aggregator [12,13]. In Ref. [14], for example, the authors propose a direct load control strategy to provide vehicle-to-grid services for three different predefined mobility patterns. In Ref. [15] the study proposes a smart charging framework to identify the benefits of nonresidential EV charging to the load aggregators and the distribution grid. With the assumption of knowing charging behaviors and direct load controlling of EVs, a case study of over 2000 nonresidential EV supply equipments (EVSEs) shows a reduction of up to 24.8% in the monthly bill where the demand charge accounts for half of the utility bill. In Ref. [16] the authors conduct a simulation study for 3000 EVs parked at a municipal parking lot and evaluate the real-time performance of a direct control approach, which maximizes the expected state of charge (SOC) of the EV aggregation in the next time step subject to mobility constraints. In Ref. [17] the authors develop an optimal direct control scheme based on global charging costs. The authors compare the proposed direct control scheme to the local scheduler in a simulation environment, including 100–400 EVs. The authors in Ref. [18] propose a DR strategy by directly controlling EVs charging start time and develop a severity index defined as the longest EV charging time delay in percentage. In Refs. [9,19–22] the authors propose coordinated charging strategies which are applicable for real-time coordinated charging control of charging stations. The results indicate that the charging load of EVs is flattened without sacrificing the charging station profits and customers' service quality.

In indirect control approaches the EV owner manages the control authority through a decentralized strategy. These strategies often make use of a broadcasted exogenous price signal. The cost of energy is minimized at each EV charging station considering the local mobility and charging constraints. An iterative cost minimal charging framework based on game theory is presented in and a similar strategy is given in Refs. [23,24]. In addition, EV charging problem is modeled as a convex optimization problem, with proof of the existence of optimal solution. However, these approaches do not include the impacts or additional costs that can be induced on the distribution network due to increased demand during low-cost periods and often assume that the supply and non-EV demand is known. Given a supply curve for EV charging, price-based control strategies may lead to significant oscillations driven by interactions between energy price and demand for a large population size of EVs [2].

Many researchers have investigated the benefits of EV charging and different grid-level services that can be provided by an aggregation of EV population using different control approaches. Note that various services can be provided by EVs, including peak shaving, regulation, voltage control, and reserves, and many studies have quantified the benefits of smart charging from various stakeholder perspectives [25]. In Ref. [25] the authors estimate that smart charging will reduce the daily electricity costs of a plug-in hybrid EV by $0.23. They also identify daily profits for the individual driver when the charging of the vehicles can be regulated. The economic benefits of fleets that participate in specific markets have also been extensively studied. In Ref. [26] the authors use historical market data and charging data collected from an EV located in a residential household to investigate financial savings and peak demand reduction. The authors conclude that the peak EV demand can be reduced by up to 56%.

In order to further explore the potential of EVs in reducing the total microgrid operation cost, this chapter will present a comprehensive modeling and decision-making framework for EVs under multiple DR markets in California, such as proxy demand response (PDR), ancillary service, and demand-based bid program (DBP) markets. It will provide combined strategies to maximize the revenues via dynamic market participations. Specifically, we analyze cost-saving performance of fleet EVs, considering the EV fleet properties and market rules, including minimum consecutive commitment, regulation market threshold, and the option to set baseline profile. A multistage operation model is established for cost-effective microgrid energy management, that is, day-ahead planning in the first stage and adaptive operations in the later stage(s). The preliminary results indicate that these approaches can considerably reduce the total monthly energy cost while satisfying the energy requests from both public and fleet EV drivers.

Asynchronous and decentralized algorithms [23,27] have been proposed in the final control stage to allocate the prescheduled energy to individual EVs, while preserving driver privacy and the overall robustness of the decision-making framework. Heterogeneous energy requests, charging schedules are considered in the framework, including arriving time and estimated leaving. In the decentralized strategies, each EV agent will make asynchronous local decisions, while coordinating with one centralized scheduler with minimum amount of information exchange. In cases of communication blackout or critical deviations of system states, the proposed decentralized method can adaptively converge to the optimal solutions considering the new inputs and updated system states. We investigate the robustness of the distribute algorithms with asynchronous coordination based on the alternating direction method of multipliers (ADMM) [28].

The overall system architecture is illustrated in Section 16.2. Deterministic problem formulation is provided in Section 16.3, followed by the more implementable day-by-day operations in Section 16.4. Asynchronous and decentralized algorithms are introduced in Section 16.5. Finally, Section 16.6 concludes this chapter with future work.

16.2 System overview

EV load is modeled as deferrable load that can be shifted to different time windows to achieve various grid objectives in different energy markets. Accordingly, optimization-based strategies were developed that allow the EV fleet manager to coordinate the integration of EVs with multiple different market strategies in order to minimize the energy cost for serving the mobility required from the fleet EVs. The overall system architecture is shown in Fig. 16.1. The aggregate EV controller will retrieve day-ahead pricing info from multiple DR markets from California Independent System Operator (CAISO) servers and collect the EV usage info, including energy demand and itineraries, from individual EV drivers. This communication has already been enabled in the demonstration project. During the next-day operation, each EV follows the day-ahead schedule in each time step to fulfill its own energy demand.

16.2.1 Smart electric vehicle charging control system

As presented in Fig. 16.2, the smart charging control system for public EVs includes the following main components: (1) an optimizer for computing the optimal charging power sequence; (2) a database for storing all the charging session data, the facility meter data, and the charging request message; (3) a web service for interacting with the pilot study participant to handle the charging request; (4) an EV charging data exchange application programming interface (API) for bridging the EV charging optimization server end

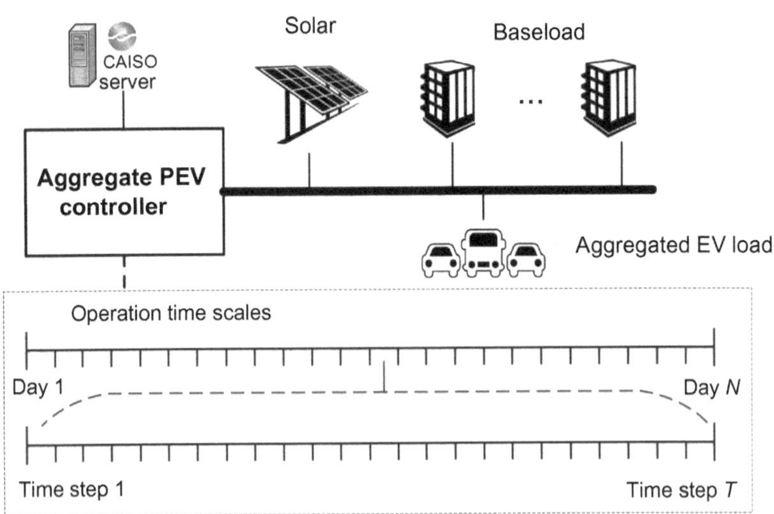

FIGURE 16.1 Overall system architecture.

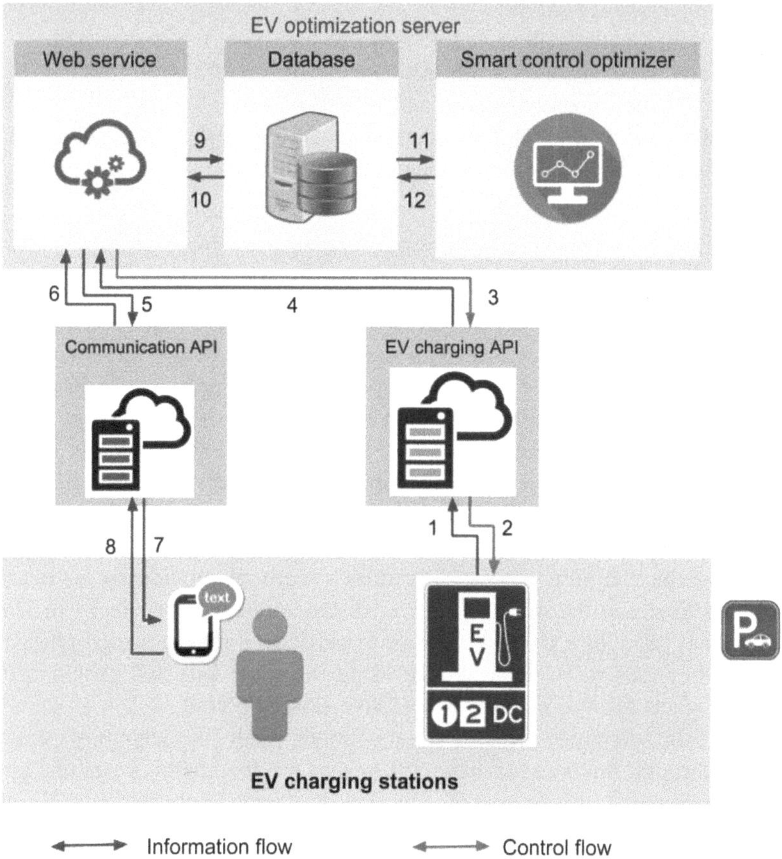

FIGURE 16.2 Smart electric vehicle (EV) charging control system architecture.

and the charging station; and (5) an API hosted by the vendor of EV charging station for managing the charging station's power settings. Among those components, the database, the web service, and the optimizer are hosted on a stand-alone server. A third-party vendor develops a function to bridge the EV optimization server and the vendor API of the charging station for the data exchange. The main capabilities of the EV optimization server and the EV data exchange server are listed as follows:

- EV optimization server:
  - Web service: to (1) handle smart charging requests; (2) interact with users; (3) data collection; (4) issue control commands
  - Database: storage for all session data, meter data, smart charging requests
  - Smart control optimizer: charging schedule optimization

- EV data exchange server:
  - communicates with each EVSE via vendor charging station API,
  - sends the charging session info (including user ID) from the EVSE to the EV optimization server, and
  - sends and implements the optimal charging power settings from the EV optimization server to the controlled EVSE.

16.2.2 Communication information exchange

In the smart charging system for public EVs, as shown in Fig. 16.2, a mobile app is deployed for interacting with the pilot study participants. When the driver plugs in the vehicle and activates the charging session, the Kisensum server will capture the active charging session(s) and pass the session data to the LBNL server. Then the web service checks if the active charging session belongs to one of the pilot study participants in the database. If it is true, the server will send a "Welcome" text message to the driver through the web service. In the text message, we provide a link to a webpage for submitting/modifying the charging request. The driver needs to provide the estimated departure time and the estimated travel distance in miles or energy in kWh for the next trip. Once the drive agrees to participate into the smart charging control, they can go to the webpage using the same link and check in the charging status at anytime during the active session period.

The data exchange server communicates with the charging station (EVSE) through the web services API to manage the charging stations connected to vendor's cloud network. The data exchanged between the EV optimization server and the data exchange API includes the charging session and meter data, which is in JSON format. In the communication system for public EVs a short message service (SMS) provider is deployed to send/receive notifications and charging request to/from drivers, as presented in Fig. 16.3.

Regarding the SMS interactive system between the EV optimization server and the participant, the driver will only receive the text message from the EV optimization server when plugging the vehicle into the controlled charging stations on the demonstration site. The charging will start as usual until the participant submits the charging request on a webpage. On the same webpage, the participant can change the departure time and/or the energy charge needed at anytime during the charging session. When participants submit their charging requests to the EV optimization server through the web service, the optimizer in the EV optimization server will first detect if this message is new or not. If it is a new message, the optimizer will initiate the optimization immediately and update the optimized the charging power sequence for all the controlled charging sessions in the database. Meanwhile, the EV optimization web service will send all the optimized charging power sequences to the data exchange server in the JSON format. Last, the data

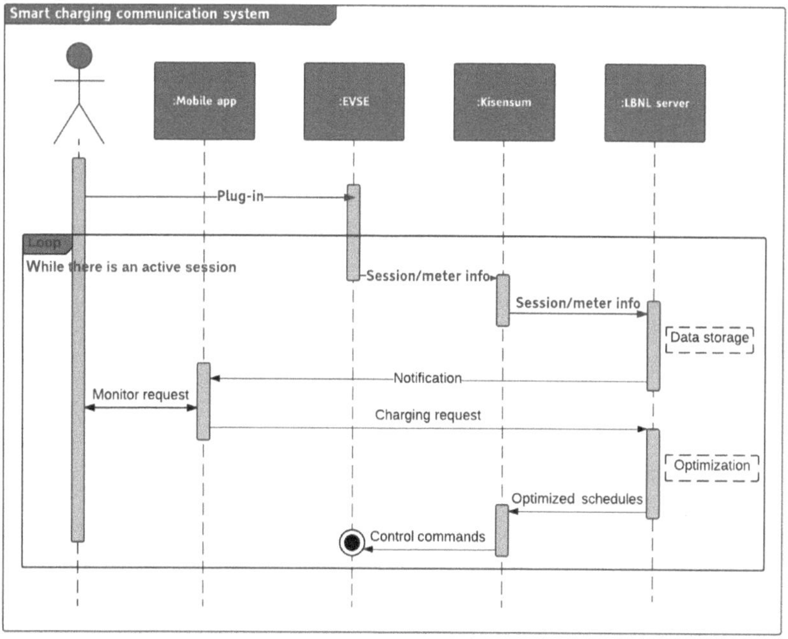

FIGURE 16.3 Smart charging communication system for electric vehicle (EVs).

exchange server implements the optimized charging power sequences into the charging stations to which the participants' vehicles plug.

16.3 Deterministic problem formulation

In this section, we introduce the EV-grid integration modeling strategies for privately owned and fleet vehicles with the objective of reducing aggregate energy and demand costs for each billing month period. The AlCoPark garage has a typical electric load pattern on weekdays with peaks in early morning and late afternoon caused by privately owned and fleet vehicle charging, respectively. Using the actual charging demand data from each charging station and the whole facility electric meter demand data, we will demonstrate the ability to shift discrete charging demand segments for individual charging sessions that reduce the aggregated peak demand in each billing month without impacting the amount of charge received by each vehicle.

For fleet vehicles, we propose a series of control strategies to control the availability of the charging stations for the fleet EVs. The goal is to manage the aggregated charging power of the fleet from time of peak demand or high-cost periods to lower cost periods. For privately owned vehicles, the goal of the proposed smart charging framework is to reschedule the power time series measured in discrete time slots $[1, \ldots, k]$ for any charging session on the same day. Our previous work describes the framework with the following key structures/assumptions: (1) The order of the measured power in time series is preserved (because the power that EVSEs draw is dependent on the SOC of the EV being charged) and (2) the charging is preemptive, which means that the rescheduled charging load is equal to the original and the charging tasks are interruptible without any decrease in the SOC of the EV.

16.3.1 Tariff and demand response markets

The AlCoPark garage is in Pacific Gas and Electric's (PG&E's) E-19 time-of-use (TOU) service rate shown in Table 16.1. This rate plan is suitable for customers who can be more flexible with their power usage between the maximum peak, maximum part-peak, and off-peak periods. Electric demand is averaged over 15-minute periods to determine peak values in each monthly billing period type (e.g., maximum peak maximum part peak). Typically, the demand charge cost accounts for about 50% of the total electric utility cost for nonresidential facilities. For example, there would be $114 savings by shifting the demand of a single 6.6 kW level 2 charging

TABLE 16.1 PG&E E-19 rate schedule.

	$/kW	Time period
Demand charges		
Maximum peak demand summer	$18.74	12:00–6:00 p.m.
Maximum part-peak demand summer	$5.23	8:30 a.m.–12:00 p.m. and 6:00–9:30 p.m.
Maximum demand summer	$17.33	Anytime
Maximum part-peak demand winter	$0.13	8:30 a.m.–9:30 p.m.
Maximum demand winter	$17.33	Anytime
Energy charges		
Peak summer	$0.15	12:00–6:00 p.m.
Part-peak summer	$0.11	8:30 a.m.–12:00 p.m. and 6:00–9:30 p.m.
Off-peak summer	$0.08	Anytime
Part-peak winter	$0.10	8:30 a.m.–9:30 p.m.
Off-peak winter	$0.09	Anytime

session from the billed maximum demand period to an off-peak period in a summer month (summer is May 1 to October 31; winter is November 1 to April 30). It is noted that the AlcoPark garage is placed on PDP (peak-day pricing) rate, which provides customers the opportunity to manage their electric costs by reducing load during high-cost periods or shifting load from high-cost periods to lower cost periods. There are 9−15 PDP event days per year. On event days, a surcharge is added between 2:00 and 6:00 p.m., which is $1.2/kWh for E-19 tariff rate.

16.3.2 Aggregation of electric vehicles

For each individual vehicle n on day d, the following constraints should be satisfied:

$$b_n^d(t)\cdot \underline{p} \le p_n^d(t) \le b_n^d(t)\cdot \overline{p} \tag{16.1}$$

$$e_n^d(t+1) = e_n^d(t) + p_n^d(t)\cdot \eta_c \cdot \Delta t \tag{16.2}$$

$$e_n^d\left(t_n^{d,l}\right) \ge e_{n,\mathrm{req}}^d \tag{16.3}$$

$b_n^d(t)$ in Eq. (16.1) is the indicator of whether vehicle n is charging at time t. Note that, the feasible charging range is not continuous so as to model the real-world EV chargers. When $b_n^d(t)$ is set to 0, both the left and right-hand sides are 0, constraining the charging power to 0, that is, the inactive state. For the active state, the charging power threshold $\overline{p}$, that is, minimal charging power, is set to 1.5 kW, which corresponds to the limit of the chargers used in the demonstration project. Eq. (16.2) indicates the accrual of energy consumption for each vehicle and the energy consumption value at the time of charging session finish time t_n^f, that is, $e_n^d\left(t_n^l\right)$, should be larger than the requested amount $e_{n,\mathrm{req}}^d$. Note that energy requests for vehicles are collected by a driver−charger interface.

In order to reduce the number of decision variables in the optimization problem, modeling approaches from Ref. [20] are adapted to aggregate numerous individual EVs as one single virtual battery with power and energy boundaries, hereby improving the computational efficiency. According to this approach, any trajectory that falls between the power and energy boundaries can be achieved by controlling each EV's charging power. The approach is summarized as follows:

$$E_d^{+/-}(t) = \sum\nolimits_{n\in N_p(t)} e_{n,d}^{+/-}(t), \quad t\in[0,T] \tag{16.4}$$

$$E_d^-(t) \le \sum_{\tau=0}^{t} P^d(\tau)\cdot \Delta t \le E_d^+(t), \quad \forall t\in[0,T] \tag{16.5}$$

$$\overline{P}^{d} = \sum\nolimits_{n \in N_p^d(t)} \overline{p} \cdot \eta_c \tag{16.6}$$

$$b_{\text{agg}}^{d} \cdot \underline{p} \le P(t) \le b_{\text{agg}}^{d} \cdot \overline{P}^{d}, \quad \forall t \in [0, T] \tag{16.7}$$

The aggregate energy boundaries, that is, $E_d^{+/-}(t)$, are obtained by summing up $e_{n,d}^{+}(t)$, which is the accumulated energy from the as-fast-as-possible charging pattern, and $e_{n,d}^{-}(t)$, which is from the as-late-as-possible charging pattern. In addition, the total power consumption value should be lower than the aggregated power from all available vehicles at time t. Note that the discontinuity of the aggregated power is also modeled, similar to that in Eq. (16.1). The optimal power consumption profiles for day-ahead operations can be used as the reference for EVs to follow during the real-time operations in distributed and asynchronous fashions, which are, however, not the focus of this chapter.

16.3.3 Time-of-use tariff structure

For commercial sites in California TOU markets, two categories of costs are generally applied to customers' bills, that is, energy charge and demand charge. Energy charges are calculated by the product of amount of electricity, measured in kilowatt-hours (kWh) used per time period, and the per-kWh rate for those respective time periods. Demand charge is calculated by using the maximum load measurement in each demand period, multiplied by the corresponding demand charge rate, in \$/kW. Thus the total monthly cost of energy charge is modeled by Eq. (16.8), where the cost of energy consumption in different time periods is also included. Eq. (16.9) models the total monthly demand charges, where I denotes the set of the demand charge periods. In the case of the E-19 tariff in the PG&E territory, there are three demand charge periods for summer months, that is, peak, part-peak, and anytime max periods, while two periods in the winter, that is, part-peak and anytime max periods.

$$C_{\text{EC}} = \sum\nolimits_{d \in D} \sum\nolimits_{t \in T} \left(L^d(t) + \sum\nolimits_{n=1}^{N_p^d(t)} P^d(t)\right) \cdot \Delta t \cdot \lambda(t) \tag{16.8}$$

$$C_{\text{DC}} = \sum_{T_i \in \{\text{T}_\text{p}, \text{T}_{pp}, \text{T}_\text{M}\}} \max_{\substack{t \in T_i \\ d \in D}} \left(L^d(t) + \sum\nolimits_{n}^{N_p^d(t)} P^d(t)\right) \cdot \omega_i \tag{16.9}$$

Thus to minimize the monthly energy bills by considering only the energy charge and demand charges, a deterministic optimization problem is formulated as follows:

Problem 1—TOU charges (energy charge + demand charge)

Objective	minimize($C_{\text{EC}} + C_{\text{DC}}$)
Subject to	(16.1)–(16.9)

16.3.4 Integration with peak-day pricing plan

PDP is an optional rate that offers businesses a discount on regular summer electricity rates in exchange for higher prices during 9–15 peak pricing event days per year, typically occurring on the hottest days of the summer (PG&E defines summer as May 1 to October 31). When utilities observe or anticipate high wholesale market prices, high demand, or a power system emergency, they call critical events during a specified time period (e.g., 2–6 p.m. on summer weekdays). The price for electricity during these time periods is substantially higher, as shown in Fig. 16.4.

The customer has to submit a capacity reservation value, that is, $P_{\mathrm{PDP}}^{\mathrm{CR}}$, to the load serving entity, in this case, the utility company. The demand peaks in different demand periods that exceed the capacity reservation value will be protected from the demand charges by the PDP policy, that is, credits will be billed to customers for the exceeding amount. This policy is modeled by Eqs. (16.10) and (16.11). However, the total energy consumption in kWh below $P_{\mathrm{PDP}}^{\mathrm{CR}}$ during PDP events will be billed with PDP energy charge rate λ_{PDP}, which is modeled by Eq. (16.12). The optimal EV charging problem with the PDP market participation is summarized in Problem 2.

$$R_{\mathrm{PDP}}^{p} = \pi_{\mathrm{PDP}}^{p} \cdot (\max_{\substack{d \in D_{\mathrm{PDP}} \\ t \in T_{\mathrm{PDP}} \cap T_P}} \left(L^d(t) + \sum\nolimits_{n \in N_p^d(t)} p_n^d(t) \right) \tag{16.10}$$

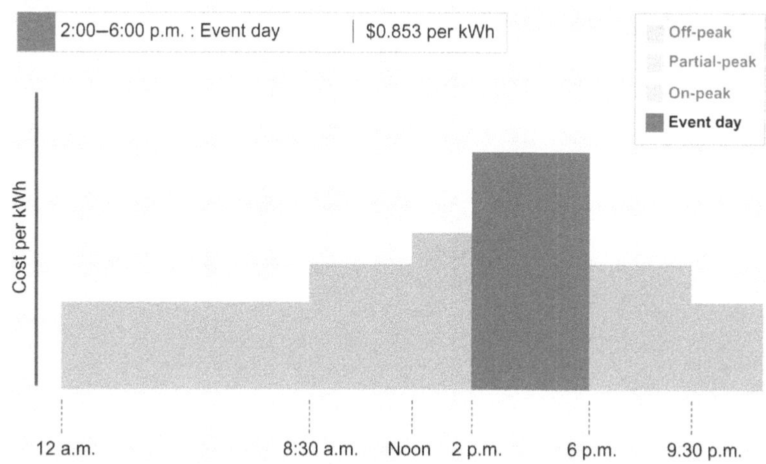

FIGURE 16.4 Peak-day pricing: event day rates.
Note: Based on A1 rates per kWh as of January 7, 2017

$$R_{\mathrm{PDP}}^{m} = \pi_{\mathrm{PDP}}^{m} \cdot \Big(\max_{\substack{d \in D_{\mathrm{PDP}} \\ t \in T_{\mathrm{PDP}} \cap T_M}} \Big(L^d(t) + \sum\nolimits_{n \in N_p^d(t)} p_n^d(t) \Big) \tag{16.11}$$

$$C_{\mathrm{PDP}} = \lambda_{\mathrm{PDP}} \cdot \sum\nolimits_{d \in D_{\mathrm{PDP}}} \sum\nolimits_{t \in T_{\mathrm{PDP}}} \max_{\substack{d \in D_{\mathrm{PDP}} \\ t \in T_{\mathrm{PDP}}}} \Big(L^d(t) + \sum\nolimits_{N_p^d(t)} p_n^d(t) - P_{\mathrm{PDP}}^{CR}, 0 \Big) \cdot \Delta t \tag{16.12}$$

Problem 2—TOU charges with PDP integration

Objective	minimize($C_{\mathrm{EC}} + C_{\mathrm{DC}} - R_{\mathrm{PDP}}^{p} - R_{\mathrm{PDP}}^{m} + C_{\mathrm{PDP}}$)
Subject to:	(16.1)–(16.12)

16.3.5 Integration with ancillary service market

To achieve instantaneous balance between the supply and demand sides of electricity transmission system, ancillary services can be utilized by calling services from various grid components, not only the traditional electricity generators but also demand-side DERs. A regulation up/down market is a representative type of ancillary service market. EVs with the capability to follow the up and downregulation signals in a short period of time can be coordinated to serve as effective and reliable resources to provide regulation services. Based on formulation of Problem 1, the EV integration with regulation market participation is modeled as follows:

$$R_{\mathrm{AS}} = \sum\nolimits_{d \in D} \sum\nolimits_{t \in T} \big(R_{\mathrm{up}}^d(t) \cdot \pi_{\mathrm{up}}^d(t) + R_{\mathrm{down}}^d(t) \cdot \pi_{\mathrm{down}}^d(t) \big) \cdot \Delta t \tag{16.13}$$

$$P^d(t) = B^d(t) + \rho_{\mathrm{up}} \cdot R_{\mathrm{up}}^d(t) + \rho_{\mathrm{down}} \cdot R_{\mathrm{down}}^d(t) \tag{16.14}$$

$$b_{\mathrm{agg}}^{d,B}(t) \cdot \underline{p} \le B^d(t) \le b_{\mathrm{agg}}^{d,B}(t) \cdot \overline{P}^d \tag{16.15}$$

$$b_{\mathrm{agg}}^{d,P}(t) \cdot \underline{p} \le P^d(t) \le b_{\mathrm{agg}}^{d,P}(t) \cdot \overline{P}^d \tag{16.16}$$

$$E_d^-(t) \le \sum\nolimits_{t_0}^{t} P^d(\tau) \cdot \Delta t \le E_d^+(t) \tag{16.17}$$

$$E_d^-(t) \le \sum\nolimits_{t_0}^{t} B^d(\tau) \cdot \Delta t \le E_d^+(t) \tag{16.18}$$

$$b_{\mathrm{agg}}^{d,B}(t) \cdot \underline{R}_{\mathrm{down}} \le R_{\mathrm{down}}^d(t) \le b_{\mathrm{agg}}^{d,B}(t) \cdot \overline{P}^d \tag{16.19}$$

$$b_{\mathrm{agg}}^{d,ru}(t) \cdot \underline{R}_{\mathrm{up}} \le R_{\mathrm{up}}^d(t) \le b_{\mathrm{agg}}^{d,ru}(t) \cdot \overline{P}^d \tag{16.20}$$

$$b_{\mathrm{agg}}^{d,bd}(t) \cdot \underline{p} \le B_d(t) + R_d(t) \le b_{\mathrm{agg}}^{d,bd}(t) \cdot \overline{P}^d \tag{16.21}$$

$$b_{\text{agg}}^{d,bu}(t)\cdot \underline{p} \leq B_d(t) - R_u(t) \leq b_{\text{agg}}^{d,bu}(t)\cdot \overline{P}^d \tag{16.22}$$

Eq. (16.16) shows the expression for calculating the total revenue from day-ahead frequency regulation markets. The revenue consists of the regulation-up capacity payment and regulation-down capacity payment. Unlike the modeling approaches in previous research where day-ahead commitments can be violated with penalties, we do not intend to violate the commitment in any circumstances due to the performance regulations in California ancillary service markets.

Due to the noncontinuity property of power boundaries, auxiliary binary decision variables are defined to indicate the options to participate in the regulation up and down markets. Given regulation signals from CAISO, an aggregate EV fleet, for example, will follow the signals, that is, increase or decrease the aggregated power consumption of the EVs. The revenue is calculated on the basis of the day-ahead bids, that is, the committed regulation up and down capacities, rather than the actual increased or decreased power consumption following real-world regulation signals, indicated by Eq. (16.17). The negative (up), ρ_{up}, and positive (down), ρ_{down}, utilization factors represent the fraction of the committed regulation dispatched by the CAISO control signal. Actual utilization factors collected in a real-world demonstration project at the Los Angeles Air Force Base were used in the simulations presented here. The baseline aggregate power $B^d(t)$ is the original power consumption profile assuming no regulation signals, whereas $P^d(t)$ is the actual power profile in Eqs. (16.18)–(16.21). Here, $B^d(t)$ is a decision variable. Eqs. (16.22) and (16.23) model the constraints so that the aggregate fleets can participate in the regulation up or down markets, or choose to stay out of the markets. We also assume that the aggregated EV fleet can follow all regulation signals, that is, the actual power consumption should always stay in the power boundaries, which is modeled by Eqs. (16.24) and (16.25).

Note that the aggregator can make regulation up and down bids for the same time periods, even one of them will not be called during implementation, but still getting benefits for the bids. In addition, the actual aggregate power and the aggregated baseline profiles should both satisfy the aggregate energy and power constraints, modeled in Eqs. (16.4)–(16.7). The problem is formulated as follows:

Problem 3—TOU charges with regulation markets

Objective	minimize $C_{\text{EC}} + C_{\text{DC}} - R_{\text{AS}}$
Subject to	(16.1)–(16.9) and (16.16)–(16.25)

16.3.6 Integration with PDR market

Aggregated EVs can also participate in the PDR market, where the fleet EVs are treated as a virtual battery with flexibility to "sell" the power in the PDR market. For EVs with V2G capabilities, "sell" operations can be achieved by discharging the vehicle batteries, while for V1G, "selling" of power would be achieved by reducing the aggregate power consumption relative to a power consumption baseline. The model is presented as follows:

$$R_{\mathrm{PDR}} = \sum\nolimits_{d \in D} \sum\nolimits_{t \in T} R^{d}_{\mathrm{sell}}(t) \cdot \pi^{d}_{\mathrm{pdr}}(t) \cdot \Delta t \tag{16.23}$$

$$M_s \cdot \left(1 - b^{\mathrm{PDR}}_{\mathrm{agg}}(t)\right) \le P^{d}(t) - B^{d}(t) + R^{d}_{\mathrm{sell}}(t) \le M_b \cdot \left(1 - b^{\mathrm{PDR}}_{\mathrm{agg}}(t)\right) \tag{16.24}$$

$$b^{\mathrm{sell}}_{\mathrm{agg}}(t) \cdot \underline{R}^{\mathrm{PDR}}_{\mathrm{sell}} \le P^{d}_{n}(t) \le b^{\mathrm{sell}}_{\mathrm{agg}}(t) \cdot \overline{P}^{d} \tag{16.25}$$

The revenue from the PDR market is a product of the virtual sell power, that is, $R^{d}_{\mathrm{sell}}(t)$, multiplied by the corresponding PDR market prices, that is, $\pi^{d}_{\mathrm{pdr}}(t)$ in Eq. (16.26). The baseline power consumption $B^{d}_{n}(t)$ is typically the averaged value of a number of previous days; thus here we model it as a known profile before optimization. In reality, PDR market participation has a requirement of a minimal threshold for virtual sell power, that is, $R\mathrm{sell}^{\mathrm{PDR}}$ in Eq. (16.28). In Eq. (16.27), $b^{\mathrm{PDR}}_{\mathrm{agg}}(t)$ is the binary indicator of whether the fleets are participating in the PDR market. When $b^{\mathrm{PDR}}_{\mathrm{agg}}(t) = 1$, that is, participating, Eq. (16.27) is reduced to

$$P^{d}(t) = B^{d}(t) - R^{d}_{\mathrm{sell}}(t) \tag{16.26}$$

where the actual power consumption value $P^{d}_{n}(t)$ equals the baseline power $B^{d}(t)$ minus the virtual sell power $R^{d}_{\mathrm{sell}}(t)$. When $b^{\mathrm{PDR}}_{\mathrm{agg}}(t) = 0$, indicating no participation, Eq. (16.20) evolves to

$$M_s \le P^{d}(t) - B^{d}(t) + R^{d}_{\mathrm{sell}}(t) \le M_b \tag{16.27}$$

where M_s is a sufficiently small number and M_b is a sufficiently big number. Eq. (16.30) remains true for all cases, making it a redundant constraint in the optimization problem, which can be effectively handled by current solvers with mixed-integer capabilities. In addition, to model the consecutive engagement constraint, some numerical approaches are applied as shown in Eqs. (16.31) and (16.32):

$$b_c(t_0) = b^{\mathrm{PDR}}_{\mathrm{agg}}(t_0) \tag{16.28}$$

$$b_c(t) \le 1 - b^{\mathrm{PDR}}_{\mathrm{agg}}(t - \Delta t) \tag{16.29}$$

$$b_c(t) \le b^{\mathrm{PDR}}_{\mathrm{agg}}(t) \tag{16.30}$$

$$b_c(t) \ge b^{\mathrm{PDR}}_{\mathrm{agg}}(t) - b^{\mathrm{PDR}}_{\mathrm{agg}}(t - \Delta t), \quad \forall t \in T \tag{16.31}$$

$$\sum_{\tau = t}^{\min(t + N_c - 1, T)} b^{\mathrm{PDR}}_{\mathrm{agg}}(t) - N_c \ge -N_b \cdot (1 - sc(t)), \quad \forall t \in T \tag{16.32}$$

An auxiliary binary decision variable, that is, $b_c(t)$, is utilized to model the consecutive participation constraint. $b_c(t) = 1$ indicates the beginning of a new block of consecutive participation at t. Eqs. (16.31)–(16.35) guarantee that the number of consecutive participating time steps is greater or equal to N_c. Incorporating binary decision variables into the optimization problems results in a mixed-integer programming problem, where sophisticated numerical solvers are needed. With PDR market integration the overall problem is formulated as follows:

Problem 4—TOU charges with PDR market participation

Objective	minimize$C_{EC} + C_{DC} - R_{PDR}$
Subject to	(16.1)–(16.9) and (16.26)–(16.35)

16.4 Cost-saving performance in different markets

Results of optimizations of EVs in DR programs and ancillary service markets described earlier are presented here. The first example optimizes charging schedules solely to minimize electric TOU costs. The second example optimizes to minimize TOU costs and maximize ancillary service regulation revenue.

First, the load shifting and cost reduction effects of smart charging programs under only TOU prices are presented. As shown in Fig. 16.5, the energy charge and demand charge rates in winter are lower than those in summer. As a result, AlCoPark Garage's actual total monthly costs for energy charges in winter were slightly lower than those in summer, indicated by the blue bars in Fig. 16.5, and the total monthly demand charges are considerably lower than those of in summer, indicated by the orange bars in Fig. 16.5.

16.4.1 Ancillary service market participation

To investigate the impact of ancillary service market integration, an additional option in the simulation is added to allow the EV fleet to modify the

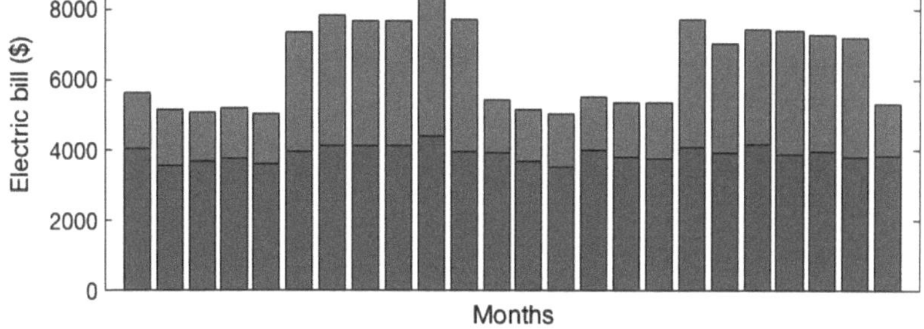

FIGURE 16.5 Total monthly electric bills from January 2015 to December 2016.

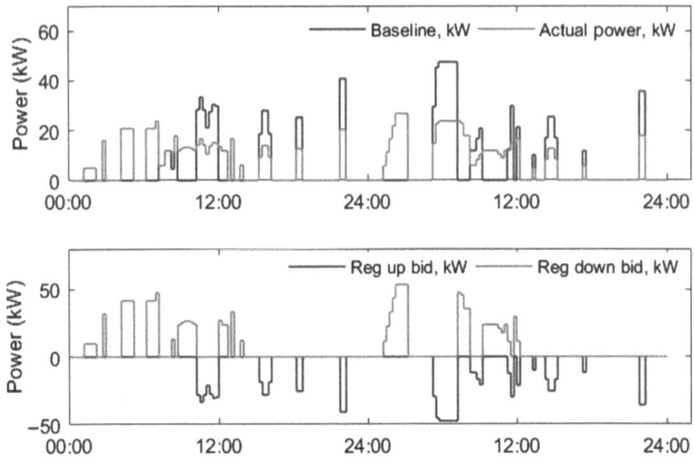

FIGURE 16.6 Details of regulation market participation.

aggregate power consumption profile in response to the regulation up and down prices from the CAISO ancillary services market, that is, Problem 4 defined in the previous section.

The details of the regulation market participation are shown in Fig. 16.6, including the baseline power, actual EV load profile (upper), and the actual regulation up/down bids (lower). EV baseline load profile and the actual EV power consumption curve are both shown in Fig. 16.6. Using both curves in the optimization, energy consumption (kWh) was held constant, constrained by Eqs. (16.20) and (16.21). Note that the duration of each regulation commitment was assumed to be 15 minutes in the optimization, within which the actual regulation signals are dispatched in every 4 seconds. A 15-minute interval was also considered the finest resolution for EV control. In addition, both regulation up and regulation down bids were allowed in the same time periods. Due to the assumption about the regulation up/down utilization rates, the regulation up/down bids were called partially, and the adjusted EV power consumption was reflected as the differences between the baseline (blue) and the actual load profile (red) in Fig. 16.6 (upper).

The monthly revenue results were collected by simulating EV management strategies for each month from January 2015 to December 2016, which was shown in Fig. 16.7 (blue circles). The highest monthly revenue was \$115 in December 2016, and the lowest revenue was \$78 in September 2016. The relationship between monthly regulation revenue and EV charging flexibility at the AlCoPark Garage is also shown in Fig. 16.7. To represent the monthly average distance between the upper and lower boundaries of the virtual battery, the flexibility index of the aggregated EVs is defined as

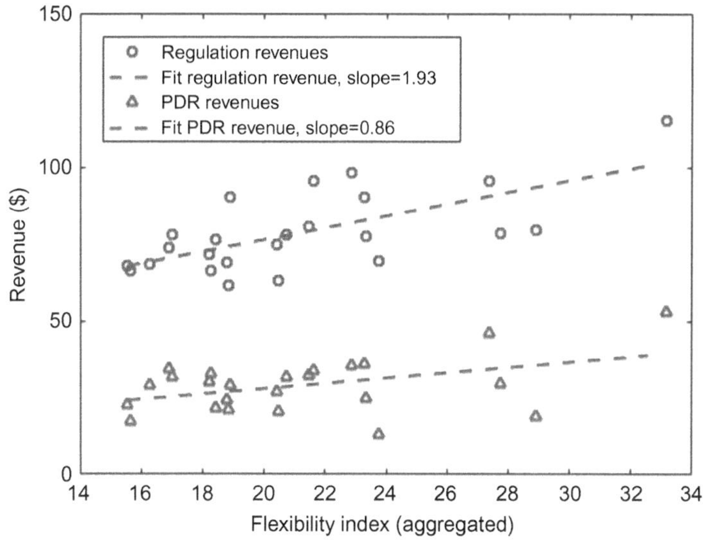

FIGURE 16.7 Monthly profits versus flexibility index.

$$f_{\text{agg}} = \frac{\sum_{d \in D} \sum_{t \in T} (E^{+}(t) - E^{-}(t))}{D \cdot T} \tag{16.33}$$

As indicated in Fig. 16.7, the ability to generate profits in regulation markets is positively correlated with the flexibility index of the aggregated virtual battery, with a correlation coefficient of 0.667.

16.4.2 PDR market participation

Problem 4 was addressed to simulate PDR market participation. CAISO requires each commitment into the PDR market to have a minimum duration of 1 hour. PDR market commitments of 1 hour were modeled with the constraints represented in Eqs. (16.31)–(16.35). As shown in Fig. 16.8 (lower), the green curve indicates the actual EV power consumption profile, whereas the red curve represents the virtual sell power of the aggregated EVs given price signals from the PDR market. Note that the total energy consumption value following the actual power consumption profile should be equal to the one that follows the baseline profile generated by Problem 2. In addition, Problem 4 models the opportunities of the EVs to participate in the PDR market as discrete options, that is, the EV aggregator does not have to stay in the market for the whole day, and it can plan to step out of market when the PDR prices are not optimal. The actual monthly revenues from PDR markets illustrated by the red triangles in Fig. 16.7 where the varying flexibilities of EV fleets to generate profits from PDR markets are shown. Note that the consecutive commitment constraint is set to 1 hour for the PDR market optimizations.

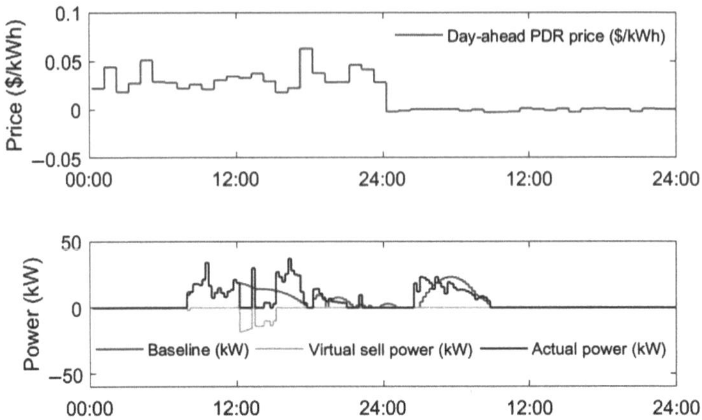

FIGURE 16.8 PDR market participation.

TABLE 16.2 Monthly revenue from demand-based bid program market.

Year	Month	Event	Revenue ($)
2016	6	5	16
2016	7	6	10
2016	8	2	0
2016	9	1	0

16.4.3 Demand-based bid program participation

The modeling approaches for the PDR market integration were similar to those for the DBP market; however, there were different requirements for commitments in the DBP market. For instance, participation in the DBP market only occurs when the DBP events are issued by the program facilitator, PG&E; however, hourly price signals in the PDR market were available on a daily basis. In PG&E's DBP program, $0.5/kW is credited to commercial customers when they reduce their demand during DBP events. It was assumed, in this analysis, that the threshold to participate was greater than or equal to 10 kW, and each commitment had to have a duration of at least two consecutive hours. The simulation results for the DBP market participation are shown in Table 16.1. Due to the 2-hour commitment constraint the existing EV resources were not qualified to participate in all of the DBP events in 2016. Thus the profit-generating capacity for EVs is not as high as that for the regulation market, considering the limits of the fleet size and V1G power and flexibility. The DBP revenues are shown in Table 16.2.

16.4.4 Peak-day pricing participation

Participating in the PDP program, the annual electric bill savings were expected to improve as the monthly peak demand and part-peak demand were partially protected by the capacity reserve level (CRL), which was required for PDP program enrollment. Specifically, as modeled by Eqs. (16.13)–(16.15), the monthly peaks above the CRL received PDP credits, whereas the energy usage not protected by CRL was billed with a fixed PDP rate. PDP events were only issued during summers, and only peak and part-peak demands were considered. The monthly PDP benefit was calculated as $R^{p}_{\mathrm{PDP}} + R^{pp}_{\mathrm{PDP}} - C_{\mathrm{PDP}}$. As shown in Fig. 16.9, PDP benefits for summer months in the year of 2016 were computed with varying CRLs. For the months with only one PDP event, that is, August and September, PDP credits dominated the total benefit, which decreased as the CRL increased. On the contrary, event energy charge became dominant in the months with more PDP events since there was less unprotected energy usage as the CRL increased between 10 and 60 kW. As CRL increased greater than 60 kW, the month benefits decreased because of the weaker protection by CRL. In addition, the annual total PDP benefit varied with the CRL with the optimal CRL value close to 40 kW.

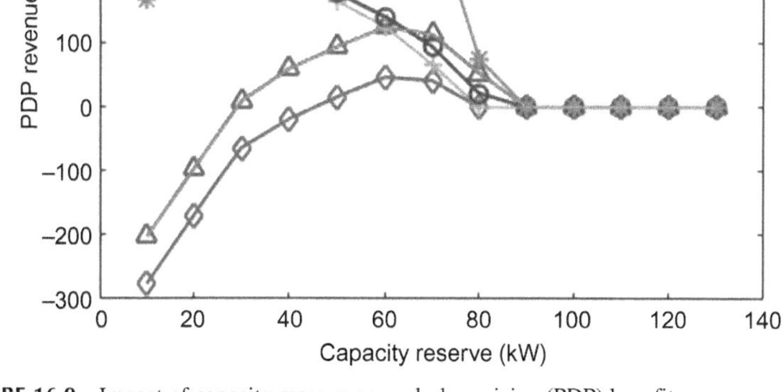

FIGURE 16.9 Impact of capacity reserve on peak-day pricing (PDP) benefits.

16.4.5 Impact of the flexibility and market participation threshold

The impacts of different factors on the revenue-generating capability of the EV fleet, including the freedom of baseline power profile selection, flexibility of individual EV, and market participation threshold, will be discussed here. Simulation results indicate that proper tuning of these factors can lead to significant improvements in revenue generation.

16.4.5.1 Impact of baseline calculation

As opposed to the regulation simulation described above where baseline charging power was a decision variable, here, instead, the charging profiles obtained by solving Problem 2 were used as baselines. As shown in Fig. 16.10, the actual power (red) generally follows the baseline power profile (blue), unlike that shown in Fig. 16.6. However, the capability of the smart charging program was limited in exploring more space to generate revenues from regulation markets. The monthly revenue from regulation in June, 2016, was reduced from \$95 to \$36. Thus with a preset baseline power profile, the flexibility was limited as well as the revenue-generating capability.

The impact of individual vehicle charging flexibility on regulation market revenues is examined here. The total connected duration of each EV was increased by multiple ratios to simulate different degrees of EV connected time flexibilities. For the months shown in Fig. 16.11, January, June, and August 2016, the revenues increased rapidly as the ratio of connected time to charging time increased from 1 to 2.5. However, as the ratio increased beyond 2.5, the total monthly revenue gradually plateaus. Note that in the simulation, the maximum connected duration for each EV was 2 days. Thus within the given time period, there exist limitations of revenue improvement by extending EV connected duration flexibility. However, for real-world

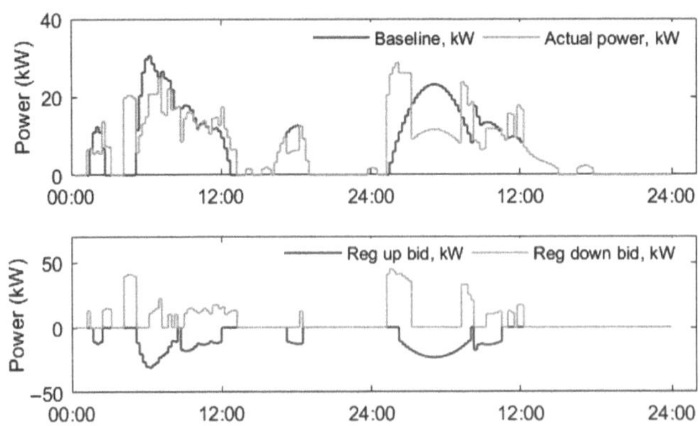

FIGURE 16.10 Fixed baseline case.

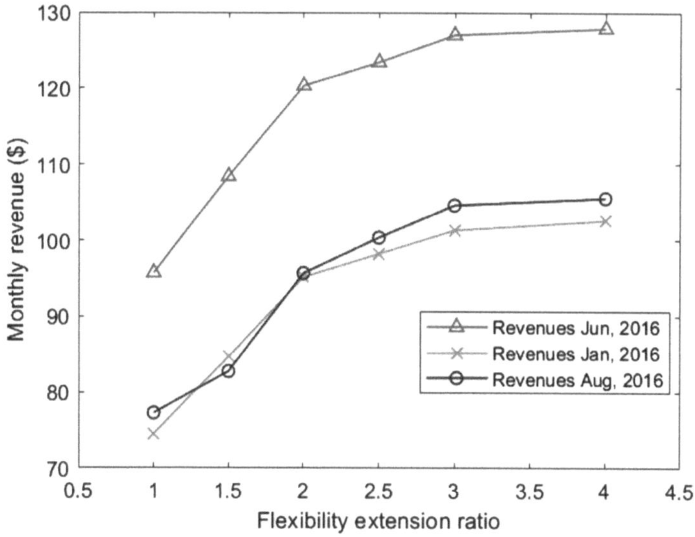

FIGURE 16.11 Impact of flexibility.

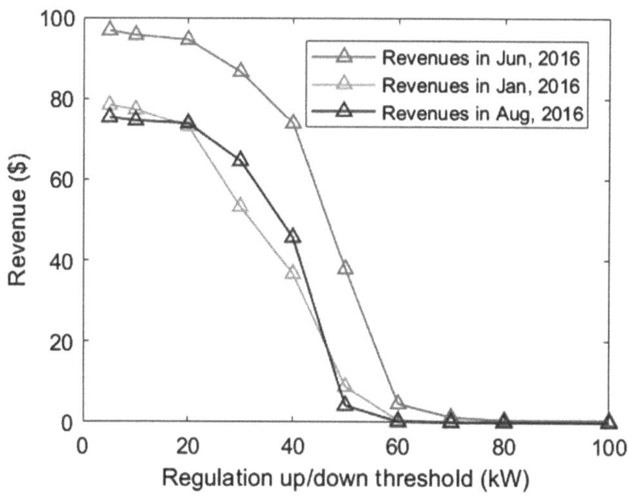

FIGURE 16.12 Impact of regulation threshold.

operations, it will be beneficial for an EV fleet manager to maximize each EV's connected duration flexibility, by either having them arrive earlier or leave later.

In the market participation simulations presented earlier, the minimum threshold to participate was 10 kW, which was appropriate the size of the

EV fleet in this study. Simulations to investigate the impact of varying threshold values on revenue are presented here. As the threshold value increases in Fig. 16.12, the initial revenue drop gets smaller; however, the revenue sharply decreases as the threshold increases from 20 to 60 kW, indicating that most commitments failed to satisfy the constraints defined by Eqs. (16.22) and (16.23) because the required power adjustments exceeded the capacity of the EV fleet.

16.5 Distributed optimization with asynchronous ADMM and V2G capabilities

Sections 16.3 and 16.4 have summarized the modeling approaches and the performance metrics of the market integration strategies of the controllable EVs within microgrid scenarios. Various types of information have been considered in the planning problems, including monthly projection of load profile, renewable generation, and the aggregate EV energy demand. However, the resultant optimal power profile, that is, $P(t), t \in [1, 2, \ldots, T]$, obtained from various market integration strategies, cannot be directly applied to control the individual vehicles, so additional steps are needed to disaggregate the aggregate power profile. In this section, we extend the previous evaluation approach on a monthly basis to an implementable day-by-day operation, whose objective is to allow multiple EV agents to follow the prescheduled aggregator's power consumption profile in a decentralized and asynchronous fashion, using ADMM.

Fig. 16.13 indicates the implementation architecture of the decentralized load following program, where each EV computes a local optimization problem but exchanges information with the aggregator by limited amount of communication. The overall problem is defined as follows:

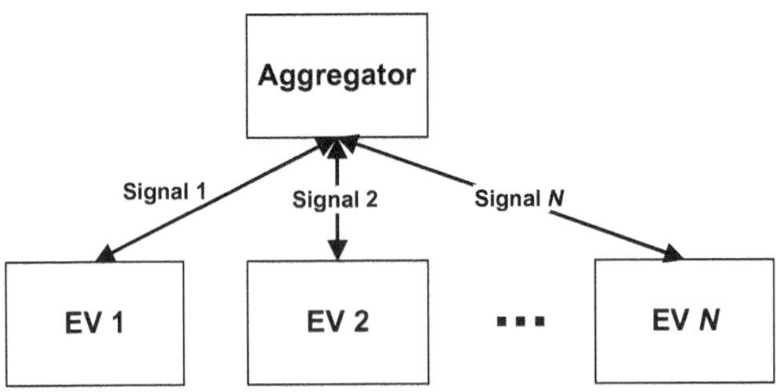

FIGURE 16.13 Distribute optimization paradigm.

Objective

$$\min_{p_a,p_n} c_a(p_a) + \gamma \cdot \sum_{n=1}^{N_{\mathrm{EV}}} c_n(p_n)$$

Subject to

$$p_a = \sum_{n=1}^{N_{\mathrm{EV}}} p_n$$

EV constraints

aggregator constraints

where $c_a(p_a)$ denotes the aggregator's cost function, and $c_n(p_n)$ is the cost function of each vehicle n. As the aggregator consists of all the EV agents, we have $p_a = \sum_{n=1}^{N_{\mathrm{EV}}} p_n$ as an additional constraint, that is, the aggregator load profile should be the summation of the load from all EVs. γ is the weight factor for EV agent's cost and is set to 1 in this case study. If we model the aggregator as one additional agent together with all the EV agents, the total number of agents, N, is equal to $N_{\mathrm{EV}} + 1$. According to Refs. [27,28], this problem can be modeled and rewritten as the exchange problem using ADMM, which is as follows:

Objective

$$\min_{p_a,p_n} \sum_{n=1}^{N} c_n(x_n)$$

Subject to

$$p_a = \sum_{n=1}^{N_{\mathrm{EV}}} p_n$$

EV constraints

aggregator constraints

Following this approach, each agent is able to compute the optimal solution of its own and exchange limited amount of information for each iteration. The optimization problem at each stage is as follows:

$$p_n^{k+1} = \min_{p_n} c_n(p_n) + \frac{\rho}{2} \cdot \left\| p_n - p_n^k + \overline{P}^k + u^k \right\|_2^2 \tag{16.34}$$

where p_n^k is the optimal power profile for EV agent n at iteration k, and $\overline{P}^k$ denotes the averaged power profiles of all EV agents at iteration k. ρ denotes the augmented Lagrangian parameter. For the aggregator the cost function is modified to minimize the difference between the real-world aggregator profile and the one from the day-ahead planning, that is, $D{:=}[P(1), P(2), \ldots, P(T)]$. Thus the updated optimization problem for the aggregator is as follows:

$$p_a^{k+1} = \min_{p_a} \left\| D - p_a \right\|_2^2 + \frac{\rho}{2} \cdot \left\| p_a - p_a^k + \overline{P}^k + u^k \right\|_2^2 \tag{16.35}$$

For each EV agent the updated optimization problem is as follows:

$$p_n^{k+1} = \min_{p_n} \alpha_n \left\| p_n \right\|_2^2 + \frac{\rho}{2} \cdot \left\| p_n - p_n^k + \overline{P}^k + u^k \right\|_2^2 \tag{16.36}$$

where α_n is a scalar to indicate the weight of battery degradation cost for each EV agent. In addition, this formulation will also ensure that each EV will have the minimal power fluctuations. For the aggregator an extra step at each iteration is taken to update the signal to be sent to EV agents:

$$u^{k+1} = u^k + P^{k+1} \tag{16.37}$$

where u^k is defined as y^k/ρ, where y^k is the dual variable for the original problem at iteration k. Note that each agent only needs to store its own copy of power profile, that is, p_n^k, to compute for the new profile p_n^{k+1}, and the signal from the aggregator agent, u^k. Thus one side effect of the decentralized algorithm is the advantage to preserve user privacy by not sharing extra information. The convergence criteria are defined by the primal feasibility, r^k, and the dual feasibility, s_n^k:

$$r^k = \overline{p}^k = \frac{1}{N_{\text{EV}}} \cdot \sum_{n=1}^{N_{\text{EV}}} p_n^k \tag{16.38}$$

$$s_n^k = -\rho N\left(p_n^k - p_n^{k-1} + \left(\overline{p}^{k-1} - \overline{p}^k\right)\right) \tag{16.39}$$

$$\left\| r^k \right\|_2 \le \epsilon^{\text{pri}} \tag{16.40}$$

$$\left\| s^k \right\|_2 \le \epsilon^{\text{dual}} \tag{16.41}$$

where s^k is defined as $\left[s_1^k, s_2^k, \ldots, s_N^k\right]$, and ϵ^{pri} and ϵ^{dual} are the primal and dual convergence criteria, respectively.

Synchronous ADMM for EV load following

Initialize N agents with 1 aggregator and $N-1$ EVs;
While $\|r^k\|_2 \ge \epsilon^{\text{pri}}$ and $\|s^k\|_2 \ge \epsilon^{\text{dual}}$:
  While aggregator receives all EV signals:
    For each EV agent $n \in N^{\text{EV}}$:
      Solve the local optimization problem and send the profile to the aggregator agent N
    Update $\overline{p}^k$ and u^k within the aggregator agent
If $\|r^k\|_2 < \epsilon^{\text{pri}}$ and $\|s^k\|_2 < \epsilon^{\text{dual}}$:
  Terminate

The detailed algorithm is illustrated in the previous table. We show the interactive load following performance in Fig. 16.14 by comparing the given aggregator's load curve to follow with aggregated EV load curves in

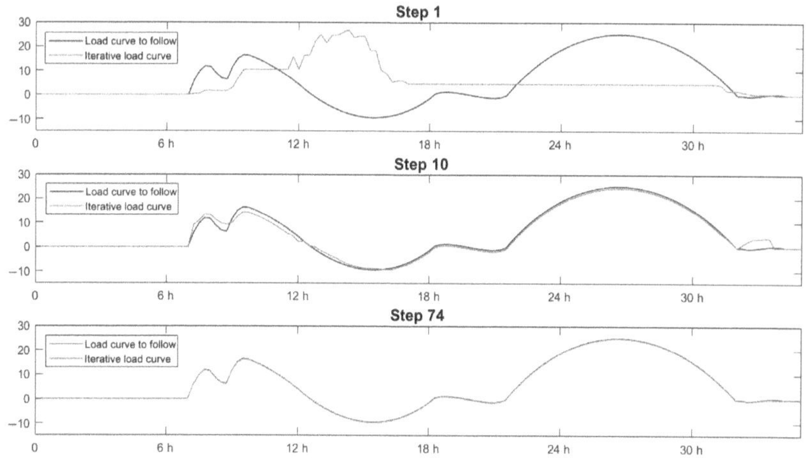

FIGURE 16.14 Iterative load following.

iterations 1, 10, and 74, respectively. Note that, in this case, we have also modified the vehicle power constraint in Eq. (16.1) to

$$\frac{\underline{p}}{\eta_{\text{dis}}} \le p_n^d(t) \le \overline{p} \cdot \eta_{\text{chg}} \tag{16.41}$$

so that discharging the energy from vehicle batteries into the grid is allowed in the day-by-day operations. In Fig. 16.14 the blue curve denotes the given aggregator load profile for the EVs to follow, which consists of both the positive part (aggregate charging) and negative part (aggregate discharging), and the optimized aggregate curve (the red curve), obtained by summing up the power profiles of all EV agents. The given load profiles are generated based on the previous models described in Sections 16.3 and 16.4 using the specific datasets, including building load and EV charging requests with 22 charging sessions, on June 8, 2018. In the iterative process the optimized aggregate power profile approaches the given curve and eventually achieves almost exactly the same curve as the given one. In an implementable environment, where communication between the aggregator and the EV agents is fully enabled, the optimality can be fulfilled by this distribute optimization process.

However, in the real-world communication networks, there can be significant communication delays and packet losses, which may cause the aggregator control to fail to collect signals from certain EV agents, making it difficult to synchronize in the iterative optimization. Thus the algorithm is expected to need more time to converge in the real-world cases. By allowing the minimal number of signals received from the EV agents, we extend the synchronous ADMM algorithm to an asynchronous version, where the EV agent will keep performing the local optimization as it does in the synchronous algorithm, whereas the aggregator will only need minimal number, that is, $N_{\min}$, of signals received from the EV agents to start the optimization and will keep records of the time lag of each EV agent, that is, L_n. In each iteration of the updated algorithm the aggregator needs to ensure that (1) the number of distinct signals from EV agents is larger than the minimal required number, that is, $N_{\text{rec}} \geq N_{\min}$, and (2) the maximum time lag among all the EV agents is less than the maximum lag allowed, that is, $\max([L_1, L_2, \ldots, L_{N-1}]) \leq L_{\max}$. The details in the asynchronous ADMM algorithm are as follows:

Asynchronous ADMM for EV load following

1 **Initialize** N agents with 1 aggregator and $N-1$ EVs, and $N-1$ copies of power profiles, i.e., $[\hat{p}_1, \hat{p}_2, \ldots, \hat{p}_{N-1}]$ within aggregator;
2 **While** $\|r^k\|_2 \geq \epsilon^{\text{pri}}$ and $\|s^k\|_2 \geq \epsilon^{\text{dual}}$:
3   **While** $\max([L_1, L_2, \ldots, L_{N-1}]) \leq L_{\max}$ and $N_{\text{rec}} \geq N_{\min}$
4   **For** each EV agent n received:
5   Set agent lag L_n to 1
6   Update the profile copy within the aggregator by $\hat{p}_n = p_n$
7   **For** each EV agent in the fleet but not received:
8   Increase the agent lag by 1, i.e., $L_n = L_n + 1$
9   Update $\overline{p}^k$ within the aggregator agent by $\overline{p}^k = \frac{\sum_{n=1}^{N_{EV}} p_n^k}{N_{EV}}$
10   If $\|r^k\|_2 < \epsilon^{\text{pri}}$ and $\|s^k\|_2 < \epsilon^{\text{dual}}$:
11   **Terminate**

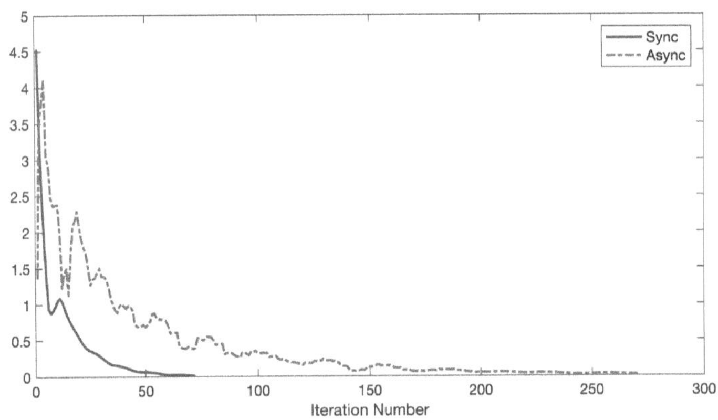

FIGURE 16.15 Primal residuals of sync-ADMM and async-ADMM.

The key difference between the asynchronous ADMM and the synchronous counterpart is that the average power update step in the asynchronous version, that is, line 9 in the previous table, does not need full information from all the EV agents in the same iteration. In other words, it allows the signals from some EV agents to be out of date and only uses the partial information to drive the system to the global optimality. However, the aggregator has to ensure the

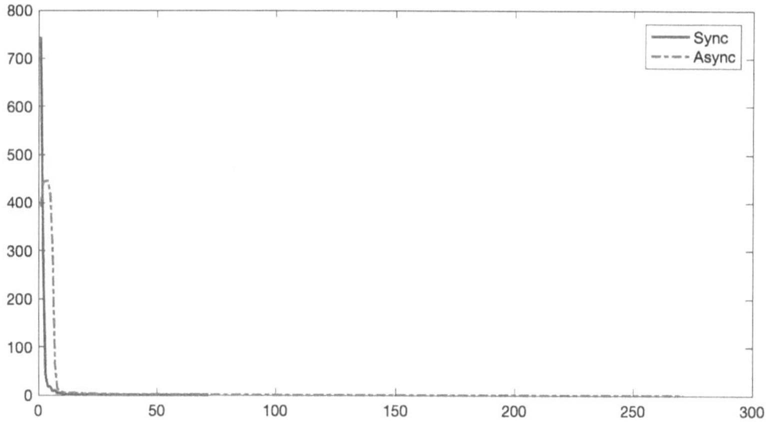

FIGURE 16.16 Dual residuals of sync-ADMM and async-ADMM.

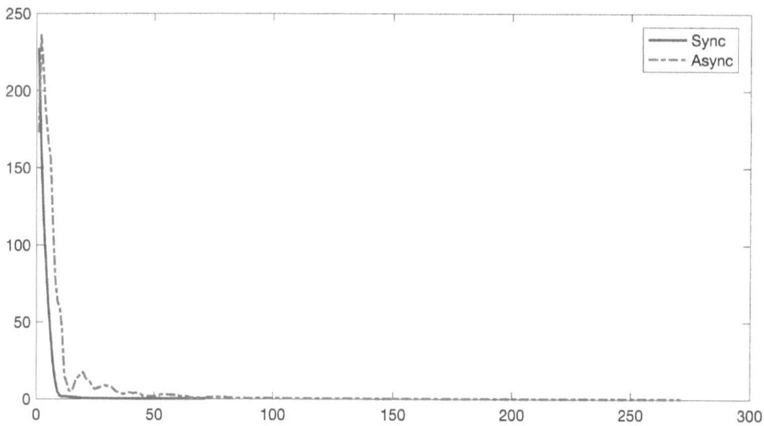

FIGURE 16.17 Objective values of sync-ADMM and async-ADMM.

maximum time lag among the EV agents to be bounded by L_{max}, which in our case is set to 5. The minimal number of agents N_{min} in this case is also set to 5. Figs. 16.15–16.17 show the convergence performance of the asynchronous ADMM algorithm compared to the sync version with regard to primal residual, dual residual, and the objective value, respectively. Though it takes more steps to converge (74 vs 271), asynchronous ADMM can still achieve the system optimality with disturbances from the communication and control systems.

16.6 Conclusion

In this chapter, we introduce the integration strategies of EVs under microgrid scenarios with the California energy, ancillary service, and DR markets.

We briefly described the implementation platform for this project with system components, communication networks, the control logics, etc. The real-world EV usage datasets collected in AlcoPark, North California, are used to the model the EV charging behaviors and quantify the benefits of EV resources to achieve various objectives. V1G cases are extended with discharging options and asynchronous ADMM algorithms to support the decentralized and asynchronous V2G operations, which preserve the user privacy and overcome the synchronization challenges in the real-world communication systems. The future work will focus on handling the shorter timescale uncertainties in energy systems, including the EV driver behaviors, fluctuating solar generation profile, and building load profile. It's also interesting to investigate more complex modeling approaches with integer variables involved, where the convergence performance and problem scalability will be of major concern.

References

[1] J.A.P. Lopes, F.J. Soares, P.M.R. Almeida, Integration of electric vehicles in the electric power system, Proc. IEEE 99 (1) (2011) 168–183.

[2] D.S. Callaway, I.A. Hiskens, Achieving controllability of electric loads, Proc. IEEE 99 (1) (2011) 184–199.

[3] B. Wang, Smart EV Energy Management System to Support Grid Services, UCLA, 2016.

[4] N. DeForest, J.S. MacDonald, D.R. Black, Day ahead optimization of an electric vehicle fleet providing ancillary services in the Los Angeles Air Force Base vehicle-to-grid demonstration, Appl. Energy 210 (2018) 987–1001.

[5] C. Marnay et al., Los Angeles Air Force Base Vehicle to grid pilot project, ECEEE 2013 Summer Study on Energy Efficiency, 2013.

[6] C. Chen, S. Duan, Optimal integration of plug-in hybrid electric vehicles in microgrids, IEEE Trans. Ind. Informat. 10 (3) (2014) 1917–1926.

[7] S. Mal, A. Chattopadhyay, A. Yang, R. Gadh, Electric vehicle smart charging and vehicle-to-grid operation, Int. J. Parallel Emergent Distrib. Syst., 28 (3), 2013, 249–265.

[8] Y. Wang, B. Wang, C.-C. Chu, H. Pota, R. Gadh, Energy management for a commercial building microgrid with stationary and mobile battery storage, Energy Build. 116 (2016) 141–150.

[9] B. Wang, Y. Wang, H. Nazaripouya, C. Qiu, C.C. Chu, R. Gadh, Predictive scheduling framework for electric vehicles with uncertainties of user behaviors, IEEE Internet Things J. 4 (1) (2017) 52–63.

[10] B. Wang et al., Predictive scheduling for electric vehicles considering uncertainty of load and user behaviors, 2016 IEEE/PES Transmission and Distribution Conference and Exposition (T&D), 2016, 1–5.

[11] M.D. Galus, M.G. Vayá, T. Krause, G. Andersson, The role of electric vehicles in smart grids, WIREs Energy Environ 2 (4) (2013) 384–400.

[12] C. Guille, G. Gross, A conceptual framework for the vehicle-to-grid (V2G) implementation, Energy Policy 37 (11) (2009) 4379–4390.

[13] B. Wang, D. Wang, C. Chan, R. Yin, D. Black, Predictive Management of Electric Vehicles in a Community Microgrid, *arXiv:1802.01512 [eess, math]*, Feb. 2018.

[14] P. Sanchez-Martin, G. Sanchez, G. Morales-Espana, Direct load control decision model for aggregated EV charging points, IEEE Trans. Power Syst. 27 (3) (2012) 1577–1584.
[15] E.C. Kara, J.S. Macdonald, D. Black, M. Bérges, G. Hug, S. Kiliccote, Estimating the benefits of electric vehicle smart charging at non-residential locations: a data-driven approach, Appl. Energy 155 (2015) 515–525.
[16] W. Su, M.-Y. Chow, Performance evaluation of an EDA-based large-scale plug-in hybrid electric vehicle charging algorithm, IEEE Trans. Smart Grid 3 (1) (2012) 308–315.
[17] Y. He, B. Venkatesh, L. Guan, Optimal scheduling for charging and discharging of electric vehicles, IEEE Trans. Smart Grid 3 (3) (2012) 1095–1105.
[18] S. Shao, M. Pipattanasomporn, S. Rahman, Grid integration of electric vehicles and demand response with customer choice, IEEE Trans. Smart Grid 3 (1) (2012) 543–550.
[19] Y. Xiong, B. Wang, C.-C. Chu, R. Gadh, Electric Vehicle Driver Clustering using Statistical Model and Machine Learning, *arXiv:1802.04193 [cs]*, Feb. 2018.
[20] Y. Xiong, B. Wang, C. Chu, R. Gadh, Distributed Optimal Vehicle Grid Integration Strategy with User Behavior Prediction, *arXiv:1703.04552 [cs, math]*, Mar. 2017.
[21] B. Wang, Y. Wang, C. Qiu, C.C. Chu, R. Gadh, Event-based electric vehicle scheduling considering random user behaviors, in: 2015 IEEE International Conference on Smart Grid Communications (SmartGridComm), 2015, pp. 313–318.
[22] Y. Xiong, B. Wang, C. Chu, R. Gadh, Vehicle grid integration for demand response with mixture user model and decentralized optimization, Appl. Energy 231 (2018) 481–493.
[23] L. Gan, U. Topcu, S.H. Low, Optimal decentralized protocol for electric vehicle charging, IEEE Trans. Power Syst. 28 (2) (2013) 940–951.
[24] B. Wang, B. Hu, C. Qiu, P. Chu, R. Gadh, EV charging algorithm implementation with user price preference, Innovative Smart Grid Technologies Conference (ISGT), 2015 IEEE Power Energy Society, 2015, 1–5.
[25] N. Rotering, M. Ilic, Optimal charge control of plug-in hybrid electric vehicles in deregulated electricity markets, IEEE Trans. Power Syst. 26 (3) (2011) 1021–1029.
[26] P. Finn, C. Fitzpatrick, D. Connolly, Demand side management of electric car charging: benefits for consumer and grid, Energy 42 (1) (2012) 358–363.
[27] J. Rivera, P. Wolfrum, S. Hirche, C. Goebel, H. Jacobsen, Alternating direction method of multipliers for decentralized electric vehicle charging control, 52nd IEEE Conference on Decision and Control, 2013, 6960–6965.
[28] S. Boyd, Distributed optimization and statistical learning via the alternating direction method of multipliers, Found. Trends Mach. Learn. 3 (1) (2010) 1–122.

Further reading

H. Zhang, Z. Hu, Z. Xu, Y. Song, Evaluation of achievable vehicle-to-grid capacity using aggregate PEV model, IEEE Trans. Power Syst. 32 (1) (2017) 784–794.

Chapter 17

Pattern-recognition methods for decision-making in protection of transmission lines

Mohammad Pazoki[1], Anamika Yadav[2] and Almoataz Y. Abdelaziz[3]
[1]*School of Engineering, Damghan University, Damghan, Iran,* [2]*Department of Electrical Engineering, National Institute of Technology, Raipur, India,* [3]*Electrical Power and Machines Department, Faculty of Engineering, Ain Shams University, Cairo, Egypt*

17.1 Introduction

Pattern recognition is a scientific procedure that focuses on the automatic discovery of regularities and relations in input patterns. Then, it assigns a label or a real-valued output to each input patterns. The input patterns can be provided based on the main goal of the procedure. They can be any type of events, measurements, processes, and captures that need to be categorized or estimated. Today the demand for practical functioning of pattern recognition has been increased in parallel with promotion of core processors. Appearance of new concepts in industrial systems and handling of large amount of information generates a novel trend in the field of pattern recognition. Development and implement of various systems of pattern recognition confirm the creation of growing trend in the application and research. The system of pattern recognition can be categorized according to the fields of applications. For example, classification of document image in the field of optical character recognition, personal identification based on the face or finger print in the field of biometric, computer-aided diagnosis in the field of medical, automatic target recognition in the field of military, sequence analysis in the field of bioinformatics, sorting objects in the field of industrial automation, power system protection in the field of power electrical engineering, etc. The problems of pattern recognition have four common solutions, which are classified into four classes: statistical, structural, hybrid methods of the statistical and structural, and artificial intelligence (AI) approaches. The last one is recommended for more complex problems where there is not a linear and direct relation between the input patterns and target vector.

Decision Making Applications in Modern Power Systems. DOI: https://doi.org/10.1016/B978-0-12-816445-7.00017-7

The main focus of this chapter is to consider the application of pattern recognition in the field of transmission line protection. The transmission lines are the vital arteries of networks that carry the generated power from power plants to the loads. Therefore the protection of power transmission lines is vital to provide the reliability and security of the system. Electromechanical, electronic, digital, numerical, and today's smart relays represent a trend in the field of power system protection and specifically protection schemes of transmission lines. Pattern-recognition methods will enable detection, prediction, classification, and decision-making, which are important functioning integrated in the implementation of the protection schemes to develop a smarter transmission system. Distance relays with different functions are used as main and backup protection systems. The model of decision-making in a distance relay contains a representation of the nonlinearity of the mapping between the input vector and output target. Fault detection, fault classification or phase selection, high-impedance fault detection, symmetrical fault detection during power swing, and power swing detection are the functions developed through pattern-recognition methods. Since the performance of some schemes of the conventional distance relays has some shortcomings, the pattern-recognition methods have potential to improve the performance of the schemes. The pattern-recognition-based functions can execute their functionality with cooperation with other routines of the conventional relays.

Since researchers investigate the application of AI methods on power systems in general forms, this chapter individually represents an overview of the application of these methods to the protection of transmission lines as follows: Section 17.2 provides more details about the pattern-recognitions methods from viewpoint of power system protection. Section 17.3 explains the most important functions of a distance relay which are based on the pattern recognition. Section 17.4 represents a general framework for the smart relays and also describes the advantages and disadvantages of pattern-recognition-based relays. Finally, Section 17.5 concludes the chapter.

17.2 Pattern recognition

Generally, there are two models for pattern recognition, that is, supervised and unsupervised models. In a supervised model, the target classes are predefined and the trained model recognizes an unknown pattern as a member of a class. In an unsupervised model, input patterns are grouped into different clusters defined as classes after this time [1]. As the pattern-recognition-based protection schemes are supervised models, this chapter only focuses on the supervised models. Fig. 17.1 at a glance provides the main parts of pattern-recognition methods in the protection of transmission lines. The feature

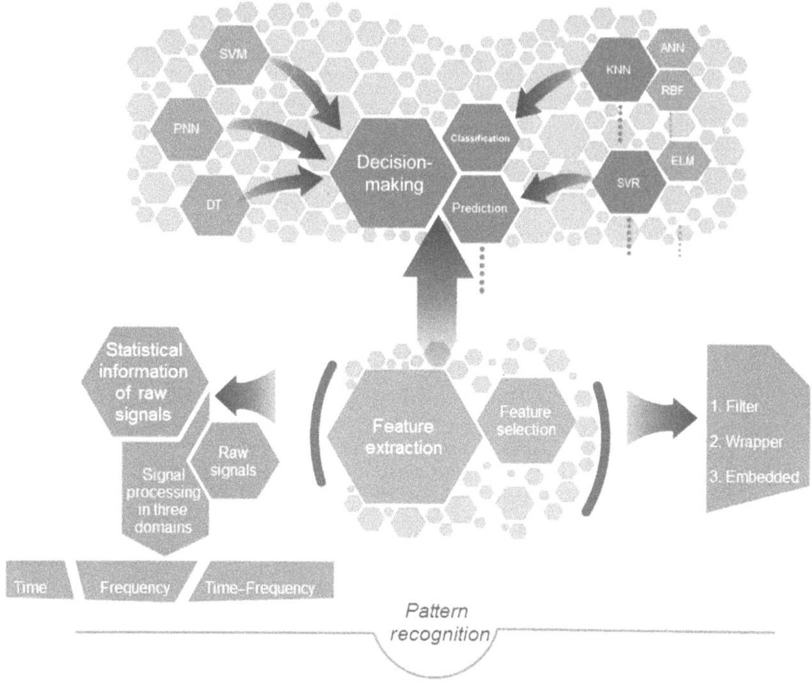

FIGURE 17.1 The structure of pattern-recognition methods.

extraction and decision-making are the primary tasks of automatic pattern recognition. The computation burden is one of the main challenges of the protection schemes, and hence, the feature selection as a reduction technique of construction of a feature vector can be considered in the procedure of the pattern recognitions.

Since distance relays are the most difficult to set in protection relays, and they need a multiplicity of settings; this section concentrates on outlining the development of pattern recognition for some functions of distance protective relays. The procedure of a pattern-recognition-based protection scheme has essential requirements to be considered:

- *Identification of function type of protection and, hence, a definition of the target of protection scheme:* A distance relay as a multitask relay has different protection functions. Some of these functions can be a candidate to replace with pattern-recognition methods. The functions are expected to issue a trip decision or supervise other performances. In addition to the existing protection functions, some other auxiliary functions can be designed to improve the performance of the distance relay. Accordingly, the target should be identified, then the pattern-recognition model can make a final decision.

- *Identification of main or backup protection:* The time operation of a pattern-recognition method can be defined based on the identifying of main or backup protection schemes. Backup protection schemes have more time to operate compared to the main schemes.
- *Identification of classification or estimation task:* Some functions of a distance relay are based on the classification purpose such as fault classification, and some functions are based on estimation purpose such as fault location. An appropriate pattern-recognition model should be designed based on classification or estimation task.
- *Identification of data which are available for protection schemes:* The power system network is an interconnected system, and hence, the measured signals, which are available through transducers at the relay location, should be identified. Nowadays more data are available for protection purposes due to developing some applications such as phase measurement units (PMUs).
- *Preprocessing of data and construction of feature vector:* Providing the appropriate input patterns for pattern-recognitions methods has a vital role to achieve desired performance. The aim of preprocessing is to extract useful features from the original measuring that are fast to calculate and also preserve suitable information.
- *Decision-making of the pattern-recognition-based protection scheme:* Finally, after the providing input patterns, the trained classifier or estimator model makes a final decision. In supervised models, the classifier or estimator model is trained using the training data and the desired output can be a class label (in classification problems) or continues variable (in estimation problems).

17.2.1 Feature extraction

Generally, a pattern-recognition method employs two basic subroutines, including feature extraction and decision-making. The process of feature extraction can be considered as an intermediary formulation to constrict a large vector of data to a small vector of attributes. In other words, based on definition in Ref. [1], "a feature of a given parameter set refers to an attribute described by one or more elements of the original pattern vector." These features construct the input vector of a pattern-recognition method. The feature vector constructed through different preprocessing techniques improves the performance of decision-making. In pattern-recognition-based schemes for protection of transmission lines, the features may include the following:

- *Raw data:* Since the voltage and current signals are available for distance relays, the instantaneous amplitudes of these signals can be used as a vector of input feature. In this category of the feature, there is no need to any preprocessing techniques. It has simple procedures, but it suffers from serious disadvantages. The instantaneous amplitude of measured voltage and current signals highly depends on the power system conditions, loading level, fault parameters, etc. The wide variation of input patterns decreases the generalization ability of trained model. For example, in a pattern-recognition-based fault classification scheme, when the input feature vector is raw signals, any sudden change of current or voltage can be misclassified as a fault. Moreover, large input data need more memory space and has more computation burden. The raw data can be useful for some simple function such as ground detection during a fault. Finding a trend and unique behavior in original signals during an events in power systems is a challenging issue.

- *Preprocessed data:* The signal transforms process the signals obtained from transducers to extract useful features. Sometimes the filtering process is also used before the signal processing. Three types of filters are used in protection functions of transmission line, that is, high-pass filter such as elliptic filters [2], low-pass filter such as Butterworth filters [3], and DC offset removal filter such as mimic filter [4]. Uncovering higher or lower frequency components by filtering of the measured voltage or current transients can improve the extracted features. Moreover, using the DC removal and low-pass filters are inevitable to extract fundamental components of power signals. Signal transforms generate representations that are useful in various protection schemes, including harmonic analysis, phasor measurement, noise reduction, filtering, frequency tracking, and feature extraction. Signal transforms are developed in three domains: frequency, time, and time–frequency. In protection schemes, the transforms in frequency and time–frequency domains are more popular compared to the transforms in the time domain. This is because of the facts that the fault signals in power transmission lines contain usefulness information in the frequency domain. Accordingly, some signal transforms in the time domain which obtain the frequency content of a signal can be interested for researchers. In this section a brief overview of signal transforms is represented from the protection viewpoint:
  1. One of the most important transforms in the frequency domain is discrete Fourier transform (DFT). The DFT is used to derive a frequency-spectrum representation of a time-varying input signal. Using the coefficients obtained by the DFT, the fundamental and non-fundamental components of input signals can be applied to protection schemes. The application of fundamental phasor components of input signals of relays is known in literature. Since faults in a power transmission line create signals having a wide frequency range, nonfundamental frequency components of input signals, as well as fundamental frequency components, are used for feature extraction in pattern-recognition-based protection schemes [3,5–7]. The DFT is a proper

tool for stationary signals, and it cannot be successfully applied to nonstationary signals such as fault signals. To overcome this problem the signal can be divided into small enough parts, then the DFT will be applied to these small parts of the signal as stationary. This improved procedure of the DFT is identified as short-term Fourier transform. Moreover, the fast Fourier transform (FFT) reduces the computation burden of the DFT considerably, and hence, the FFT as an elegant algorithm can also compute the DFT [8].

2. Since the transforms in time–frequency domain have more abundance compared to ones in time domain, time–frequency-based transforms are considered as a second category. Time–frequency transforms simultaneously analyze a time series signal in time and frequency domains. The discrete wavelet transform (DWT) as a multiresolution analysis provides simultaneous information on the time and frequency domains. The DWT decomposes an original signal on a set of functions, called the mother wavelet. The DWT produces two sets of coefficients from a given signal: approximation coefficients and detail coefficients. Approximation and detail coefficients are computed by convolving the signal with the low-pass filter and the high-pass filter, respectively. This procedure can be repeated for approximation coefficients as a new signal, and hence, the DWT can be organized for n level of decomposition. The S-transform (ST) has been introduced by RG Stockwell in 1996 with better time–frequency representation compared to the DWT [9]. In addition to original version of the ST, different versions of the ST are proposed for feature extraction purpose such as hyperbolic ST [10] and fast discrete orthonormal ST [11,12]. The output of the ST is a matrix with complex elements. The rows of the matrix represent the frequency information of the input signal, and the columns represent the time. Therefore the ST obtains the instantaneous frequency information of the signal corresponding to each time of sampling. Therefore the ST can be impressively applied to nonstationary signals such as fault signals captured from transmission lines. Since the DWT and ST contain the instantaneous frequency content of power system signals, they can be applied to extract proper features for the protection purpose.
3. Third category of signal transforms, which are recently applied to transmission line protection, is in time domain. These transforms decompose the input signal into different time domain components. Accordingly, n levels of decomposition obtain n components with specified frequency content. Empirical mode decomposition (EMD) method is based on the fact that any nonstationary signal includes different simple intrinsic mode oscillations. The EMD decomposes the input signal into intrinsic mode functions (IMFs) through sifting procedure. The decomposition obtains a group of frequency-ordered IMF

components. Each sequential of IMF components includes lower frequency oscillations compared to the previous one [13,14]. Another signal transforms in time domain that recently used for the protection of transmission lines is intrinsic time decomposition (ITD). The ITD decomposes a signal into the proper rotation components (PRCs) and a monotonic baseline signal in the time domain. The PRCs represent the inherent amplitude and frequency information of the input signal [15,16]. The EMD and ITD are self-adaptive signal transforms; hence, these signal transforms do no need basis function. They are based on the local characteristic timescales of the input signal and compute a series of data sequences with distinctive characteristic scales. Consequently, the time domain transforms provide the frequency information of a signal in the time domain; therefore they can be applied to extract useful features from the signals of power systems.

Different fault conditions considerably affect the raw signal. In case where the raw signals are insufficient as a proper feature vector, incorporating the signal transforms into the feature extraction procedure improves the distinctive ability of features to obtain better decision-making in protection schemes. To improve the overall quality of the input feature vector, in addition to signal transforms, some data processing techniques can be applied to the output of signal transforms before the classification stage. For example, the matrix methods make the features from the decomposed matrix accessible. Nonnegative matrix factorization, principal component analysis, and singular value decomposition are some known examples of matrix methods [17]. Moreover, the statistical methods can extract useful information from the output of signal transforms. This type of features can be named as hidden features that decrease the dimensionality of the output of signal transforms by eliminating the redundancy and also keeping the representative and discriminative information of the output [3]. The ending process of the feature vector construction is the normalization. The normalization techniques map the extracted features to a small specified range, such as $[-1, 1]$ or $[0, 1]$. There are some normalization techniques such as min−max normalization, Z-score normalization, and decimal scaling normalization. Specifically, the latter two techniques can change the original input patterns quite a bit, whereas the min−max normalization provides linear transformation on original range of data. The min−max normalization keeps the relationship and distance among original input patterns. Fig. 17.2 shows the possible ways of feature vector construction concisely.

17.2.2 Feature selection

After construction of feature vector and before decision-making, the feature selection may be employed. In the stage of feature extraction, different

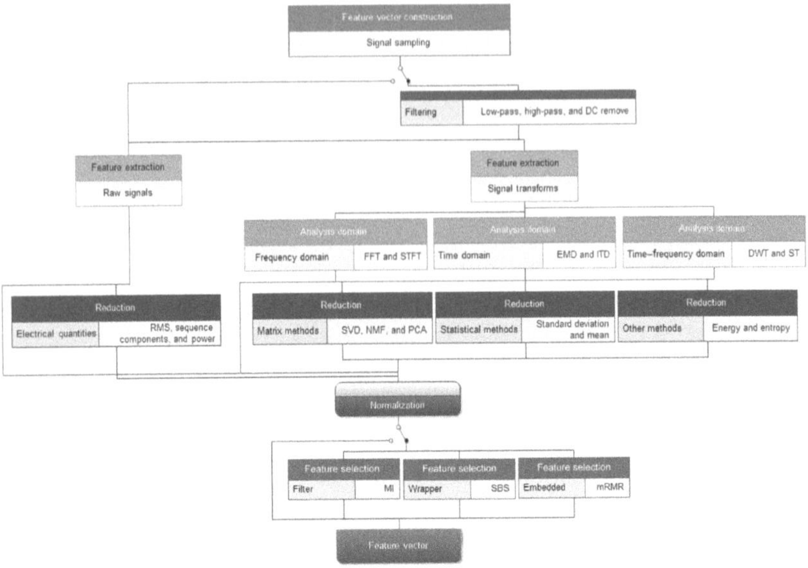

FIGURE 17.2 The details of feature vector construction.

features may be extracted from the sampled signals at the relay location. The procedure of feature selection selects a subset from the original features. The feature selection is performed to

- select informative and pertinent features and reduce the feature set,
- save memory space and decrease computation burden of the algorithm,
- improve the classification and/or predictive accuracy,
- enhance generalization ability, and
- represent knowledge about the extracted features.

In this chapter the feature selection techniques using supervised models are investigated. The techniques are classified into filter, wrapper, and embedded [18].

Filter techniques rank features to filter out the less relevant variables before classification. How to measure the relevancy of a feature to the other features or the data or the target output is one of the main challenges of filter techniques. A proper ranking criterion is employed to score the input features, and a feature that has the score below the threshold can be a candidate to remove. Filter techniques are more useful to represent the relationships between the input features, whereas they don't consider the performance of a classifier or predictor model. Many filter techniques achieve only a feature ranking based on the measuring of relationships instead of the best subset of input feature. Mutual information-based methods, correlation coefficient–based criteria, relief-based algorithms are some of the examples of filter techniques.

Wrapper techniques use the performance of the classifier or predictor model as an objective function to score the input features. In these techniques, different search algorithms can be employed to search a subset of input features, and the wrapper technique needs to train a new model for each subset, and subsequently, the performance of predictor model determines the best selection of the subset. Exhaustive search algorithms impose high-computation burden for a large size of input features. Filter techniques can also be employed as a preprocessing subroutine for wrapper techniques, and hence, it allows a wrapper technique to be applied to the large size of input features. On the other hand, the wrapper techniques can find the best subset of features corresponding to a particular type of predictor. Wrapper techniques can be classified into Sequential search algorithms such as sequential backward selection and sequential floating forward selection or evolutionary algorithms such as genetic algorithm (GA) and particle swarm optimization.

Embedded techniques incorporate the training part and feature selection part simultaneously. The training phase is not considered in filter techniques. On the other hand, wrapper techniques perform the reclassifying the subsets of input feature to find the best subset of the feature which can be combined with any classifier or predictor model. In embedded techniques, the structure of the classifier or predictor model employed in pattern recognition has an important role [19]. The max-relevancy, min-redundancy is an example of embedded techniques [20].

To sum up, based on feature selection techniques, some extracted features which don't have discriminative ability can be removed from the original feature vector while the model of pattern recognition has the highest accuracy. Finally, the performance of the classifier or predictor model demonstrates the necessity of using a feature selection technique.

17.2.3 Decision-making

As shown in Fig. 17.1, the final decision of a protection scheme based on supervised pattern recognition is made through two types of function: classification or prediction. When a protection scheme is designed, the type of its output defines the type of required function. For example, a fault classification scheme should identify the faulty phase(s) in the case of an occurrence of a fault at a transmission line. Corresponding to each phase, there is a binary classification task. Three phases should be classified according to fault or no-fault classes. In this manner the protection scheme can classify 10 types of fault which may happen in transmission lines. As another example, a fault location problem in transmission lines can be solved through pattern recognition based on the prediction function. In this problem, after a fault occurrence, a regression function estimates the fault location using the input feature vector. The output of the prediction function is a real value that shows the distance between the fault point and the relay location.

17.2.3.1 Classification

In a transmission line, when the input signals are sampled at the relay location, some preprocessing is made to construct the input feature vector, and the vector is fed into the classifier. This pattern-recognition-based relay is allowed to determine that the input signals are belong to which output class. The classification procedure contains two main sequential steps: training and testing steps. The matrix of input data set including feature vector and target vector can be divided in two parts: use onc for training and the remaining is for testing. Therefore the testing data set is unseen during the training step. In the training step, the classification model is learned through training data set, and hence, a learned classifier will be built. The testing date set is employed to evaluate the classification accuracy of a learned classifier. Using the classification function, the class labels for the testing data set are predicted. The classification accuracy is computed as percentage of correct predictions. Moreover, a confusion-matrix can be developed when more details about the performance of the classifier are required in order to discover where the classifier is failing [21]. To obtain higher accuracy of classification, the one of the best ways is to test out different classification algorithm and also examining different parameters corresponding to each algorithm. The best selection can be determined through K-fold cross-validation [1].

Different classification functions can be applied as a classifier in pattern recognition. Since the final decision of a protection scheme is made through a classifier, the selection of classifier among various classifiers is a challenging issue. On the other hand, there is no single predictor superior to its rivals for solving all the problems. However, evidence suggests and confirms that the extraction of right features is the most important factor for proper designing a pattern-recognition method. The classification algorithms can be categorized into binary and multiclass algorithms. Binary classification is the task of classifying the instances of an input data set into two classes. Multiclass classification is the task of classifying the instances of input data set into three or more classes. It is worth mentioning that the binary classification can be used in multiclass problems based on class-versus-class and one class-versus-rest techniques. Some known classifiers that are used to decide in pattern-recognition-based relays are artificial neural networks (ANN), support vector machines (SVMs), probabilistic neural network (PNN), decision tree (DT), k-nearest neighbor (k-NN), random forest (RF), etc.

17.2.3.2 Prediction

In some functions of protection relays of transmission line, the real-value estimation is required. For example, one of the solutions to find the location of the fault is pattern-recognition-based methods [22]. In classification problems, the pattern-recognition model predicts discrete outcomes, for example, "Is there a fault or not?" and "Is a stable swing or an unstable?". In regression problems, the pattern-recognition model predicts continuous outcomes. For example, "What is the distance between the fault location and relaying point?". The same as classification, some data should be used to train the algorithm, and then the learned algorithm can make a prediction. In prediction problems, prediction error of input test data set is the criterion that can be employed to evaluate how good the learned algorithm is. In prediction of continues target, prediction error can be computed based on absolute or relative error. Some known predictors that are used to estimate value in pattern-recognition-based relays are ANN, SVMs, *k*-NN, etc.

17.3 Pattern recognition application on protection of transmission line

Generally, protection of transmission line employs two subroutines including fault detection and fault-type classification. The pattern-recognition-based methods for fault detection and classification in transmission lines have been comprehensively deliberated over the last three decades. This section presents a comprehensive survey of the pattern-recognition methods for decision-making in protection of transmission lines. With the progression of the present electrical power grid to a smart grid, the significance of designing a smart relay capable of rapidly detecting and classifying the fault in a transmission line is growing interest of several researchers. The developments in signal processing and feature extraction techniques, PMU, global positioning system, and communications have facilitated the protection engineers to develop smart and adaptive relays. Recently, few review articles related to the transmission lines protection schemes [23–27] have been published. The major challenges in the protection of transmission lines are the detection of high-impedance fault (HIF), power swings, and fault during power swings, wherein the conventional distance protection scheme fails. Understanding the concept of these phenomena would assist the protection engineers to cultivate an intelligent fault detection and classification techniques based on fault pattern recognition to assist the conventional relaying scheme.

17.3.1 Fault detection, classification, and location

Fault in transmission lines is an inevitable event either due to environmental stress such as thunder storm, lightning, fog, snow fall, dust contamination, or electrical stress, for instance, internal or partial discharge in insulators causing its failure. The status of the transmission line, whether it is healthy or faulty, is usually predicted by distinctive characteristic variation in the associated voltage and current waveforms during faulty or healthy conditions. Therefore the task of protective relaying can be treated as pattern recognition or classification problem. Based on the transmission line configuration, the different pattern-recognition-based fault detection and classification schemes are categorized into two types: single-circuit transmission lines and double-circuit transmission lines (DCTLs). In the recent past, different AI and data mining techniques have been applied to solve the pattern recognition problem of fault detection and classification in a power transmission network. ANN, fuzzy logic (FL), adaptive neuro-fuzzy system (ANFIS), *k*-NN, DT, SVM, and extreme learning machine are more noticeable among different AI and data mining techniques employed to solve such pattern recognition or classification problems of transmission line protection.

17.3.1.1 Fault detection, classification, and location in single-circuit transmission line

In this section, different pattern-recognition-based schemes for fault detection and classification in the single-circuit transmission line have been reviewed in terms of two aspects; first, the feature extraction method and the classifier or predictor are employed. Owing to the different characteristics of ANN such as generalization capability, noise immunity, adaptability, fault tolerant, and fast response, it has been most widely used for fault detection and classification in transmission lines [27–36]. Different topologies of ANN have been used such as feed-forward neural network (FFNN) [28,29], self-organized map (SOM) [30], PNN [31,32], Elman recurrent network (ERN) [33], adaptive resonant theory (ART) [34], etc. In Ref. [28] the raw signals of voltage, current, and direction of current, while in Ref. [29] the fundamental components of voltage and current are used as input to multi-layered perceptron network with backpropagation as training algorithm. The fault detection, faulty phase selection, and fault distance estimation in double-end fed transmission lines are presented using fundamental components of voltage and current signals and ANN in Ref. [30]. Further, ANN has been used to detect and classify the intercircuit and cross-country faults in Ref. [31]. FFT with FFNN is presented in Ref. [32] for fault classification using SOM algorithm and fault location using Levenberg–Marquardt algorithm. ST in conjunction with PNN is employed for fault detection and classification whereas ST with BPNN for fault location in Ref. [33]. In Ref. [34], wavelet energies of the current signals are used as input to PNN classifier to distinguish the transients originating from the faults from those originating from normal switching events. Recently, a fast time-domain signal processing tool called as ITD is used with PNN to classify the fault in Ref. [16]. DWT is used to extract distinctive features of both voltage and current signals and fed to ERN for fault location estimation in Ref. [35]. Further, amplitude, phase, and sequence components of current and voltage are utilized by ART-based fault phase selector in Ref. [36]. Despite the popularity of ANN-based scheme, the major drawback of the ANN-based algorithm is trial and error procedures to find the number of layers, neurons, and activation functions, which makes the overall design process a tedious and complex task.

SVM is a statistical technique used for the purpose of computational learning in Refs. [37–40], which overcomes the drawback of ANN by giving a global solution rather than local minima. Finite impulse response filter with SVMs has been used for fault detection, classification and support vector regression (SVRs) are utilized to locate the faults in Ref. [37]. EMD is used for decomposing voltage signal into IMFs, and then Hilbert Huang Transform is used to extract the features from IMFs. Finally, SVM is proposed for fault classification in Ref. [38]. A new combination of hyperbolic ST and SVMs/SVRs is applied for fault detection, classification, and location in Ref. [10]. DWT is employed to yield the change in energy and standard deviation (SD) of fault current and voltage signal for faulty phase identification and fault location in Ref. [39]. In Ref. [40] wavelet entropy criterion is applied to detail wavelet coefficients of voltage and current signals to reduce the size of the feature vector and fed to SVM for fault classification and location estimation.

DWT-based algorithms in pattern recognition techniques are also popular in recent times. An ANN classifies the fault based on the voltage and current waveforms in the time domain using real oscillographic data in Ref. [41]. The entropies of the wavelet decomposed voltage signals have been fed to the neural networks for fault classification and location in Ref. [42]. An exhaustive survey on the application of wavelet transforms-based fault identification in high voltage (HV) transmission line has been reported in Ref. [43], which discusses different schemes utilizing DWT with different pattern recognition schemes, useful for researchers working in this field. Another scheme in Ref. [44] uses the wavelet packet transform, a generalization of DWT wherein discrete time signal passes through a series of filters, for feature extraction and fed to SVM for fault classification and location. DWT is used in conjunction with DT and *k*-NN classifiers as a means of the semisupervised machine learning in Ref. [45].

FL provides an inference mechanism that enables approximate human reasoning capabilities to be applied to knowledge-based systems such as protective relaying task. Fuzzy inference system (FIS) has been used to detect the fault (both forward and reverse), locate and also identify the faulty phase (s) in a multisection transmission system in Ref. [46]. Further, application of FIS to detect the presence of a fault and its direction using phase angle of positive sequence current only is proposed in Ref. [47].

DT-based methods have been employed for fault classification in a transmission line using the odd harmonics up to 19th of the voltage and current signals in Ref. [5]. Further, DWT-DT–based fault classification scheme with classification and regression tree analysis is proposed in Ref. [48]. ST and PNN-based fault classification and section identification schemes have been proposed for series-compensated line in Ref. [49]. In Ref. [10], hyperbolic ST and SVMs/SVRs have been used to detect, classify, and locate the fault in series-compensated transmission line. In Ref. [50], the combination of ST and logistic model trees is used to classify a fault in series-compensated lines.

Besides the hybrid approach using two different pattern recognition methods is also envisaged by authors in Refs. [51–54]. In Ref. [51], ST is used to extract statistical features, and then DT-induced–fuzzy rule base is developed for fault classification. In Ref. [52] the change in energy and SD are used an input to ANFIS method to predict the fault location. Wavelet and neuro-fuzzy-based fault location for combined overhead and underground cable transmission systems is presented in Ref. [53]. An adaptive neuro-fuzzy approach is presented for fault direction estimation in three-section transmission lines using fundamental components of current signals in Ref. [54].

17.3.1.2 Fault detection, classification, and location in double-circuit transmission lines

Since the last two decades, most of the new transmission lines constructed are double-circuit lines owing to the advantage of carrying the double amount of power as compared to single-circuit lines. This section presents a brief review of the pattern-recognition-based methods employed for the protection of the DCTLs. ANN is used to calculate the appropriate tripping impedance for fault detection in a DCTL in Ref. [55]. Another hybrid approach comprising DWT and ANN for fault section identification and distance location is presented in Ref. [56] which offers primary protection as well as backup protection the adjacent forward and reverse line sections. Further, the first zone reach setting is improved using ANN-based protection system for DCTLs in Ref. [57]. It is based on discrete Fourier coefficients of the three-phase voltage and current signals measured at one end of the line. An adaptive distance protection scheme using Kohonen neural network for different switching and operating modes of DCTLs is reported in Ref. [58]. But it considered only single-line-to-ground fault in one circuit. DWT and back-propagation neural network (BPNN) based on Clarke's transformation are proposed for fault detection and classification in parallel lines in Ref. [59]. A fuzzy system is used to classify the series, shunt, and simultaneous series-shunt faults and locate the fault in a DCTLs in Ref. [60] using synchronized two ends measurement. A complete protection scheme is proposed for series capacitor-compensated DCTLs using DWT and *k*-NN in Ref. [61].

17.3.2 High-impedance fault detection

HIF occurs when the transmission line live conductors make contact with a surface which offers high impedance, such as asphalt, concrete roads, dry and high resistance ground, gravel, or when the live conductor touches ground through a tree limb. The conventional distance protection scheme is unable to detect the HIF. As a result, designing of a fast, accurate, and reliable algorithm for HIF detection is an open problem in protection of transmission lines, up till now. The foremost pattern-recognition methods that have been applied for HIF detection is ANN [62]. Further combined DWT and ANN in Refs. [63,64] and multiresolution morphological gradient with ANN in Refs. [65] were reported for HIF detection in distribution lines. Recently, mathematical morphological filter is used in Ref. [66] for preprocessing of signal, and data mining–based DT is used for classification of HIF or non-HIF. Although there are numerous other pattern-recognition methods discussed earlier, they have not been reported for HIF detection in a transmission line. However, different pattern-recognition methods could still exhibit a high level of effectiveness in detecting the HIF if accompanied by appropriate preprocessing and feature extraction techniques discussed in Section 17.2.

17.3.3 Power swing detection

An abrupt variation in loading or configuration of the power system network such as a change in mechanical power input to the turbine, tripping of one of the parallel lines due to a fault and sudden application of heavy load; originates power swing in the network. The characteristic behavior of the power swing is similar to three-phase symmetrical fault in the network which may cause undesirable tripping of the distance relay. Therefore it is essential to quickly and reliably discriminate between the fault and power swing conditions to prevent the instability of the power system network. The conventional power swing detection needs a lot of stability study in a power system, and also they don't have accurate setting [67]. In Ref. [68], fault phase selector has been proposed for unbalanced fault during a power swing based on series multiresolution morphological gradient transform. It uses superimposed components of modular current to identify the faulted phase during power swing. Further, summation of the sixth-level detail coefficients for current signals of the three phases has been used to classify the fault type, and ANFIS is used to locate the fault with power swing in Ref. [69]. An ANFIS-based scheme is proposed in Refs. [70,71] to block the tripping of the relay during power swings using the variation of positive sequence impedance, positive and negative sequence currents, and power swing center voltage. Detection of unstable power swings during distance relay operation using the ST and ANN is proposed in Ref. [72]. In Ref. [73], SVMs have been used to classify the power system events into three categories: faults, power swings, and voltage instability, to assist in supervisory control and operation of the conventional distance relays. Further, k-NN has been applied to detect the power swing in Ref. [74].

17.3.4 Symmetrical fault detection during power swing

The distance relay uses power swing blocking element to prevent the false tripping in case of power swing. But if a symmetrical fault occurs during power swing, distance relay is unable to detect this fault, being disabled by power swing blocking element. Since the nature of both symmetrical fault and power swing event is balanced and has similar characteristics, it poses serious concern for distance protection scheme. There are very few papers that deal with detection of symmetrical faults during power swing using pattern-recognition-based methods [75–80]. An intelligent scheme based on SVM and PNN to detect the symmetrical three-phase fault during power swing and voltage collapse is reported in Ref. [76]. Here, ST has been used to extract the features. SVM has been used to detect faults during power swing using current signals of half-cycle only in Ref. [76]. Another protection scheme is proposed to identify fault during power swing based on combination of PNN and ST in Ref. [77] using only half cycle data of current signal. Two pattern-recognition-based methods such as DT and RF were proposed in Ref. [78] for detection of power swing and fault during power swing in series-compensated line using 20 differential features. Thus this scheme depends upon remote end data retrieval. The scheme in Ref. [79] utilizes DT for detection and classification of both symmetrical and unsymmetrical faults during power swing using fundamental components of voltages and currents of one end only. Another paper [80] reports an ultrahigh-speed scheme to detect fault during a power swing in the series-compensated line using multiresolution morphological gradient and SVM classifier.

Table 17.1 shows abovementioned papers used the pattern-recognition-based protection schemes concisely. Moreover, the table represents the purpose of the schemes, signal processing tools, extracted features, and type of classifiers or predictors.

17.4 Decision-making based on smart relays

Electromechanical protective relaying is the first generation of relays, and they were introduced in the 1910s. The solid state or static relays is the next generation of relays after electromechanical relays. The second generation was first introduced in the 1960s. With the development of microprocessors and microcontrollers, in around the 1980s, the digital relays as the third generation of relays were produced. In digital relays, analog signals are replaced with digital signals. Finally, in the trend of improvement in the technology of relays, numerical relays, as the newest generation, are introduced due to advances in digital signal processors (DSP). They were introduced in 1985 and developed in 1990. In this chapter the relays that use pattern-recognition-based scheme are named smart relays. Since the numerical relays are multifunction, the pattern-recognition methods can be added to these

TABLE 17.1 Recent publications used the pattern-recognition-based protection schemes.

Ref.	Purpose	Method					
		Feature			Decision		
		DSP	Input			Classification	Prediction
[29]	Fault location	DFT	• Magnitudes of fundamental components	Voltage and current	ANN	✓	-
[30]	Fault classification and location	DFT	• Fundamental components	Voltage and current	ANN	✓	✓
[31]	Fault detection and classification	DFT	• Fundamental components	Voltage and current	ANN	✓	-
[32]	Fault classification and location	DFT	• DC component • Fundamental frequency component • Seventh, ninth, harmonic frequency	Voltage and current	ANN	✓	✓
[33]	Fault classification and location	ST	• Phase angle corresponding to every peak of the S-matrix • The frequency components of the S-matrices	Voltage and current	PNN (classification); BPNN (prediction)	✓	✓
[34]	Fault classification	WT	• Wavelet energies calculated	Current	DT, hidden Markov model, PNN	✓	-
[35]	Fault location	WT	• Six detail coefficients of three-level decomposition	Voltage and current	Elman recurrent network	-	✓
[36]	Fault classification	-	• Fault components of	Voltage and current	ART	✓	-
[37]	Fault detection, classification, and location	FIR	• Two stage FIR filter	Voltage and current	SVM	✓	✓
[38]	Fault classification	EMD	• Energy distribution of instantaneous amplitude. • Standard deviation of amplitude. • Standard deviation of phase	Voltage	SVM	✓	-
[39]	Fault classification and location	WT	• Energy • Standard deviation	Voltage and current	SVM (classification); RBFNN (prediction)	✓	✓
[40]	Fault classification and location	WT	• Detail coefficients of the first three-level decomposition	Voltage and current	SVM	✓	✓
[41]	Fault detection and classification	WT	• Energy of wavelet coefficient	Voltage and current	ANN	✓	-
[42]	Fault classification and location	WT	• Entropies of the wavelet decompositions	Voltage and current		PNN	Recursive neural network

(Continued)

TABLE 17.1 (Continued)

						✓	✓
[44]	Fault classification and location	WT	• Energy and entropy of the decomposed coefficients	Voltage and current	SVM	✓	✓
[45]	Fault classification	WT	• Energies of the approximation and detail coefficients of the four decomposition levels	Voltage and current	*k*-NN and DT	✓	
[46]	Directional relaying, fault classification and fault location	DFT	• Phase angle of positive sequence current signal • Fundamental components of three phase current, voltage and zero sequence current signal • Positive sequence current and voltage signals	Voltage and current	Fuzzy inference system	✓	✓
[47]	Fault direction detection	-	• Phase angle of positive sequence current	Current	Fuzzy inference system	✓	-
[48]	Fault classification	WT	• Calculate the seventh level detailed coefficients	Current	CART	✓	-
[49]	Fault section Identification and fault classification	ST	• Energy and standard deviation of ST contours	Current	PNN	✓	-
[50]	Fault section identification and fault classification	ST	• Square height value of level 1 obtained from frequency contour • Sum of square height values of levels 1–9 obtained from frequency contour • The area between 0 and 200 Hz of locus of maximum magnitude of frequency • The area between 200 and 400Hz of locus of maximum magnitude of frequency • The area between 400 and 1000 Hz of locus of maximum magnitude of frequency • The standard deviation of maximum element from each column of S-matrix	Current	Logistic model tree	✓	-
[51]	Fault classification	ST	• Energy content of the frequency contour • Variance of the S-matrix (amplitude and frequency) • Auto-correlation of the absolute value of S-matrix • Standard deviation of the S-matrix at contour level 1 (amplitude and frequency)	Current	Fuzzy and DT	✓	-
[52]	Fault location	Hyperbolic ST	• Change in the signal energy • Standard deviation of the hyperbolic ST contour	Current	ANFIS	-	✓
[53]	Fault section and location	WT	• First step of the detail coefficients • Second-level of approximation coefficients	Voltage and current	Neuro-fuzzy	✓	✓
[54]	Fault direction detection	DFT	• Fundamental component	Current	ANFIS	✓	-
[55]	Compensate for the influence of the variable power system conditions	-	• Bus voltages • Branch currents • Breaker statuses	-	ANN	-	✓
[56]	Fault section identification and location	WT	• Three level approximate coefficients	Voltage and current	ANN	✓	✓
[57]	Fault detection, direction detection, and classification	DFT	• Fundamental frequency component	Voltage and	ANN		

(Continued)

TABLE 17.1 (Continued)

				current		✓	-
[58]	Reducing mutual coupling between two lines	-	• Voltage • Current • Switching mode • Operation mode	-	Kohonen NN	-	✓
[59]	Fault classification	WT	• Four levels of detail coefficients • Wavelet energy	Current	ANN	✓	-
[60]	Fault detection/classification and location	DFT	• Fundamental component	Voltage and current	ANN	✓	✓
[61]	Fault detection, classification, and location	WT	• Approximate coefficient up to first level • Approximate coefficient up to third level • Standard deviation of wavelet coefficients	Voltage and current	*k*-NN	✓	✓
[62]	High-impedance fault detection	DFT	• Residual current and voltage • Magnitude and angle of second and third harmonics	Voltage and current	ANN	✓	-
[63]	High-impedance fault detection	WT	• Details at levels 3, 4 and 5 • Approximations at level 5	Current	ANN	✓	-
[64]	High-impedance fault detection	WT	• First, second, third, and forth detail level • Fourth approximation level	Voltage	ANN	✓	-
[65]	High-impedance fault detection	MMG	• Time-based feature using MMG	Current	ANN	✓	-
[66]	High-impedance fault detection	MM	• Mean: • Standard deviation • Energy • Entropy • Kurtosis • Skewness	Current	DT	✓	-
[69]	Fault location	WT	• Summation of coefficients for the 6th and 11th levels	Current	ANFIS	-	✓
[70]	Power swing detection	-	• Positive sequence impedance • Positive and negative sequence currents • Power swing center voltage	Voltage and current	ANFIS	✓	-
[71]	Power swing detection	-	• Rate of change of positive sequence current, • Active power • Reactive power	Voltage and current	ANFIS	✓	-
[72]	Detection of unstable power swing	ST	• Derivative of bus voltage, • Bus voltage and bus active power processed by the ST	Voltage and current	ANN	✓	-
[73]	Fault classification and power swing detection	-	• Positive-sequence complex voltage • Positive-sequence magnitude of current • Real power flow • Reactive power flow	Voltage and current	SVM	✓	-
[74]	Power swing detection	DFT	• Areas of triangles constituted by phase angles of instantaneous power phasors	Voltage and	*k*-NN		

(*Continued*)

TABLE 17.1 (Continued)

				current		✓	-
[75]	Power swing detection	ST	• Derivative of bus voltage • Bus voltage and bus active power processed by the ST	Voltage and current	PNN	✓	-
[76]	Fault detection during power swing	-	• Current signal	-	SVM	✓	-
[77]	Fault detection during power swing	ST	• Frequency spectrum obtained from S-matrix	Current	PNN	✓	-
[78]	Fault and power swing detection and classification	-	• Rate of change of active power • Rate of change of reactive power • Rate of change of sequence components of current • Rate of change of sequence components of voltage • Rate of change of current • Rate of change of voltage • Rate of change of phase angle • Rate of change of Load angle	Voltage and current	DT	✓	-
[79]	Fault detection during power swing	DFT	• Fundamental components of currents and voltages and zero sequence currents	Voltage and current	DT	✓	-
[80]	Fault detection during power swing	MMG	• Traveling wave	Voltage and current	SVM	✓	-

ANFIS, Adaptive neuro-fuzzy system; *ANN*, artificial neural networks; *ART*, adaptive resonant theory; *CART*, classification and regression tree; *DFT*, discrete Fourier transform; *DSP*, digital signal processors; *DT*, decision tree; *FIR*, finite impulse response; *k-NN*, *k*-nearest neighbor; *MM*, Multiresolution morphological; *MMG*, multiresolution morphological gradient; *NN*, Neural network; *RBFF*, Radial basis function neural network; *ST*, *S*-transform; *WT*, Wavelet transform.

relays or they can be replaced with some existing functions. Consequently, the smart relays are new development of numerical relays. Fig. 17.3 shows the trend of completion in relay design and also represents the main specifications of all generations of relays concisely. Regarding the advantages and disadvantages of these relays are discussed in literature [81,82].

Nowadays, power transmission systems are more sophisticated due to growing interconnected systems, FACTS technology, high voltage direct current (HVDC) transmission line, and tapped transmission lines. To deliver secure and reliable electricity to customers, the protection of such complex systems is inevitable. At the same time, using the meters, PMUs, intelligent electronic devices, communication systems, and automation systems make available information required about the whole power system. To protect the power transmission systems from abnormal conditions, a protective relay which has an accurate operation and also smart function is necessary. Some numerical relays even have a PMU module to enhance the capability of decision-making of protection schemes and hence increase the reliability and robustness of the protection system. Moreover, most of the new protection

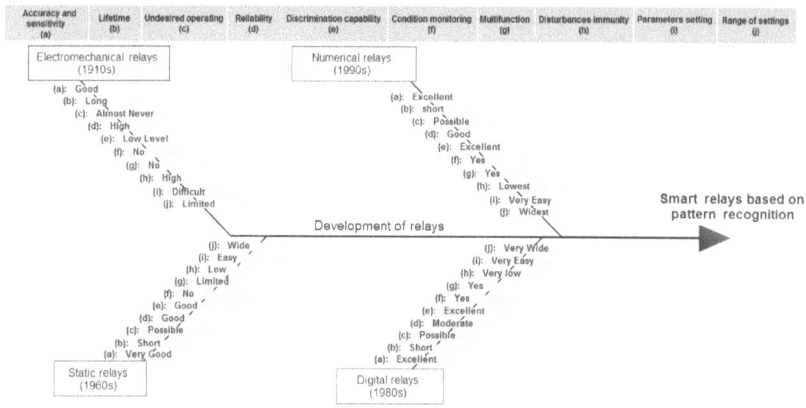

FIGURE 17.3 Trends of completion in relay design.

schemes are established based on communication channels, which increase both the protection function capability and also the risk of a communication failure. For example, any communication failure causes to fail most differential-and pilot-based relays against abnormal conditions. Therefore since the main algorithms of numerical relays are also based on the conventional relays, they may fail to make a correct decision sometimes [83]. In this manner the conventional numerical relays fail to cover all operating conditions of a transmission line, despite the fact that they employ wide range of protection functions. For example, in a compensated transmission line, the conventional phase selection functions, which are based on the relationship between sequence components or the magnitude of phase currents, cannot operate accurately [11]. Therefore the protection systems of transmission lines should be more accurate and also adaptive to provide reliable operation. Changing the fundamental logic of protection schemes to make final decision should be the main motivation to suggest intelligent-based protection schemes.

From a theoretical viewpoint, as discussed in Section 17.3, many papers have proposed different protection schemes based on pattern recognition. Moreover, some papers have devoted to implement the pattern-recognition-based protection scheme on a hardware platform and test the performance of the framework against different scenarios [84]. In 2004 a relay developed by a commercial company has been addressed the problem of HIF detection using a scheme based on a pattern-recognition algorithm. It can be introduced as the first commercial use of pattern recognition for protection purpose [85,86].

In the smart relaying task, the pattern recognition should decide

- to issue a trip signal or retrain itself from tripping,
- to issue a block or unblock signal,

- to classify the ongoing transients as faults or other conditions,
- to estimate the value to monitor the information such as fault location, etc.

17.4.1 Structure of smart relays

Generally, there are two types of protection functions in a numerical distance relay: basic functions and additional functions. For example, the basic function in a distance relay is a full scheme distance protection with six impedance measuring units for all types of faults in forward and reverse impedance zones. The phase selection function is an example of additional functions in a distance relay that provides the option of single-pole tripping of the circuit breaker during single-phase-to-ground faults. As another example of additional functions is fault locator as a necessary complement to the distance relay. Pattern recognition can be employed as basic or additional protection functions in a transmission line protection.

Fig. 17.4 shows the structure of a smart relay based on pattern recognition. In this figure, analog-to-digital converters are utilized to measure signals from the database at a preidentified sampling rate. The development of digital signal processing board that holds the field programmable gate logic (FPGA) provides high-speed mezzanine card (HSMC) interfaces. Since the measured data are stored in a memory module, one sampling clock in the HSMC interface triggers one dynamic random-access memory writing cycle via the dual-in-line memory module interface. The implementation of a protection function in FPGA is described in Ref. [84]. The data used to train the classifier or estimator model(s) came from tests conducted at a real or simulated transmission network. Then the developed model(s) will be employed to test and implement the protection function.

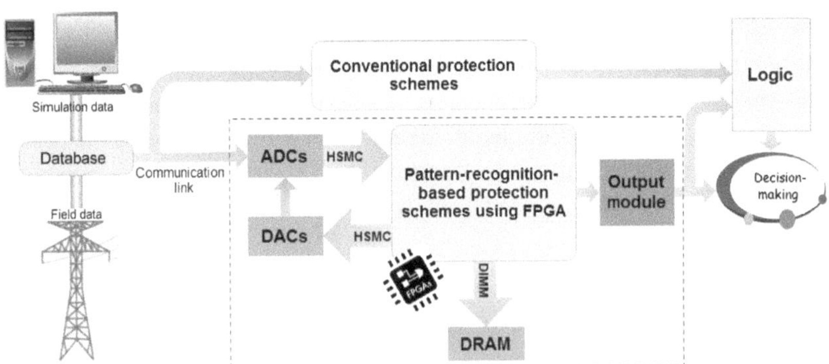

FIGURE 17.4 A general overview of the structure of a smart relay based on pattern recognition.

The design of the new pattern-recognition-based function can be operated in two ways:

1. Stand-alone functions that consist of independent classification or estimation mechanism. In this type of function the input is a predetermined constructed feature vector, and the final decision made by the function does not interact with other functions. For example, the smart fault locator proposed in Ref. [3] is a stand-alone function, and the output of the locator only identifies the fault point in a transmission line to restore the power system as soon as possible.
2. Integrated functions that are a part of other protection schemes. This type of function can communicate with other functions to enhance the performance of them. The input of integrated functions contains the output of other protection functions, and/or the input is selected or activated by the decision of other protection functions. Moreover, the output of integrated functions can be used to improve the performance of other protection function. For example, in Ref. [16] the input of proposed pattern-recognition-based fault classifier is dependent on disturbance detection module. When disturbance detection module detects a disturbance in three-phase current signal, the activator switch actives the fault classifier. Then, the output of fault classifier can select one of the six impedance measuring units to protect the protected zone. The structure of this example of an integrated function is shown in Fig. 17.5. The decision made by an integrated function also can be used individually. As shown in Fig. 17.5, the output of the fault classifier module is used for fault report.

An integrated pattern-recognition-based function can also be operated in a voting configuration. The pattern-recognition-based function makes a

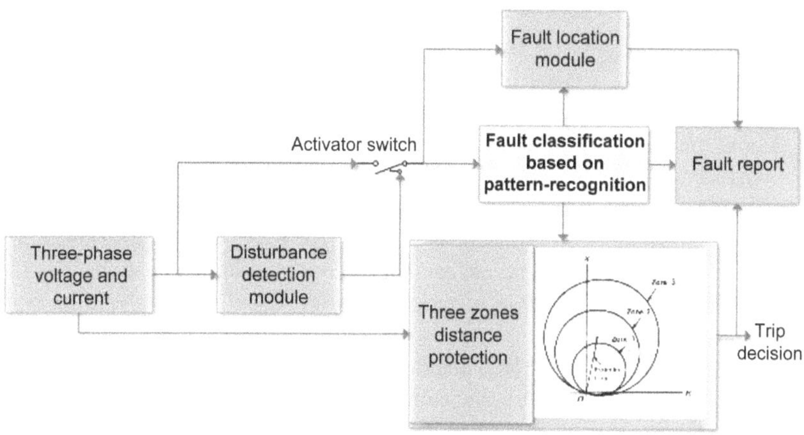

FIGURE 17.5 Integrated pattern-recognition-based function.

decision in parallel with other functions that have the same outputs to aggregate the information. Finally, the number of votes determines the winner and final decision.

Since numerical distance relays are the most difficult to set and need a multiplicity of settings, pattern recognition helps to develop smarter distance protective relays. Further, regarding the applicability of the pattern-recognition-based methods practically, some papers have devoted to implement the protection scheme on a hardware platform and test the performance of the framework against different scenarios. Field programmable neural arrays (FPNA) make it possible to adapt the definition of large and effective neural architectures to constraints of hardware topological. The concept of an FPNA is based on FPGA principles. Therefore the complex and large functions can be realized using a set of simple programmable resources. Since the numerical relays are multifunction, the pattern-recognition methods can be added to these relays, or they can be replaced with some existing functions. The FPNA-based scheme makes a bridge between the complex architecture of neural networks and hardware devices. In hardware architecture of the FPNA configurable, predefined blocks are available for each neural resource, and therefore the FPNA is mapped onto the hardware device through completely modular implementation. Consequently, the smart relays based on pattern-recognition methods can be a new development of numerical relays based on FPGA/FPNA technology [85,87].

17.4.2 Advantages

The performance of distance relays may be substantially improved if the decision-making of some protection functions is based on pattern recognition. There are several advantages to use pattern recognition in the protection of transmission lines [11,84,88,89]:

- Tripping decisions can be based on several criteria to improve the accuracy, reliability, and selectivity.
- Pattern recognition functions can be used in parallel with other conventional protection functions to enhance the performance of the relay, whereas they can also operate as a stand-alone function.
- The function can be implemented without/a little information on the parameters of the protected line.
- There is no limitation to select several input features, and hence, the function can make a more accurate decision based on using multidimensional input matrix.
- In power system protection, there are large items of vagueness and uncertainty due to the complex and nonlinear relationships between the different numbers of power system parameters, and also there are different types of transmission line topology. Moreover, in some cases, there is no

straight relationship between the input and output of a protection scheme. These impose a considerable difficulty when employing conventional protection schemes. Some of these uncertainties and complexities are as follows: different power system operating conditions such as changes in load demand and generation type (renewable energy source, distributed generation, conventional energy resource, DC storages, etc.); different power system topology and configuration such as single/double line, untransposed/transposed line, uncompensated/compensated line, two-terminal/multiterminal line, and overhead/combined line; different pre-fault and fault conditions such as fault type, fault inception angle, fault resistance, fault location, loading level, short-circuit level, and references values of compensator device; different phenomena in transmission lines such as fault, power swing, fault during power swing, and transients of lightning strike; different access point to relaying data such as only local data and synchronized/unsynchronized local and remote data; error caused by CT saturation or CCVT transients; noisy conditions caused by various interferences. In addition, the random nature of some parameters and conditions increase the complexity of the problem. When some conditions of the power system have been changed, a pattern-recognition-based function can adapt itself to the new condition through considering a new training data set in the learning phase of classifier or estimator model(s).
- Since different training patterns can be employed in the training phase, the functions obtain high generalization ability.
- The hardware implementation of pattern-recognition-based functions confirms the robust performance of the smart function from practical viewpoint.
- The pattern-recognition-based functions are efficient in terms of complexity and speed. Consequently, they can stand as a candidate for a protection function.

17.4.3 Disadvantages

There are some challenges to apply pattern recognition in numerical distance relay and generally protective relays. There are many papers proposed the smart protection functions based on pattern recognition theoretically, but there are a few papers that consider the smart protection functions from practical viewpoint. In this regard a basic question may be raised: "what are the challenging issues to employ pattern-recognition-based protection function in commercial relays?". The response to this question can be considered as the most disadvantages and shortcomings of smart functions [16,84]:

- Before commercial application the smart functions should be investigated in more details from a perspective of hardware implementation, whereas there are a few papers in this area. These papers implement the pattern recognition on the FPGA board and confirm the practical ability of pattern-recognition-based functions.

- Most classifier or estimator model is a black box. Some types of classifiers or estimators such as DT are not a black box. However, engineers are not interested in smart function, because pattern recognition is not easily explainable. Therefore they prefer to use conventional schemes which are based on only power system quantities.
- Providing a train data set is a challenging task. Nowadays, different train data sets can be generated through simulation of a real power system in the software environment. The electromagnetic transient program can simulate transient phenomena with high accuracy.
- One of the main concerns of researchers is whether a trained classifier or estimator model for a specified sample system is applicable to other power systems or not. The normalization technique may overcome this shortcoming, but it is still a concern for the smart functions. Moreover, some researchers propose combined processing on the raw signals to extract more useful features which have the same behavior corresponding different sample systems.
- To achieve high generalization ability of trained classifier or estimator model, large training data are required. On the other hand, considering all scenarios in a sample power system to generate train data set is only possible by simulation software. It is worth mentioning that some classifiers such as SVM have good generalization ability even without large training data set. Due to growing of simulation software, it is not a critical issue.
- One of the main stages of the smart function is the training of a classifier or estimator model. All classifiers or estimators have different setting parameters which should be set to enhance the performance of a trained model. There is no straight-forward way to find the best parameters of a trained model.

17.5 Conclusion

This chapter presents a general overview of pattern recognition to solve some problems in distance protection. From viewpoint of transmission line protection, the detail structure of a pattern recognition model is demonstrated. Moreover, the literature review confirms that pattern recognition functions are effective in fault detection, fault classification, fault location, HIF detection, power swing detection, and symmetrical fault detection during power swing. These are even more important in view of the fact that the waveforms measured at relaying point are affected by complicated and nonlinear relationships that exist between many power system parameters. Further, these parameters are influenced by different system configurations, topologies, compensation devices, DC grids, and fault conditions. From this chapter, it can be concluded that pattern-recognition-based functions provide effective and, therefore, interesting alternatives to conventional protection function or supplementary functions in conventional distance relays for solving different complex protection problems. Researches represent that the performance of pattern-recognition-based functions can be considerably improved by the combination of different preprocessing techniques rather than the employment of only raw signals. Consequently, the generalization capability of learned classifier or estimator models can increase. Recently, the hardware implementation of pattern-recognition-based functions illustrates that these functions can be effectively employed in protection of transmission line practically. However, to apply the functions to commercial relays, more consideration is required. Recent advances in the hardware implementation of ANN-based approaches make smart relay possible for the future technology of relays. Moreover, an ANN can be implemented in a reconfigurable orthogonal memory multiprocessor (REOMP) as a reconfigurable parallel computer architecture using FPGA. Rapid FPGA technology enables the modular design concept based on the REOMP implementation in a system-on-chip approach. Integration of pattern-recognition functions thus develops a way toward the new generation of relays in smart protection systems.

References

1. E. Micheli-Tzanakou, Supervised and Unsupervised Pattern Recognition: Feature Extraction and Computational Intelligence, CRC Press, 1999.
2. R. Salat, S. Osowski, Accurate fault location in the power transmission line using support vector machine approach, IEEE Trans. Power Syst. 19 (2) (2004) 979–986.
3. Z. Moravej, M. Pazoki, M. Khederzadeh, New smart fault locator in compensated line with UPFC, Int. J. Electr. Power Energy Syst. 92 (2017) 125–135.
4. G. Benmouyal, Removal of DC-offset in current waveforms using digital mimic filtering, IEEE Trans. Power Delivery 10 (2) (1995) 621–630.
5. A. Jamehbozorg, S. Mohammad Shahrtash, A decision-tree-based method for fault classification in single-circuit transmission lines, IEEE Trans. Power Delivery 25 (4) (2010) 2190–2196.
6. A. Jamehbozorg, S.M. Shahrtash, A decision tree-based method for fault classification in double-circuit transmission lines, IEEE Trans. Power Delivery 25 (4) (2010) 2184–2189.
7. M. Farshad, J. Sadeh, Accurate single-phase fault-location method for transmission lines based on k-nearest neighbor algorithm using one-end voltage, IEEE Trans. Power Delivery 27 (4) (2012) 2360–2367.
8. P.F. Ribeiro, C.A. Duque, P.M. Ribeiro, A.S. Cerqueira, Power Systems Signal Processing for Smart Grids, Wiley, 2013.
9. R.G. Stockwell, L. Mansinha, R.P. Lowe, Localization of the complex spectrum: the *S* transform, IEEE Trans. Signal Process. 44 (4) (1996) 998–1001.
10. Z. Moravej, M. Khederzadeh, M. Pazoki, New combined method for fault detection, classification, and location in series-compensated transmission line, Electr. Power Compon. Syst. 40 (9) (2012) 1050–1071.

11. Z. Moravej, M. Pazoki, M. Khederzadeh, New pattern-recognition method for fault analysis in transmission line with UPFC, IEEE Trans. Power Delivery 30 (3) (2015) 1231–1242.
12. P.K. Mishra, A. Yadav, M. Pazoki, A Novel Fault Classification Scheme for Series Capacitor Compensated Transmission Line Based on Bagged Tree Ensemble Classifier, IEEE Access, 2018.
13. N.E. Huang, Z. Shen, S.R. Long, M.C. Wu, H.H. Shih, Q. Zheng, et al., The empirical mode decomposition and the Hilbert spectrum for nonlinear and non-stationary time series analysis, Proc. R. Soc. Lond. Ser. A: Math. Phys. Eng. Sci. 454 (1971) (1998) 903–995.
14. Z.H.U. Zhihui, Y. Sun, Fault classification for power transmission line using EMD-approximate entropy and SVM, Electr. Power Autom. Equip. 7 (2008) 023.
15. M.G. Frei, I. Osorio, Intrinsic time-scale decomposition: time–frequency–energy analysis and real-time filtering of non-stationary signals, Proc. R. Soc. Lond. Ser. A: Math. Phys. Eng. Sci. 463 (2078) (2007) 321–342.
16. M. Pazoki, A new fault classifier in transmission lines using intrinsic time decomposition, IEEE Trans. Ind. Inf. 14 (2) (2018) 619–628.
17. B. Ghoraani, Selected topics on time-frequency matrix decomposition analysis, J. Pattern Recognit. Intell. Syst. 1 (3) (2013) 64–78.
18. G. Chandrashekar, F. Sahin, A survey on feature selection methods, Comput. Electr. Eng. 40 (1) (2014) 16–28.
19. T.N. Lal, O. Chapelle, J. Weston, A. Elisseeff, Embedded methods, Feature Extraction, Springer, Berlin, Heidelberg, 2006, pp. 137–165.
20. H. Peng, F. Long, C. Ding, Feature selection based on mutual information criteria of max-dependency, max-relevance, and min-redundancy, IEEE Trans. Pattern Anal. Mach. Intell. 27 (8) (2005) 1226–1238.
21. B.D. Ripley, Pattern Recognition and Neural Networks, Cambridge University Press, 2007.
22. M. Kezunovic, Smart fault location for smart grids, IEEE Trans. Smart Grid 2 (1) (2011) 11–22.
23. K. Chen, C. Huang, J. He, Fault detection, classification and location for transmission lines and distribution systems: a review on the methods, High Voltage 1 (1) (2016) 25–33.
24. V.H. Ferreira, R. Zanghi, M.Z. Fortes, G.G. Sotelo, R.B.M. Silva, J.C.S. Souza, et al., A survey on intelligent system application to fault diagnosis in electric power system transmission lines, Electr. Power Syst. Res. 136 (2016) 135–153.
25. A. Ghaderi, H.L. Ginn III, H.A. Mohammadpour, High impedance fault detection: a review, Electr. Power Syst. Res. 143 (2017) 376–388.
26. D.P. Mishra, P. Ray, Fault detection, location and classification of a transmission line, Neural Comput. Appl. 30 (2018) 1377–1424.
27. A. Yadav, Y. Dash, An overview of transmission line protection by artificial neural network: fault detection, fault classification, fault location, and fault direction discrimination, Adv. Artif. Neural Syst. 2014 (2014) 12.
28. S.A. Khaparde, N. Warke, S.H. Agarwal, An adaptive approach in distance protection using an artificial neural network, Electr. Power Syst. Res. 37 (1) (1996) 39–44.
29. A.J. Mazon, I. Zamora, J.F. Minambres, M.A. Zorrozua, J.J. Barandiaran, K. Sagastabeitia, A new approach to fault location in two-terminal transmission lines using artificial neural networks, Electr. Power Syst. Res. 56 (3) (2000) 261–266.
30. A. Jain, V.S. Kale, A.S. Thoke, Application of artificial neural networks to transmission line faulty phase selection and fault distance location, in: Proceedings of the IASTED International Conference "Energy and Power System", Chiang Mai, Thailand, 2006, pp. 262–267.

31. A. Jain, A.S. Thoke, R.N. Patel, E. Koley, Intercircuit and cross-country fault detection and classification using artificial neural network, India Conference (INDICON), 2010 Annual IEEE, IEEE, 2010, 1–4.
32. O.O. Kalu, T.C. Madueme, Application of artificial neural network (ANN) to enhance power systems protection: a case of the Nigerian 330 kV transmission line, Electr. Eng. 100 (3) (2018) 1467–1479.
33. N. Roy, K. Bhattacharya, Detection, classification, and estimation of fault location on an overhead transmission line using *S*-transform and neural network, Electr. Power Compon. Syst. 43 (4) (2015) 461–472.
34. N. Perera, A.D. Rajapakse, Recognition of fault transients using a probabilistic neural-network classifier, IEEE Trans. Power Delivery 26 (1) (2011) 410–419.
35. S. Ekici, S. Yildirim, M. Poyraz, A transmission line fault locator based on Elman recurrent networks, Appl. Soft Comput. 9 (1) (2009) 341–347.
36. Y. Yang, N.L. Tai, W.Y. Yu, ART artificial neural networks based adaptive phase selector, Electr. Power Syst. Res. 76 (1-3) (2005) 115–120.
37. H. Fathabadi, Novel filter based ANN approach for short-circuit faults detection, classification and location in power transmission lines, Int. J. Electr. Power Energy Syst. 74 (2016) 374–383.
38. N. Ramesh Babu, B. Jagan Mohan, Fault classification in power systems using EMD and SVM, Ain Shams Eng. J. 8 (2) (2017) 103–111.
39. S.R. Samantaray, P.K. Dash, G. Panda, Distance relaying for transmission line using support vector machine and radial basis function neural network, Int. J. Electr. Power Energy Syst. 29 (7) (2007) 551–556.
40. S. Ekici, Support Vector Machines for classification and locating faults on transmission lines, Appl. Soft Comput. 12 (6) (2012) 1650–1658.
41. K.M. Silva, B.A. Souza, N.S.D. Brito, Fault detection and classification in transmission lines based on wavelet transform and ANN, IEEE Trans. Power Delivery 21 (4) (2006) 2058–2063.
42. A. Dasgupta, S. Nath, A. Das, Transmission line fault classification and location using wavelet entropy and neural network, Electr. Power Compon. Syst. 40 (15) (2012) 1676–1689.
43. A.R. Adly, R.A. El Sehiemy, A.Y. Abdelaziz, N.M.A. Ayad, Critical aspects on wavelet transforms based fault identification procedures in HV transmission line, IET Gener. Transm. Distrib. 10 (2) (2016) 508–517.
44. P. Ray, D.P. Mishra, Support vector machine based fault classification and location of a long transmission line, Eng. Sci. Technol. 19 (3) (2016) 1368–1380.
45. T.S. Abdelgayed, W.G. Morsi, T.S. Sidhu, Fault detection and classification based on co-training of semisupervised machine learning, IEEE Trans. Ind. Electron. 65 (2) (2018) 1595–1605.
46. A. Yadav, A. Swetapadma, Enhancing the performance of transmission line directional relaying, fault classification and fault location schemes using fuzzy inference system, IET Gener. Transm. Distrib. 9 (6) (2015) 580–591.
47. P.K. Mishra, A. Yadav, A non-unit protection scheme for series-compensated transmission system using fuzzy inference system, Smart Computing and Informatics, Springer, Singapore, 2018, pp. 453–466.
48. J. Upendar, C.P. Gupta, G.K. Singh, Statistical decision-tree based fault classification scheme for protection of power transmission lines, Int. J. Electr. Power Energy Syst. 36 (1) (2012) 1–12.

49. S.R. Samantaray, P.K. Dash, Pattern recognition based digital relaying for advanced series compensated line, Int. J. Electr. Power Energy Syst. 30 (2) (2008) 102–112.
50. Z. Moravej, M. Pazoki, A.A. Abdoos, A new approach for fault classification and section detection in compensated transmission line with TCSC, Int. Trans. Electr. Energy Syst. 21 (1) (2011) 997–1014.
51. S.R. Samantaray, A systematic fuzzy rule based approach for fault classification in transmission lines, Appl. Soft Comput. 13 (2) (2013) 928–938.
52. P.K. Dash, S.R. Samantaray, A novel distance protection scheme using time-frequency analysis and pattern recognition approach., Int. J. Electr. Power Energy Syst. 29 (2) (2007) 129–137.
53. C.K. Jung, K.H. Kim, J.B. Lee, B. Klöckl, Wavelet and neuro-fuzzy based fault location for combined transmission systems, Int. J. Electr. Power Energy Syst. 29 (6) (2007) 445–454.
54. A. Swetapadma, A. Yadav, High-speed directional relaying using adaptive neuro-fuzzy inference system and fundamental component of currents, IEEJ Trans. Electr. Electron. Eng. 10 (6) (2015) 653–663.
55. A.G. Jongepier, L. Van Der Sluis, Adaptive distance protection of double-circuit lines using artificial neural networks, IEEE Trans. Power Delivery 12 (1) (1997) 97–105.
56. A. Yadav, A. Swetapadma, A single ended directional fault section identifier and fault locator for double circuit transmission lines using combined wavelet and ANN approach, Int. J. Electr. Power Energy Syst. 69 (2015) 27–33.
57. A. Yadav, A. Swetapadma, Improved first zone reach setting of artificial neural network-based directional relay for protection of double circuit transmission lines, IET Gener. Transm. Distrib. 8 (3) (2013) 373–388.
58. S. Skok, A. Marusic, S. Tesnjak, I. Pavic, Double-circuit line adaptive protection based on Kohonen neural network considering different operation and switching modes, Power Engineering 2002 Large Engineering Systems Conference on, LESCOPE 02, IEEE, 2002, 153–157.
59. A.A.M. Zin, M. Saini, M.W. Mustafa, A.R. Sultan, New algorithm for detection and fault classification on parallel transmission line using DWT and BPNN based on Clarke's transformation, Neurocomputing 168 (2015) 983–993.
60. A. Swetapadma, A. Yadav, Fuzzy inference system approach for locating series, shunt, and simultaneous series-shunt faults in double circuit transmission lines, Comput. Intell. Neurosci. 2015 (2015) 79.
61. A. Swetapadma, P. Mishra, A. Yadav, A.Y. Abdelaziz, A non-unit protection scheme for double circuit series capacitor compensated transmission lines, Electr. Power Syst. Res. 148 (2017) 311–325.
62. H.K. Zadeh, An ANN-based high impedance fault detection scheme: design and implementation, Int. J. Emerg. Electr. Power Syst. 4 (2) (2005) 1.
63. M.-T. Yang, J.-C. Gu, Detecting high impedance faults utilizing combined phase voltages with neutral line current, Int. J. Emerg. Electr. Power Syst. 2 (2) (2005) 1051.
64. T. Hubana, M. Saric, S. Avdakovic, Approach for identification and classification of HIFs in medium voltage distribution networks, IET Gener. Transm. Distrib. 12 (5) (2018) 1145–1152.
65. M. Sarlak, S.M. Shahrtash, High impedance fault detection using combination of multi-layer perceptron neural networks based on multi-resolution morphological gradient features of current waveform, IET Gener. Transm. Distrib. 5 (5) (2011) 588–595.

66. K. Sekar, N.K. Mohanty, Data mining-based high impedance fault detection using mathematical morphology, Comput. Electr. Eng. 69 (2018) 129–141.
67. P.J. Mooney, N. Fischer, Application guidelines for power swing detection on transmission systems, Power Systems Conference: Advanced Metering, Protection, Control, Communication, and Distributed Resources, 2006. PS'06, IEEE, 2006, 159–168.
68. L. Zou, Q. Zhao, X. Lin, P. Liu, Improved phase selector for unbalanced faults during power swings using morphological technique, IEEE Trans. Power Delivery 21 (4) (2006) 1847–1855.
69. M.J. Reddy, D.K. Mohanta, Adaptive-neuro-fuzzy inference system approach for transmission line fault classification and location incorporating effects of power swings, IET Gener. Transm. Distrib. 2 (2) (2008) 235–244.
70. H.K. Zadeh, L. Zuyi, A novel power swing blocking scheme using adaptive neuro-fuzzy inference system, Electr. Power Syst. Res. 78 (7) (2008) 1138–1146.
71. A. Esmaeilian, S. Astinfeshan, A novel power swing detection algorithm using adaptive neuro fuzzy technique, 2011 International Conference on Electrical Engineering and Informatics (ICEEI), IEEE, 2011, 1–6.
72. A.F. Abidin, A. Mohamed, H. Shareef, Intelligent detection of unstable power swing for correct distance relay operation using *S*-transform and neural networks, Expert Syst. Appl. 38 (12) (2011) 14969–14975.
73. K. Seethalekshmi, S.N. Singh, S.C. Srivastava, A classification approach using support vector machines to prevent distance relay maloperation under power swing and voltage instability, IEEE Trans. Power Delivery 27 (3) (2012) 1124–1133.
74. İ.G. Tekdemir, B. Alboyacı, Improvement of power swing detection performance of a distance relay by using k-NN algorithm, 2015 9th International Conference on Electrical and Electronics Engineering (ELECO), IEEE, 2015 541–545.
75. A.F. Abidin, A. Mohamed, H. Shareef, Power swing and voltage collapse identification schemes for correct distance relay operation in power system, J. Cent. South Univ. 20 (4) (2013) 988–1000.
76. N.G. Chothani, B.R. Bhalja, U.B. Parikh, New support vector machine-based digital relaying scheme for discrimination between power swing and fault, IET Gener. Transm. Distrib. 8 (1) (2014) 17–25.
77. Z. Moravej, J.D. Ashkezari, M. Pazoki, An effective combined method for symmetrical faults identification during power swing, Int. J. Electr. Power Energy Syst. 64 (2015) 24–34.
78. R. Dubey, S.R. Samantaray, B.K. Panigrahi, V.G. Venkoparao, Data-mining model based adaptive protection scheme to enhance distance relay performance during power swing, Int. J. Electr. Power Energy Syst. 81 (2016) 361–370.
79. A. Swetapadma, A. Yadav, Data-mining-based fault during power swing identification in power transmission system, IET Sci., Meas. Technol. 10 (2) (2016) 130–139.
80. M. Daryalal, M. Sarlak, Fast fault detection scheme for series-compensated lines during power swing, Int. J. Electr. Power Energy Syst. 92 (2017) 230–244.
81. V. Gurevich, Digital Protective Relays: Problems and Solutions, CRC Press, 2016.
82. A. Abdelmoumene, H. Bentarzi, A review on protective relays' developments and trends, J. Energy South. Afr. 25 (2) (2014) 91–95.
83. Y. Zhang, Mitigating Future Blackouts via Smart Relays: A Machine Learning Approach (Ph.D. Diss.), Carnegie Mellon University, 2010.
84. J.-A. Jiang, C.-L. Chuang, Y.-C. Wang, C.-H. Hung, J.-Y. Wang, C.-H. Lee, et al., A hybrid framework for fault detection, classification, and location—Part II: Implementation and test results, IEEE Trans. Power Delivery 26 (3) (2011) 1999–2008.

85. J. Stoupis, M. Maharsi, R. Nuqui, S.A. Kunsman, R. Das, Ground alert, ABB Rev. 1 (2004) 28–31.
86. S.A. Kunsman, I.I. Jouny, S. Kaprielian, High Impedance Fault Detection, U.S. Patent 7,069,116, Issued June 27, 2006.
87. A.R. Omondi, J.C. Rajapakse, *FPGA Implementations of Neural Networks*", Springer, 2006.
88. J. Gracia, A.J. Mazon, I. Zamora, Best ANN structures for fault location in single-and double-circuit transmission lines, IEEE Trans. Power Delivery 20 (4) (2005) 2389–2395.
89. No. 22 K. Warwick, A. Ekwue, R. Aggarwal (Eds.), Artificial Intelligence Techniques in Power Systems, IET, 1997.

Chapter 18

A reliable decision-making algorithm for fault during power swing in 400 kV double-circuit transmission line: a case study of Chhattisgarh state power system network

V. Ashok[1], Anamika Yadav[1], C.C. Anthony[2], K.K. Yadav[3] and Umakant Yadav[4]

[1]*Department of Electrical Engineering, National Institute of Technology, Raipur, India,* [2]*Transmission, Chhattisgarh State Power Transmission Company Ltd., Raipur, India,* [3]*Meter, Relay and Testing Division, 400 kV Sub-Station, Bhilai-3, Chhattisgarh State Power Transmission Company Ltd., Bhilai, India,* [4]*Extra High Tension: Construction and Maintenance Division, Gudhiyari, Chhattisgarh State Power Transmission Company Ltd., Raipur, India*

18.1 Introduction

Generally, the power swing (PS) is defined as oscillation in the power angle of the two area that occurs owing to abrupt change in system configuration such as removal of faulted line due to clearance of fault by opening of circuit breakers, loss of synchronism, or changes in direction of power flow because of switching-in/out of various equipment/elements [1]. PS can be characterized into two types such as stable PS or unstable PS. In case of a stable PS the power oscillations will die down after some time, whereas in case of unstable PS, there is progressive parting of power angle between the two areas causing large oscillations in both active and reactive power flows, voltages, and currents, thereby leading to loss of synchronism of generators in the system [2,3]. The reports of the blackouts in India endorse the decisive role played by protection relays [4–6]. The recent blackout in India during July 30 and 31, 2012 was originated by the tripping of zone-3 distance relay

Decision Making Applications in Modern Power Systems. DOI: https://doi.org/10.1016/B978-0-12-816445-7.00018-9

of the 400 kV Bina–Gwalior line due to PS and load encroachment [7,8]. Further, the absence of real-time data and of synchronized controls are as well certain major reasons of power failure [9].

There are various conventional techniques available for the detection of fault and power swing in protective relays to ascertain the PS [10]. Concentric characteristic and double blinder techniques were derived using the rate of change of apparent impedance of the transmission line at relaying point [11,12]. These schemes entail wide range of off-line stability studies to obtain the threshold settings [13]. However, these schemes are unable to detect the faults during a PS being blocked by power swing blocking (PSB) function. Subsequently, resistance factor of the apparent impedance measured by the protective relay oscillates during PS and remains persistent during the common shunt fault period except at the instant of fault inception; the rate of change of resistance is a worthy pointer to ensure a shunt fault condition [13]. The response time becomes an apprehension because the rate of change of resistance varies considerably slow during the transient condition. By merging the concentric characteristic with constant monitoring of apparent impedance, a PS detection scheme was proposed in [14]. Further, the PS was detected by swing center voltage in [15] using more than two-cycle data. A modified distance protection method which is invulnerable to the PS effect was discussed in [16]. In [17,18], few learning-based methods, such as support vector machine (SVM) and adaptive neuro-fuzzy inference system, were used to design the PSB function for distance protection relay by employing feature from numerous input signals. These schemes need extensive simulation studies to get erudite the model for wide range of faults during PS, and it suffers owing to high computational complexity. In order to discriminate the PS from the fault in [19], they used the discrete wavelet transform (DWT) processed three-phase voltage and current signals. Detection of fault based on superimposed current signals was recommended in [20]. The derivative of the three-phase active and reactive power was employed to identify symmetrical faults during PS in [21]. In [22], by using the virtual existence of decaying dc in the current signals during PS, a symmetrical fault detection scheme was reported. A data-mining model based on decision tree (DT) technique had been demonstrated to improve distance relay performance during PS [23] and DT-based fault identification during PS in power transmission system was reported in [24]; however, single DT model–based scheme are subjected to data over-fit problem. The Teager–Kaiser energy operator was used to detect fault during PS in [25], and the features of fundamental frequency phasors had been used for fault detection during PS in [26]; nevertheless, the effect of variation in source impedance was not considered. A variety of Lissajous figures had been used to detect PS and fault during PS in [27], but they does not considered the effect of current transformer (CT)–saturation and capacitor coupling voltage transformer (CCVT) transients.

In view of the above, there is a necessity to develop an intelligent scheme that can provide adaptive protection to prevent the malfunction of distance relay during PS and fault during PS in a DCTL. The proposed algorithm helps in minimizing the possibility of hidden failures and budding cascading events by adopting proper decision-making algorithm, thereby concerning the sensitivity and security equilibrium of distance relays at predominant system conditions. The main contribution of the proposed research work is to demonstrate the effectiveness of an intelligent decision-making algorithm based on wavelet packet energy (WPE) and bagged DT (BDT) (WPE_BDT) for detection and classification of shunt faults during PS condition.

In this chapter a double-circuit line of Chhattisgarh state power transmission system is considered to perform various simulation studies for evaluation of proposed decision-making algorithm based on WPE_BDT for protection of transmission lines during PS condition. In Section 18.2, discrimination of fault scenario from PS and fault during PS is emphasized. In Section 18.3, modeling and simulation of Chhattisgarh state power transmission network is described. In Section 18.4 an application of data preprocessing technique such as wavelet packet transform (WPT) and an ensemble classifier based on BDT is illustrated. Section 18.5 deals with the performance assessment of the proposed decision-making algorithm and its validation on real-time digital simulation (RTDS) simulator. In Section 18.6 the summarization of final outcomes of proposed decision-making algorithm and its adaptability and reliability is highlighted. In Section 18.7 a comparative assessment with existing schemes is reported. Finally, Section 18.8 concludes the chapter.

18.2 Protection challenges in case of a fault during power swing

Whenever the power transmission system experiences an abnormal event or a fault, the system is subjected to severe fluctuations in voltage, current, power flow, and frequency. These fluctuations are incepted at the generator rotor angle and start evolving into severe power flow oscillations across the power system network. The occurrence of PS condition in a power transmission network should be detected to avoid malfunction of the appropriate protection scheme. There is a need of an intelligent decision-making algorithm to discriminate different power system operating conditions such as fault, PS (stable and unstable), and fault during PS.

During PS condition, the load point impedance trajectory will navigate toward the operating region of a distance protection scheme. In this condition the operation of distance protection scheme needs to be blocked using PSB function to avoid undesirable tripping of distance protection scheme by improving decision-making capability. As these PSs are phase symmetric, discriminating them from faults is another protection challenge. Besides in the case of an unstable PS, the out-of-step relaying scheme is used to split the network at specified locations/lines to maintain the system within stability margin. The out-of-step tripping (OST) scheme achieved by locally measured relaying quantities such as rate of change of frequency, load angle, and swing voltage center has a drawback of setting a time delaying function since the PS frequency is not consistent all the time it may vary as low as 1–3 Hz and as high as 4–7 Hz. Moreover, due to the PSB and OST function, the distance relaying algorithm does not respond to the appropriate shunt faults during the PS. In this regard, based on the above protection challenges, there is a need of an intelligent decision-making algorithm which should be more reliable and adaptable to overcome maloperation of protection scheme at stressed condition, thereby improve the reliability and stability of the power system.

18.3 Modeling and simulation of Chhattisgarh state power transmission network

A 400 kV, 50 Hz India power transmission network of Chhattisgarh state as shown in Fig. 18.1 is considered with corresponding network parameters are given in Table 18.1. The network consists of two power stations at bus-4 (KSTPS/NTPC), Station I with 4×500 MW and Station II with 3×210 MW, four double-circuit transmission lines (DCTLs) (toward Vindyachal with 215 km, Raipur/Power Grid Corporation of India Limited (PGCIL)-B with 220 km, toward Birsinghpur with 231 km, and Bhilai/Khedamara with 198 km), and three single-circuit transmission lines (toward Korba-West with 17 km, Bhatpara with 100 km, from Sipat with 60 km). At bus-3 (Bhilai/Khedamara), one triple circuit line (from Raipur/Raita with 65.68 km), three feeders connected to Bhilai 220 kV substation through step down transformer, one double-circuit line (from KSTPS/NTPC 198 km), and seven single-circuit lines (from Korba-West with 212 km, Bhatpara with 90 km, toward Raipur/PGCIL-A with 20 km, Seoni with 250 km, Koradi with 272 km, Bhadrawati with 322 km, and from Marwa with 170 km) are connected. Here the DCTL connected between bus-3 and bus-4 (KSTPS/NTPC and Bhilai/Khedamara) of 198 km length is considered, and its configuration is depicted in Fig. 18.2. The existing 400 kV power transmission network of Chhattisgarh state has been modeled in MATLAB/Simulink software to perform simulation studies for various types of shunt faults. The proposed reliable decision-making algorithm is employed at bus-4 (KSTPS/NTPC), by simulating shunt faults during the PS condition at a wide range of fault parameters. PS condition has been initiated by isolating the line between bus-4 and bus-11 after the occurrence of a three-phase to ground fault.

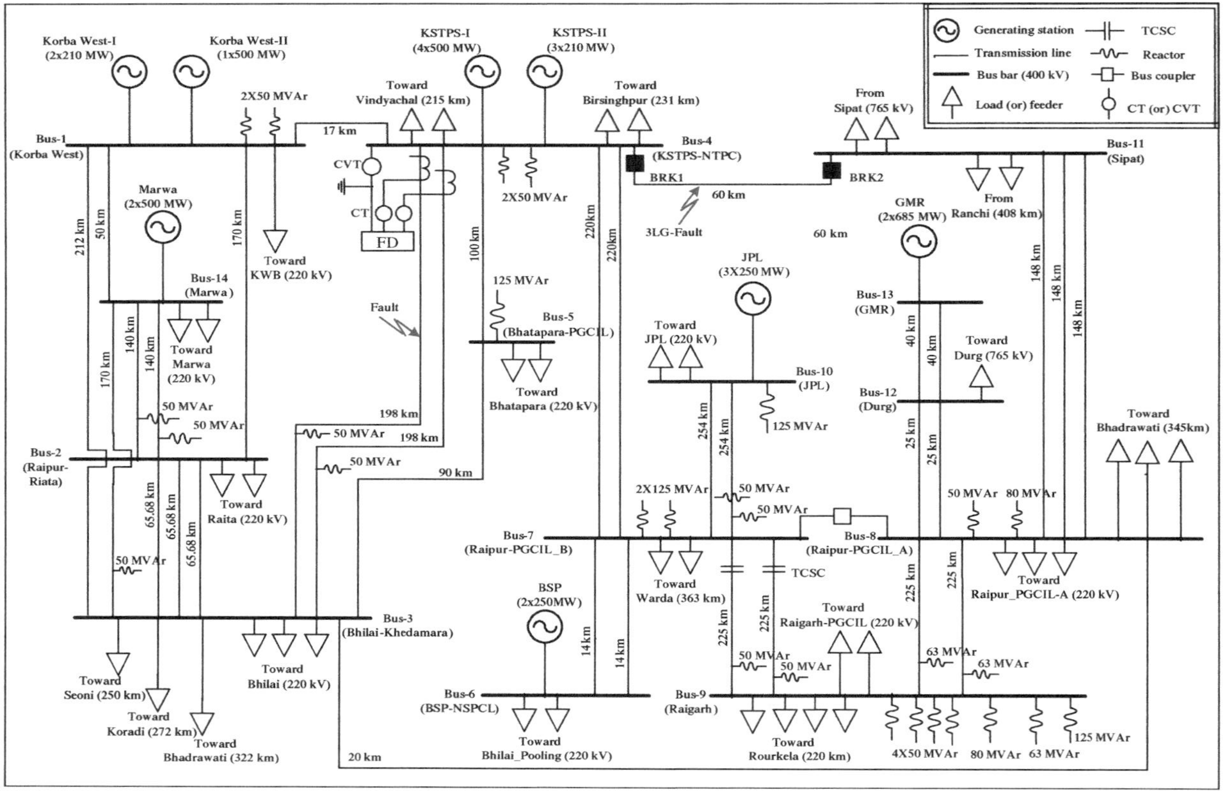

FIGURE 18.1 400 kV Power transmission network of Chhattisgarh state.

TABLE 18.1 Generator and load/feeder data of 400 kV double-circuit line of Chhattisgarh state network.

Generator			Load/feeder connected at bus-3 and bus-4			
Parameter	KSTPS-I	KSTPS-II	Feeders	P (MW)	Q (MVAr)	Bus
Rating (MVA)	600	250	Korba-West-I	166.7	5.1	4
X_l (pu)	0.15	0.116	Vindyachal (D/C)	467.5	27.6	4
R_a (pu)	0.00192	0.0016	Birsinghpur (D/C)	250.7	68.8	4
X_d (pu)	2.31	1.7	Sipat	134.8	4.2	4
X_q (pu)	2.19	1.62	PGCIL-B (D/C)	200.1	39.7	4
X_d^1 (pu)	0.267	0.256	Bhatapara	407.2	150.3	4
X_q^1 (pu)	0.37	0.245	Bhilai/Khedamara (D/C)	232.4	9.0	4
X_d^{11} (pu)	0.212	0.185	Korba-West-I	229.2	78.6	3
X_q^{11} (pu)	0.233	0.147	Marwa	244.2	20.4	3
T_{d0}^1 (pu)	6.69	4.8	Raipur/Raita (T/C)	134.7	15.2	3
T_{q0}^1 (pu)	2.5	0.5	KSTPS/NTPC (D/C)	230.4	33.7	3
T_{d0}^{11} (pu)	0.038	0.0437	Bhatapara	48.6	73.4	3
T_{q0}^{11} (pu)	0.05	0.141	PGCIL-A	420.7	72.1	3
H (s)	3	4.129	Bhilai (T/C)	226.3	119.7	3
–	–	–	Bhadrawati	371.8	19.8	3
–	–	–	Koradi	473.2	27.7	3
–	–	–	Seoni	283.1	100.8	3

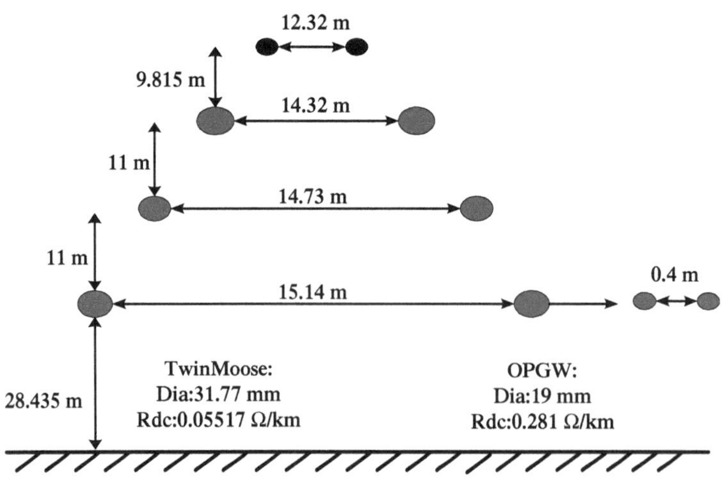

FIGURE 18.2 400 kV Double-circuit transmission line configuration.

18.4 Proposed wavelet packet energy and bagged decision tree decision-making algorithm

18.4.1 Wavelet packet energy

The WPT is a generalized form of DWT. In WPT the signal (S) is passed through low-pass and high-pass filter, and concurrently, the signal is decomposed into high-frequency (d) and low-frequency (a) coefficients. Subsequently, the low-frequency (a) and high-frequency (d) coefficients are decomposed further recursively at different levels to form a tree structure called as a wavelet packet tree which is illustrated in Fig. 18.3. In DWT, at each level, only the approximate coefficients signal is decomposed into low and high-frequency components, that is, it discards the detail coefficients in each level. Also in the case of WPT, from Fig. 18.3, it can be observed that, after the decomposition process for each and every level, there will be two wavelet packet nodes available [28]. Therefore WPT provides more numbers of attributes, improved frequency resolution, offers more information content in the high-frequency component, and gives a comprehensive view of the decomposed signal. In the present work, sampling frequency considered is 1 kHz, for the 50 Hz system; thus the number of samples per cycle is 20. Using WPT, decomposition up to third level is performed using "db6" as mother wavelet that is a well-known and suitable wavelet for transient signal analysis. According to this decomposition process, at each level, precise wavelet coefficients of the signal are calculated in order to find optimal decomposition at each and every node. These decomposed coefficients of each node will explore different attributes. Subsequently, a statistical feature, that is, WPE, is obtained from the decomposed coefficient of WPT at each subband. The value of the transient energy signal is more as compared to the value of the undistorted or normal signal. Total eight nodes of decomposed WPT coefficients are generated; these decomposed coefficients of the signal are having different WPE in each and every subband. The plot of variation in the standard deviation (SD) of the wavelet energy packets of current and

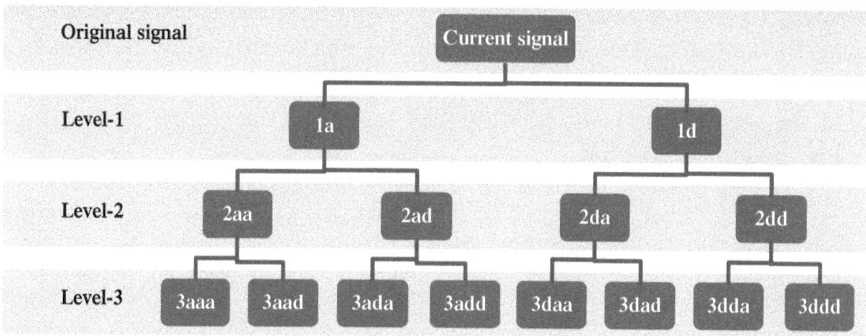

FIGURE 18.3 Structure of wavelet packet tree.

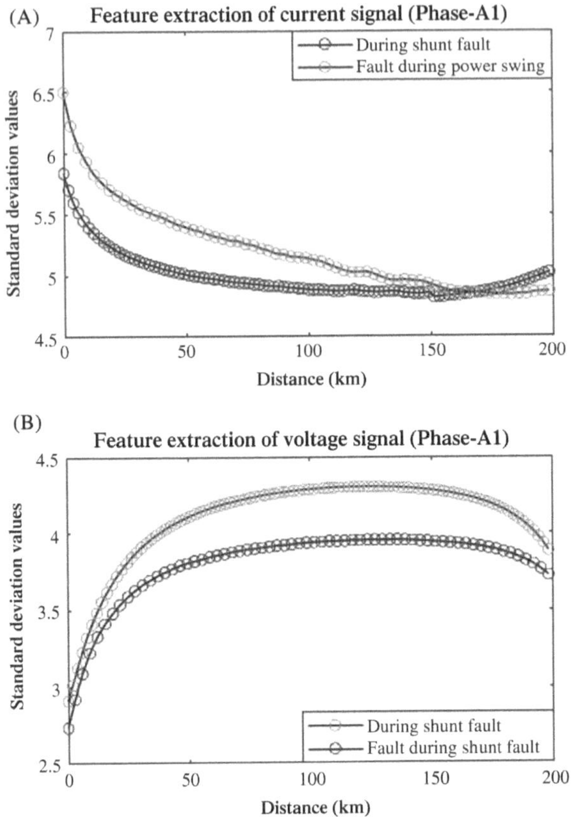

FIGURE 18.4 Variation of SD of WPE with fault distance (A) SD values of the current of phase-A1 during shunt fault and fault during power swing (B) SD values of voltage of phase-A1 during shunt fault and fault during power swing. *SD*, Standard deviation; *WPE*, wavelet packet energies.

voltage respectively during PS and fault during PS with respect to varying fault distance is shown in Fig. 18.4A and B. It is clear from Fig. 18.4 that SD values of faulted phase current and voltage have a distinct variation with respect to fault distance along the complete line length. Thus SD values of all three-phase voltages and currents have been taken as input feature set to design the input matrix for training and testing of the proposed intelligent decision-making algorithm.

18.4.2 Bagged decision tree

In the perspective of machine learning, tree-based ensemble techniques are commonly used for prediction and classification purposes in many research fields such as protection of transmission lines for instance. One

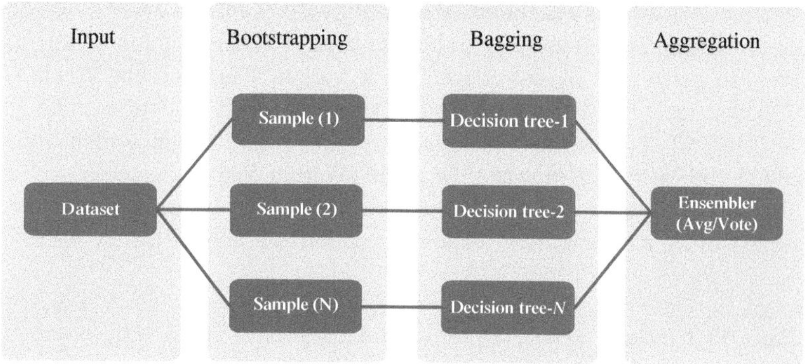

FIGURE 18.5 Architecture of a BDT model. *BDT*, Bagged decision tree.

such tree-based ensemble technique is BDT that builds, by randomization, numerous DTs and then aggregates their predictions as exemplified in Fig. 18.5. From an ensemble of trees, one can derive an importance score for each variable of the problem that assesses its relevance for predicting the output. Every tree in the ensemble is grown on an independently drawn bootstrap replica of input data [29]. Observations not included in this replica are "out-of-bag" for this tree [30]. The bagging is nothing but a bootstrap aggregation of a set of DTs. Individual DTs tend to over-fit. BDT (Bootstrap–aggregated) combines the results of many DTs, thereby overwhelming the effects of over-fitting and improves generalization. Tree Bagger grows the DTs in the ensemble using bootstrap samples of the data. The estimated out-of-bag error variances can be reduced by setting a more balanced misclassification cost matrix or a less skewed prior probability vector. BDT generates in-bag samples by oversampling classes with large misclassification costs and undersampling classes with small misclassification costs. Consequently, out-of-bag samples have fewer observations from classes with large misclassification costs and more observations from classes with small misclassification costs. To train a BDT ensemble classifier using a small data set, a highly skewed cost matrix is considered, and then the number of out-of-bag observations per class might be very low. In both bagging and random forest, many individual DTs are built on a bootstrapped version of the original dataset and are ensemble together. The difference is that, in a random forest, a random subset of variables are considered during node split while in bagging of DTs, all the variables are considered in a node split [31].

18.4.3 Wavelet packet energy and bagged decision tree–based decision-making algorithm

In order to train and test the proposed WPE_BDT based decision-making algorithm for detection and classification of shunt faults during PS condition, an extensive dataset is generated as exemplified in Table 18.2–18.5. To generate an appropriate dataset for training/testing, extensive simulation studies have been performed by considering wide variation in operating conditions as follows: fault resistance (R_f) from 0 to 100 Ω, fault inception angle (ϕ_f) from 0 to 360 degrees, fault location (L_f) from 1 to 197 km, source impedance (Z) within ± 10%, power flow angle (δ) from 5 to 20 degrees.

In this proposed protection scheme, exclusive datasets have been designed using an illustrative feature extraction technique, that is, WPE. To

TABLE 18.2 Dataset generated for a wide range of variation in different parameters for wavelet packet energy and bagged decision tree-1 modules.

Parameter	Common shunt faults	Power swing (stable and unstable)
Fault on line	NTPC/KSTPS-to-Bhilai/ Khedamara	NTPC/KSTPS-to-Sipat
Fault clearing time (cycle)	5	2, 4, 6, 8, 10, 12, 14, 16, 18, 20
Fault location (L_f)	198 km in steps of 2 km	60 km in steps of 3 km
Fault inception angle (ϕ_f) (degrees)	0, 90, and 270	0, 90, and 270
Fault resistance (R_f) (Ω)	0, 50, and 100	0, 50, and 100
No. of fault cases	*LG*: 2646, *LLG*: 2646	*LLLG*: 1710
	LL: 2646, *LLL*: 0882	
Total no. of cases	9020 = 8820 + 200 (no fault)	1710

TABLE 18.3 Dataset generated for a wide range of variation in fault parameters for wavelet packet energy and bagged decision tree-2 modules.

Parameter	Stable power swing	Unstable power swing
Fault on line	NTPC/KSTPS-to-Sipat	NTPC/KSTPS-to-Sipat
Fault clearing time (cycle)	2, 4, 6, 8	10, 12, 14, 16
Fault location (L_f)	60 km in steps of 3 km	60 km in steps of 3 km
Fault inception angle (ϕ_f) (degrees)	0, 90, and 270	0, 90, and 270
Fault resistance (R_f) (Ω)	0	0
Total no. of cases	*LLLG*: 228	*LLLG*: 228

TABLE 18.4 Dataset generated for a wide range of variation in fault parameters for wavelet packet energy and bagged decision tree-3 modules.

Parameter	Shunt faults during power swing
Fault on line	NTPC/KSTPS-to-Bhilai/Khedamara
Fault clearing time	5-cycle
Fault location (L_f)	198 km in steps of 3 km
Fault inception angle (ϕ_f)	0, 90, and 270 degrees
Fault resistance (R_f)	0, 50, and 100 Ω
No. of fault cases	*LG*: 1755
	LLG: 1755
	LL: 1755
	LLL: 585
Total no. of cases	9850 = 5850 + 4000 (no fault)

TABLE 18.5 Dataset generated for a wide range of variation in fault parameters for wavelet packet energy and bagged decision tree-4 modules.

Parameter	Common shunt faults	Shunt faults during power swing
Fault on line	NTPC/KSTPS-to-Bhilai/ Khedamara	NTPC/KSTPS-to-Bhilai/ Khedamara
Fault type	*LG*: A1G, B1G, C1G	*LG*: A1G, B1G, C1G
	LLG: A1B1G, B1C1G, A1C1G	*LLG*: A1B1G, B1C1G, A1C1G
	LL: A1B1, B1C1, A1C1	*LL*: A1B1, B1C1, A1C1
	LLL: A1B1C1	*LLL*: A1B1C1
Fault location (L_f)	198 km in steps of 2 km	198 km in steps of 3 km
Fault inception angle (ϕ_f) (degrees)	0, 90, and 270	0, 90, and 270
Fault resistance (R_f) (Ω)	0, 50, and 100	0, 50, and 100
No. of fault cases	*LG*: 2646	*LG*: 1755
	LLG: 2646	*LLG*: 1755
	LL: 2646	*LL*: 1755
	LLL: 0882	*LLL*: 585
Total no. of cases	9020 = 8820 + 200 (no fault)	9850 = 5850 + 4000 (no fault)

design a reliable decision-making algorithm, total 15 features are selected which are very distinctive to discriminate different operating scenarios of the power system network such as no-fault, fault, PS, and fault during PS conditions. The distinct features that have been considered to design input datasets to the WPE_BDT modules are calculated using Eqs. (18.1)–(18.6) and given hereunder:

- $S1$ = Rate of change of frequency at the sending-end bus (Δf)
- $S2$ = Rate of change of load angle at the sending-end bus ($\Delta\delta$)
- $S3$ = Rate of change of voltage at the sending-end bus (ΔV)
- $S4$ = Rate of change of active power at relaying point (ΔP)
- $S5$ = Rate of change of reactive power at relaying point (ΔQ)
- $S6$ = SD of wavelet packet energy of Ia1
- $S7$ = SD of WPE of Ib1
- $S8$ = SD of WPE of Ic1
- $S9$ = SD of WPE of va1
- $S10$ = SD of WPE of vb1
- $S11$ = SD of WPE of vc1
- $S12$ = SD of WPE of Ig1
- $S13$ = SD of WPE of Ia2
- $S14$ = SD of WPE of Ib2
- $S15$ = SD of WPE of Ic2

$$\Delta f_s = \left|f_s(i) - f_s(i-k)\right| \tag{18.1}$$

$$\Delta V_s = \left|V_s(i) - V_s(i-k)\right| \tag{18.2}$$

$$\Delta \delta_s = \left|\delta_s(i) - \delta_s(i-k)\right| \tag{18.3}$$

$$\Delta P_s = \left|P_s(i) - P_s(i-k)\right| \tag{18.4}$$

$$\Delta Q_s = \left|Q_s(i) - Q_s(i-k)\right| \tag{18.5}$$

$$\sigma_s = \left\{ \sqrt{\frac{1}{(N-1)} \sum_{i=0}^{N-1} (x_i - \mu)^2} \right\} \tag{18.6}$$

where i is the current sample count, k is the number of samples/cycle, N is the total number of samples, μ is the mean of samples, σ is the SD, x is the current sample, f is the operating frequency, V is the operating voltage at the sending-end bus, δ is the power flow angle at the sending-end bus, P is the active power, Q is the reactive power.

The flowchart of the proposed protection scheme has been described in Fig. 18.6, which constitutes six stages and four different WPE_BDT modules. Different stages of proposed reliable decision-making algorithm has

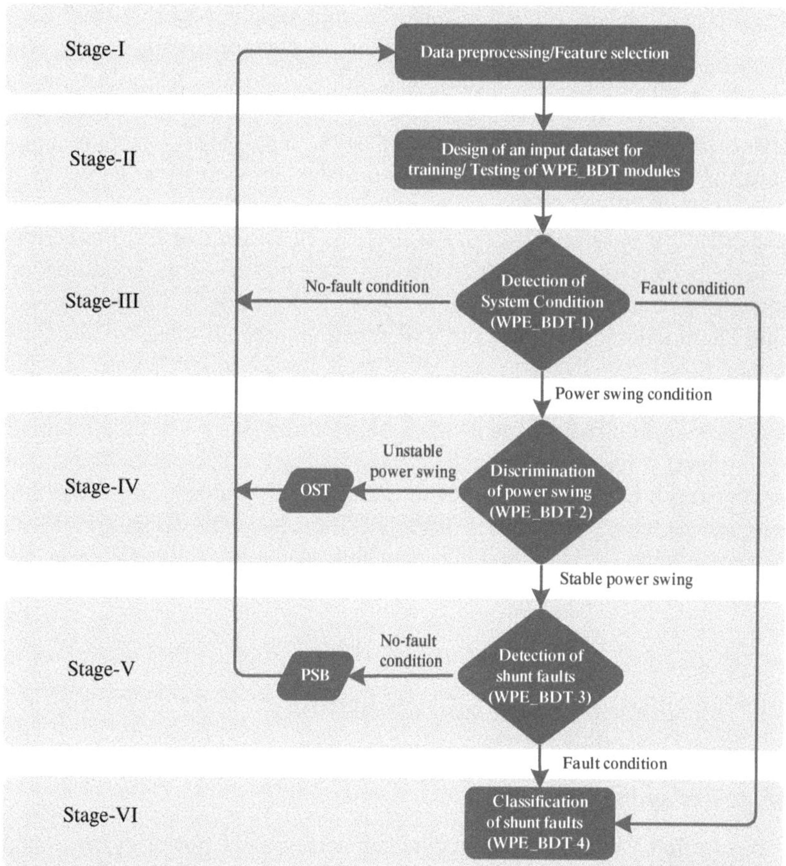

FIGURE 18.6 Flowchart of proposed reliable decision-making algorithm using WPE_BDT modules. *WPE_BDT*, Wavelet packet energy and bagged decision tree.

been instigated as follows: (1) Stage I: data preprocessing and feature selection by applying appropriate signal processing technique, (2) Stage II: design of input dataset for training/testing of different WPE_BDT modules considering wide range of variation in different parameters as per Tables 18.2–18.5, (3) Stage III: WPE_BDT-1 for detection of power system network condition such as no-fault, fault, PS conditions, (4) Stage IV: WPE_BDT-2 for discrimination of stable and unstable PS, (5) Stage V: WPE_BDT-3 for detection of faults during PS, and (6) Stage VI: WPE_BDT-4 for classification of shunt faults during PS.

Four different WPE_BDT modules have been developed using distinct and suitable features, such as S1–S5 are used for WPE_BDT-1 and -2 module, S6–S15 are used for WPE_BDT-3 and -4 modules. The output of

WPE_BDT-1 module for detection of power system condition will be of three classes: 0, 1, 2 for no fault, fault, and PS condition, respectively. Then if it is a no-fault condition, it goes back to the first stage, while if a PS condition is detected, it activates the stage IV, else, if the fault is detected, it activates stage VI. Moreover, in case of a PS, WPE_BDT-2 module is used to identify whether the PS is stable or unstable. In case of unstable PS the output of the WPE_BDT-2 module is "1," then the OST function of conventional MHO relay is enabled, else if it is "0," it is a stable PS. Furthermore in order to identify that whether there is only power swing condition or a fault during stable PS, WPE_BDT-3 module is used. If the system is in stable PS condition, it enables the PSB function, while if there is fault during PS, then it activates the last stage VI of fault classification. In case of a fault during stable PS, WPE_BDT-4 module classifies the fault type into 10 classes pertaining to different shunt faults. Henceforth, there will be a multiclass target/output; 0 for no-fault, 1–10 corresponding to 10 types of shunt faults consecutively. Exclusive datasets as tabulated in Tables 18.2–18.5 have been used to train the proposed WPE_BDT modules for designing of accurate and reliable decision-making algorithm to avoid maloperation at fault during the PS condition.

18.5 Simulation results and discussions

18.5.1 Detection of system condition (no-fault, fault, and power swing)

In this section, WPE_BDT-1 module for detection of power system condition is tested with a versatile dataset consisting of distinct features (S1–S5). It is very significant to train and test the module to get an exact condition of the system to prevent maloperation of the proposed protection scheme. In line with this, WPE_BDT-1 module has been trained and tested for different combinations/segmentation of datasets based on cross-validation philosophy, that is, simultaneously training and testing with dataset segmentation such as (90–10), (80–20), (70–30), (60–40), and (50–50) for training and testing, respectively. The confusion matrix has been reported in Table 18.6, which illustrates the actual and predicted class and accuracy during testing for a particular combination of the dataset. It can be seen the 100% accuracy in detection of the system condition is achieved in different percentage of dataset segmentation using the proposed WPE_BDT-1 module. The accuracy of the detection/classification capability of WPE_BDT-1 module depends on the selection of suitable feature and appropriate combination/segmentation of datasets for training and testing collectively, which confirms the suitability of the proposed scheme for discrimination of different power system conditions.

TABLE 18.6 Test results and confusion matrix of wavelet packet energy and bagged decision tree-1 module for detection of system condition (no-fault, power swing).

Case study	Data segmented (%)	Predicted class	Actual class			Accuracy (%)
			0	1	2	
Case-1	90–10	0	2000			100
		1		8820		
		2			1710	
Case-2	80–20	0	2000			100
		1		8820		
		2			1710	
Case-3	70–30	0	2000			100
		1		8820		
		2			1710	
Case-4	60–40	0	2000			100
		1		8820		
		2			1710	
Case-5	50–50	0	2000			100
		1		8820		
		2			1710	

18.5.2 Discrimination of power swing condition (stable and unstable)

A stable PS is initiated by creating a fault in the adjacent line connected to the monitored bus and clearing the fault before the critical clearing time (CCT), whereas if the fault is cleared after the CCT then it will initiate unstable PS in the system. For instance, a three-phase to ground fault has been incepted at 1.001 seconds on a single-circuit transmission line of 60 km length which is connected between NTPC and Sipat in Fig. 18.1 and the same has been cleared by tripping the circuit breakers of both the ends at 1.15 seconds, that is, fault clearing time is 7.5-cycles. In order to identify the state of the system during this stressed condition, that is, the PS is stable or unstable; WPE_BDT-2 module is tested, and the test results are demonstrated in Fig. 18.7. The three-phase current and voltages of the monitored double-circuit line are

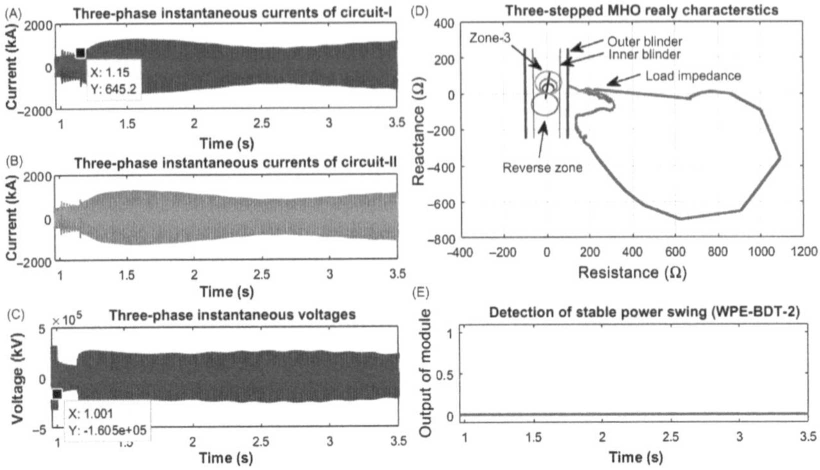

FIGURE 18.7 Performance of proposed WPE_BDT-2 module in case of a stable power swing. (A) Three phase instantaneous currents of circuit-I (B) Three phase instantaneous currents of circuit-II (C) Three phase instantaneous voltages at sending-end bus (D) Three stepped MHO relay characteristics (E) Output of WPE_BDT-2 module during stable PS. *WPE_BDT*, Wavelet packet energy and bagged decision tree.

depicted in Fig. 18.7A–C, respectively, and Fig. 18.7D the performance of the three-stepped MHO relay characteristic with double blinder PS detection technique in R–X plane, and Fig. 18.7E shows the output of proposed WPE_BDT-2 module output, which remains at zero level confirming the system state in stable PS. Furthermore, an unstable PS is initiated and illustrated in Fig. 18.8, when the same fault is cleared after 9-cycles from the fault inception at 1.18 seconds (PS has been initiated). From Fig. 18.8A–C, it can be appreciated from the high swing rate of the voltage and current waveforms, that it is an unstable PS condition. Therefore the impedance seen by the MHO relay enters in its third zone in Fig. 18.8D, and correspondingly, the output of WPE_BDT-2 module in Fig. 18.8E rises up to high (1) level at 1.243 seconds, confirming it to be an unstable PS.

18.5.3 Detection of faults during power swing condition

The conventional distance relays fail to detect the fault during PS owing to blocking through PS blinders. However, if fault during PS remains undetected, it may damage the equipment and create a severe economic loss. It is very important to detect the fault during the PS condition to prevent maloperation of distance relaying scheme, thereby neutralizing the PSB function. In the proposed protection scheme, WPE_BDT-3 module is used for detection of shunt faults during the PS has been trained and tested for different combinations of training and testing datasets using cross-validation

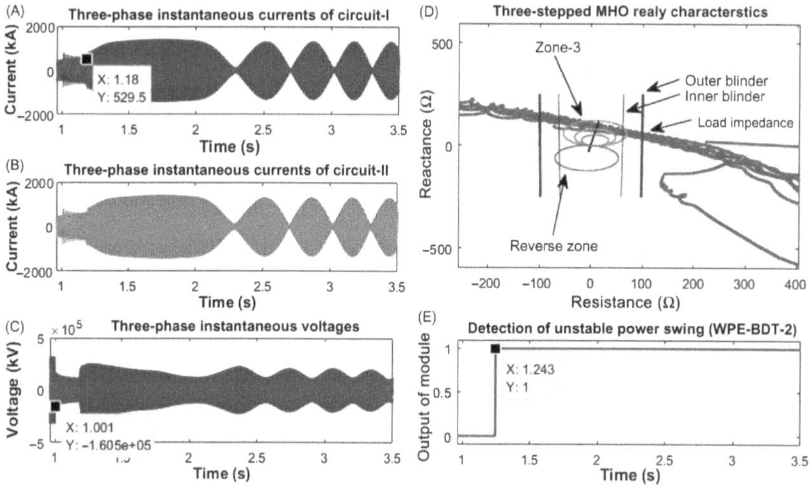

FIGURE 18.8 Performance of WPE_BDT-2 module in case of unstable power swing: (A) Three phase instantaneous currents of circuit-I (B) Three phase instantaneous currents of circuit-II (C) Three phase instantaneous voltages at sending-end bus (D) Three stepped MHO relay characteristics (E) Output of WPE_BDT-2 module during unstable PS. *WPE_BDT*, Wavelet packet energy and bagged decision tree.

philosophy and the confusion matrix obtained for the tested cases has been reported in Table 18.7 which illustrate the actual and predicted class for different combination of dataset. It can be observed that the proposed scheme offers 99.8% accuracy in prediction of the fault during PS. The output of WPE_BDT-3 module will be a two class target/output: 0 for no-fault during the PS and 1 for fault during the PS. If the output class is "0," the relay will initiate a PSB function, and there is a no-fault condition. Further, if the output class is "1," the relay will initiate a trip command/signal. For instance, phase-A1 to ground fault during PS is created at 3.018 seconds, and resultant three-phase currents of both the circuit and voltages are depicted in Fig. 18.9A–C. The performance of three-stepped distance MHO relay with a double blinder technique is reported in Fig. 18.9D wherein the apparent impedance of the transmission line passes through the zone-1 for a phase to ground fault (A1G). The output of WPE_BDT-3 module for detection of a fault during PS has been demonstrated in Fig. 18.9E where the fault has been detected at 3.035 seconds.

18.5.4 Performance evaluation of proposed wavelet packet energy and bagged decision tree-4 module for classification of faults

If normal shunt fault is detected in the system in stage III or V of the proposed algorithm depicted in Fig. 18.6, the next task is to classify the fault

TABLE 18.7 Test results and confusion matrix of wavelet packet energy and bagged decision tree-3 module for detection of a fault condition during the power swing.

Case study	Data segmented (%)	Predicted class	Actual class		Accuracy (%)
			0	1	
Case-1	90–10	0	3948	52	99.37
		1	10	5840	
Case-2	80–20	0	3947	53	99.41
		1	5	5845	
Case-3	70–30	0	3972	28	99.70
		1	1	5849	
Case-4	60–40	0	3974	26	98.22
		1	2	5848	
Case-5	50–50	0	3972	28	99.68

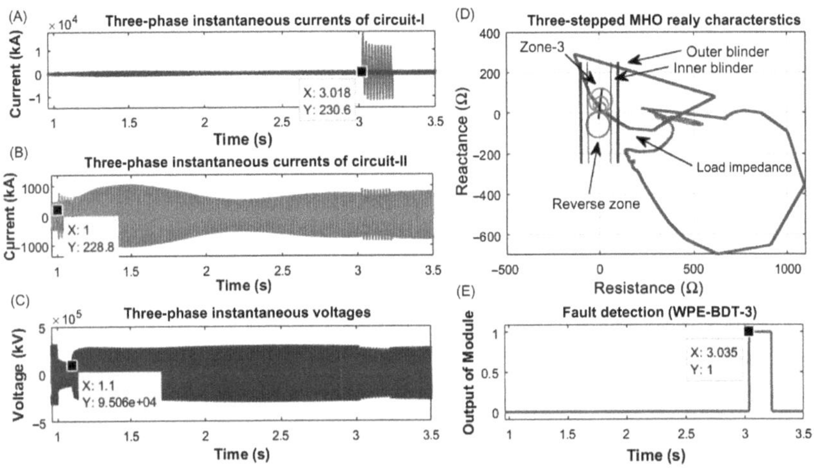

FIGURE 18.9 Performance of WPE_BDT-3 module in case of shunt fault (phase-A1 to ground) during a stable power swing: (A) Three phase instantaneous currents of circuit-I (B) Three phase instantaneous currents of circuit-II (C) Three phase instantaneous voltages at sending-end bus (D) Three stepped MHO relay characteristics (E) Fault detection using WPE_BDT-3 module. *WPE_BDT*, Wavelet packet energy and bagged decision tree.

and identify the faulted phase(s). To examine the efficacy of proposed decision-making algorithm, WPE_BDT-4 module for fault classification has been tested here with new fault cases involving normal shunt faults and faults during PS by varying different fault parameters as given in Table 18.5.

TABLE 18.8 Test results and confusion matrix of wavelet packet energy and bagged decision tree-4 module for classification of shunt fault during the power swing (stable).

Case study	Data segmented (%)	Predicted class	Actual class											Accuracy (%)
			0	1	2	3	4	5	6	7	8	9	10	
Case-1	80–20	0	3999		1									99.96
		1		585										
		2	1		584									
		3				585								
		4					585							
		5						585						
		6							585					
		7					1			584				
		8									585			
		9										585		
		10											585	
Case-2	70–30	0	3999									1		99.95
		1		585										
		2	1		584									
		3				585								
		4					585							
		5						584			1			
		6							585					
		7					1			584				
		8									585			
		9										585		
		10											585	
Case-3	60–40	0	3998		1				1					99.97
		1		585										
		2			585									
		3				585								
		4					585							
		5						585						
		6							585					
		7								585				
		8									585			
		9										585		
		10											585	
Case-4	50–50	0	3998		1			1						99.94
		1		585										
		2	1		584									
		3				585								
		4					585							
		5						585						
		6							584			1		
		7					1			585				
		8									585			
		9										585		
		10											585	

The confusion matrix for the tested cases has been reported in Table 18.8, which confirms high accuracy of prediction of the fault classification module. A detailed discussion and analysis of appropriate case studies are reported in further subsections.

18.5.4.1 Performance in case of varying fault parameters

In real-time situation, fault may occur at any time, that is, at varying fault inception angle (ϕ_f) in the transmission line and in general the ground fault resistance (R_f) is in the range of 10–20 Ω; however, in some cases, namely, conductor falling on a high-resistance path such as rock, sand, asphalt, it may vary from 0 to 100 Ω approximately. Therefore in order to investigate its impact on proposed relaying scheme an extensive simulation studies have been performed on a 400 kV, 198 km DCTL of Chhattisgarh state transmission network by creating faults at diverse locations with different fault inception angle and fault resistance. As the signals are of periodic nature the impact of $\phi_f = 0$, 90, and 270 degrees and $R_f = 0$, 50, 100 Ω have been considered. From the test results shown in Table 18.9, it can be witnessed that performance of WPE_BDT-based scheme is not affected by variation of fault parameters such as fault inception angle, fault resistance, and fault location. To investigate the performance of proposed WPE_BDT-4 module, let us consider a double phase to ground fault (A1B1G) occurred at 3.037 seconds as depicted in Fig. 18.10A–C which shows the currents of faulted circuit I, II, and voltages. In this case the output of WPE_BDT-3 module for fault detection is exemplified in Fig. 18.10D and WPE_BDT-4 module for fault classification in Fig. 18.10E, which detects and classifies the fault type as A1B1G at 3.058 seconds, that is, after a delay of 21 ms.

18.5.4.2 Performance during variation in power flow angle

It is essential to examine the variation in prefault power flow angle on proposed relaying algorithm which determines not only the amount of active power flow through the transmission line but also provides the magnitude of the voltage and current signals at relaying point. In this regard, proposed WPE_BDT modules have been tested with variation of power flow angle from 5 to 20 degrees in steps of 5 degrees and presuming the fault parameters as $L_f = 50$ km, $R_f = 30\Omega$, $\phi_f = 90$degrees. Some of the test cases have been described in Table 18.10. The competency of the proposed scheme is verified for detection and classification of a shunt fault (A1B1G depicted as bold in Table 18.10) with varying power flow angle as depicted in Fig. 18.11A–C which occurred at 3.021 seconds. It has been detected and classified as A1B1G fault after 24 ms as shown in Fig. 18.11D and E, that is, within 1.5 cycles from fault inception time.

18.5.4.3 Performance during variation in source impedance ratio

During a fault condition, the source impedance of the system plays a vital role, because it determines whether the system is a weak or strong system in case of protection sensitivity, and also it will define the short-circuit level or fault level at particular fault condition. In this context the source impedance of both the generating units of KSTPS-I and KSTPS-II has been varied within a range of $\pm 10\%$ as illustrated in Table 18.11 by assuming fault parameters $L_f = 100$, $R_f = 40\ \Omega$, $\phi_f = 270$degrees to test the efficiency of WPE_BDT modules for detection and classification of shunt faults during the PS. The effectiveness of proposed WPE_BDT-3 and -4 modules have been illustrated in Fig. 18.12 for a single phase to ground fault (A1G), which is incepted at 3.028 seconds, and the same has been detected at

TABLE 18.9 Response of wavelet packet energy and bagged decision tree-3 and 4 modules for detection and classification of shunt faults during the power swing.

	Type of fault	L_f(km)	$R_f(\Omega)$	ϕ_f(degrees)	Fault detection time (ms)	Fault classification time (ms)						
						A1	B1	C1	G	A2	B2	C2
Shunt faults	AG	30	0.001	45	16.0	16	–	–	16	–	–	–
	ABG	**60**	**25**	**135**	**21.0**	**21**	**21**	–	**21**	–	–	–
	AB	120	75	225	11.0	11	11	–	–	–	–	–
	ABC	180	100	315	12.0	12	12	12	–	–	–	–
Shunt faults during power swing	AG	30	0.001	45	16.0	16	–	–	16	–	–	–
	ABG	60	25	135	22.0	22	22	–	22	–	–	–
	AB	120	75	225	07.0	07	07	–	–	–	–	–
	ABC	180	100	315	12.0	12	12	12	–	–	–	–

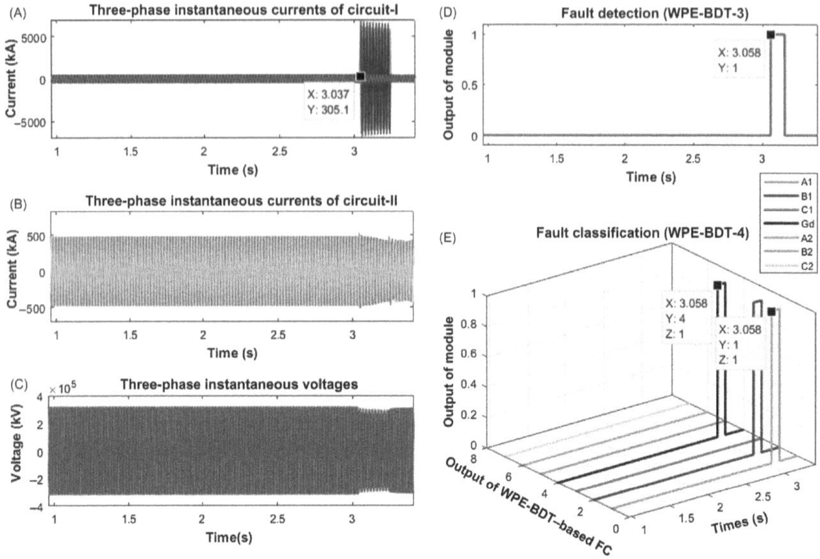

FIGURE 18.10 Performance of WPE_BDT-3 and -4 modules for detection and classification of shunt fault (double phase to ground fault-A1B1G): (A) Three phase instantaneous currents of circuit-I (B) Three phase instantaneous currents of circuit-II (C) Three phase instantaneous voltages at sending-end bus (D) Fault detection using WPE_BDT-3 module (E) Fault classification using WPE_BDT-4 module. *WPE_BDT*, Wavelet packet energy and bagged decision tree.

TABLE 18.10 Response of wavelet packet energy and bagged decision tree-3 and 4 modules for detection and classification of shunt faults during the power swing.

	Type of fault	$\Delta\delta$ (degrees)	Fault detection time (ms)	Fault classification time (ms)						
				A1	B1	C1	G	A2	B2	C2
Shunt faults	AG	+05	14.0	14	–	–	14	–	–	–
	ABG	+10	22.0	22	22	–	22	–	–	–
	AB	+15	22.0	22	22	–	–	–	–	–
	ABC	+20	18.0	18	18	18	18	–	–	–
Shunt faults (power swing)	AG	+05	14.0	14	–	–	14	–	–	–
	ABG	+ 10	24.0	24	24	–	24	–	–	–
	AB	+15	17.0	17	17	–	–	–	–	–
	ABC	+20	19.0	19	19	19	–	–	–	–

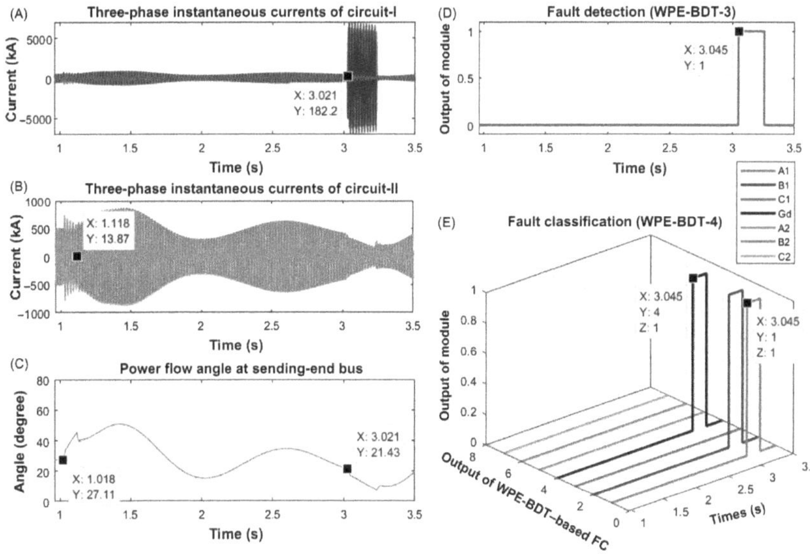

FIGURE 18.11 Performance of WPE_BDT-3 and -4 modules for detection and classification of shunt fault during PS, with variation in power flow angle: (A) Three phase instantaneous currents of circuit-I (B) Three phase instantaneous currents of circuit-II (C) Power flow angle at sending-end bus (D) Fault detection using WPE_BDT-3 module (E) Fault classification using WPE_BDT-4 module. *WPE_BDT*, Wavelet packet energy and bagged decision tree.

TABLE 18.11 Response of wavelet packet energy and bagged decision tree-3 and 4 modules for detection and classification of shunt faults during the power swing.

	Type of fault	ΔZ_{S1} (%)	ΔZ_{S2} (%)	Fault detection time (ms)	Fault classification time (ms)						
					A1	B1	C1	G	A2	B2	C2
Shunt faults	AG	−10	–	12.0	12	–	–	12	–	–	–
	ABG	−5	–	16.0	16	16	–	16	–	–	–
	AB	–	+5	18.0	18	18	–	–	–	–	–
	ABC	–	+10	11.0	11	11	11	–	–	–	–
Shunt faults (power swing)	*AG*	−10	–	25.0	25	–	–	25	–	–	–
	ABG	−5	–	16.0	16	16	–	16	–	–	–
	AB	–	+5	21.0	21	21	–	–	–	–	–
	ABC	–	+10	29.0	29	29	29	–	–	–	–

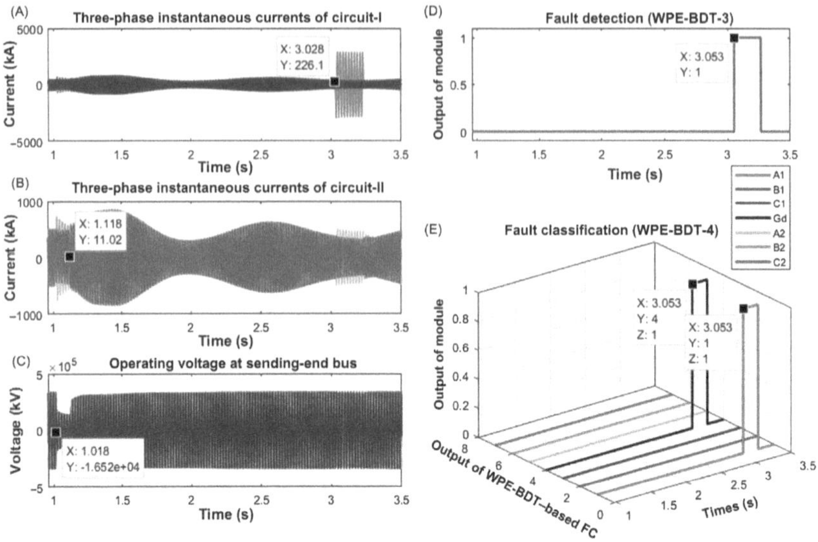

FIGURE 18.12 Performance of WPE_BDT-3 and -4 modules for detection/classification of phase-A1 to ground fault during PS: (A) Three phase instantaneous currents of circuit-I (B) Three phase instantaneous currents of circuit-II (C) Three phase instantaneous voltages at sending-end bus (D) Fault detection using WPE_BDT-3 module (E) Fault classification using WPE_BDT-4 module. *PS*, Power swing; *WPE_BDT*, wavelet packet energy and bagged decision tree.

3.053 seconds. From Fig. 18.12 and test results in Table 18.11, the efficacy of the proposed scheme has been confirmed for variation of the source impedance.

18.5.4.4 Performance during variation in sampling frequency

It is imperative to test the impact of variation in sampling frequency on the proposed scheme. In this regard, different type of case studies have been performed with variation in sampling frequency at 1.2, 5, 10, and 20 kHz, and the test results are summarized in Table 18.12 by assuming fault parameters $L_f = 25$ km, $R_f = 10\ \Omega$, $\phi_f = 90$degrees. The results from Table 18.12 show that the proposed WPE_BDT modules are not affected by variation in the sampling frequency.

18.5.4.5 Performance during current transformer–saturation and capacitor coupling voltage transformer transient

While measuring high fault currents at the relaying point, it is imperative to consider the effect of measuring CT saturation due to large DC offset current component in the faulty signal which may lead to an error in measurements and creates overreach problem as well. In order to assess the influence of CT

TABLE 18.12 Response of wavelet packet energy and bagged decision tree-3 and 4 modules for variation in sampling frequency.

	Type of fault	Sampling frequency (kHz)	Fault detection time (ms)	Fault classification time (ms)						
				A1	B1	C1	G	A2	B2	C2
Shunt faults	AG	1.2	13.0	13	–	–	13	–	–	–
	ABG	5	10.0	10	10	–	10	–	–	–
	AB	10	5.0	05	05	–	–	–	–	–
	ABC	20	4.0	04	04	04	–	–	–	–
Shunt faults (during power swing)	AG	1.2	17.0	17	–	–	17	–	–	–
	ABG	5	12.0	12	12	–	12	–	–	–
	AB	10	9.0	09	09	–	–	–	–	–
	ABC	20	6.0	06	06	06	–	–	–	–

saturation and CCVT transients on the proposed WPE_BDT scheme the standard CT and CCVT models as given in [32] have been selected and a wide-ranging simulation studies have been performed by creating different types of fault by assuming the $\phi_f = 0$degrees, $L_f = 3$ km, $R_f = 0.001\ \Omega$. Three-phase current and voltage waveforms during double phase to ground fault (A1B1G) occurred at 3.018 seconds in the presence of CTs and CCVTs are shown in Fig. 18.13A and B, respectively, wherein it can be observed that the CT secondary current multiplied by CT ratio does not imitate the primary current owing to CT saturation. The effectiveness of the proposed WPE_BDT-based scheme has been established on the current and voltage samples derived from the modified model. It can be seen from the Fig. 18.13D and E that the performance of the proposed scheme is not affected by CT saturation and CCVT transient as the fault is detected and identified correctly at 3.024 seconds as a double phase to ground fault involving phases A1 and B1 of circuit-I.

18.5.4.6 Performance in case of a fault in circuit-II

As the two circuits are operating in parallel with mutual coupling between the two circuits, a fault in one circuit affects the corresponding healthy phase of the second circuit too. Therefore it is important to investigate the performance of the proposed scheme for faults in the second circuit as well. For instance, consider a phase to ground fault (C2G) occurred at 3.024 seconds on circuit-II at $L_f = 60$ km, $R_f = 10\ \Omega$, and $\phi_f = 270$degrees. Three-phase current waveforms of circuit-I and circuit-II and voltage waveforms during a fault in the second circuit are shown in Fig. 18.14A–C, respectively,

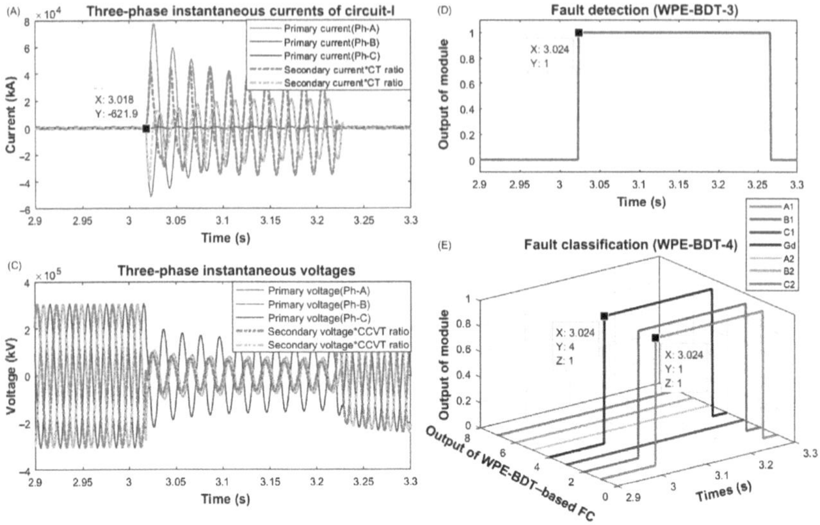

FIGURE 18.13 Performance of WPE_BDT-3 and -4 modules in presence of CT saturation and CCVT transients: (A) Three phase instantaneous currents of circuit-I (B) Three phase instantaneous voltages at sending-end bus (C) Fault detection using WPE_BDT-3 module (D) Fault classification using WPE_BDT-4 module. *CCVT*, Capacitor coupling voltage transformer; *CT*, current transformer; *WPE_BDT*, wavelet packet energy and bagged decision tree.

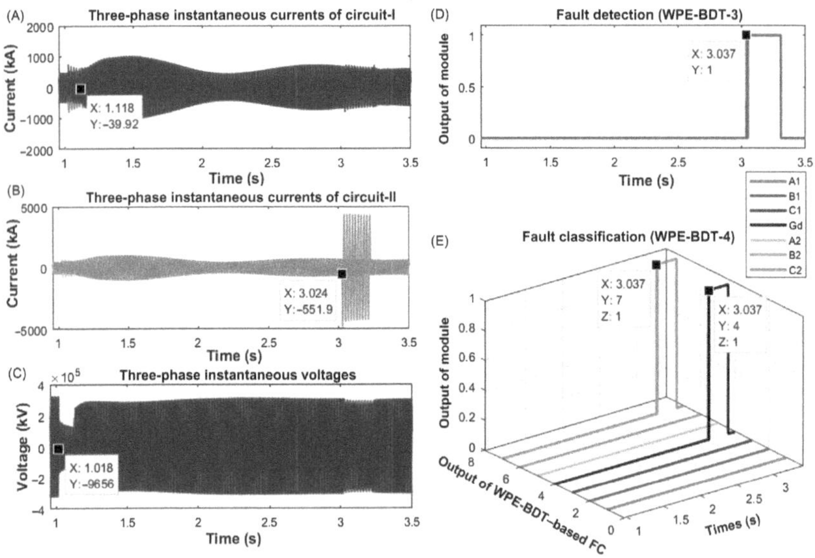

FIGURE 18.14 Performance of WPE_BDT-3 and -4 modules in case of a fault in circuit-II: (A) Three phase instantaneous currents of circuit-I (B) Three phase instantaneous currents of circuit-II (C) Three phase instantaneous voltages at sending-end bus (D) Fault detection using WPE_BDT-3 module (E) Fault classification using WPE_BDT-4 module. *WPE_BDT*, Wavelet packet energy and bagged decision tree.

wherein it can be observed that the currents of the second circuit are quite high after the occurrence of a fault, and the corresponding healthy phase-A1 current of circuit-I is also affected. The output of the proposed WPE_BDT scheme can be seen from the Fig. 18.14 D and E, which detects the fault type as phase to ground fault (C2G) of circuit-II at 3.037 seconds. Thus it can be concluded that the proposed WPE_BDT scheme is correctly identifying the faults in both the circuits of the double-circuit line.

18.5.4.7 Real-time validation using real-time digital simulation

To evaluate the performance of the proposed scheme in real-time, an experimental simulation has been performed in RTDS laboratory at CPRI, Bangalore. In this regard, the 400 kV power system network of Chhattisgarh state has been modeled in power system simulation software package for real-time simulation studies (RSCAD)/RTDS software to imitate fault transients on the real power system network. RSCAD/RTDS laboratory setup and respective real-time control window have been shown in Fig. 18.15. Some of the fault events have been simulated on the selected double-circuit line between NTPC/KSTPS-to-Bhilai/Khedamara and corresponding voltage and current signals are saved in COMTRADE format and then transferred from master control PC to a slave control PC which is installed with MATLAB/Simulink software wherein the proposed WPE_BDT scheme is coded. For example, a double phase to ground fault (A1B1G) is created at 0.41 seconds, $L_f = 1$ km, $R_f = 0\ \Omega$, and$\phi_f = 0$degrees using RTDS setup; the real-time voltage and current signals captured of circuit I and II are illustrated in Fig. 18.16A–C, from which it can be seen that the circuit-I faulted phase currents are significantly high while the corresponding healthy circuit-II phase currents are also increased due to the effect of zero sequence mutual coupling between the two circuits. This real-time simulated data has been concurrently tested with already trained WPE_BDT modules in MATLAB/Simulink, and its performance is depicted in Fig. 18.16D and E. It can be seen that the fault

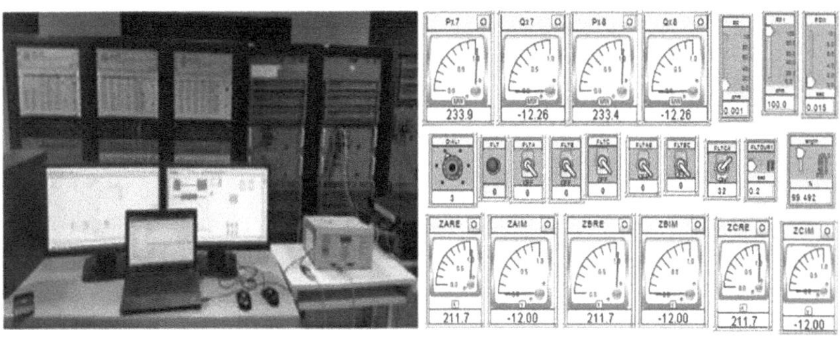

FIGURE 18.15 RTDS setup at CPRI, Bangalore, and real-time control/monitoring window. *RTDS*, Real-time digital simulation.

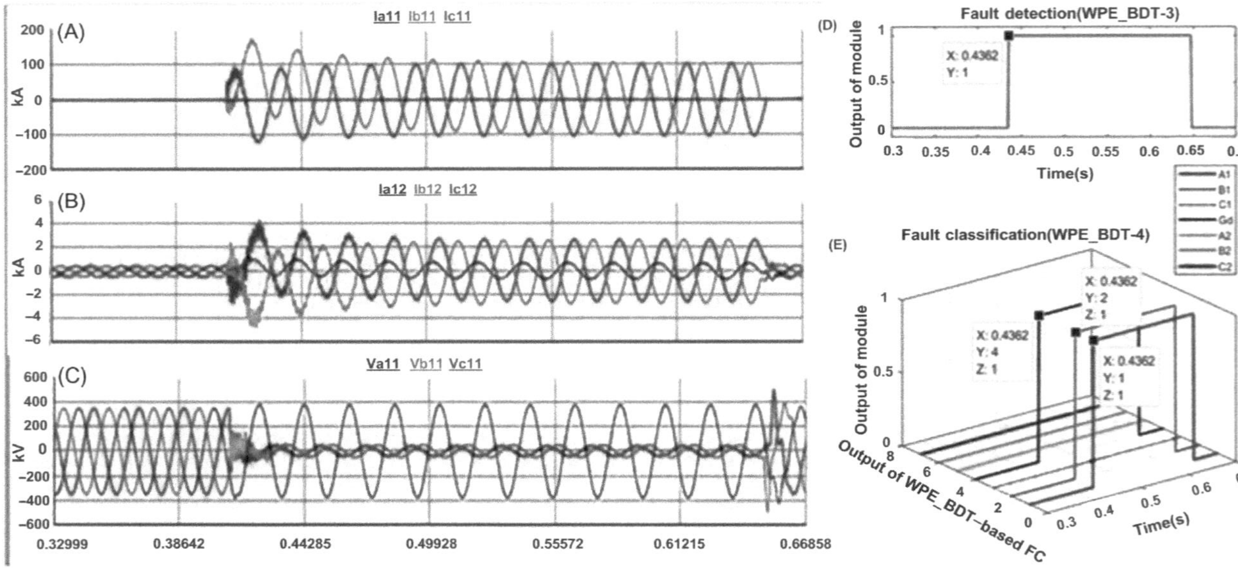

FIGURE 18.16 Performance in case of real-time signals captured during A1B1G fault at $L_f = 1$ km, $R_f = 0\ \Omega$, and $\phi_f = 0$ degrees using RTDS setup: (A) three-phase currents of circuit-I, (B) three-phase currents of circuit-II, (C) three-phase voltages, (D) fault detection of WPE_BDT-3 module, and (E) fault classification of WPE_BDT-4 module. *RTDS*, Real-time digital simulation; *WPE_BDT*, wavelet packet energy and bagged decision tree.

has been detected and classified as A1B1G fault at 0.4362 seconds. Thus the proposed scheme is validated using RTDS for real-time implementation. Further regarding the applicability of the proposed method practically for any given power network, some papers have devoted to implement the pattern-recognition-based protection scheme on a hardware platform and test the performance of the framework against different scenarios [33–36]. Since the numerical relays are multifunction, the pattern-recognition methods can be added to these relays or they can be replaced with some existing functions. Consequently, the smart relays based on pattern-recognition methods are new development of numerical relays. Some companies such as ASEA Brown Boveri (ABB) implemented pattern recognition-based methods in the field of protection [33]. Moreover, neural network-based classifiers also can be implemented by field programmable gate array (FPGA) [34]. In 2004 a relay developed by a commercial company has been addressed the problem of high impedance faults (HIF) detection by means of a scheme based on a pattern-recognition algorithm. It can be introduced as first commercial use of pattern recognition for protection purpose [35,36]. The smart relays can be introduced as next generation of relays.

18.6 Overall performance assessment of proposed decision-making scheme

The performance assessment has been done with various performance metrics, thereby comparing different decision-making algorithms such as BDT, AdaBoost DT (ABDT), and RUSBoost DT (RBDT), which are described in Table 18.13. Although ABDT, RBDT show similar performance compared with BDT, BDT is exploring good accuracy, sensitivity, and specificity. The main objective is to prevent the likely cascaded tripping events at power system network under the stressed condition such as PS, thereby regulating the sensitivity, security/dependability of protection system, and hidden failures of an existing power system. It is also worth to mention computer system configuration when dealing with machine learning techniques such as decision-making algorithm, SVM, and artificial neural network (ANN), since the CPU speed and its internal memory capacity play a vital role while preprocessing of raw data, training, and testing as well. This research work has been carried out on Intel Xeon CPU E3-1225 v5 at 3.30 GHz processor with 8GB RAM 64-bit Windows 7 operating system. The preprocessing of the raw signals using WPE and decision-making using the BDT enhances the ability of the proposed scheme without any maloperation.

18.7 Comparative assessment

The proposed WPE_BDT based scheme has been compared with other methods available in the literature for detection/classification of shunt faults

TABLE 18.13 Performance assessment of various decision-making algorithms (bagged decision tree, AdaBoost decision tree, and RUSBoost decision tree).

Modules	BDT			ABDT			RBDT		
	Accuracy (%)	TPR	FPR	Accuracy (%)	TPR	FPR	Accuracy (%)	TPR	FPR
WPE_BDT-1	100	1	1	70.4	3.3	2.3	96.2	8.9	9.73
WPE_BDT-2	75.00	7.5	8.35	74.88	7.48	8.31	74.54	7.45	8.21
WPE_BDT-3	99.82	9.92	9.92	99.58	9.9	9.9	99.54	9.9	9.9
WPE_BDT-4	99.96	9.96	9.95	98.08	9.75	9.71	97.92	9.77	9.68

FPR, *False Positive Rate; TPR*, *True Positive Rate; WPE_BDT*, Wavelet packet energy and bagged decision tree.

TABLE 18.14 Comparative assessment of proposed scheme with other methods/schemes.

References	[17]	[18]	[23]	Proposed Scheme
Schemes → Parameters ↓	ANFIS-based power swing detection	SVM-based classification of faults during power swing	Decision tree–based data-mining model for detection of shunt faults during power swing	WPE_BDT–based detection/classification of shunt faults during power swing
Fault type	Yes	Yes	Yes	Yes
Fault location (L_f)	Yes	Yes	Yes	Yes
Fault inception angle	0–360 degrees	Yes	0–90 degrees	0–270 degrees
Fault resistance (R_f) (Ω)	0–100	0–100	0–300	0–100
CT saturation	–	–	–	Yes
CCVT transients	–	–	–	Yes
Power flow angle	Yes	–	–	5–20 degrees
Source impedance	Yes	–	Yes	±10%
Sampling frequency (kHz)	1	1.92	–	1
Response time (min.–max.) (cycles)	1–4	1.5–2.5	1.75	0.25–1.5

Note: "Yes"—indicates that the variation in the parameters is considered and "–" indicates not-considered/mentioned. *ANFIS*, Adaptive Neuro Fuzzy Inference System; *CT*, Current transformer; *CCVT*, capacitor coupling voltage transformer; *SVM*, support vector machine; *WPE_BDT*, wavelet packet energy and bagged decision tree.

during PS. A comparative assessment with respect to different fault parameters and at various operating scenarios is tabulated in Table 18.14. Most of the previously reported protection schemes have not considered the effect of CT saturation and CCVT transients. However, the proposed scheme has been extensively tested for both the stable and unstable PS that are the more concerned cause of cascading tripping. Besides the effect of variation in the power flow angle is also considered. Furthermore, the proposed scheme also deals with classification of faults during PS which is not covered by other schemes except [18], and in [17,23], the effect of CT-saturation and CCVT transients have not been considered/analyzed to evaluate the performance of the protection scheme. Moreover, the proposed scheme is capable to detect/classify all types of faults within minimum 0.25 cycle to maximum 1.5-cycles time. Thus from the comparative analysis, it is clear that the proposed WPE_BDT scheme outperforms over other schemes.

18.8 Conclusion

A reliable decision-making algorithm based on WPE_BDT modules for detection and classification of shunt faults during PS has been reported in an existing 400 kV DCTL network of Chhattisgarh state. The proposed algorithm has been designed in such a way that first the detection of power system network condition is done whether there is no-fault, fault, or PS, followed by discrimination of PS (stable and unstable) and the detection and classification of faults during PS with a fault classifier at the end. All the WPE_BDT modules have been trained/tested with exclusive datasets based on cross-validation philosophy, and their performance assessment has been demonstrated. The proposed scheme has been assessed with the conventional three-stepped MHO relay characteristics, and the comparison shows that the proposed scheme outperforms over existing schemes. The RTDS studies confirm the security/reliability of the proposed scheme. The proposed method rapidly detects and classifies the faulty phase(s) using one terminal/single-end data only. Furthermore, the proposed scheme accurately discriminates stable and unstable PS and also detects and classifies faults during the PS, whereas the conventional distance protection scheme suffers due to OST and PSB functions during PS condition. The results confirm the suitability of the proposed decision-making algorithm for fault during PS condition.

Acknowledgments

The authors acknowledge the financial support of Central Power Research Institute, Bangalore, for funding the project no. RSOP/2016/TR/1/22032016, dated: 19.07.2016. The authors are grateful to the local power utility (Chhattisgarh State Power Transmission Company Limited) for their cooperation in providing valuable data to execute research work.

The authors also indebted to the Power Systems Division of Central Power Research Institute, Bangalore, for permitting to carry simulation studies in real-time digital simulator (RTDS) laboratory.

References

[1] S.H. Horowitz, A.G. Phadke, Power System Relaying, second ed., Wiley, New York, 1995.
[2] A.G. Phadke, J.S. Thorp, Computer Relaying for Power Systems, second ed., Wiley, Hoboken, NJ, 2009.
[3] D. Novosel, G. Bartok, G. Henneberg, P. Mysore, D. Tziouvaras, S. Ward, IEEE PSRC report on the performance of relaying during wide area stressed conditions, IEEE Trans. Power Deliv. 25 (1) (2010) 3–16.
[4] NYISO, "Blackout August 14, 2003" Final Rep, 2005.
[5] UCTE, Final Report of the Investigation Committee on the 28 September 2003 Blackout in Italy, 2004.
[6] UCTE, Final Report: System Disturbance on 4 November 2006, 2006.
[7] Tech Rep. Power-Grid, Report of the Inquiry Committee on Grid Disturbance in Northern Region on 30th July 2012 and in Northern, Eastern and North-Eastern Region on 31st July 2012, Tech Rep. Power-Grid, New Delhi, India, 2012.
[8] S.H. Horowitz, A.G. Phadke, Third zone revisited, IEEE Trans. Power Del. 21 (1) (2006) 23–29.
[9] P. Pouyan, S. Kundur Prabha, W. Taylor Carson, The anatomy of a power grid blackout, IEEE Power Energy Mag. 4 (5) (2006) 22–29.
[10] D. Tziouvaras, D. Hou, Out-of-Step Protection Fundamentals and Advancements, 2003.
[11] IEEE Power System Relaying Committee of the IEEE Power Engineering Society, "Power Swing and Out-of-Step Considerations on Transmission Line" Rep PSRCWG D6, July 2005.
[12] J.P.E. Mooney, N. Fischer, Application guidelines for power swing detection on transmission systems, in: Proc Power System Conf, 2006, pp. 159–168.
[13] H. Khoradshadi-Zadeh, Evaluation and performance comparison of power swing detection algorithms, in: Proc IEEE Power Eng Soc Gen Meeting, vol. 2, 2005, pp. 1842–1848.
[14] L. Martuscello, E. Krizauskas, J. Holbach, Y. Lu, Tests of distance relay performance on stable and unstable power swings reported using simulated data of the August 14th 2003 system disturbance, in: Proc IEEE 62nd Annu Conf Protect Relay Eng, 2009, pp. 236–255.
[15] G. Benmouyal, D. Hou, D. Tziouvaras, Zero-Setting Power Swing Blocking Protection, 2004.
[16] X. Lin, Z. Li, S. Ke, Y. Gao, Theoretical fundamentals and implementation of novel self-adaptive distance protection resistant to power swings, IEEE Trans. Power Del. 25 (3) (2010) 1372–1383.
[17] H.K. Zadeh, Z. Li, A novel power swing blocking scheme using adaptive neuro-fuzzy inference system, Electr Power Syst. Res. 78 (2008) 1138–1146.
[18] K. Seethalekshmi, S.N. Singh, S.C. Srivastava, A classification approach using support vector machines to prevent distance relay mal operation under power swing and voltage instability, IEEE Trans. Power Del. 27 (3) (2012) 1124–1133.
[19] S.M. Brahma, Distance relay with out-of-step blocking function using wavelet transform, IEEE Trans. Power Del. 22 (3) (2007) 1360–1366.

[20] A.P. Apostolov, D. Tholomier, S.H. Richards, Superimposed components based sub-cycle protection of transmission lines, in: Proc IEEE Power Eng Soc Power Syst Conf Expo, vol. 1, 2004, pp. 592–597.

[21] X. Lin, Y. Gao, P. Liu, A novel scheme to identify symmetrical faults occurring during power swings, IEEE Trans. Power Del. 23 (1) (2008) 73–78.

[22] S. Lotfifard, J. Faiz, M. Kezunovic, Detection of symmetrical faults by distance relays during power swings, IEEE Trans. Power Del. 25 (1) (2010) 81–87.

[23] R. Dubey, S.R. Samantaray, B.K. Panigrahi, V.G. Venkoparao, Data-mining model based adaptive protection scheme to enhance distance relay performance during power swing, J. Power Energy Syst. 81 (2016) 361–370.

[24] A. Swetapadma, A. Yadav, Data-mining based fault during power swing identification in power transmission system, IET Sci. Meas. Technol. 10 (2) (2016) 130–139.

[25] J. Kumar, P. Jena, Solution to fault detection during power swing using teager-kaiser energy operator, Arab J. Sci. Eng. 42 (2017) 5003–5013.

[26] B. Patel, P. Bera, Detection of power swing and fault during power swing using lissajous figure, IEEE Trans. Power Deliv. 33 (6) (2018) 3019–3027.

[27] S.M. Hashemi, M. Sanaye-Pasand, M. Shahidehpour, Fault detection during power swings using the properties of fundamental frequency phasors, IEEE Trans. on Smart Grid, 10 (2), 2019,1385–1394. https://doi.org/10.1109/TSg.2017.2765200.

[28] P. Ray, D.P. Mishra, SVM based fault classification and location of a long transmission line, Int. J. Eng. Sci. Technol. 19 (2016) 1368–1380.

[29] L. Breiman, Bagging predictors, Mach. Learn. 24 (02) (1996) 123–140.

[30] R. Polikar, Ensemble based systems in decision making, IEEE Circuits Syst. Mag. 06 (03) (2006) 21–45.

[31] TreeBagger class. Available at: <https://in.mathworks.com/help/stats/treebagger.html>.

[32] B. Kumar, A. Yadav, A backup protection scheme for transmission line compensated with UPFC during high impedance faults and dynamic situations, IET Sci. Meas. Technol. 11 (6) (2017) 703–712.

[33] J. Stoupis, M. Maharsi, R. Nuqui, S.A. Kunsman, R. Das, Ground alert, ABB Rev. 1 (2004).

[34] A.R. Omondi, J.C. Rajapakse, FPGA Implementations of Neural Networks, Springer, 2006.

[35] J.-A. Jiang, C.-L. Chuang, Y.-C. Wang, C.-H. Hung, J.-Y. Wang, C.-H. Lee, et al., A hybrid framework for fault detection, classification, and location—Part II: implementation and test results, IEEE Trans. Power Deliv. 26 (3) (2011) 1999–2008.

[36] S.A. Kunsman, I.I. Jouny, S. Kaprielian, High Impedance Fault Detection, U.S. Patent 7,069,116, 2006.

Chapter 19

Modeling and processing of smart grids big data: study case of a university research building

Lênio O. Prado, Jr[1], Paulo Fernando Ribeiro[2], Carlos Augusto Duque[3] and Shady H.E. Abdel Aleem[4]

[1]IF—Federal Institute of Science and Technology Education, Minas Gerais, Brazil, [2]Institute of Electrical System and Energy, Federal University of Itajuba, UNIFEI, Itajubá, Brazil, [3]UFJF—Federal University of Juiz de Fora, Juiz de Fora, Brazil, [4]Mathematical, Physical and Engineering Sciences Department, 15th of May Higher Institute of Engineering, Cairo, Egypt

19.1 Introduction

Smart grids (SG) are known as the next evolutionary generation of the energy grid plant [1]. A SG is an intelligent electricity/communication two-way network that integrates utilities, distributed energy resources (DERs), and the actions of all stakeholders connected to them, making use of advanced information, control, and communication technologies to save energy and reduce costs, and increase reliability and transparency [2]. Although having many complementary definitions, it is a consensus that SGs need new technology development and standards for effective, reliable, and robust information and communication technologies (ICTs) infrastructure [3].

The technological and business advances in the energy systems are making the design and development of SG possible. These advances are related to some technologies such as the use of advanced metering infrastructure (AMI), DERs, business models, organizational structures, regulations and roles, and operating practices, along with the ICTs improvements [1].

With the implementation of an AMI [4], a large amount of data needs to flow from smart meters to a central or concentrator where this data must be processed within a time limit, and then some fast action or decision should be taken. The analysis of collected data may allow consumers' power usage to be better observed, which may lead to a more conscious use of power, either by reducing the usage or even by changing the hours of use of certain appliances. Data analysis is a key role of SG.

Decision Making Applications in Modern Power Systems. DOI: https://doi.org/10.1016/B978-0-12-816445-7.00019-0

In the past few years, new ways of data analysis became available [5,6]. Some of these techniques require high computational power, high rate of data exchange between the assets and sensors involved, and short response times, which imposes high demands on the SG communication [7].

These techniques of data analysis are called big data analytics (BDA) and provide ways to process data from many sources, making data available instantly, with a granular access. BDA allows processing of massive volume of data with increasing storage capacity and parallel processing [5,8].

Among the benefits brought by data analysis techniques such as BDA, we can highlight the management of supply and demand, the core benefit of the SG [9], which allow one to efficiently handle the use of equipment at peak times, scheduling a better time for use out of peak periods when power prices are higher [10,11]. Also, BDA provides the means of identifying, collecting, visualizing, and analyzing critical parameters, among others [12–14]. Price negotiation is another benefit brought, allowing consumer to sell generated surplus power with great prices [15].

Fig. 19.1 shows an overview of the architecture required to collect, transmit, and process data, as well as the benefits brought by the data analysis.

A very important point that needs special attention when data is collected and processed refers to how this data will be used. The main questions are, what can be done with massive data created by this new energy scenario

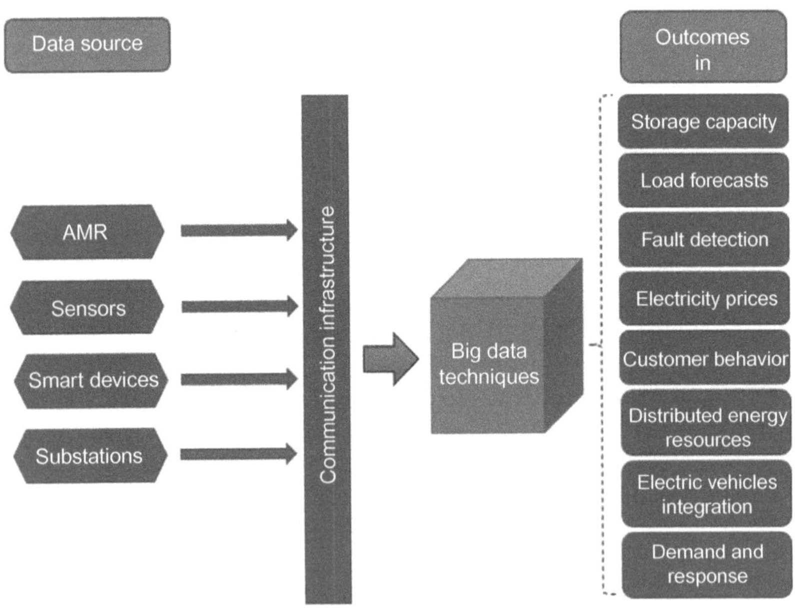

FIGURE 19.1 Big data overview.

brought by the SG? What benefits can be delivered both to consumers and utilities by data analysis? The answers lead to big data techniques.

It's important to highlight which benefits can be obtained from the extraction of knowledge from the data collected, thus justifying the creation, development, and use of such analysis techniques. The correlation of data is made to find some hidden characteristics that could provide more comprehensive understanding of power consumption and generation.

In this chapter, data collected from a university research laboratory at the Federal University of Itajubá (UNIFEI) was analyzed, and the solar generation profile is evidenced through the generated graphs. The analyzed data was collected in the period between January 01 and March 31, 2018, at every 15 minutes. The analyzed characteristics were current phase, power factor, frequency, apparent power, active power, reactive power, energy production, and phase voltage.

In addition to data analysis, a social analysis was carried out regarding the use of collected data in order to highlight the ethical question about the use of information. Because SGs use detailed user consumption data, this social analysis is very important. The problem is discussed, and some points of attention are listed. It presented an approach on the social impacts of the use of collected data in relation to user privacy in cases where collected information may reveal users' habits and behavior.

19.2 Big data

Big data can be considered a large amount of data of numerous formats, diverse structures, and many origins. The data collected can be images, texts, sounds, hyperlinks, sensor measurements, social media content, videos recorded by users themselves and many other types, and traditional data-processing programs are unable to work with this massive amount of data. Some important characteristics are used in the literature to define the term big data [5,16,17].

Although there are many definitions in the literature about the term big data, what is consensus is the definition of the Vs [18].

- *Volume*: Information generated by sensors, equipment, smart devices, people, social networks, security cameras, emails, among others, results in a large volume of data to be processed and analyzed to generate results [19].
- *Variety*: Big data techniques must handle data from multiple sources, with varying formats, types, and sizes. In addition to getting the data from your home locations, storing this large amount of information creates challenges as to data availability. The needed computational power to process audios, images, videos, text documents, and sensory data is also a technological challenge [20].

- *Velocity*: The data arrives for analysis at a very high speed. In addition to the speed of arrival, it is also important to consider the processing time. The sooner results are obtained, the more useful the information will be [19].
- *Veracity*: The reliability of the data is very important as it ensures the veracity of the information. Big data techniques should be able to detect, ignore, or correct erroneous information such as incorrect sensor readings, errors caused by data transmission, or even corrupted data [21].
- *Value*: It refers to the impact that the information can bring or even the insight obtained through the results obtained from the data [22].

A huge amount of data generated is a point which the term big data fits properly. With the advent of the Internet of Things, sensors inserted at most diverse scenarios produce an expressive amount of information [23]. Sensors are used to collect data related to temperature, gas presence, pressure, tracking individuals, equipment, and others. Every day more and more sensors are added to aid for sensing equipment health, vehicle telemetry, energy consumption, climate, and many others tasks [24,25].

This large use of sensors generates data at an impressive speed, and because of this, the speed at which data is created and need to be processed also stands out as closely related to BDA [6,16]. Data size collected by some sensors turn around a few kilobytes, but the frequency in which they are collected in some cases is too high.

The crucial point is that, on a number of occasions, the data will only be useful if processed in a short time after being collected. This situation requires a robust communication infrastructure to send the information and receive the result, real-time processing to enable the analysis of the data immediately, and storage space to keep the data received and the result obtained, all this in an efficient and reliable way [4].

Big data needs innovative and efficient techniques to process data, and the use of techniques previously intended for other types of use is welcome as long as they can add value to the data.

Several areas of knowledge are used in big data, and there is no clear definition of where one area ends and another begins, as they are intrinsically related. Neural networks, pattern recognition, visualization techniques, statistics, and machine learning are some of the many areas that make up the complex scenario of techniques that can be used with big data [8,26].

19.3 Big data requirements

Big data is not a new subject, and there are current applications working on this topic [6,27]. Even so, there are still gaps to be filled and methods to be improved. Although there are already big data applications, there are still open questions about scalability, storage, security, information handling format, and evaluation of results of data [16,28].

Applications need to meet some requirements to work the data efficiently and reliably. One of these requirements is a network with high bandwidth. In the scenarios of SGs, optical fibers are required as the data communication medium along transmission and distribution power lines scopes [29].

Another observation when considering some real-time characteristics of big data applications, even in a private network, is that protocols such as transmission control protocol would not be a good alternative, due to its reliable delivery, communication model (unicast), retransmission mechanism, window management, and timeout strategies, that can cause network congestion due to the protocol control data [30].

A very important requirement regarding the use of big data for SGs refers to the fact that the old meters used in consumers' house cannot provide information about the energy usage in a detailed way, by day and time of consumption. Therefore it would be necessary to exchange the old one by a smart meter one [29]. As stated in Ref. [31], there are some countries that already replaced old meters by smart meters.

Once smart meters are present in the network and the communication environment is available, information can be collected and analyzed, and a new set of applications should be created to meet the needs arising from data analysis and the market around SGs [9].

19.3.1 Big data analytics for smart grid

Current power lines have an energy flow in a single direction, from utilities to consumers, and almost no communication occurs between consumer and utilities about the consumption of power [31].

SGs brought a challenge related to data exchanging. There is a huge amount of data that needs to flow in a fast, reliable, and secure way. All this data is generated by sensors, smart meters, and equipment in the grid [6]. One of the key requirements of SGs is a communication network able to handle data flow in the grid [3].

In addition to the implementation of a robust and secure communication environment, applications and methods related to big data, cloud storage, and data mining must be used together to allow predicting future observations related to consumption profile, fault location, self-healing, and to enable the development of a robust and effective environment where utilities, grid operators, investors, and consumers can enjoy the benefits of an SG [9,15].

The information traffic between devices in the network increases as the number of appliances grows, reaching a number of hundreds of items per house. The amount of data being transferred in the grid follows this growth. The matter is the storage capacity, processing response time, and use of network bandwidth to keep and process all this data [32,33]. BDA fits properly in this scenario.

The computers processing power and memory (CPU and RAM) currently available is extremely high, and the speed in which these hardware components are improved is becoming higher and higher [16]. Despite the high processing power currently available, it is not a trivial task to analyze the huge amount data and then delivery measurable results.

Although data analysis tools already exist, SG brings new challenges, the main one being the high volume of information to be analyzed within a short time. Even with abundant processing and storage power, many of the existing analytical techniques do not adequately apply to the SGs scenario [34].

Transforming raw data into useful information is one of the best advantages of BDA usage for SG. With appropriate tools, utilities can improve generation park usage, reducing the power demand peak and efficiently managing generation schedule during the day. On the customer side, detailed consumption information can enable more energy-conscious use. The sale of surplus renewable energy generated is also an excellent benefit for the consumer, since it allows the reduction of expenses [3].

One of the SG applications that play an important role is dynamic energy management (DEM). As emphasized in [35], DEM in the conventional power grid has already consolidated some maturity. In an SG scenario, the situation is quite different. The complexity of the environment, with distributed resources and smart meters with bidirectional data flow, requires strict processing and communication requirements.

Energy management systems must properly handle the data collected, so that it can deliver useful information. Based on the information resulting from the collected data processing, it should be possible to perform protection and corrective actions in an assertive way.

19.3.2 Big data analytics—challenges and trends

Although the BDA concept is extensively used in the literature as a viable alternative for solving SG demands, it is important to note that there is no established and consolidated standard or government regulation on how all parts will work [36].

The big data usage for storing, analyzing, and providing results from data collected in the context of SGs has a great complication. All the information generated is available in a large demographic area, either located in the smart meters in the consumers home or in electrical equipment and devices located in remote environments, spread throughout the utilities range [2,3].

Since the data is spread and needs to be transmitted through the communication channel to the control center of the utility, some questions are raised:

- Are all the information collected really important?

  Before being sent for analysis, the collected data can be locally verified according to a commercial rule or by specified limits, or even if the data is similar to the last one sent. If, for example, a limit check is being used, only data outside that limit must be sent for analysis; otherwise, the data has no useful value and doesn't need to be sent, reducing the volume of traffic data [37].

- Is there any method to make the information more meaningful?

  In [38], it is proposed a unified index for power quality evaluation in distributed generation systems. Such technique facilitates the analysis of energy quality through a single index and also decreases the amount of data to be sent to the data center.
- How to integrate existing tools, solutions, and data with the new big data tools?

Integration with currently used environment such as supervisory control and data acquisition (SCADA) systems, distribution and transmission automation, enterprise asset management, and historical data stored by such systems is a great challenge to transform the current grid in a smart one.

Big data technologies along with cloud storage and machine learning are part of the new structure of an SG. To make possible more assertive forecasts, existent meteorological data must also be incorporated in order to assist the forecasting of wind and photovoltaic generation. Create financial incentives to encourage investors to embrace SG-oriented projects and government laws to define rules about the operation of the SG are also needed.

BDA should be used in the context of SG to provide solutions through services that add value, such as the following:

- *Billing*: Smart meters measurements make possible usage not only for billing energy, but also to identify the customers' consumption class and the detailed consumption profile per equipment.
- *Marketing*: With consumers being segmented by classes, it is easy to conduct promotions and offers to groups with specific needs. The electricity commercialization will also be a dynamic market open to customers, utilities, and investors.
- *Design and planning*: For forecasting power generation, energy consumption, estimating loads and prices, analyzing customer behaviors, identifying system faults, and minimizing errors during the design and planning steps.
- *In-home displays, controls, and other devices*: The customer will have access to devices and tools that allow real-time energy consumption reports; they will also obtain data related to generated renewable energy in their residence and have a more direct communication channel with the utility, among others.

19.3.3 Security (cyber and physical)

A set of technological solutions has been developed to address security problems in SG communication networks. Even a methodology to assessing SG security was proposed [39]. The main security technology standards for SG communication networks are summarized in Ref. [40].

For privacy issues, we can cite solutions based on

1. a method for securely anonymizing metering data from the smart meter (the data does not need to be attributable to a specific consumer smart meter) [41];
2. an aggregation method with two different solutions, one using a trusted third party as a data aggregation proxy and the other that adds random value to the data from a particular probability distribution [42];
3. the use of homomorphic (Paillier) encryption and additive secret sharing to make the total consumption readings visible at the neighborhood area network level, without revealing meter data at home area network level [43];
4. a cooperative state estimation technique for the privacy preservation of personal behavior [44];
5. a privacy-preserving smart metering with multiple data consumers, an approach that allows utilities and third parties (data consumers) to collect measurement data with different levels of spatial and temporal aggregation from smart meters without revealing the individual measurements to any single node of the architecture [45]; and
6. a location-based encryption security for SG applications, where a message is encrypted such that it can only be decrypted at a specified location and time by using a location-dependent decryption key [46].

For more approaches on privacy protection, management and architectures, authentication and access control, and intrusion detection systems, please refer to Ref. [47].

For the physical layer [48], presents a three-layer, not individually, framework, with a cross-layer distributed, smart, and real-time defenses to simultaneously address security of the cyber (servers, workstations, mobile devices), network (routers, switches, firewalls), and physical (smart meters, power line communications (PLCs), remote terminal units (RTUs)) domains against pervasive and persistent attacks.

The amounts of data traffic in SG have inspired approaches to monitor and prevent DDoS attacks. A proposed method uses software-defined networking (SDN) architecture to detect and mitigate the congestion attacks by exploiting the consumer quality of experience measurements as well as the traditional quality of service parameters [49]. Another research makes an analysis about the SG resilience under SDN controller failures [50].

A key management architecture and associated protocols to support encrypted SG communications (unicast, multicast, and broadcast) is proposed by Ref. [51]. The proposed key management schema supports the hierarchical structure of the SG control mechanisms and takes its resource and electronic/physical security differences into account.

19.4 Big data social impact

The data collected in the SG must be worked on to generate useful data. In addition to the current demand for applications, methods, and computational resources to deal with this data, there is another very important point that needs to be debated and analyzed: user privacy.

Since customers consumption profile can be detailed by reading the data from smart meters, some very important points come into question: Who will have access to this data? What information will be extracted and for what purposes? What actions will be taken from the extracted information? Is the environment where information is collected and processed safe and enables effective protection mechanisms [52]?

These and other issues permeate the relationship between technology and privacy. It is not unusual to have situations in which user data is stolen, or even collected without the express authorization of the user [28]. Even when the user authorizes data usage, the data is worked in an unknown way, without the user exact knowledge of what will be done with the information collected and with the resulting information and how it will be used.

It is necessary to make clear which data will be collected and which purpose will be given for the information processed. In order to establish rules that effectively guide data collection and processing, an accurate and rigorous data policy should be in place to prevent user privacy from being harmed, as well as to ensure accountability in cases of information theft, including fines and penalties in case of breach of security and privacy parameters.

Despite the numerous benefits brought by the use of big data techniques, it is important to note that data collection from smart meters gives to utility a full knowledge of the power usage habits from their clients. Is it safe to send detailed information such as which equipment is being used, at what time, and for how long? A bad use of this information can lead to security problems, as well as a breach of confidentiality for all the utility's customers.

It is necessary to establish regulatory norms that inhibit the creation of energy cartels, the marketing of customer power consumption information, theft of information, and the prevention of automated software that could buy cheap energy and resell with abusive prices.

Debates about the user data privacy should be stimulated by regulators. It is a fact that intelligent grids are the future of the energy grid, but special attention must be paid to the social impacts inherent in its constitution.

19.5 Laboratory building data and analysis

At the university, sensors were installed in the inverters of the solar panels, and the data collected were used to trace the generation profile and also analyze the following aspects: current phase, power factor, frequency, apparent power, active power, reactive power, energy production, and phase voltage (Fig. 19.2).

The data collection interval was 15 minutes, starting at 0645 or 0700, depending on the start of the power generation, and ending at 1800 or extending

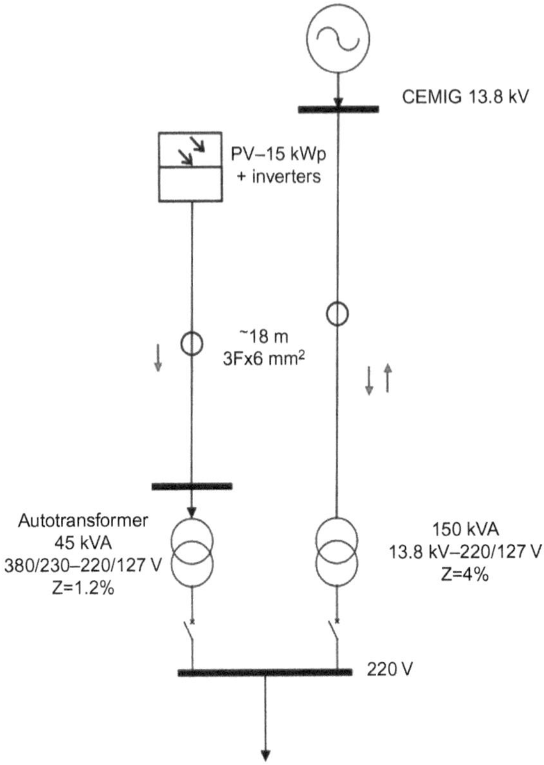

FIGURE 19.2 CERIn uniform topology [53].

until 1900, when there is still generation until this time. The following graphs and tables shown were intentionally divided into periods of 10 days in order to facilitate the detailed visualization of the data. The following table shows the lowest and the highest values found for each analyzed characteristic.

At the end of each section, graphs are displayed showing the values of each analyzed characteristic during a typical weekday and also the weekend profile, in order to verify differences in the profile on different days. In the graphs, information is displayed in a way that allows the visualization of the daily profile of the data collected.

Fig. 19.3 demonstrates the process of obtaining input values through measurements of electrical data and the output information obtained through the use of methods and filters.

19.5.1 Phase current

In the analyzed period the lowest phase current value was *0.09*, found on 22 March at 1845. The highest figure was **17.36**, found on 20 March at 1245 (Figs. 19.4–19.8).

January to March—Phase current

Interval	Low	Day—time	High	Day—time
01–10 January	0.35	03—1845	17.10	03—1315
11–20 January	0.12	15—1845	16.53	17—1245
21–31 January	0.60	23—1845	16.49	28—1415
01–10 February	0.30	07—0700	17.34	07—1245
11–20 February	0.28	20—1830	16.57	17—1300
21–28 February	0.10	27—1845	17.16	25—1145
01–10 March	0.16	03—1830	17.07	03—1130
11–20 March	0.12	20—1845	**17.36**	20—1245
21–31 March	*0.09*	22—1845	17.26	26—1015

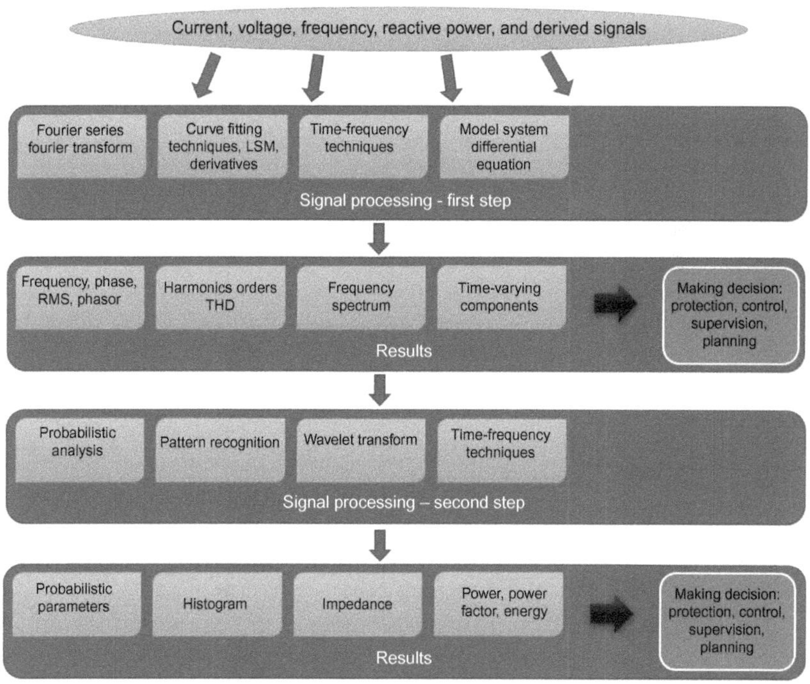

FIGURE 19.3 Processed signals and parameters. *Adapted from P.F. Ribeiro, C.A. Duque, P.M. Ribeiro, A.S. Cerqueira, Power Systems Signal Processing for Smart Grids, John Wiley & Sons, 2013 [54].*

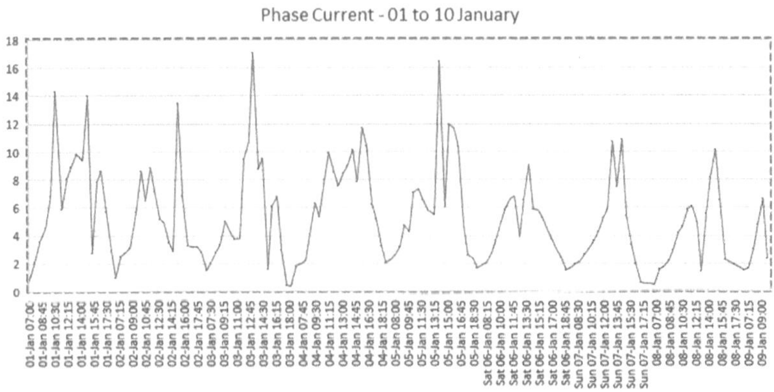

FIGURE 19.4 Phase current 01–10 January.

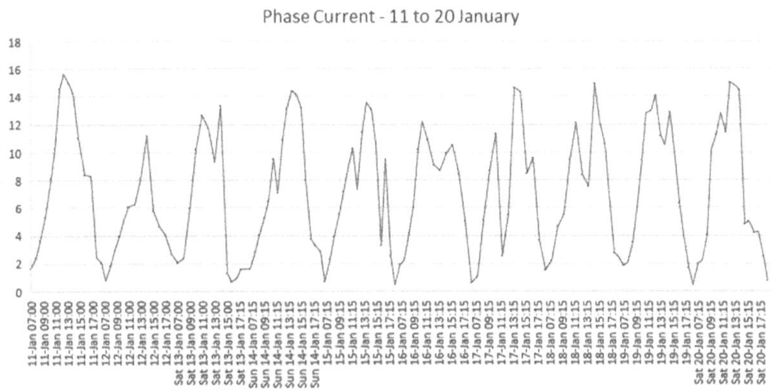

FIGURE 19.5 Phase current 11–20 January.

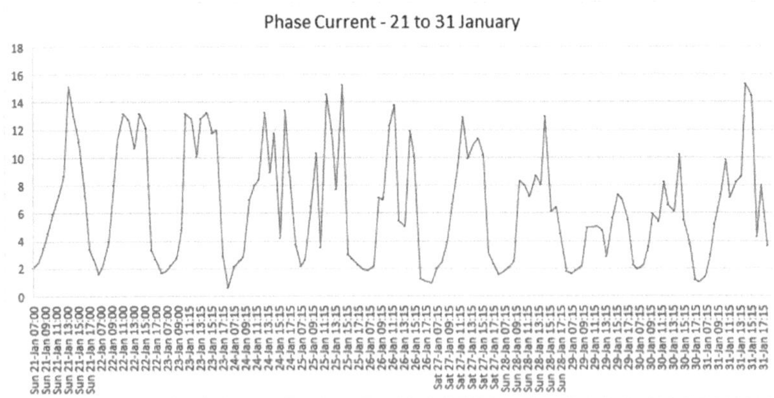

FIGURE 19.6 Phase current 21–31 January.

Week and weekend profile of phase current.

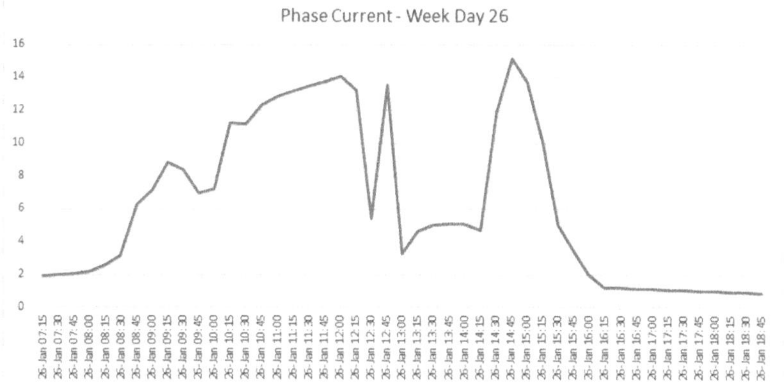

FIGURE 19.7 Phase current week day.

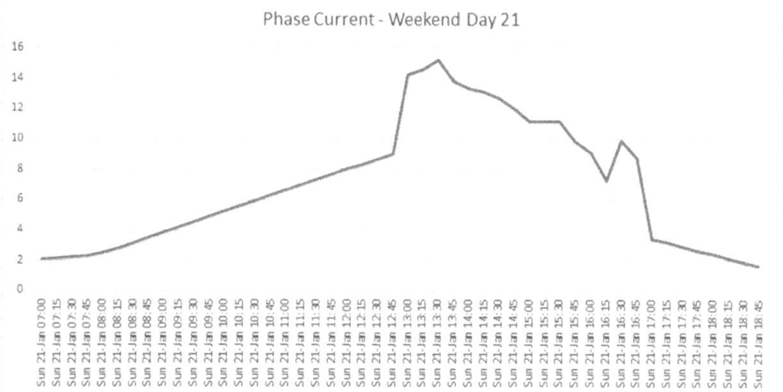

FIGURE 19.8 Phase current weekend day.

19.5.2 Power factor

In the analyzed period, the lowest power factor value was *− 58.49* found on 28 February at 1815. The highest figure was **−99.99** found on 17 February at 1330 (Figs. 19.9–19.13).

January to March—Power factor

Interval	Low	Day—time	High	Day—time
01–10 January	−75.67	03—1930	−99.9	01—1500
11–20 January	−84.31	15—1930	−99.9	17—1300
21–31 January	−77.98	21—1930	−99.9	21—1445
01–10 February	−76.82	05—1930	−99.9	01—1500
11–20 February	−76.28	11—1930	**99.9**	17—1330
21–28 February	*− 58.49*	28—1815	−99.9	25—1400
01–10 March	−87.96	07—1815	−99.9	02—1415
11–20 March	−91.47	12—1815	−99.9	14—1515
21–31 March	−85.03	22—1800	−99.9	27—0945

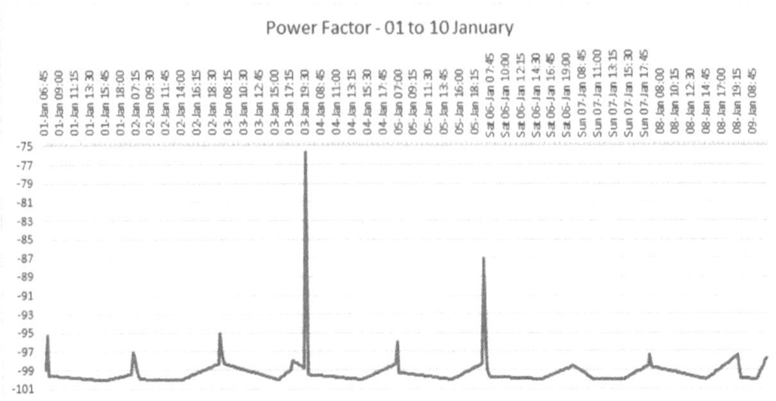

FIGURE 19.9 Power factor 01–10 January.

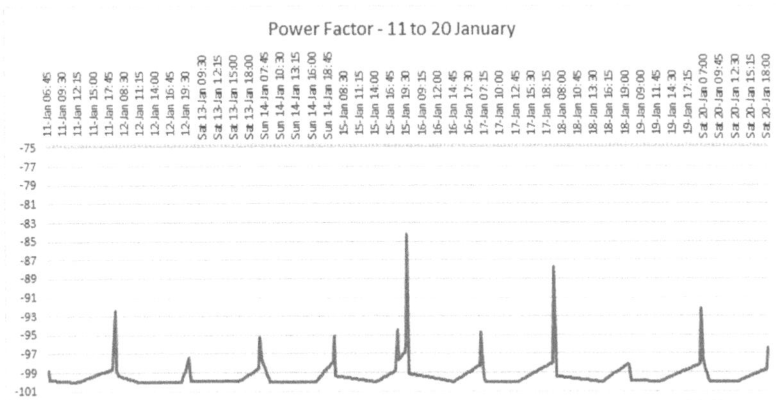

FIGURE 19.10 Power factor 11–20 January.

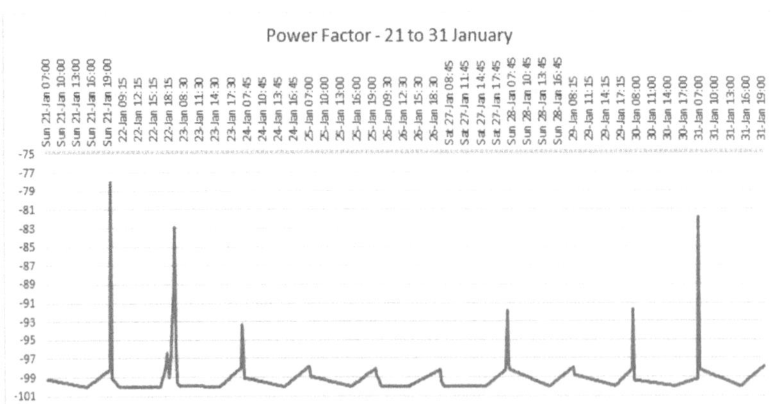

FIGURE 19.11 Power factor 21–31 January.
Week and weekend profile of power factor.

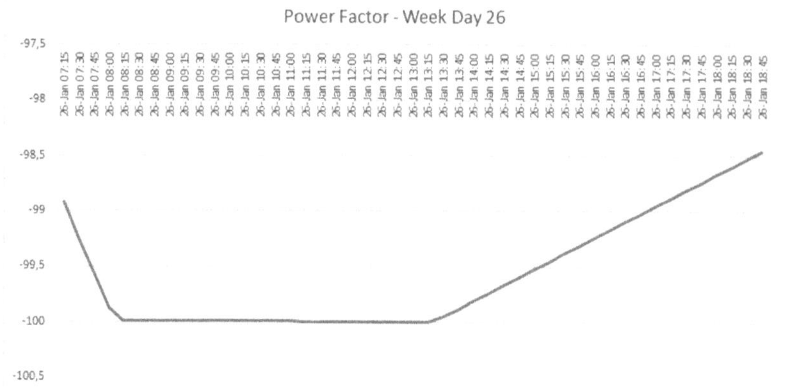

FIGURE 19.12 Power factor week day.

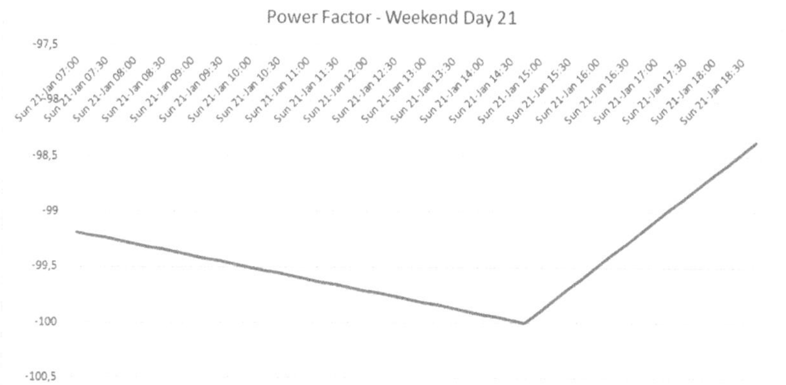

FIGURE 19.13 Power factor weekend day.

19.5.3 Frequency

In the analyzed period, the lowest frequency value was *1.34*, found on 29 March at 1800. The highest figure was **60.11**, found on 04 January at 0645 (Figs. 19.14–19.18).

January to March—Frequency

Interval	Low	Day—time	High	Day—time
01–31 January	23.53	04—0630	**60.11**	04—0645
01–28 February	10.07	18—0630	60.09	04—0700
01–31 March	*1.34*	29—1800	60.06	29—1745

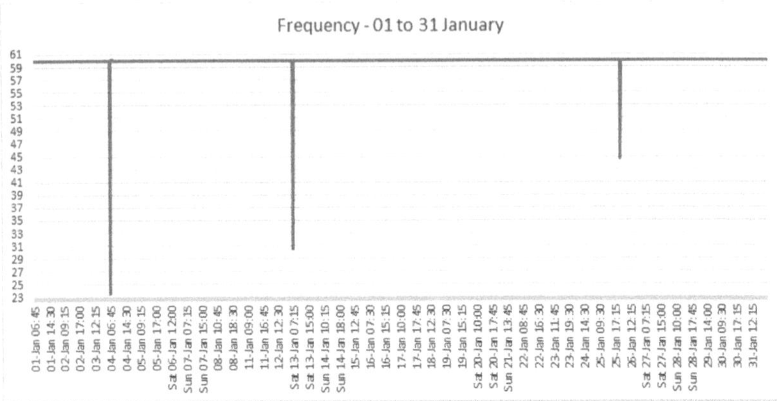

FIGURE 19.14 Frequency 01–31 January.

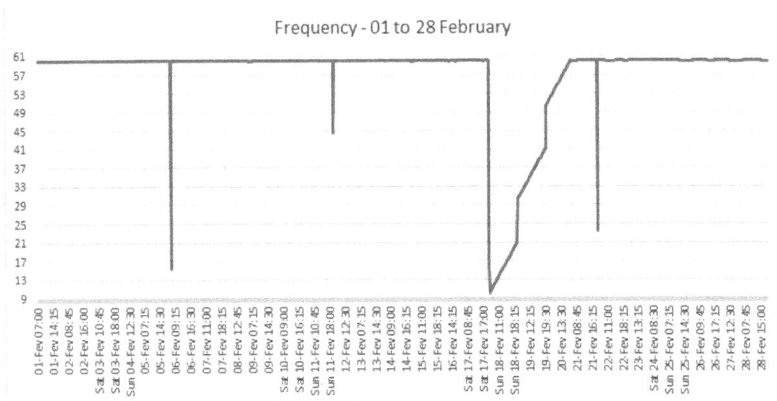

FIGURE 19.15 Frequency 01–28 February.

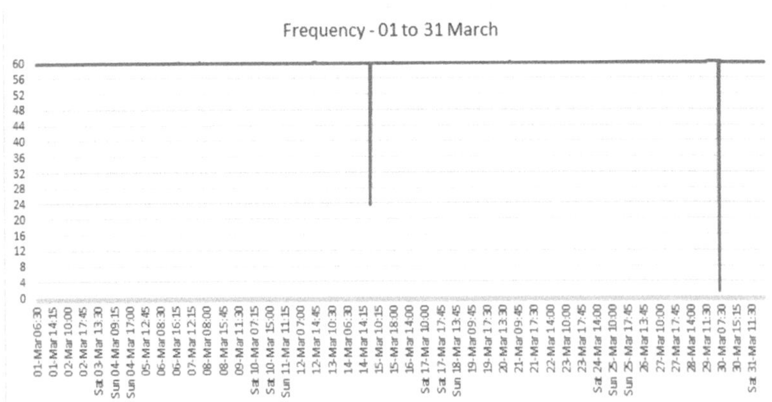

FIGURE 19.16 Frequency 01–31 March.

Week and weekend profile of frequency.

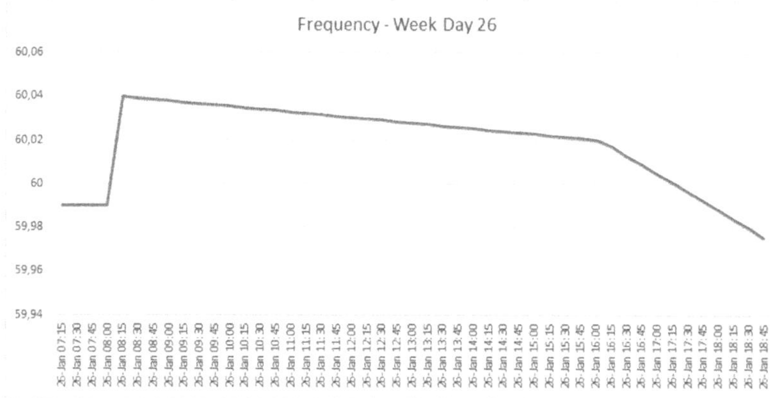

FIGURE 19.17 Frequency week day.

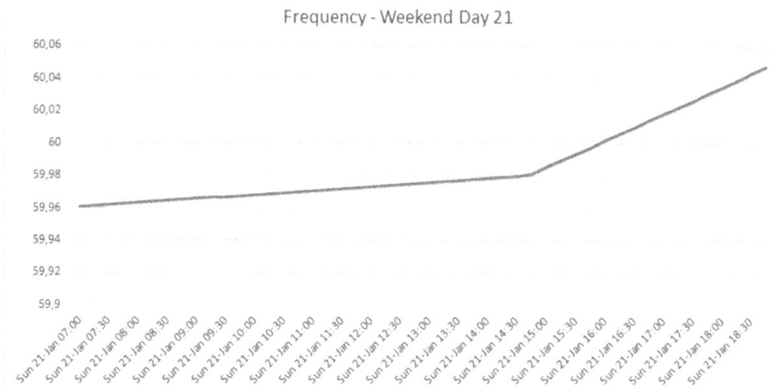

FIGURE 19.18 Frequency weekend day.

19.5.4 Apparent power

In the analyzed period, the lowest apparent power value was *26.47*, found on 07 March at 1815. The highest figure was **12126.33**, found on 25 February at 1200 (Figs. 19.19–19.23).

January to March—Apparent power

Interval	Low	Day—time	High	Day—time
01–10 January	50.25	01—0700	11,540.74	03—1315
11–20 January	50.38	11—1830	11,167.83	19—1330
21–31 January	41.63	22—1845	10,514.65	22—1400
01–10 February	41.74	04—0700	11,620.83	04—1245
11–20 February	34.22	20—1800	11,358.85	14—1345
21–28 February	26.63	21—1845	**12126.33**	25—1200
01–10 March	*26.47*	07—1815	11,640.02	10—1145
11–20 March	34.48	12—1815	11,821.87	20—1245
21–31 March	27.23	22—1800	11,962.18	26—1015

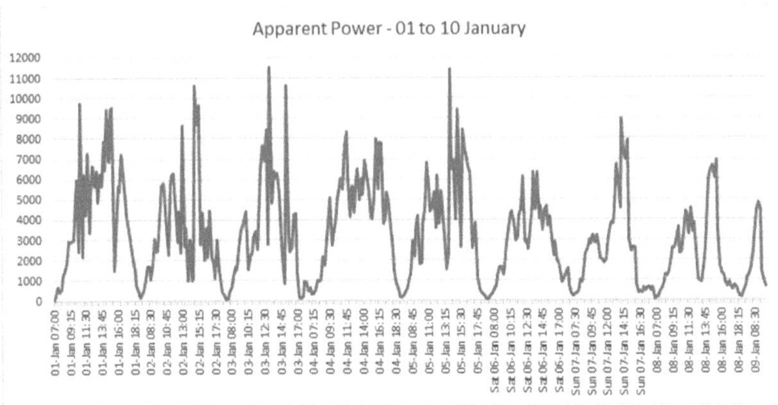

FIGURE 19.19 Apparent power 01–10 January.

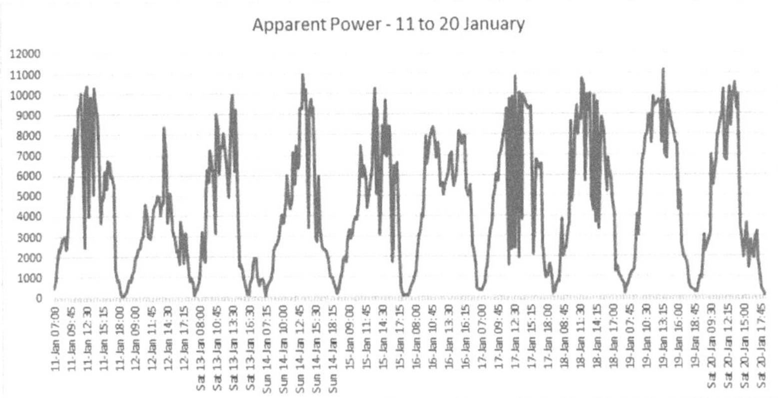

FIGURE 19.20 Apparent power 11–20 January.

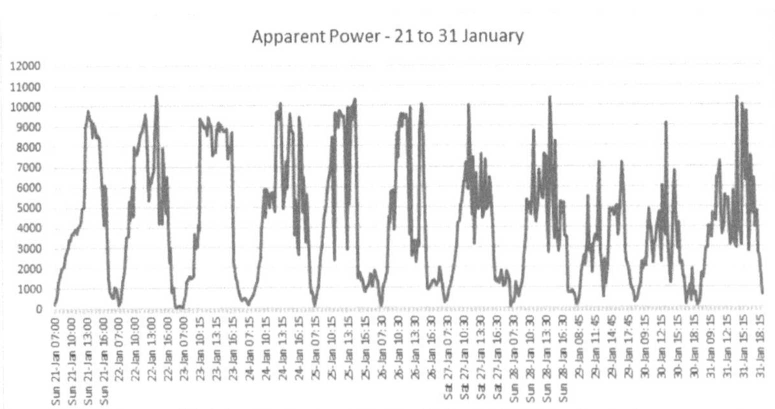

FIGURE 19.21 Apparent power 21–31 January.

Week and weekend profile of apparent power *apparent power week day.*

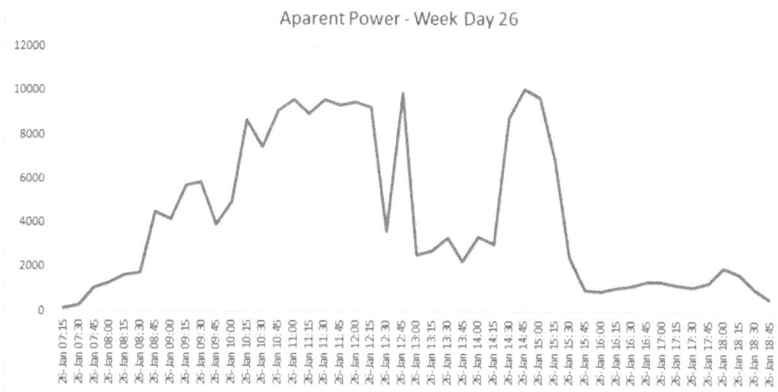

FIGURE 19.22 Apparent power week day.

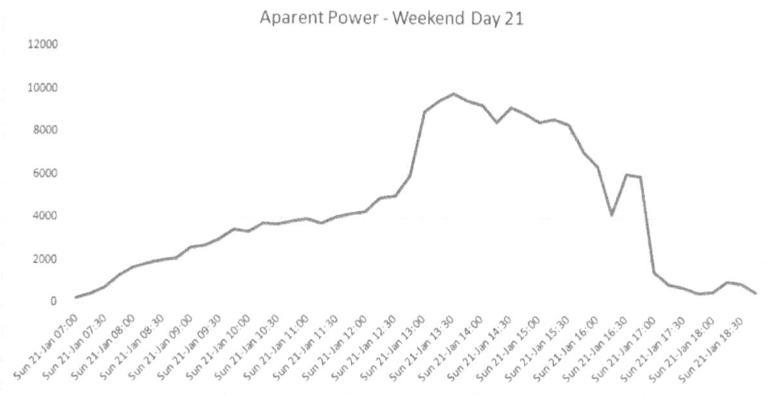

FIGURE 19.23 Apparent power weekend day.

19.5.5 Active power

In the analyzed period, the lowest active power value was *0.72*, found on 29 March at 1800. The highest figure was **12126.24**, found on 25 February at 1200 (Figs. 19.24–19.30).

January to March—Active power

Interval	Low	Day—time	High	Day—time
01–10 January	47.94	01—0700	11,540.69	03—1315
11–20 January	48.01	11—1830	11,167.75	19—1330
21–31 January	25.98	26—0700	10,514.58	22—1400
01–10 February	39.20	04—0700	11,620.74	04—1245
11–20 February	22.26	15—1800	11,358.75	14—1345
21–28 February	8.58	21—1845	**12126.24**	25—1200
01–10 March	23.55	07—1815	11,639.95	10—1145
11–20 March	31.79	12—1815	11,821.78	20—1245
21–31 March	*0.72*	29—1800	11,962.13	26—1015

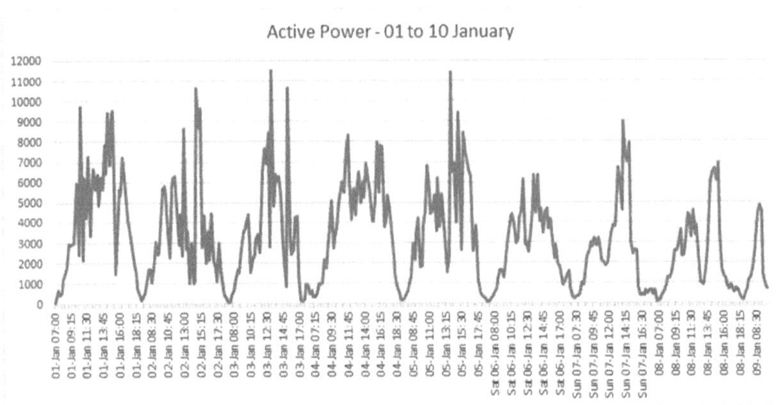

FIGURE 19.24 Active power 01–10 January.

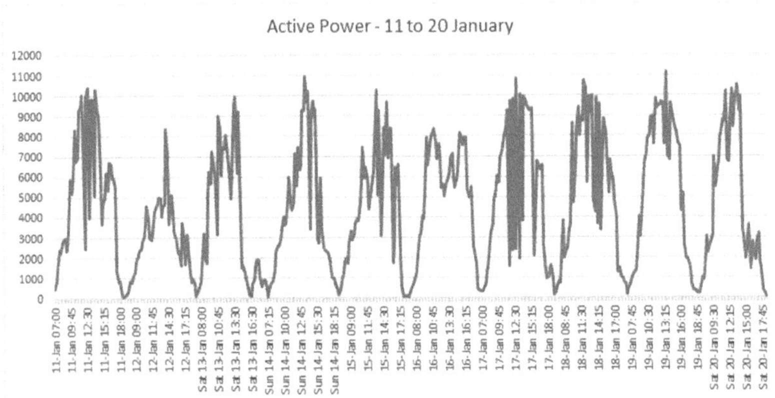

FIGURE 19.25 Active power 11–20 January.

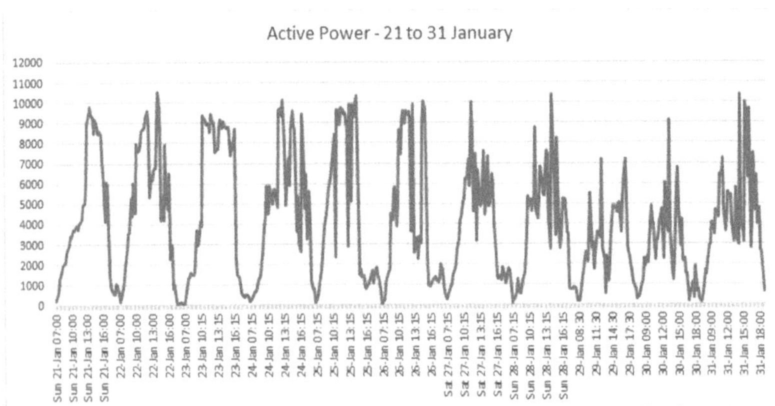

FIGURE 19.26 Active power 21–31 January.
Week and weekend profile of active power.

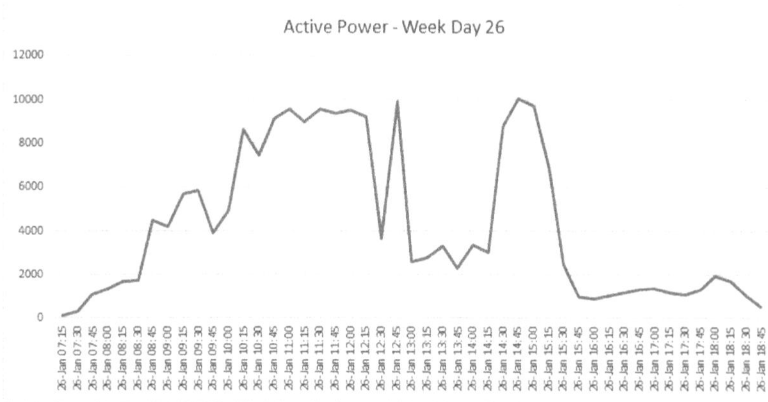

FIGURE 19.27 Active power week day.

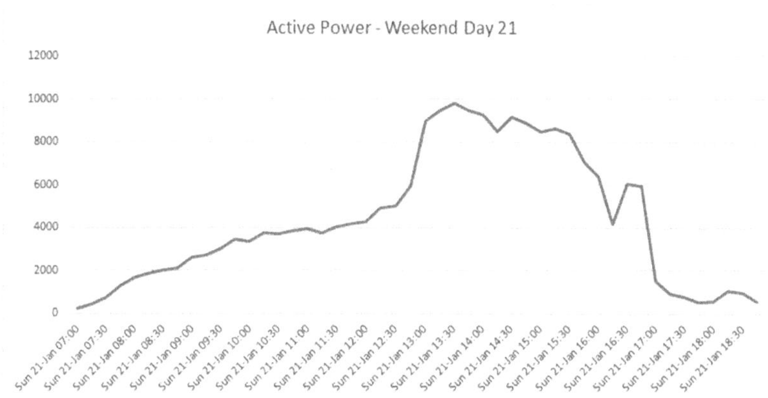

FIGURE 19.28 Active power weekend day.

Comparative between week and weekend profile of active power *Active Power Week Days Comparative.*

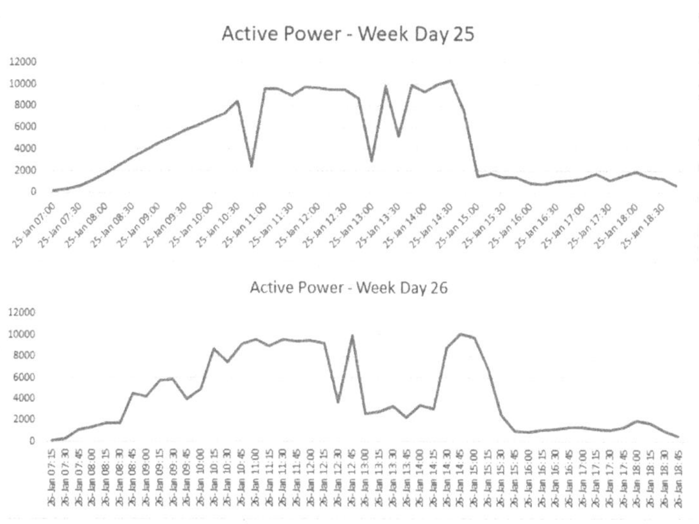

FIGURE 19.29 Active power week days comparative.

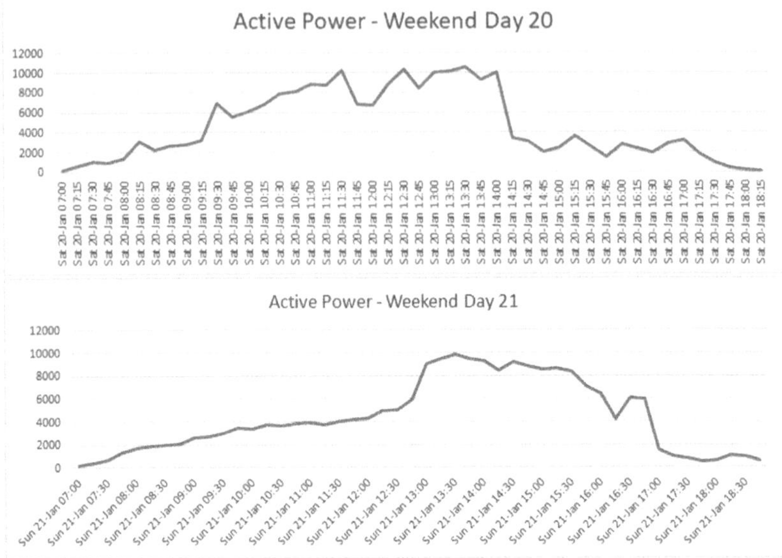

FIGURE 19.30 Active power weekend days comparative.

19.5.6 Reactive power

In the analyzed period, the lowest reactive power value was *5.41*, found on 31 March at 1630. The highest figure was **60.80**, found on 12 March at 1000 (Figs. 19.31–19.35).

January to March—Reactive power

Interval	Low	Day—time	High	Day—time
01–10 January	12.40	08—1715	41.25	05—1400
11–20 January	9.86	19—1745	49.82	18—1515
21–31 January	8.11	30—1615	46.24	23—1345
01–10 February	6.00	03—1830	53.97	09—1345
11–20 February	8.60	13—1500	52.96	15—0945
21–28 February	8.48	24—0615	56.36	22—1415
01–10 March	8.73	05—1645	49.40	09—1515
11–20 March	8.00	11—0900	**60.80**	12—1000
21–31 March	*5.41*	31—1630	57.45	26—1415

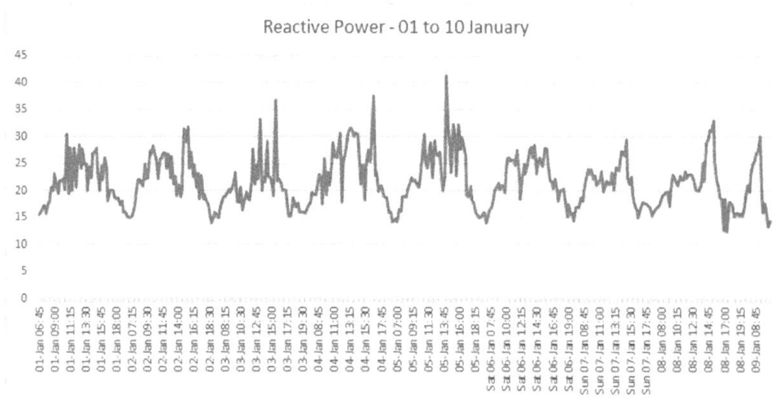

FIGURE 19.31 Reactive power 01–10 January.

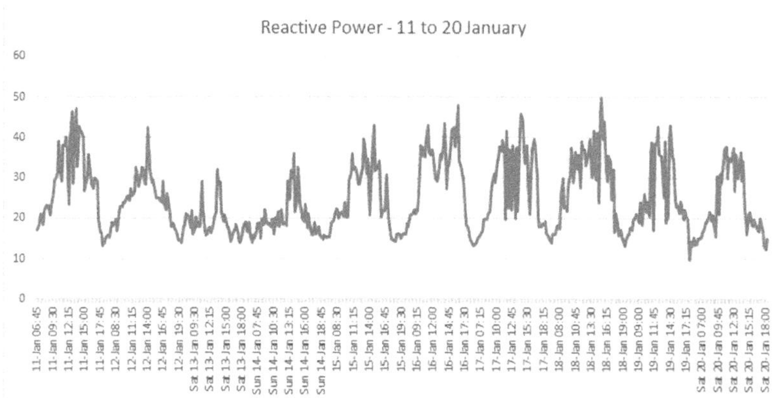

FIGURE 19.32 Reactive power 11–20 January.

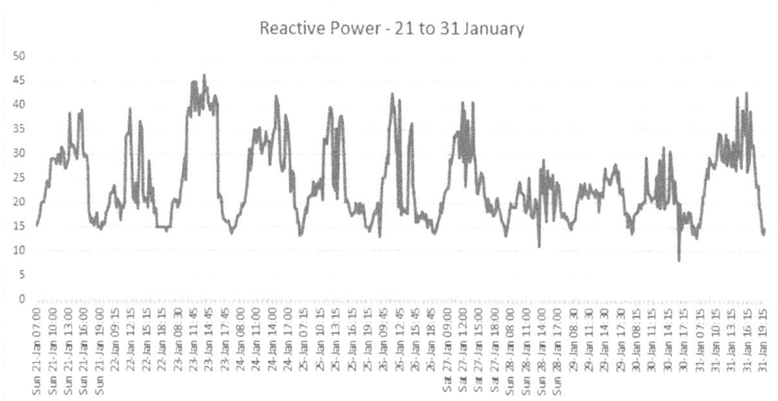

FIGURE 19.33 Reactive power 21–31 January.
Week and weekend profile of reactive power.

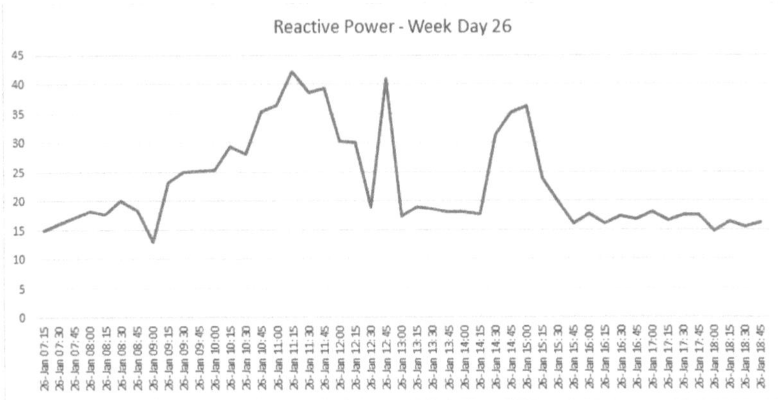

FIGURE 19.34 Reactive power week day.

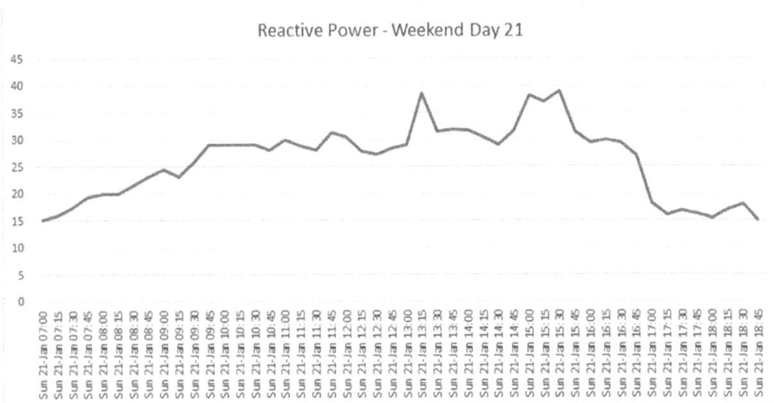

FIGURE 19.35 Reactive power weekend day.

19.5.7 Energy production

In the analyzed period, the lowest energy production value was *26.00*, found on 07 February at 0745. The highest figure was **3618.00** found on 21 February at 1200 (Figs. 19.36–19.40).

January to March—Energy production

Interval	Low	Day—time	High	Day—time
01–31 January	52	15—1830	2664	02—1500
01–28 February	*26*	07—0745	**3618**	21—1200
01–31 March	38	17—0630	2700	29—0930

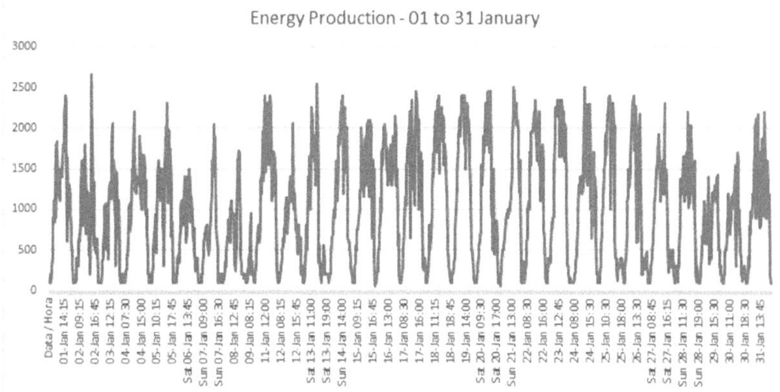

FIGURE 19.36 Energy production 01–31 January.

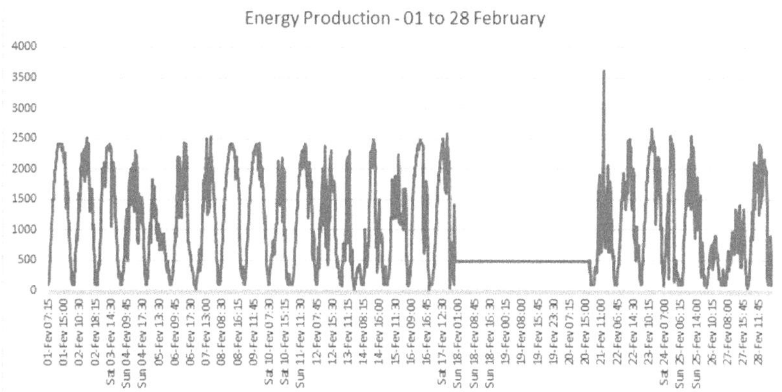

FIGURE 19.37 Energy production 01–28 February.

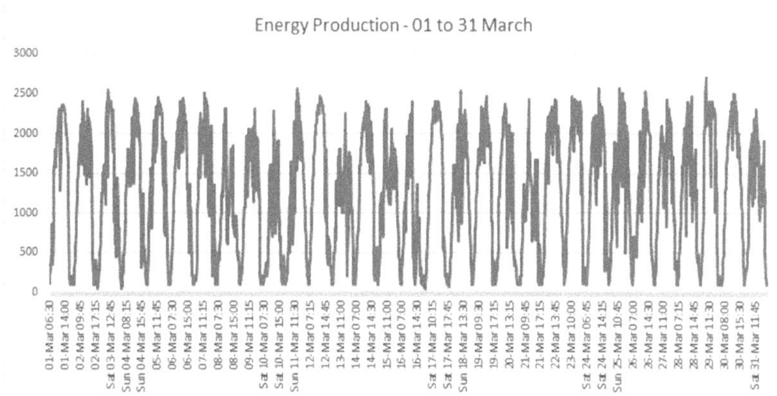

FIGURE 19.38 Energy production 01–31 March.
Week and weekend profile of energy production.

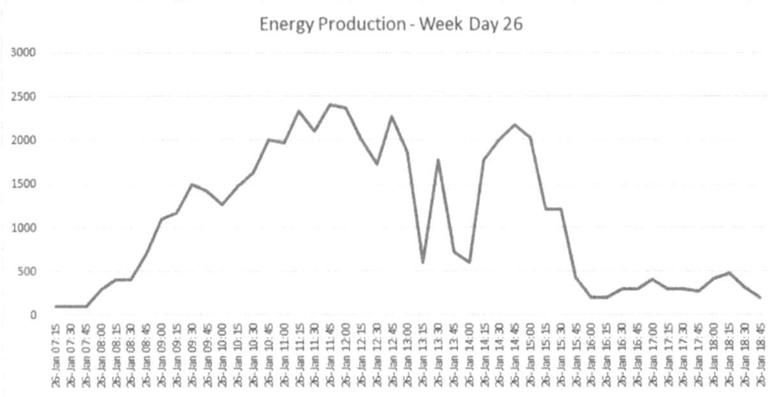

FIGURE 19.39 Energy production week day.

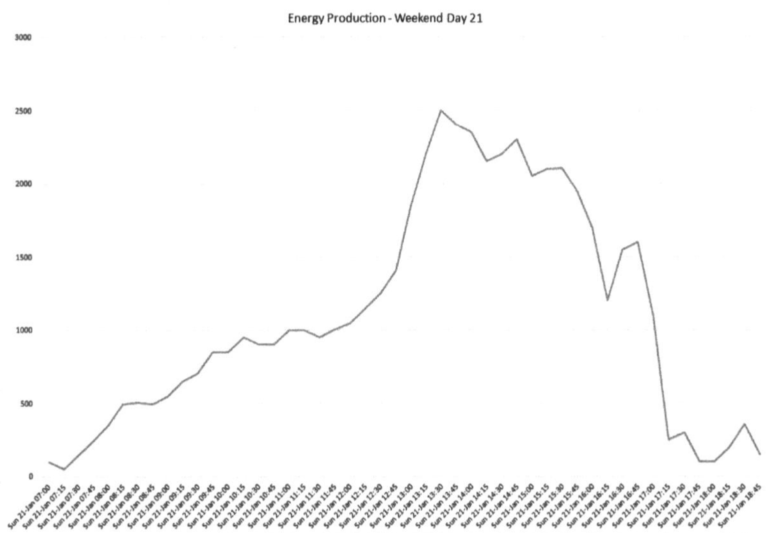

FIGURE 19.40 Energy production weekend day.

19.5.8 Voltage phase

In the analyzed period, the lowest voltage phase value was *1.38* found on 17 February at 1945. The highest figure was **401.54**, found on 01 January at 1430 (Figs. 19.41–19.45).

January to March—Voltage phase

Interval	Low	Day—time	High	Day—time
01–10 January	385.67	02—1345	**401.54**	01—1430
11–20 January	385.25	15—1815	399.29	13—1100
21–31 January	385.05	26—0800	398.53	23—1230
01–10 February	385.11	07—1930	399.32	09—1230
11–20 February	*1.38*	17—1945	398.08	17—1215
21–28 February	379.02	27—1215	399.51	22—1100
01–10 March	382.84	07—1815	400.18	09—1230
11–20 March	382.80	13—1400	400.73	19—1345
21–31 March	384.69	30—0615	399.08	25—1115

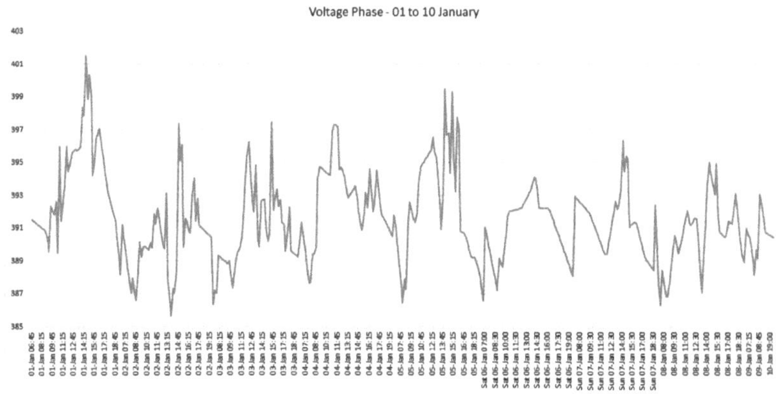

FIGURE 19.41 Voltage phase 01–10 January.

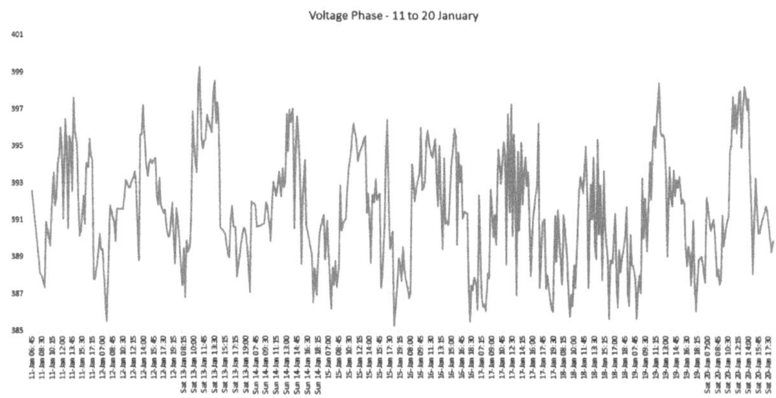

FIGURE 19.42 Voltage phase 11–20 January.

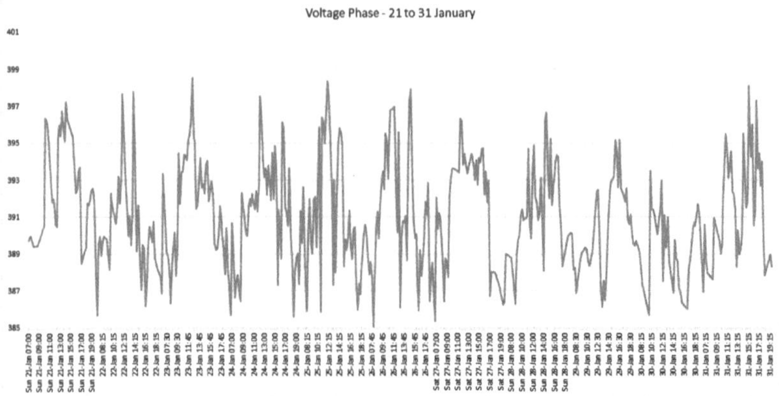

FIGURE 19.43 Voltage phase 21–31 January.
Week and weekend profile of voltage phase.

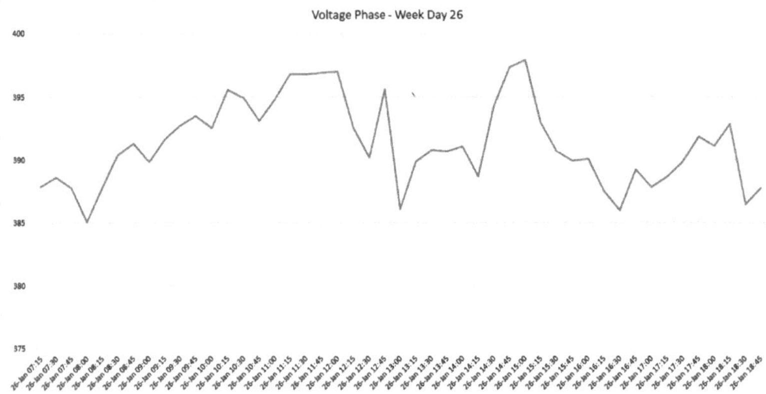

FIGURE 19.44 Voltage phase week.

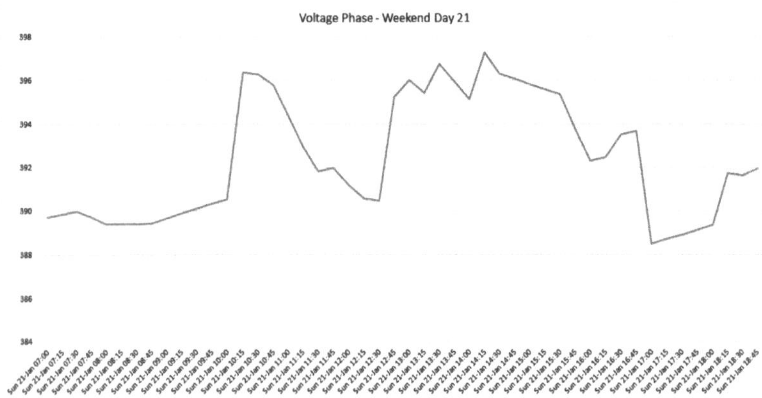

FIGURE 19.45 Voltage phase weekend.

19.6 Conclusion

The analysis of the electrical data, including the photovoltaic generation, helped one to highlight the existing potential of the BDA for better handling a large amount of data.

With current and voltage data collected and analyzed, a huge range of options opens up to turn the lab grid into a smart one. Signal analysis techniques can be applied on data already demonstrated for monitoring and control of power systems.

The analysis of a period of 3 months confirmed the fact that there is a little variation of the parameters analyzed among the analyzed months, which encourages the use of the generated energy as an alternative source of power.

No voltage fluctuation was observed in the analyzed data, which shows that the system successfully withstood the reactive loads of the components present in the system. Harmonic and interharmonic information were not collected by the meters; however, in the future, there is an intention to collect such information in order to verify distortions, monitor, and supervise the harmonic levels of the grid.

In the analyzed data, there is substantial energy generation between 0900 and 1700 as can be seen in Figs. 19.36–19.40. Small variations in energy production were observed due to clouds or even solar radiation decrease. There were changes in the peak generation time, but the period of greater generation is concentrated mainly between 1200 and 1500, at which time the solar radiation is higher.

The frequency did not suffer considerable oscillations during the analyzed period, remaining stable at 60 Hz between 0900 and 1700 as can be seen in Figs. 19.14–19.18.

Both in January (Fig. 19.14) and in March (Fig. 19.16), it is possible to notice that the frequency dropping is due to the data collection time, when there is no electricity generation. The frequency dropping point(s) in January were at 0630, 0645, and 0700 and in March it was at 1800.

The data collected in the night period until sunrise, when there is still no solar generation, shows a drop-in frequency. As soon as the solar generation began, the frequency stabilized. The same happens in the late afternoon, when the generation of solar energy comes to an end and, consequently, the frequency falls.

A voltage dip is shown in Fig. 19.15 in the interval from 0630 of 18 February to 0630 of 20 February. As 18 February is a Sunday, the voltage dip is probably related to some problem or maintenance on the campus power grid or even a sensor data reading error. The problem was solved on 19 February, since in the first hour of 20 February the problem disappeared.

A comparative of the active power is shown in Figs. 19.29 and 19.30. It is possible to verify that on the weekend the values oscillation is smoother than in a weekday. Since the inverter is tied with the campus grid, and there is no massive usage on weekends, the values of the active power are stable.

A more comprehensive analysis of photovoltaic generation data could be made with the inclusion of climatic data collected by the meteorological station installed in the university. This could allow the prediction of the amount of energy to be generated and injected into the grid. The use of big data techniques would be required to store the collected data, as well as to process and extract information from that data.

References

[1] L.M. Camarinha-Matos, Collaborative smart grids: a survey on trends, Renew. Sustain. Energy Rev. 65 (Supplement C) (2016) 283–294 [Online]. Available. Available from: http://www.sciencedirect.com/science/article/pii/S1364032116303264.

[2] Z. Fan, P. Kulkarni, S. Gormus, C. Efthymiou, G. Kalogridis, M. Sooriyabandara, et al., Smart grid communications: overview of research challenges, solutions, and standardization activities, IEEE Commun. Surv. Tutorials 15 (1) (2013) 21–38. First.

[3] T. Eleftherios, B. Dimitrios, E. Charalambos, S. Lambros, Communication challenges in smart grid, in: MATEC Web of Conferences, vol. 41, 2016, p. 01004. [Online]. Available from: <https://doi.org/10.1051/matecconf/20164101004>.

[4] US DOE, Communications Requirements of Smart Grid Technologies, US Department of Energy, Tech. Rep. (2010) 1–69.

[5] A. Oussous, F.-Z. Benjelloun, A.A. Lahcen, S. Belfkih, Big data technologies: a survey, J. King Saud Univ.-Comput. Inf. Sci. 30 (4) (2018) 431–448. Available from: https://www.sciencedirect.com/science/article/pii/S1319157817300034.

[6] C.P. Chen, C.-Y. Zhang, Data-intensive applications, challenges, techniques and technologies: a survey on big data, Inf. Sci. 275 (2014) 314–347.

[7] M.L. Tuballa, M.L. Abundo, A review of the development of smart grid technologies, Renew. Sustain. Energy Rev. 59 (2016) 710–725.

[8] L. Zhou, S. Pan, J. Wang, A.V. Vasilakos, Machine learning on big data: opportunities and challenges, Neurocomputing 237 (2017) 350–361.

[9] A. El Khaouat, L. Benhlima, Big data based management for smart grids, Renewable and Sustainable Energy Conference (IRSEC), 2016 International, IEEE, 2016, pp. 1044–1047.

[10] J.-S. Chou, N.-T. Ngo, Smart grid data analytics framework for increasing energy savings in residential buildings, Autom. Constr. 72 (2016) 247–257.

[11] K. Zhou, C. Fu, S. Yang, Big data driven smart energy management: from big data to big insights, Renew. Sustain. Energy Rev. 56 (2016) 215–225.

[12] Y. Wang, Q. Chen, C. Kang, Q. Xia, Clustering of electricity consumption behavior dynamics toward big data applications, IEEE Trans. Smart Grid 7 (5) (2016) 2437–2447.

[13] C. Kang, Y. Wang, Y. Xue, G. Mu, R. Liao, Big data analytics in china's electric power industry: modern information, communication technologies, and millions of smart meters, IEEE Power Energy Mag. 16 (3) (2018) 54–65.

[14] R.J. Hyndman, X.A. Liu, P. Pinson, Visualizing big energy data: solutions for this crucial component of data analysis, IEEE Power Energy Mag. 16 (3) (2018) 18–25.

[15] T. Khalifa, K. Naik, A. Nayak, A survey of communication protocols for automatic meter reading applications, IEEE Commun. Surv. Tutorials 13 (2) (2011) 168–182.

[16] X. Jin, B.W. Wah, X. Cheng, Y. Wang, Significance and challenges of big data research, Big Data Res. 2 (2) (2015) 59–64.
[17] L. Qing, Z. Boyu, L. Qinqian, Impact of big data on electric-power industry, in: 2017 IEEE Second International Conference on Big Data Analysis (ICBDA), March 2017.
[18] S. Pradeep, J.S. Kallimani, A survey on various challenges and aspects in handling big data, in: Electrical, Electronics, Communication, Computer, and Optimization Techniques (ICEECCOT), 2017 International Conference on, IEEE, 2017, pp. 1–5.
[19] A. Gandomi, M. Haider, Beyond the hype: big data concepts, methods, and analytics, Int. J. Inf. Manage. 35 (2) (2015) 137–144.
[20] G. Wang, A. Gunasekaran, E.W. Ngai, T. Papadopoulos, Big data analytics in logistics and supply chain management: certain investigations for research and applications, Int. J. Prod. Econ. 176 (2016) 98–110.
[21] S.F. Wamba, A. Gunasekaran, S. Akter, S.J.-F. Ren, R. Dubey, S.J. Childe, Big data analytics and firm performance: effects of dynamic capabilities, J. Bus. Res. 70 (2017) 356–365.
[22] Y. Sun, H. Song, A.J. Jara, R. Bie, Internet of things and big data analytics for smart and connected communities, IEEE Access 4 (2016) 766–773.
[23] C. Bekara, Security issues and challenges for the IoT-based smart grid, Procedia Comput. Sci. 34 (2014) 532–537.
[24] V.C. Gungor, B. Lu, G.P. Hancke, Opportunities and challenges of wireless sensor networks in smart grid, IEEE Trans. Ind. Electron. 57 (10) (2010) 3557–3564.
[25] M. Erol-Kantarci, H.T. Mouftah, Wireless sensor networks for cost-efficient residential energy management in the smart grid, IEEE Trans. Smart Grid 2 (2) (2011) 314–325.
[26] M. Amjad, F. Iradat, A brief review of methods and approaches proposed in existing literature to address issues in wireless big data, Information and Computer Technologies (ICICT), 2018 International Conference on, IEEE, 2018, pp. 13–17.
[27] H. Zou, Y. Yu, W. Tang, H.-W.M. Chen, Flex analytics: a flexible data analytics framework for big data applications with i/o performance improvement, Big Data Res. 1 (2014) 4–13.
[28] C. Tu, X. He, Z. Shuai, F. Jiang, Big data issues in smart grid—a review, Renew. Sustain. Energy Rev. 79 (2017) 1099–1107.
[29] A. Aggarwal, S. Kunta, P.K. Verma, A proposed communications infrastructure for the smart grid, Innovative Smart Grid Technologies (ISGT), 2010, IEEE, 2010, pp. 1–5.
[30] C.H. Hauser, D.E. Bakken, I. Dionysiou, K.H. Gjermundrod, V. Irava, J. Helkey, et al., Security, trust, and QoS in next-generation control and communication for large power systems, Int. J. Crit. Infrastruct. 4 (1-2) (2008) 3–16.
[31] N.S. Nafi, K. Ahmed, M.A. Gregory, M. Datta, A survey of smart grid architectures, applications, benefits and standardization, J. Netw. Comput. Appl. 76 (2016) 23–36.
[32] K. Wang, Y. Wang, X. Hu, Y. Sun, D.J. Deng, A. Vinel, et al., Wireless big data computing in smart grid, IEEE Wirel. Commun. 24 (2) (2017) 58–64.
[33] V.C. Gungor, D. Sahin, T. Kocak, S. Ergut, C. Buccella, C. Cecati, et al., Smart grid technologies: communication technologies and standards, IEEE Trans. Ind. Inf. 7 (4) (2011) 529–539.
[34] A.A. Munshi, A.-R.M. Yasser, Big data framework for analytics in smart grids, Electr. Power Syst. Res. 151 (2017) 369–380.
[35] P.D. Diamantoulakis, V.M. Kapinas, G.K. Karagiannidis, Big data analytics for dynamic energy management in smart grids, Big Data Res. 2 (3) (2015) 94–101.
[36] S. Sagiroglu, R. Terzi, Y. Canbay, I. Colak, Big data issues in smart grid systems, in: Renewable Energy Research and Applications (ICRERA), 2016 IEEE International Conference on, IEEE, 2016, pp. 1007–1012.

[37] C.J. Deepu, C. Heng, Y. Lian, A hybrid data compression scheme for power reduction in wireless sensors for IoT, IEEE Trans. Biomed. Circuits Syst. 11 (2) (2017) 245–254.

[38] G.S. Elbasuony, S.H.A. Aleem, A.M. Ibrahim, A.M. Sharaf, A unified index for power quality evaluation in distributed generation systems, Energy 149 (2018) 607–622.

[39] A.J. McBride, A.R. McGee, Assessing smart grid security, Bell Labs Tech. J. 17 (3) (2012) 87–103.

[40] Y. Wang, D. Ruan, D. Gu, J. Gao, D. Liu, J. Xu, et al., Analysis of smart grid security standards, in: Computer Science and Automation Engineering (CSAE), 2011 IEEE International Conference On, IEEE, vol. 4, 2011, pp. 697–701.

[41] C. Efthymiou, G. Kalogridis, Smart grid privacy via anonymization of smart metering data, in: Smart Grid Communications (SmartGridComm), 2010 First IEEE International Conference on, IEEE, 2010, pp. 238–243.

[42] J.-M. Bohli, C. Sorge, O. Ugus, A privacy model for smart metering, in: Communications Workshops (ICC), 2010 IEEE International Conference on, IEEE, 2010, pp. 1–5.

[43] F.D. Garcia, B. Jacobs, Privacy-friendly energy-metering via homomorphic encryption, in: International Workshop on Security and Trust Management, Springer, 2010, pp. 226–238.

[44] Y. Kim, E.C.-H. Ngai, M.B. Srivastava, Cooperative state estimation for preserving privacy of user behaviors in smart grid, in: Smart Grid Communications (SmartGridComm), 2011 IEEE International Conference on, IEEE, 2011, pp. 178–183.

[45] C. Rottondi, G. Verticale, A. Capone, Privacy-preserving smart metering with multiple data consumers, Comput. Netw. 57 (7) (2013) 1699–1713.

[46] E. Khan, B. Adebisi, B. Honary, Location based security for smart grid applications, Energy Procedia 42 (2013) 299–307.

[47] E. Ancillotti, R. Bruno, M. Conti, The role of communication systems in smart grids: architectures, technical solutions and research challenges, Comput. Commun. 36 (17) (2013) 1665–1697.

[48] M.M. Farag, M. Azab, B. Mokhtar, Cross-layer security framework for smart grid: physical security layer, in: Innovative Smart Grid Technologies Conference Europe (ISGTEurope), 2014 IEEE PES, IEEE, 2014, pp. 1–7.

[49] Y. Wu, J. Wei, Towards attack-resilient communications for smart grids with software defined networking, in: IEEE PES GM 2017: Cybersecurity e Communications Talks, Tutorials, and Panels, 2017.

[50] U. Ghosh, X. Dong, R. Tan, Z. Kalbarczyk, D.K. Yau, R.K. Iyer, A simulation study on smart grid resilience under software-defined networking controller failures, in: Proceedings of the Second ACM International Workshop on Cyber-Physical System Security, ACM, 2016, pp. 52–58.

[51] X. Long, D. Tipper, Y. Qian, A key management architecture and protocols for secure smart grid communications, Secur. Commun. Netw. 9 (16) (2016) 3602–3617.

[52] C. Zhao, J. He, P. Cheng, J. Chen, Privacy-preserving consensus-based energy management in smart grid, in: Power & Energy Society General Meeting, 2017 IEEE, IEEE, 2017, pp. 1–5.

[53] T.E.C.D. Oliveira, Estudo da capacidade de hospedagem de fontes de geração distribuída no sistema elétrico de um campus universitário, 2015.

[54] P.F. Ribeiro, C.A. Duque, P.M. Ribeiro, A.S. Cerqueira, Power Systems Signal Processing for Smart Grids, John Wiley & Sons, 2013.

Index

Note: Page numbers followed by "*f*" and "*t*" refer to figures and tables, respectively.

B

C

D

F

G

H

K

L

M

N

T

FSC
www.fsc.org
MIX
Papier | Fördert gute Waldnutzung
FSC® C083411

Kolibri 360 GmbH
Ferdinand-Jühlke-Straße 7
99095 Erfurt, Deutschland
produktsicherheit@kolibri360.de